Environmental Science and Technology

A Sustainable Approach
to Green Science and Technology

SECOND EDITION

Environmental Science and Technology

A Sustainable Approach to Green Science and Technology

SECOND EDITION

STANLEY E. MANAHAN

Taylor & Francis
Taylor & Francis Group
Boca Raton   London   New York

CRC is an imprint of the Taylor & Francis Group,
an informa business

CRC Press
Taylor & Francis Group
6000 Broken Sound Parkway NW, Suite 300
Boca Raton, FL 33487-2742

© 2007 by Taylor & Francis Group, LLC
CRC Press is an imprint of Taylor & Francis Group, an Informa business

No claim to original U.S. Government works
Printed in the United States of America on acid-free paper
10 9 8 7 6 5 4 3 2 1

International Standard Book Number-10: 0-8493-9512-7 (Hardcover)
International Standard Book Number-13: 978-0-8493-9512-3 (Hardcover)

Visit the Taylor & Francis Web site at
http://www.taylorandfrancis.com

and the CRC Press Web site at
http://www.crcpress.com

PREFACE

Throughout the brief period that humankind has populated planet Earth, the species has faced challenges to its survival. Human ingenuity and science have been remarkably effective in meeting these. Diseases that once virtually wiped out entire populations have been conquered. Modern agriculture has enabled the support of a global population several times larger than would have been possible without it. Enough water has been coaxed from often-scarce sources to support large human populations in arid regions. The growth of human population has slowed to an extent that predictions of runaway population growth from just a few decades ago have proven to be unduly pessimistic.

However, despite its remarkable powers of adaptation, humankind is on a collision course with the carrying capacity of planet Earth, which, in the extreme, raises questions of human survival on Earth, at least with anything like the standard of living that we have come to expect. Peak production levels of petroleum, a resource upon which modern economic systems are based have now been reached and wrenching adjustments must occur as this resource dwindles to insignificant levels over the next several decades. During the last 50 years, a mere moment in the life span of human existence on Earth, atmospheric carbon dioxide levels have increased by 15%, well on their way to doubling from preindustrial levels during the next century. The potential effects of this greenhouse warming gas on global climate and all that implies for Earth's carrying capacity, are many and profound. Many other examples can be cited of trends that must change if we are to continue to exist comfortably on Earth.

So, the enormous challenge facing humankind can be summarized in one word: **sustainability**. The definition of sustainability is essentially self-evident; achieving it is a challenge of enormous proportions. In 1987 the World Commission on Environment and Development (the Bruntland Commission) defined **sustainable development** as "industrial progress that meets the needs of the present without compromising the ability of future generations to meet their own needs." The achievement of sustainable development is the central challenge facing the present generations and those that immediately follow. The need is urgent, and time is short.

Environmental pollution has long been recognized as a problem and measures have been taken to alleviate it. Dating somewhat arbitrarily to the 1960s, various laws and regulations have been implemented to deal with environmental pollution.

These have concentrated on a "command-and-control" approach mandating maximum amounts of pollutants that can be released to water, the atmosphere, and other parts of the environment. Measures taken to control pollution have largely been "end-of-pipe" measures that remove pollutants from exhaust gases or wastewater before they are released and that deal with solid wastes by burying them in a (hopefully) secure location.

In more recent times the limitations of "end-of-pipe" measures have become obvious and emphasis has shifted to pollution prevention. An even more sophisticated approach has been the evolution of **green science**, as exemplified by the green chemistry movement and its engineering counterpart, **green technology**. Green science and green technology are designed to carry out science, engineering, manufacturing, and other areas of human endeavor in ways that are oriented toward minimal environmental and resource impact with the highest degree of sustainability.

Although there are excellent basic books in the areas of green chemistry and green engineering, little is available at a very basic level in the general area of green science and technology. *Environmental Science and Technology: A Sustainable Approach to Green Science and Technology*, second edition, is designed to provide a general overview of green science and technology and their essential role in ensuring sustainability and sustainable development. The book is designed to be useful for individuals who need to know the principles of green science.

This book differs in a fundamental way from the other standard environmental science textbooks in that it recognizes a fifth distinct sphere of the environment, the anthrosphere, that has developed into a huge part of Earth's environment made and operated by humans. In so doing, the book recognizes that humans simply will modify and manage Earth to their own perceived self-benefit. Therefore, we must recognize that reality and, to the best of our ability, manage Earth in a positive way, avoiding those measures that are unsustainable and certain to do environmental harm on a large scale, doing things in ways that minimize environmental impact, and even using anthrospheric activities to enhance the environment as a whole and to maintain sustainability. With the anthrosphere in mind as a major environmental sphere, the book is organized into six major sections as outlined below.

Chapter 1 to Chapter 3 are written to provide the essential background for understanding green science and technology. Chapter 1, "Sustainability Through Green Science and Technology" is an introduction to green science and technology and how they relate to sustainability. It recognizes natural capital, consisting of Earth's resources and its capacity to support life and human activities. Chapter 2, "The Five Environmental Spheres," defines and explains the four traditionally recognized environmental spheres — the hydrosphere, atmosphere, geosphere, and biosphere (water, air, earth, and life) — as do all common works on environmental science. Additionally, it recognizes the fifth environmental sphere, the anthrosphere, which is defined above and has an enormous influence on the environment as a whole and that must be considered as an integral part of Earth's environment. Chapter 3, "Green Chemistry, Biology, and Biochemistry," is a brief overview of these disciplines that are essential to understanding green science and technology. These topics are covered at a fundamental level in recognition of the fact that many of the users will have minimal backgrounds in the sciences.

Chapter 4 to Chapter 6 deal with the hydrosphere. Chapter 4, "Water: A Unique Substance Essential for Life," explains the special physical and chemical characteristics of water and bodies of water which determine its crucial role in the environment. Chapter 5, "Aquatic Biology, Microbiology, and Chemistry," discusses the chemical and biochemical processes that occur in water. Chapter 6, "Keeping Water Green," covers the essential role of water in green science and technology and the preservation of this valuable resource.

The next three chapters cover the atmosphere and air. Chapter 7, "The Atmosphere: A Protective Blanket Around Us," is a discussion of the properties of air and the atmosphere emphasizing the protective role of the atmosphere for life on Earth. Atmospheric chemical processes and their effects on air pollution are discussed in Chapter 8, "Environmental Chemistry of the Atmosphere." Protection of the atmosphere as a green resource is discussed in Chapter 9, "Sustaining an Atmosphere Conducive to Life on Earth."

Chapter 10 to Chapter 12 deal with the geosphere. Chapter 10, "The Geosphere," introduces the geosphere, or solid earth, as one of the major environmental spheres and includes discussion of the geosphere as an essential source of minerals. The thin layer of soil on the surface of the geosphere consisting of weathered minerals and organic matter essential for plant growth is outlined in Chapter 11, "Soil, Agriculture, and Food Production," which also discusses the role of soil in producing food required for life on Earth. Preservation of the quality of the geosphere as a life support system is the topic of Chapter 12, "Geospheric Hazards and Sustaining a Green Geosphere." This chapter also discusses ways in which regions of the geosphere may suddenly and sometimes without warning turn treacherous, resulting in earthquakes, tsunamis, mudslides, and destructive volcanoes.

Chapter 13 to Chapter 15 are a discussion of the biosphere. Ecology and the relationship of organisms to their environment and to each other are discussed in Chapter 13, "The Biosphere: Ecosystems and Biological Communities." Chapter 14, "Toxic Effects on Organisms and Toxicological Chemistry," discusses how organisms handle and metabolize toxic substances and the ill effects that may occur from exposure to toxic substances. Chapter 15, "Bioproductivity for a Greener Future," addresses the key issue of production of biomass by photosynthesis and the critical role of biomass, not only for food, but for raw materials as well, in sustaining the future needs of humans and other organisms on Earth.

The final major section of the book deals with the anthrosphere as a distinct part of the environment. The anthrosphere and its major aspects are the topic of Chapter 16, "The Anthrosphere as Part of the Global Environment." The chapter begins with a section on the "Earth as Made by Humans" that divides the anthrosphere into (1) anthrospheric constructs, such as dwellings made by humans; (2) anthrospheric flows of materials, energy, communications, and people; and (3) anthrospheric conduits through which these flows move. Chapter 17, "Industrial Ecology for Sustainable Resource Utilization," outlines the rapidly evolving area of industrial ecology in which industrial enterprises process materials and energy, interacting in ways somewhat analogous to natural ecosystems. Chapter 18, "Adequate, Sustainable Energy: Key to Sustainability," emphasizes the importance of ample supplies from sustainable sources of energy that can be used by humans to sustain themselves and their environment.

Reader input and suggestions are welcome. They should be addressed to the author at manahans@missouri.edu.

THE AUTHOR

Stanley E. Manahan is professor of chemistry at the University of Missouri–Columbia, where he has been on the faculty since 1965. He received his B.A. degree in chemistry from Emporia State University, Emporia, Kansas in 1960 and his Ph.D. in analytical chemistry from the University of Kansas in 1965. Since 1968 his primary research and professional activities have been in environmental chemistry, toxicological chemistry, and waste treatment. His classic textbook, *Environmental Chemistry*, eighth edition (CRC Press, Boca Raton, Florida, 2004) has been in print continuously in various editions since 1972 and is the longest-standing title on this subject in the world. Other books that he has written are *Green Chemistry and the Ten Commandments of Sustainability*, third edition (ChemChar Research, Inc., 2006), *Toxicological Chemistry and Biochemistry*, third edition (CRC Press/Lewis Publishers, 2001), *Fundamentals of Environmental Chemistry*, 2nd edition (CRC Press/Lewis Publishers, 2001), *Industrial Ecology: Environmental Chemistry and Hazardous Waste* (CRC Press/Lewis Publishers, 1999), *Environmental Science and Technology* (CRC Press/Lewis Publishers, 1997), *Hazardous Waste Chemistry, Toxicology and Treatment* (Lewis Publishers, 1992), *Quantitative Chemical Analysis*, (Brooks/Cole, 1986), and General Applied Chemistry, second edition (Willard Grant Press, 1982). He has lectured on the topics of environmental chemistry, toxicological chemistry, waste treatment, and green chemistry throughout the United States as an American Chemical Society Local Section Tour Speaker and has presented plenary lectures on these topics in international meetings in Puerto Rico; the University of the Andes in Mérida, Venezuela; Hokkaido University in Japan; the National Autonomous University in Mexico City; Italy; and France. He was the recipient of the Year 2000 Award of the Environmental Chemistry Division of the Italian Chemical Society. His research specialty is gasification of hazardous wastes.

CONTENTS

1. SUSTAINABILITY THROUGH GREEN SCIENCE AND TECHNOLOGY

"If we do not change direction, we are likely to end up where we are headed," (old Chinese proverb).

"If we make the effort to learn its language, the Earth will speak to us and tell us what we must do to survive."

1.1. SUSTAINABILITY

The old Chinese proverb certainly applies to modern civilization and its relationship to world resources that support it. Evidence abounds that humans are degrading the Earth life support system upon which they depend for their existence. The emission to the atmosphere of carbon dioxide and other greenhouse gases is almost certainly causing global warming. Discharge of pollutants has degraded the atmosphere, the hydrosphere, and the geosphere in industrialized areas. Natural resources including minerals, fossil fuels, freshwater, and biomass have become stressed and depleted. The productivity of agricultural land has been diminished by water and soil erosion, deforestation, desertification, contamination, and conversion to nonagricultural uses. Wildlife habitats including woodlands, grasslands, estuaries, and wetlands have been destroyed or damaged. About 3 billion people (half of the world's population) live in dire poverty on less than the equivalent of U.S. $2 per day. The majority of these people lack access to sanitary sewers and the conditions under which they live give rise to debilitating viral, bacterial, and protozoal diseases. At the other end of the standard of living scale, a relatively small fraction of the world's population consumes an inordinate amount of resources with lifestyles that involve living too far from where they work in energy-wasting houses that are far larger than they need, commuting long distances in large "sport-utility vehicles" that consume far too much fuel, and overeating to the point of unhealthy obesity with accompanying problems of heart disease, diabetes, and other obesity-related maladies.

As We Enter the Anthropocene

Humans have gained an enormous capacity to alter Earth and its support systems. Their influence is so great that we are now entering a new epoch, the **anthropocene**, in which human activities have effects that largely determine conditions on the planet. The major effects of human activities on Earth have taken place within a miniscule period of time relative to the time that life has been present on the planet or, indeed, relative to the time that modern humans have existed. These effects are largely unpredictable, but it is essential for humans to be aware of the enormous power in their hands — and of their limitations if they get it wrong and ruin Earth and its climate as life-support systems.

Achieving Sustainability

Although the condition of the world and its human stewards outlined in the preceding paragraphs sounds rather grim and pessimistic, this is not a grim and pessimistic book. That is because the will and ingenuity of humans that have given rise to conditions leading to deterioration of Planet Earth can be — indeed, are being — harnessed to preserve the planet, its resources, and its characteristics that are conducive to healthy and productive human life. The key is **sustainability** or **sustainable development** defined by the Bruntland Commission in 1987 as industrial progress that meets the needs of the present without compromising the ability of future generations to meet their own needs.[1] A key aspect of sustainability is the maintenance of Earth's **carrying capacity**, that is, its ability to maintain an acceptable level of human activity and consumption over a sustained period of time. Although change is a normal characteristic of nature, sudden and dramatic change can cause devastating damage to Earth support systems. Change that occurs faster than such systems can adjust can cause irreversible damage to them. The purpose of this book is to serve as an overview of the science and technology of sustainability — green science and green technology. This chapter is an introduction to green science and technology and their relationship to sustainability.

Figure 1.1 illustrates the evolution leading from early attempts to control pollution to the current emphasis upon sustainability. Until approximately 1980, **pollution control** was almost exclusively driven by regulations. Pollutants were produced, but efforts were concentrated on so-called end-of-pipe measures to prevent their release to water, air, or land. As it became more difficult to meet increasingly stringent regulations, it was realized that a better approach was **pollution prevention**, reducing the amounts of pollutants and wastes at the source and employing recycle and reuse to lower levels of release while using less materials. Pollution prevention led to **design for environment** that went beyond simple compliance and was proactive in reducing pollutants, waste, and material consumption. Design for environment recognized that responsibility for products extended beyond the point of sale and made use of life-cycle analysis and eco-efficiency (see Section 1.9) in reducing adverse

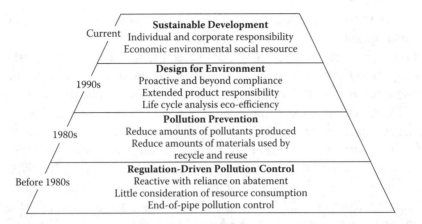

Figure 1.1. Evolution from regulation-driven pollution control to current systems emphasizing sustainable development.

environmental and resource impacts. Since the 1990s, **sustainable development** has come into vogue. Although the concept has taken until recently to become widely accepted as the best means of doing business, it dates back to the previously mentioned 1987 Bruntland Commission report entitled "Our Common Future" resulting from a United Nations commission chaired by the Prime Minister of Norway, Gro Harlem Brundtland. Sustainable development makes use of the concepts of *green science* and *green technology*. It emphasizes individual and corporate responsibility and considers economic, environmental, social, and resource impacts.

The Economics of Sustainability

Humans obtain food, shelter, health, security, mobility, and other necessities through economic activities carried out by individuals, businesses, and government entities. By their nature, all economic systems utilize resources (renewable and non-renewable) and all tend to produce wastes. With these characteristics in mind, it is possible to define three key characteristics of a sustainable economic system operating within Earth's carrying capacity.[2]

- The usage of renewable resources is not greater than the rates at which these resources are regenerated.

- The rates of use of nonrenewable resources do not exceed the rates at which renewable substitutes are developed.

- The rates of pollution emission or waste production do not exceed the capacity of the environment to assimilate these materials.

Although they are useful guidelines, these rules cannot be followed exactly. Certainly, it should be possible to keep usage of renewable resources at levels that are sustainable, and there are many cases of economic systems that have suffered grievously when such resources are not renewed at a sufficient rate. For example, the consumption of firewood in Haiti has greatly exceeded the rates at which the wood resource is replenished, and the population has suffered grievously as a result. With regard to the second point, it is not always possible to use substitutes for non-renewable resources, such as essential metals in some applications, although greatly reduced levels of usage can often be achieved, and recycling can reduce consumption of some resources extracted from the Earth almost to the point of renewability. The third point above suffers from uncertainty regarding the capacity of the environment to assimilate wastes and pollutants. For example, until the early 1970s, there was no concern regarding known emissions of chlorofluorocarbons (freon gases) to the atmosphere because the quantities were small, the substances among the least toxic known, and their reactivities in the lower atmosphere were negligible. Then it was found that they caused destruction of the essential protective stratospheric ozone layer and, as a result, the issue of their discharge into the atmosphere became very important.

The challenge of attaining global sustainability is enormous. The total burden on Earth's carrying capacity is a product of population times demand per person. This leads to the conclusion that most of the increase in the burden on Earth's carrying capacity will come from the populations of developing countries. This fact also provides an opportunity, however, in that sustainable systems are easier to introduce into developing regions in which the infrastructure and economic systems are less developed and therefore more amenable to development along lines of greater sustainability.

1.2. NATURAL CAPITAL AND THE QUALITY OF LIFE

As the industrial revolution developed, natural resources were abundant relative to needs. In the earlier years of the industrial revolution, production was limited by factors other than resources, such as labor. Now population is in surplus and automated production continues to reduce the need for human labor. Increasingly, production is becoming limited by the Earth's natural environment including the availability of natural resources, the vital life-support ability of ecological systems, and the capacity of the natural environment to absorb the byproducts of industrial production including wastes and most notably carbon dioxide, which, released to the atmosphere, is the major cause of global warming.

Traditionally, the success of economies has been measured in material factors including financial assets, income, and real estate. The achievement of sustainability requires a broader view of assets. As shown in Figure 1.2, there are three forms of capital required for a high quality of life. **Economic capital** consists of the traditional economic assets including money, property, and possessions. **Social capital** consists of opportunity, freedom, health, healthy households, and well-functioning societies. **Natural capital** consists of resources, including minerals and fuels; biological productivity; capacity to absorb pollutants; and other assets normally thought

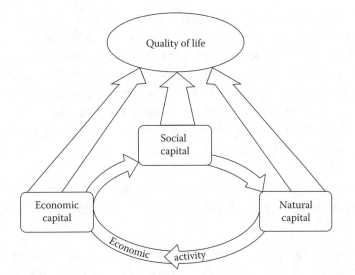

Figure 1.2.   Economic activity determining quality of life depends upon three major categories of capital. Sustainable development requires maintenance and enhancement of natural capital.

of as "the environment" potentially used in economic and social systems. Other kinds of capital can be defined as well that are parts of or overlap with the three categories above. **Intellectual capital** refers to the information base, management systems, knowledge and education of people and related areas, and similar areas. Closely related to intellectual capital are such things as management structures, laws and regulations, computer software and hardware, and related areas comprising **organizational capital**.

Natural capital is particularly pertinent to sustainability.[3] Natural capital can be reduced to two major areas: **natural resources** and **ecosystem or environmental services**. In addition to providing natural resources, such as metal ores, natural capital values such natural assets as diverse as the protective stratospheric ozone layer, the capacity of ecosystems interacting with the natural environment to maintain conditions conducive to human life and comfort, and even the plant pollination function of bees and other insects. An appreciation of human effects on natural capital is illustrated by the fact that human activities have drastically altered and transformed as much as one-half of Earth's land surface by activities such as cultivation. The limits of the carrying capacity of land and freshwater resources are being approached globally and are already exceeded in some areas. It is important to realize that above certain **sustainable yield threshold** levels, which may not be obvious, overuse or abuse of natural capital can lead to its irreversible loss.

Until recently, capitalists from the traditional business community and environmentalists were often in opposition with regard to economic development. However, the recognition of natural capital as an essential part of economic systems has led to the development of an economic activity termed **natural capitalism**. Such a system properly values natural and environmental resources increasing well-being, productivity, wealth, and capital while reducing waste, consumption of resources, and adverse environmental effects. Such a system takes advantage of the individual and corporate incentives that have made the traditional capitalist economic system so powerful in delivering consumer goods and services. In so doing, it seeks to have businesses emulate biological systems by recycling wastes back into the raw material stream and emphasizing the provision of services rather than just material goods (see the discussion on industrial ecology in the following text and in Chapter 17).

For a system of natural capitalism to function properly, the following changes in business practices are required:

1. Implement changes in technology that enable significantly higher productivity with greatly reduced use of minerals, energy, water, and biomass products such as wood.

2. Take advantage of the models provided by closed-loop biological systems in maximizing the recycling of materials such that waste products from one sector become raw materials for another and the most efficient possible use is made of energy.

3. Move from a business model that emphasizes selling goods to one that provides services. A particularly pertinent example would be a shift from selling automobiles to the provision of transportation, including public transportation.

4. Emphasize reinvestment in natural capital to increase production of ecosystem services. For example, new housing developments should include investment in neighborhood parks and natural areas with the idea that such amenities are just as important as dishwashers and multicar garages in providing a high-quality residential life.

1.3. SUSTAINABILITY AND THE COMMON GOOD

Natural capital is something that was described in a classic work as the "commons."[4] This term was used centuries ago in England to describe a common pasture used by most residents of a village for grazing cattle, sheep, and horses. Each family could gain wealth (more meat, milk, or horsepower) by putting more animals into the commons. For example, a family with one cow could acquire a second one and double its wealth (in cows). Because the commons might accommodate perhaps 100 head of livestock, this individual action would detract from the commons by only 1%. The natural tendency was for each of many families to seek to increase its wealth by adding more animals and, over time, in the aggregate, the carrying capacity of the commons became grossly exceeded, and the pasture was ruined from overgrazing. During the 14th century, this practice became so widespread that the economies of many villages collapsed with whole populations no longer able to provide for their basic food needs.

Examples abound of the counterproductive attitudes of people toward resources and of their disregard for the commons upon which, ultimately, their own livelihoods depend. During the 1880s, ranchers in Edwards County, Texas, became concerned with the number of settlers who began cultivating the grasslands used by the ranchers for their herds. At a stockmen's meeting, they came up with the following resolution: "Resolved that none of us know, or care to know, anything about grasses, native or otherwise, outside of the fact that for the present, there are lots of them, the best on record, and we are getting the most of them while they last."[5] Within a few short years, overgrazing and drought drastically decreased the yields from the grasslands that provided the ranchers' livelihoods. Unfortunately, the attitude expressed by these individuals persists in different guises even today. Examples of modern tragedies of the commons include vast amounts of land unwisely cultivated and turned to desert (desertification), ongoing destruction of the Amazon rain forest, severe deterioration of the global ocean fisheries resource, freeways that at times become great linear parking lots (residents of Houston fleeing inland from Hurricane Rita in 2005 were stuck in a 100-mi-long traffic jam on I-35 for up to 24 h), and, of much direct concern to many university students and faculty, parking facilities that have become so crowded by excess demand that their utility is seriously curtailed.

The idea of the commons can be applied to modern civilizations in which the global commons consist of the air humans must breathe, water resources, agricultural lands, mineral resources, capacity of the natural environment to absorb wastes, and all other facets of natural capital. And the logic of the commons still prevails. According to this logic, each consumer unit has the right to acquire a unit of natural capital, the cost of which is distributed throughout the commons and shared by all. Some consumer units accumulate resources to a greater extent than others, making them relatively wealthier. However, if enough consumer units use relatively large amounts of natural capital, it becomes exhausted and unsustainable, therefore unable to support the society as a whole, so that everybody suffers.

Automotive transportation can be used to illustrate a modern tragedy of the commons. When an individual acquires an automobile, it adds to that person's possessions and mobility. The single automobile makes a relatively small impression on the environment in terms of materials required to make the automobile, fuel to run it, and pollution from exhausts. However, as more and more people acquire vehicles, material resources to manufacture them and fossil fuels to keep them running become strained, traffic becomes so heavy that the automobile loses its convenience as a mode of transportation, and, in some places at some times, the whole transportation system collapses.

Various "tragedies of the commons," such as those described in the preceding text, make a strong case for collective actions in the public sector to ensure the well-being of humankind and the preservation of the support systems on which humans depend, and they illustrate the limitations of unregulated "free-for-all" capitalist economic systems in achieving sustainable development. However, the dismal abandoned factories, seriously deteriorated environments, and relatively low living standards of nations that have emerged from communist domination since the fall of the "iron curtain" around 1990 give testimony to the failures of economies in which private enterprise is discouraged. A major challenge facing modern and developing economies is to devise systems in which enlightened regulations act to preserve the support systems on which these economies ultimately depend while harnessing the tremendous power of human ingenuity, initiative, and even greed in developing and maintaining sustainable economic systems.

1.4. THE MASTER EQUATION

Environmental impact of human economic activities has been described by a **master equation** relating population and gross domestic product (GDP, a measure of economic activity) per person.[6] The master equation is expressed as

$$\text{Environmental impact} + \text{population} \times \frac{\text{GDP}}{\text{Person}} \times \frac{\text{environmental impact}}{\text{unit of GDP}} \tag{1.4.1}$$

This relationship is sometimes called the *IPAT equation,*

$$I = P \times A \times T \tag{1.4.2}$$

where I is environmental impact, P is population, A is affluence (GDP/person), and T stands for technology (impact/unit GDP). This equation has been used to estimate that by mid/late in the 21st century, making the reasonable assumptions that the global environmental burden (I) should be halved while population (P) doubles and wealth per person (A) increases 5-fold, environmental efficiency, the inverse of T, must increase 20-fold![7] Such is the challenge facing the practitioners of green technology during the next several decades.

To consider the possibilities for increasing the ratio of GDP/(environmental impact), the inverse of T in the IPAT equation, it is useful to examine a hypothetical plot of resource impact as a function of economic development as shown in Figure 1.3.

As expected, preindustrial environmental effects of human activities were very low (though not zero; primitive humans did impact the environment by activities such as burning forests so that grass would grow to support higher populations of game animals in the cleared areas). As the industrial revolution got well under way around 1800, its environmental impacts rose sharply. Eventually, pollution and waste problems became painfully obvious and laws and regulations were enacted which, especially during the latter 1900s, had perceptible, if uneven, positive effects on environmental and resource impacts. The regulatory approach focused on preventing discharges, largely by mandating so-called end-of-pipe measures in which pollutants were produced, but were removed from wastewater and exhaust gas streams before release. The regulation and cleanup of hazardous waste sites was also undertaken

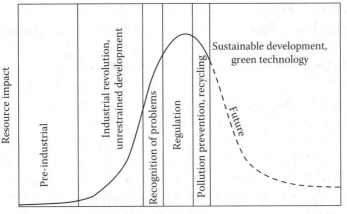

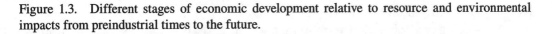

Figure 1.3.   Different stages of economic development relative to resource and environmental impacts from preindustrial times to the future.

to remediate often long-standing problems with hazardous wastes usually discarded in landfills. As the costs and limitations of the regulatory (command-and-control) approach were realized, emphasis shifted to pollution prevention and recycling by emphasizing products and processes that do not produce pollutants and that utilize recycling of materials; by the year 2000, much emphasis was placed on such measures. Now the emphasis is shifting toward technologies that are inherently safe, nonpolluting, and non-resource-intensive. These are the green technologies based upon green science as described in the following section.

1.5.   THE GOALS AND PRIORITIES OF GREEN SCIENCE AND TECHNOLOGY

Sustainable development requires setting of prioritized goals. Graedel and Allenby have termed these the "grand objectives."[6] They are discussed briefly here as goals for the practice of sustainable science and technology.

Because humans are obviously concerned with their own survival, the first of the grand objectives is **maintenance of the human species**. Exclusive of some catastrophic cosmic event, there are few things that would wipe out the human species completely, but a variety of conditions could greatly diminish human populations and make the existence of the survivors unpleasant. Major global climate change, either a new Ice Age or major global warming, could render large parts of Earth unsuitable for human habitation and drastically reduce food supply. A large-scale nuclear war could kill millions directly and from delayed effects of radionuclide contamination, or could even cause changes in climate (a "nuclear winter" from particles blasted into the atmosphere is one scenario). Depletion of water, land, and mineral resources could seriously lower Earth's capacity to support human life. Direct effects on humans, such as damage to human DNA from toxic substances or an epidemic of some new disease (in 2006, "bird flu" was regarded as a potential threat to the lives of millions of people) for which there is no cure or vaccine could occur.

A second grand objective is **sustainable development**, the subject of much of this book. Many factors could prevent sustainable development. Included are climate change, lack of adequate water supplies, mineral depletion, fuel depletion, and even exhaustion of available landfill space.

Maintenance of **biodiversity** is a third grand objective. Species may be lost as a consequence of habitat destruction from factors such as loss of water availability and quality, changed land use, and deforestation. Some animal species have been hunted to extinction and fishery stocks lost from overfishing and stream diversion (damming). Aquatic species can be lost from acid deposition and thermal pollution. The loss of protective stratospheric ozone can result in species loss due to damage from ultraviolet radiation.

A fourth grand objective is maintenance of **esthetic richness**. Air pollution, water pollution, and uncontrolled urban development can seriously damage esthetics. In congested urban areas one of the greatest detractors to esthetics is photochemical smog. Oil spills may ruin beaches. Odors from various sources may make some areas unpleasant.

1.6. GREEN SCIENCE

Although *science* is a widely used word having somewhat different meanings in different contexts, it can generally be regarded as a body of knowledge or system of study dealing with an organized body of facts verifiable by experimentation that are consistent with a number of general laws. In its purest sense, science avoids value judgments; it involves a constant quest for truths whether they be good, such as the biochemical basis of a cure for some debilitating disease, or bad, such as the nuclear physics behind the development of nuclear bombs. However, in defining green science, it is necessary to modify somewhat the view of "pure" science. *Green science* is science that is oriented strongly toward the maintenance of environmental quality, the reduction of hazards, the minimization of consumption of nonrenewable resources, and overall sustainability.

When the public thinks of environmental pollution, exposure to hazardous substances, consumption of resources such as petroleum feedstocks, and other unpleasant aspects of modern industrialized societies, chemical science (the science of matter) often comes to mind. So, it is fitting that to date the most fully developed green science is *green chemistry* defined as the practice of chemistry in a manner that maximizes its benefits while eliminating or at least greatly reducing its adverse impacts.[8] Green chemistry is based upon "twelve principles of green chemistry" and, since the mid-1990s, has been the subject of a number of books, journal articles, and symposia. In addition, centers and societies of green chemistry and a green chemistry journal have been established.

1.7. GREEN TECHNOLOGY

Technology refers to the ways in which humans do and make things with materials and energy directed toward practical ends. In the modern era, technology is to a large extent the product of engineering based on scientific principles. Science deals with the discovery, explanation, and development of theories pertaining to interrelated natural phenomena of energy, matter, time, and space. Based on the fundamental knowledge of science, engineering provides the plans and means to achieve specific practical objectives. Technology uses these plans to carry out the desired objectives. Technology obviously has enormous importance in determining how human activities affect Earth and its life support systems.

Technology has been very much involved in determining levels of human population on Earth, which has seen three great growth spurts since modern humans first appeared. The first of these, lasting until about 10,000 years ago, was enabled by the primitive, but remarkably effective tools that early humans developed, resulting in a global human population of perhaps 2 or 3 million. For example, the bow and arrow enabled early hunters to kill potentially dangerous game for food at some (safer) distance without having to get very close to an animal and stab it with a spear or club it into submission. Then, roughly 10,000 years ago, humans who had existed as hunter/gatherers learned to cultivate plants and raise domesticated animals, an effort that was aided by the further development of tools for cultivation and food production. This development ensured a relatively dependable food supply in smaller areas. As a result, humans were able to gather food from relatively small agricultural fields rather than having to scout large expanses of forest or grasslands for game to kill or berries to gather. This development had the side effect of allowing humans to remain in one place in settlements and gave them more free time in which humans freed from the necessity of having to constantly seek food from their natural surroundings could apply their ingenuity in areas such as developing more sophisticated tools. The agricultural revolution allowed a second large increase in numbers of humans and enabled a human population of around 100 million, 1000 years ago. Then came the industrial revolution, the most prominent characteristic of which was the ability to harness energy other than that provided by human labor and animal power. Wind and water power enabled mills and factories to use energy in production of goods. After about 1800, this power potential was multiplied manyfold with the steam engine and later the internal combustion engine, turbines, nuclear energy, and electricity, enabling current world population of around 6 billion to grow (though not as fast as some of the more pessimistic projections from past years).

There is ample evidence that new technologies can give rise to unforeseen problems. According to the **law of unintended consequences**, whereas new technologies can often yield predicted benefits, they can also cause substantial unforeseen problems. For example, in the early 1900s, visionaries accurately predicted the individual freedom of movement and huge economic boost to be expected from the infant automobile industry. It is less likely that they would have predicted millions of deaths from automobile accidents, unhealthy polluted air in urban areas, urban sprawl, and depletion of petroleum resources that occurred in the following century. The tremendous educational effects of personal computers were visualized when the first such devices came on the market. Less predictable were the mind-numbing hours that students would waste playing senseless computer games. Such unintended negative consequences have been called **revenge effects**.[9] Such effects occur because of the unforeseen ways in which new technologies interact with people.

Avoiding revenge effects is a major goal of **green technology** defined as technology applied in a manner that minimizes environmental impact and resource consumption and maximizes economic output relative to materials and energy input. During the development phase, people who develop green technologies, now greatly aided by sophisticated computer methodologies, attempt to predict undesirable consequences of new technologies and put in place preventative measures before revenge effects have a chance to develop and cause major problems.

A key component of green technology is **industrial ecology**, which integrates the principles of science, engineering, and ecology in industrial systems through which goods and services are provided, in a way that minimizes environmental impact and optimizes utilization of resources, energy, and capital. In so doing, industrial ecology considers every aspect of the provision of goods and services from concept, through production, and to the final fate of products remaining after they have been used. It is above all a sustainable means of providing goods and services. It is most successful in its application when it mimics natural ecosystems, which are inherently sustainable by nature. Industrial ecology works through groups of industrial concerns, distributors, and other enterprises functioning to mutual advantage, using each others' products, recycling each others' potential waste materials, and utilizing energy as efficiently as possible. By analogy with natural ecosystems, such a system comprises an **industrial ecosystem**.

1.8. LIFE-CYCLE ANALYSIS

An important component of green technology is **life-cycle analysis** (assessment) which considers process and product design in the management of materials from their source through manufacturing, distribution, use, reuse (recycle), and ultimate fate. The objective of life-cycle analysis is to determine, quantify, and minimize adverse resource, environmental, economic, and social impacts.

Figure 1.4 shows a generalized life cycle to which a life-cycle analysis can be applied. Initially, product manufacture requires acquisition of energy and materials. Usually the material has to be refined and components are then fabricated, followed by assembly into the final product. After product use, there are several possible recycling loops in the life cycle. Wider loops are indicative of less "green" life cycles. Most efficient is simple product reuse as shown by the innermost loop. In many cases, such as with several kinds of automotive parts, the product or its components are remanufactured and reenter the cycle at the point of product assembly. When the

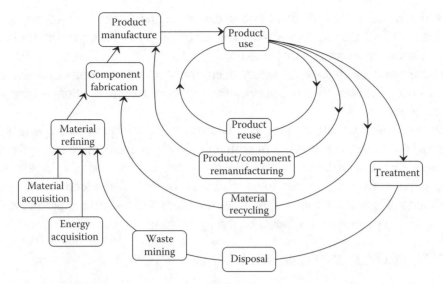

Figure 1.4.   A generalized diagram of life cycles showing various levels of material use. The inner loops are most desirable from the viewpoint of sustainable development.

product or its components cannot be recycled, materials may be recycled; aluminum from cans remelted to produce aluminum metal for manufacture is an example of such a material. Finally, in some cases, wastes are disposed, often after treatment to reduce hazards, and at a later time materials may be extracted from wastes, a process sometimes called **waste mining**.

A life-cycle analysis has four major components:[10]

1. Determination of the scope of the assessment

2. Inventory analysis of materials mass and energy

3. Analysis of impact on the environment, human health, and other potentially impacted areas

4. Improvement analysis

An important early step in life-cycle assessment is to determine the boundaries of time, space, materials, processes, and products to be considered, a process called **scoping**. On a relatively narrow level of scoping, consideration might be given to the life cycle of batteries used in hybrid internal-combustion engine/electric automobiles. The scope would be confined to the battery, itself, with questions raised such as its suitability for recycling over a time period confined to its normal lifetime of several years. A more broadly-based scope could consider alternatives to the battery, such as ultra-high-speed flywheel assemblies to provide for temporary energy storage. Such a scope would have to consider a broader base of technologies that would take some time to develop. An even broader scope would evaluate the need for the automobile and consider alternatives, such as public transportation.

In the **inventory analysis** of life-cycle assessment, the flows of materials are quantified. Usually, energy flows are measured as well. This information enables development of mass and energy balances.

The **impact analysis** has largely been confined to environmental and human health impacts. However, it is important to consider resource, economic, and even societal impacts as well.

Once the above factors have been determined, an **improvement analysis** can be carried out to determine ways in which adverse impacts can be minimized. Several major factors can be considered in an improvement analysis. In some cases, alternate materials can be selected to minimize wastes. Consideration can be given to the kinds of materials that can be reused or recycled. Alternate pathways for the manufacturing process or segments of it may be considered.

1.9. THE ECO-ECONOMY AND ECO-EFFICIENCY

Traditionally, economists have viewed Earth's environment as part of the broader economy. In this view, the environment is regarded as a source of economic wealth — minerals, food, forests, and land on which to place buildings and other anthrospheric structures. These were looked upon as assets to be exploited, not necessarily as precious attributes to be used sustainably and preserved insofar as possible.

Originally coined in a publication entitled "Changing Course" from The World Business Council for Sustainable Development,[11] **eco-efficiency** refers to the affordable provision of goods and services that satisfy human material and quality of life needs with the least possible use of resources and energy while staying within Earth's carrying capacity. Eco-efficient economic systems emphasize the delivery of services, not simply more material goods. By doing more with less and by producing less waste materials with their inherent costs of control and disposal, eco-efficiency is ideally the most profitable way of doing business. By limiting consumption of resources and discharge of pollutants, eco-efficient firms ideally require less regulation, a feature that always appeals to the business community. On a broader scale, eco-efficiency is part of the concept of **sustainable production and consumption**, which seeks to modify both production and consumption patterns so that they are consistent with sustainable use of natural capital.

Eco-efficiency has three major aspects as illustrated in Figure 1.5. Eco-efficient resource use maximizes the value produced per unit of material and energy used. This requires that a manufacturer coordinate closely with raw materials suppliers and processors. It also requires careful consideration of customers. For example, a customer's needs might well be satisfied with an alternative product or substitution of a service for a product that requires less material and energy. By facilitating recycle of consumer products back into the materials stream, less material may be required. Eco-efficient processes emphasize provision of products and services with minimum waste and pollution. In many cases, a product that causes little harm to

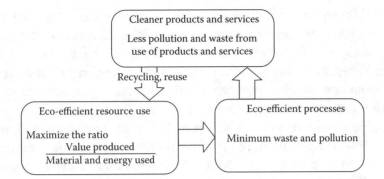

Figure 1.5.   The major aspects of eco-efficiency.

the environment in its use is undesirable because it is produced by processes that tend to generate large quantities of wastes and pollutants. The products and service provided should be designed to generate minimal pollution and waste and to have minimal adverse impact. Products should be designed for reuse and recycling of materials to the maximum extent possible. The World Business Council for Sustainable Development has identified the following essential aspects of eco-efficiency:

- **Dematerialization** by fulfilling economic needs with minimum amounts of materials, especially those from nonrenewable sources (minerals)

- **Substitution** of service and knowledge flows for material flows

- **Closing production loops**, for which natural ecosystems provide excellent models

- **Service extension** by shifting from a supply-driven to a demand-driven economy

- **Functional extension** by manufacturing "smarter" products with enhanced functionality and selling services to increase product functionality.

Eco-efficiency seeks to reduce both the material intensity and the energy intensity of goods and services while increasing the service intensity of goods and services. Dispersion of toxic materials is minimized or eliminated in eco-efficient systems. Eco-efficient products are designed to be as durable as possible, consistent with their intended uses and maximum lifetimes, and are designed for ease of recycling of components and materials.

1.10. DESIGN FOR ENVIRONMENT

A key aspect of eco-efficiency and economic sustainability is **design for environment** consisting of a systematic consideration of environmental performance and potential environmental impacts at the earliest stages of product design and development.[12] Design for environment considers environmental impact at all stages of a product lifetime including raw materials acquisition, manufacturing, packaging, distribution, installation, operation, and ultimate fate at the end of the useful product lifetime. Whereas earlier efforts in pollution prevention focused on incremental improvements on existing processes and products to minimize environmental impact, design for environment concentrates on the entire cycle of manufacturing products and providing services. Therefore, design for environment is much more effective and less costly than more primitive pollution prevention measures.

Design for environment is composed of two broad areas. One of these is **design for sustainability**, which seeks to minimize uses of energy, mineral, material, water, and other resources and aims to preserve natural capital. A second major category is **design for health and safety**. This area seeks to reduce risks from toxic substances, pollutants, and wastes as well as preventing losses from accidents to workers, in transportation, and in use of products.

A number of specific design considerations go into design for environment. Material substitution uses more readily available materials accessible, if possible, from renewable sources and that are more recyclable and require less energy in their production. Materials that are environmentally and toxicologically undesirable and unduly consumptive of resources are avoided. Packaging is minimized. Energy use throughout the cycle of a product from its manufacture, use, and disposal, or recycling is minimized. Products are designed to have a long life. To promote recycling and reuse, products are designed for separability, disassembly, reuse, remanufacture, and recyclability. Items that require disposal are designed for disposal, such as is the case with biodegradable plastics. Components and substances that require disposal preferably are made from combustible materials so that they can be burned for energy, if recycling is not practical. As examples, plastics should not contain chlorine or toxic heavy metals, which cause major problems in emissions and ash when they are burned.

1.11. GREEN PRODUCTS AND SERVICES

A *green product* is one that uses smaller amounts of materials that are less hazardous to produce and that has a lower potential to expose people or the environment to hazardous substances, pollutants, and wastes in its use and disposal. A green service is one that fulfills these criteria in providing a service. For example, a hybrid fuel/electric automobile is a relatively green product with minimal environmental impact, whereas a well-utilized public transportation system based on buses and rail is a green service. Green products and services can improve profitability because of lower requirements for materials and lower costs of disposal and environmental cleanup.

Green products are generally highly durable so long as their durability does not pose undue disposal problems. They are generally reusable, reparable, and remanufacturable. Green products come with minimal, recyclable packaging. In the case of materials used in consumer applications, green products are relatively more concentrated meaning that they have minimum inert ingredients and are more economical to transport. An example is a concentrated liquid laundry detergent requiring only half as much detergent per load of laundry compared to washing powders that contain a large fraction of "filler" ingredients. Green products have minimal toxicities. The extent to which products are green can depend largely upon business or governmental services related to their use. The repairability of a product requires that replacement parts be available. Electrical batteries may be relatively green if recycling facilities are maintained in which they may be collected.

An effective tool in promoting green products and services are **product take back laws**, in which Germany has taken the lead. Product take back includes packaging and packing, which often makes up much of the potential waste involved with marketing a product. By requiring that a vendor takes back a product at the end of its useful life, such laws put much of the responsibility for proper disposition of products and packaging on producers instead of customers.

1.12. TWELVE PRINCIPLES OF GREEN SCIENCE AND TECHNOLOGY

Green chemistry has been guided by Twelve Principles of Green Chemistry later extended to Twelve Principles of Green Engineering. Similarly, it is possible to list **Twelve Principles of Green Science and Technology**, which are the following:

1. With their present activities, humans will deplete Earth's resources and damage Earth's environment to an extent that conditions for human existence on the planet will be seriously compromised or even become impossible. In the past, civilizations have declined and entire populations have died out because they have degraded key environmental systems.

2. The equation below describes burden on, and degradation of Earth's support system; both factors must be addressed:

$$Burden = (number\ of\ people) \times (demand\ per\ person) \qquad (1.12.1)$$

3. Even at the risk of global catastrophe, technology will be used in attempts to meet human needs; therefore, technologies must be designed with a goal of zero environmental impact and maximum sustainability.

4. In the recognition of the reality of Principle 3, it is essential to recognize the anthrosphere as one of five basic spheres of the environment.

5. A key to sustainability is the development of efficiently-used abundant sources of energy that have little or no environmental impact; such sources will require hard decisions and compromise.

6. Climate conducive to life on Earth must be maintained.

7. Earth's capacity for biological and food productivity must be maintained and enhanced; this will require consideration of the interactions of all five environmental spheres.

8. Material demand must be drastically reduced and materials must come from sustainable sources, be recyclable, and, for those that get into the environment, degradable.

9. The production and use of toxic, dangerous, persistent substances should be minimized and such substances should never be discarded to the environment.

10. Human welfare must be measured in terms of quality of life, not just acquisition of material possessions. Economics, governmental systems, creeds, and personal lifestyles must consider environment and sustainability.

11. The risks of not taking risks must be acknowledged.

12. In a word, the goal must be to achieve *sustainability*, a concept in which students and the public must be educated. Although sustainability will require major changes in societal systems, scientists, engineers, and, ultimately, enlightened citizens must take the lead; there is not time for politicians and nonscientists to make up their minds.

Depletion of Earth's Resources and Environmental Destruction

The first step in the achievement of sustainability is to recognize that, on their present course, humans will deplete and damage Earth's crucial support systems to the extent that human existence on the planet with living standards anything like those that prevail today will become impossible.

In 1968, the Stanford University biologist Paul Ehrlich published a book entitled *The Population Bomb*,[13] a pessimistic work that warned Earth had reached its population carrying capacity sometime in the past and that catastrophe loomed. Ehrlich predicted rapid resource depletion, species extinction, grinding poverty, starvation, and a massive dying of human populations in the relatively near future. "Not so," retorted Julian Simon (deceased) a University of Maryland economist writing in a number of books, the most recent of which is titled *Hoodwinking the Nation*.[14] Ehrlich hedged his views by stating that he might be wrong and that "some miraculous change in human behavior" or a "totally unanticipated miracle" might "save the day." Simon expressed the view that Ehrlich's doom and gloom views were nonsense and that human ingenuity would overcome the problems foreseen by him.

The debate between Ehrlich and Simon led to a famous wager by Simon in 1980 that $200 worth of each of five raw materials chosen by Ehrlich — copper, chromium, nickel, tin, and tungsten — would actually decrease in price over the next 10 years in 1980 dollars. Each did in fact decrease in price and Ehrlich paid.

Simon then offered to raise the ante to $20,000, a proposition that Ehrlich declined. This incident is often cited by antienvironmentalists as evidence that we will never run out of essential resources and that a way will always be found to overcome shortages.

However, common sense dictates that Earth's resources are finite. Whereas unexpected discoveries, ingenious methods for extracting resources, and uses of substitute materials will certainly extend resources, a point will inevitably be reached at which no more remains and modern civilization will be in real trouble.

Unfortunately, the conventional economic view of resources often fails to consider the environmental harm done in exploiting additional resources. Fossil fuels provide an excellent example. As of 2006, there was ample evidence that world petroleum resources were strained as prices for petroleum reached painfully high levels. This has resulted in a flurry of exploration activities including even drilling in some cemeteries! Natural gas supplies have been extended by measures such as tapping coal seams for their gas content, often requiring pumping of large quantities of alkaline water from the seams and release of the polluted water to surface waters. There is no doubt that liquid and gaseous fossil fuel supplies could be extended by decades using coal liquefaction and gasification and extraction of liquid hydrocarbons from oil shale. But these measures would cause major environmental disruption from coal mining and processing, production of salt-laden oil shale ash, and release of greenhouse gases.

The sad fact is that on its present course, humankind will deplete Earth's resources and damage its environment to an extent that conditions for human existence on the planet will be seriously compromised or even become impossible. There is ample evidence that in the past civilizations have declined, and entire populations have died out because key environmental support systems were degraded.[15] A commonly cited example is that of the Easter Islands where civilizations once thrived, and the people erected massive stone statues that stand today. The populations of these islands vanished and it is surmised that the cause was the denuding of once abundant forests required to sustain human life on the islands. A similar thing happened to pre-Columbian Viking civilizations in Greenland, where 3 centuries of unusually cold weather and the Vikings' refusal to adopt the ways of their resourceful Inuit neighbors were contributing factors to their demise. Iceland almost suffered a similar fate, but the people learned to preserve their support systems so that Iceland is now a viable country.

Number of People Times Demand Per Person

Equation 1.12.1 shows that both the number of people and the demand that each puts on Earth's resources must be considered in reducing the impact of humans on Earth. Both must be addressed to achieve sustainability.

As of 2005, Earth's human population stood at approximately 6.5 billion people and in July 2006, the U.S. at population passed 300 million. These are staggering numbers to be sure. However, the good news is that these numbers are not nearly as high as those from projections made 40 or 50 years earlier. Even in developing countries, birth rates have fallen to much lower levels than expected earlier. Particularly in Italy, Spain, France, and other nations in Europe, birth rates have fallen to much below the replacement level, and there is concern over depopulation and the social and economic impacts of depleted, aging populations. Even in the U.S., the birth rate has fallen below replacement levels and population growth that is taking place is the result of immigration. The increase in world population that has occurred over the last half century has been more due to decreasing death rates than to increasing birth rates. One U.N. official opined that, "It is not so much that people started reproducing like rabbits that they stopped dying like flies!" Although these trends do not provide room for complacency — explosive population growth could resume — they are encouraging and give hope that the first factor in Equation 1.12.1 may be controlled.

The second factor in Equation 1.12.1, demand per person, may prove to be more intractable. Examination of almost any measure of demand per person, such as consumption of fossil fuel per capita, shows that the highest values of this parameter are found in the more developed countries — the U.S., Canada, Australia, Europe, and Japan. Demand per capita is much less in the highly populous countries of China and India. As the economies of these two giants grow, however, demand for material goods and energy-consuming services will grow as well. For example, if the living standard of the citizens of China were to reach the average of those of Mexico — not considered by most Mexicans to be very high — world petroleum consumption would have to double under conventional economic systems. Were the average person in China to live like the average person in the U.S., an impossible burden would be placed on Earth's carrying capacity. Obviously, ways must be found to meet the basic resource needs per person in more developed countries and means found to deliver a high quality of life to residents of less developed countries without placing unsupportable demands on Earth's resources.

Another point regarding the relationship of population and consumption per capita is that an increase in population in more developed countries has a much greater impact on resources than it does in less developed nations. The addition of one person to the U.S. population has at least 10 times the impact as adding one person to India's population. It may be inferred that immigration into the U.S. and other developed countries from less highly developed nations has an inordinate impact upon resources as the immigrants attain the living standards of their new countries.

Technology Will Be Used

One of the most counterproductive attitudes of some environmentalists is hostility to technology and to technological solutions to environmental problems. Humans are simply not going to go back to living in caves and teepees. Technology is here to stay. And even recognizing that the misuse of technology could result in catastrophe, it will be used to attempt to fulfill human needs. To deny that is unrealistic and foolish.

Therefore, a challenge for modern humankind is to use technology in ways that do not irreparably damage the environment and deplete Earth's resources. The application of technology sustainably is one of the basic tenets of green science and engineering, green chemistry, and industrial ecology. It requires recognition of the anthrosphere as one of the fundamental environmental spheres as discussed below.

The Anthrosphere

In using technology sustainably, it is essential to recognize the *anthrosphere* — structures and systems in the environment designed, constructed, and modified by humans — as one of the five main spheres of the environment. A key to sustainability is reorientation of the anthrosphere so that (1) it does not detract from sustainability and (2) it makes a contribution to sustainability. There is enormous potential for improvement in both of these areas.

Much is already known about designing and operating the anthrosphere so that it does not detract from sustainability. This goal can be accomplished through applications of the principles of industrial ecology discussed in Chapter 16 to Chapter 18. Basically, the anthrosphere must be operated so that maximum recycling of materials occurs, the least possible amount of wastes are generated, the environment is not polluted, and energy is used most efficiently. Furthermore, to the maximum extent possible, materials and energy must come from renewable sources.

The anthrosphere can be designed and operated in a positive way to improve and enhance the other environmental spheres. Much of this endeavor has to do with reversing damage to the environment by previous human activities. Some examples are putting natural meandering pathways into rivers that had been straightened, restoring wetlands that had previously been drained, reforestation of previously deforested lands, and production of desalinated water from saltwater for irrigation.

Energy: Key to Sustainability

With enough energy from sources that are sustainable and nonpolluting almost anything is possible. Toxic organic matter in hazardous waste substances can be totally destroyed and any remaining elements can be reclaimed or put into a form in which they cannot pose any hazards. Wastewater from sewage can be purified to a form in which it can be reused as drinking water. Seawater can be desalinated to provide freshwater for domestic use, industrial use, and irrigation. Pollutants can be removed from stack gas. Essential infrastructure can be constructed.

The accomplishment of sustainability is impossible without the development of efficient, sustainable, nonpolluting sources of energy. Here lies the greatest challenge to sustainability, because the major energy sources used today and based on fossil fuels are inefficient, unsustainable, and, because of the threat to world climate from greenhouse gases, threaten Earth with a devastating form of pollution. Alternatives must be developed.

Fortunately, alternatives are available to fossil fuels, given the will to develop them. Most renewable energy sources are powered ultimately by the sun. The most direct use of solar energy is solar heating. Solar heating of buildings and of water has been practiced increasingly in recent decades and should be employed wherever possible. The conversion of solar energy to electrical energy with photovoltaic cells is feasible and also practiced on an increasing scale. At present, electricity from this source is more expensive than that from fossil fuel sources, but solar electricity is gradually coming down in price and is already competitive in some remote locations far from power distribution grids. A tantalizing possibility is direct solar conversion of water to hydrogen and oxygen gases. Hydrogen can be used in fuel cells and oxygen has many applications, such as in gasification of biomass discussed in Chapter 17. With modern nuclear power reactors and reprocessing of spent fuel, nuclear energy can safely and sustainably provide base-load electrical power for generations to come.

Protection of Climate

The most likely way for humans to ruin the global environment is by modifying the atmosphere such that global warming on a massive scale occurs. The most common cause of such a greenhouse effect is release of carbon dioxide into the atmosphere from fossil fuel combustion as discussed in Chapter 7 and Chapter 9. Human activities are definitely increasing atmospheric carbon dioxide levels and there is credible scientific evidence that global warming is taking place. These phenomena and the climate changes that will result pose perhaps the greatest challenge for human existence, at least in a reasonably comfortable state, on the planet.

The majority of increase in atmospheric carbon dioxide levels is tied with energy and fossil fuel use. Other factors are involved as well. Destruction of forests removes the carbon dioxide-fixing capacity of trees, and the decay of biomass residues from forests releases additional carbon dioxide to the atmosphere. Methane is also a greenhouse gas. It is emitted to the atmosphere by flatulent emissions of ruminant animals (cows, sheep, moose), from the digestive tracts of termites attacking wood, and from anoxic bacteria growing in flooded rice paddies. Some synthetic gases, particularly virtually indestructible fluorocarbons, are potent greenhouse gases as well. The achievement of sustainability requires minimization of those practices that result in greenhouse gas emissions, particularly the burning of fossil fuels.

Unfortunately, if predictions of greenhouse gas warming of Earth's climate are accurate, some climate change inevitably will occur. Therefore, it will be necessary to adapt to warming and the climate variations that it will cause. Some of the measures that will have to be taken are listed as follows:

- Relocation of agricultural production from drought-plagued areas to those made more hospitable to crops by global warming (in the Northern Hemisphere agricultural areas will shift northward)

- Massive irrigation projects to compensate for drought

- Development of heat-resistant, drought-resistant crops

- Relocation of populations from low-lying coastal areas flooded by rising sea levels caused by melted ice and expansion due to warming of ocean water

- Construction of sea walls and other structures to compensate for rising sea levels

- Water desalination plants to produce freshwater to compensate for reduced precipitation in some areas

Maintenance and Enhancement of Biological and Food Productivity

The loss of Earth's biological productivity would certainly adversely affect sustainability and, in the worst case, could lead to massive starvation of human populations. A number of human activities have been tending to adversely affect biological productivity, but these effects have been largely masked by remarkable advances in agriculture such as by increased use of fertilizer, development of highly productive hybrid crops, and widespread irrigation. Some of the factors reducing productivity are the following:

- Loss of topsoil through destructive agricultural practices

- Urbanization of land and paving of large amounts of land area

- Desertification in which once productive land is degraded to desert

- Deforestation

- Air pollution that adversely affects plant growth

Biological productivity is far more than a matter of proper soil conditions. To preserve and enhance biological productivity, all five environmental spheres must be considered. Obviously, in the geosphere, topsoil must be preserved; once it is lost, the capacity of land to produce biomass is almost impossible to restore. Deforestation must be reversed and reforestation of areas no longer suitable for crop production promoted. (Reforestation is happening in parts of New England where rocky, hilly farmland is no longer economical to use for crop production.) In more arid regions where trees grow poorly, prairie lands should be preserved, desertification from overgrazing and other abuse prevented, and marginal crop lands restored to grass.

The hydrosphere may be managed in a way to enhance biological productivity. Measures such as terracing of land to minimize destructive rapid runoff of rainfall and to maximize water infiltration into groundwater aquifers may be taken. Watersheds, areas of land that collect rainwater and which may be areas of high biological productivity should be preserved and enhanced.

Management of the biosphere, itself, may enhance biological productivity. This has long been done with highly productive crops. The production of wood and wood pulp on forest lands can be increased — sometimes dramatically — with high-yielding trees, such as some hybrid poplars. Hybrid poplars from the same genus as cottonwoods or aspen trees grow faster than any other tree variety in northern temperate regions, so much so that for some applications they may be harvested annually. They have the additional advantage of spontaneous regrowth from stumps left from harvesting, which can be an important factor in conserving soil from erosion, particularly on sloping terrain. Furthermore, it may be possible to genetically engineer these trees to produce a variety of useful products in addition to wood, wood pulp, and cellulose.

Proper management of the anthrosphere is essential to maintaining biological productivity. The practice of paving large areas of productive land should be checked. Factories in the anthrosphere can be used to produce fertilizers for increased biological productivity. Massive water desalination plants can be operated that are powered by renewable energy sources (solar and wind) to provide irrigation water for crops.

Reduction of Material Demand

Reduced material demand, particularly that from nonrenewable sources, is essential to sustainability. Fortunately, much is being done to reduce material demand and the potential exists for much greater reductions. Nowhere is this more obvious than in the communications and electronics industries. Old photos of rail lines from the early 1900s show them lined with poles holding 10 or 20 heavy copper wires, each for carrying telephone and telegraph communications. Now far more information than that carried by 10 to 20 wires can be carried by a single thread-sized strand of fiber-optic material. The circuitry of a bulky 1948-vintage radio with its heavy transformers and glowing vacuum tubes has been replaced by circuit chips smaller than a fingernail. These are examples of **dematerialization** and also illustrate **material substitution**. For example, fiber-optic cables are made from silica extracted from limitless supplies of sand whereas the conducting wires that they replace are made from scarce copper.

Wherever possible, materials should come from renewable sources. This favors wood, for example, over petroleum-based plastics for material. Wood and other biomass sources can be converted to plastics and other materials. From a materials sustainability viewpoint, natural rubber is superior to petroleum-based synthetic rubber, and it is entirely possible that advances in genetic engineering will enable growth of rubber-producing plants in areas where natural rubber cannot now be produced.

Materials should be recyclable insofar as possible. Much of the recyclability of materials has to do with how they are used in fabricated products. For example, binding metal components strongly to plastics makes it relatively more difficult to recycle metals. Therefore, it is useful to design apparatus, such as automobiles or electronic devices, in a manner that facilitates disassembly and recycling.

Some materials, by the nature of their uses, have to be discarded to the environment. An example of such a material is household detergent, which ends up in wastewater. Such materials should be readily degradable, usually by the action of micoorganisms. Detergents provide an excellent example of a success story with respect to degradability. The household detergents that came into widespread use after World War II contained ABS surfactant (which makes the water "wetter") that was poorly biodegradable such that sewage treatment plants and receiving waters were plagued with huge beds of foam. The ABS surfactant was replaced by LAS surfactant which is readily attacked by bacteria and the problem with undegradable surfactant in water was solved.

Minimization of Toxic, Dangerous, Persistent Substances

The most fundamental tenet of green chemistry is to avoid the production and use of toxic, dangerous, persistent substances and to prevent their release to the environment. With the caveat that it is not always possible to totally avoid such substances, significant progress has been made in this aspect of green chemistry. Much research is ongoing in the field of chemical synthesis to minimize involvement with toxic and dangerous substances. In cases where such substances must be used because no substitutes are available, it is often possible to make minimum amounts of the materials on demand so that large stocks of dangerous materials need not be maintained.

Many of the environmental problems of recent decades have been the result of improperly disposed hazardous wastes. Current practice calls for placing hazardous waste materials in secure chemical landfills. There are two problems with this approach. One is that, without inordinate expenditures, landfills are not truly "secure" and the second is that, unlike radioactive materials that do eventually decay to nonradioactive substances, some refractory chemical wastes never truly degrade to nonhazardous substances. Part of the solution is to install monitoring facilities around hazardous waste disposal facilities and watch for leakage and emissions. But problems may show up hundreds of years later, not a good legacy to leave to future generations.

Therefore, any wastes that are disposed should first be converted to nonhazardous forms. This means destruction of organics and conversion of any hazardous elements to forms that will not leach into water or evaporate. A good approach toward this goal is to cofire hazardous wastes with fuel in cement kilns; the organics are destroyed and the alkaline cement sequesters acid gas emissions and heavy metals. Ideally, hazardous elements, such as lead, can be reclaimed and recycled for useful purposes. Conversion of hazardous wastes to nonhazardous forms may require expenditure of large amounts of energy.

Quality of Life

One of the greatest challenges in the achievement of sustainability has to do with human attitudes that are hostile to sustainability. It appears to be a natural human instinct to want more material possessions — more "toys," more spacious dwellings, more land, more energy to use for various purposes. Such desires are fed by advertising campaigns, real estate interests that profit by building residential and commercial structures on formerly productive farmlands, and an unceasing quest for more money and higher profits. Nor is the quest for material possessions confined to wealthier societies. The drive for more things is found at all levels of society and in virtually every society on Earth. The problem is that a single-minded quest for the materialistic without regard to environmental effects and sustainability eventually becomes self-destructive and will ultimately destroy the economic systems that have profited from it.

Obviously, human welfare and happiness require certain levels of material possessions and activities that use materials and energy. People need comfortable homes with adequate room, safe and comfortable transportation, adequate nourishing food, and comfortable clothing. But they do not have to have huge homes on enormous lots far from where they work with all of the unsustainable aspects that such dwellings entail, such as large amounts of energy to heat and cool largely unused living space, loss of productive farmland to dwelling lots, consumption of scarce water to keep decorative lawns healthy, and the vehicles and fuel required to commute long distances. They do not have to have monster sport-utility vehicles and pickup trucks for routine transportation when smaller — but still safe — automobiles and minivans using half the fuel of SUVs are perfectly adequate (and more comfortable to drive). These things are not required for happiness and satisfaction and in some cases may even be detrimental to it. They do not require the artery-clogging rich food that has resulted in an epidemic of obesity and its accompanying illnesses now afflicting more affluent societies.

The things most important for true happiness and satisfaction in human existence include adequate, comfortable, conveniently located dwellings; the right amounts of healthy food; satisfying social relationships, good education; good recreational activities; satisfying cultural activities; the best possible health care; access to creed and belief systems that are satisfying to the individual and consistent with sustainability — all supported by strong physical, societal, and governmental infrastructure systems. Fortunately, all of these things can be had sustainably and with minimal consumption of nonrenewable materials and energy. Furthermore, provision of these attributes of a satisfying life can be profitable and consistent with profit-driven systems that have been so successful in providing material goods. The question is, "How do we get there from here?" It is hoped that part of the answer will be found in the pages of this book.

The Risks of No Risks

Some things for which there are no suitable substitutes are inherently dangerous. We must avoid becoming so risk adverse that we do not allow dangerous, but necessary activities (some would put sex in this category) to occur. A prime example is nuclear energy. The idea of using a "controlled atom bomb" to generate energy is a very serious one. But the alternative of continuing to burn large amounts of greenhouse-gas-generating fossil fuels, with the climate changes that almost certainly will result, or of severely curtailing energy use, with the poverty and other ill effects that would almost certainly ensue, indicates that the nuclear option is the best approach.

Therefore, it is necessary to manage risk and to use risky technologies in a safe way. As discussed above, with proper design and operation, nuclear power plants can be operated safely. Modern technology and applications of computers can be powerful tools in reducing risks. Computerized design of devices and systems can enable designers to foresee risks and plan safer alternatives. Computerized control can enable safe operation of processes such as those in chemical manufacture. Redundancy can be built into computerized systems to compensate for failures that may occur. The attention of computers does not wander, they do not do drugs, become psychotic, or do malicious things (although people who use computers are not so sure). Furthermore, as computerized robotics advance, it is increasingly possible for expendable robots to do dangerous things in dangerous areas where in the past humans would have been called upon to take risks.

Although the goal of risk avoidance in green chemistry and green technology as a whole is a laudable one, it should be kept in mind that without a willingness to take some risks, many useful things would never get done. Without risk-takers in the early days of aviation, we would not have the generally safe and reliable commercial aviation systems that exist today. Without the risks involved in testing experimental pharmaceuticals, many life-saving drugs would never make it to the market. Although they must be taken judiciously, a total unwillingness to take risks will result in stagnation and a lack of progress in important areas required for sustainability.

The Achievement of Sustainability

To provide for human needs without a catastrophic collapse of Earth's support systems, *sustainability* must be achieved. The achievement of sustainability will require a massive commitment on the part of governments, industry, and Earth's population. Key roles must be played by educational systems and by the media in educating the populace on the importance of sustainability. An especially important responsibility resides with scientists and technical people in getting information to educators, the media, and common citizens regarding the meaning of sustainability, its importance, and how it may be achieved.

LITERATURE CITED

1. World Commission on Environment and Development, *Our Common Future*, Oxford University Press, New York, 1987.

2. Daly, Herman, *Beyond Growth: The Economics of Sustainable Development*, Beacon Press, Boston, MA, 1996.

3. Hawken, Paul, Amory Lovins, and L. Hunter Lovins, *Natural Capitalism: Creating the Next Industrial Revolution*, Back Bay Books, Boston, MA, 2000.

4. "The Tragedy of the Commons," Garrett Hardin, *Science*, **162**, 1243, 1968.

5. Duncan, Dayton, *Miles from Nowhere*, Viking, New York, 1994, p. 145.

6. Graedel, Thomas E. and Braden R. Allenby, *Industrial Ecology*, 2nd ed., Prentice Hall, Upper Saddle River, NJ, 2003, p. 5.

7. Quist, Jaco, Marjolijn Knot, William Young, Ken Green, and Philip Vergragt, Strategies towards sustainable households using stakeholder workshops and scenarios, *International Journal of Sustainable Development*, **4**, 75–89 2001.

8. Manahan, Stanley E., *Green Chemistry and The Ten Commandments of Sustainability*, 2nd ed., ChemChar Research, Inc., Columbia, MO, 2006.

9. Tenner, Edward, *Why Things Bite Back*, Vantage, New York, 1996.

10. Lankey, Rebecca L. and Paul T. Anastas, Life-cycle approaches for assessing green chemistry technologies, *Industrial and Engineering Chemistry Research*, **41**, 4498–4502 (2002).

11. Schmidheiny, Stephan, *Changing Course: A Global Business Perspective on Development and the Environment*, The MIT Press, Cambridge, MA, 1992.

12. Fiksel, Joseph, Measuring sustainability in eco-design, in *Sustainable Solutions: Developing Products and Services for the Future*, M. Charter and U. Tischner, Eds., Greenleaf Publishing Co., Surrey, U.K., 2000, chap. 9.

13. Ehrlich, Paul R., *The Population Bomb*, Ballantine Books, New York, 1968.

14. Simon, Julian, *Hoodwinking the Nation*, Transaction Publishers, Somerset, NJ, 1999.

15. Diamond, Jared, *Collapse: How Societies Choose to Fail or Succeed*, Viking, New York, 2005.

SUPPLEMENTARY REFERENCES

Allenby, Braden, *Reconstructing Earth: Technology and Environment in the Age of Humans*, Island Press, Washington, D.C., 2005.

Binder, Manfred, Martin Jänicke, and Ulrich Petschow, *Green Industrial Restructuring: International Case Studies and Theoretical Interpretations*, Springer-Verlag, Berlin, 2001.

Caldararo, Niccolo, *Sustainability, Human Ecology, and the Collapse of Complex Societies: Economic Anthropology and a 21st Century Adaptation*, Edwin Mellen Press, Lewiston, New York, 2004.

Clark, James and Duncan MacQuarrie, Eds., *Handbook of Green Chemistry and Technology*, Blackwell Science, Malden, MA, 2002.

Committee on Sustainability of Technical Activities, *Sustainable Engineering Practice: An Introduction*, American Society of Civil Engineers, Reston, VA, 2004.

Cunningham, William P., Mary Ann Cunningham, and Barbara Woodworth Saigo, *Environmental Science: A Global Concern,* 8th ed., McGraw-Hill Higher Education, Boston, MA, 2005.

Doering, Don S., *Designing Genes: Aiming for Safety and Sustainability in U.S. Agriculture and Biotechnology*, World Resources Institute, Washington, D.C., 2004.

Easton, Thomas, *Taking Sides: Clashing Views on Controversial Environmental Issues,* 11th ed., McGraw-Hill, Boston, MA, 2005.

Enger, Eldon D., Bradley F. Smith, and Anne Todd Bockarie, *Environmental Science: A Study of Interrelationships*, 10th ed., McGraw-Hill, Boston, MA, 2006.

Gallopín, Gilberto and Paul D. Raskin, *Global Sustainability: Bending the Curve*, Routledge, New York, 2002.

Goldie, Jenny, Bob Douglas, and Bryan Furnass, Eds., *In Search of Sustainability*, CSIRO Publishing, Collingwood, Victoria, Australia, 2005.

Ikerd, John E., *Sustainable Capitalism: A Matter of Common Sense*, Kumarian Press, Bloomfield, CT, 2005.

Jha, Raghbendra and K.V. Bhanu Murthy, *Environmental Sustainability: A Consumption Approach*, Routledge, New York, 2006.

Kant, Shashi and R. Albert Berry, Eds., *Economics, Sustainability, and Natural Resources*, Springer-Verlag, Berlin, 2005.

Kibert, Charles J., Jan Sendzimir, and G. Bradley Guy, Eds., *Construction Ecology: Nature as the Basis for Green Buildings*, Spon Press, New York, 2002.

Lempert, Robert, *Transition Paths to a New Era of Green Industry: Technological and Policy Implications*, RAND, Santa Monica, CA, 2002.

Mawhinney, Mark, *Sustainable Development: Understanding the Green Debates*, Blackwell Science, Malden, MA, 2002.

Miller, G. Tyler, *Environmental Science: Working with the Earth*, 11th ed., Brooks Cole, Belmont, MA, 2005.

National Academies, *Sustainability in the Chemical Industry: Grand Challenges and Research Needs*, National Academies Press, Washington, D.C., 2005.

Norton, Bryan G., *Sustainability: A Philosophy of Adaptive Ecosystem Management*, University of Chicago Press, Chicago, IL, 2005.

Ooi, Giok Ling, *Sustainability and Cities: Concept and Assessment*, World Scientific Publishing, Hackensack, NJ, 2005.

Olson, Robert and David Rejeski, Eds., *Environmentalism and the Technologies of Tomorrow: Shaping the Next Industrial Revolution*, Island Press, Washington, D.C., 2005.

Redclift, Michael, *Sustainability: Critical Concepts in the Social Sciences*, Routledge, New York, 2005.

Schaper, Michael, Ed., *Making Ecopreneurs: Developing Sustainable Entrepreneurship*, Ashgate, Burlington, VT, 2005.

Schellnhuber, Hans Joachim, Paul J. Crutzen, and William C. Clark, *Earth System Analysis for Sustainability*, MIT Press, Cambridge, MA, 2004.

Schmidt, Gerald, *Positive Ecology: Sustainability and the "Good Life,"* Ashgate, Burlington, VT, 2005.

Sernau, Scott R., *Global Problems: The Search for Equity, Peace and Sustainability*, Allyn and Bacon, Boston, MA, 2006.

Spellman, Frank R. and Nancy E. Whiting, *Environmental Science and Technology: Concepts and Applications*, 2nd ed., Government Institutes, Lanham, MD, 2006.

Steger, Ulrich, *Sustainable Development and Innovation in the Energy Sector*, Springer-Verlag, Berlin, 2005.

Swaan Arons, Jakob, Hedzer van der Kooi, and Krishnan Sankaranarayanan, *Efficiency and Sustainability in the Energy and Chemical Industries*, Taylor & Francis, London, 2004.

Tilbury, Daniella and David Wortman, Engaging People in Sustainability, IUCN, World Conservation Union, Gland, Switzerland, 2004.

Wallace, Bill, *Becoming Part of the Solution: The Engineer's Guide to Sustainable Development,* American Council of Engineering Companies, Washington, D.C., 2005.

Wright, Richard T., *Environmental Science: Toward a Sustainable Future*, 9th ed., Pearson/Prentice Hall, Upper Saddle River, NJ, 2004.

QUESTIONS AND PROBLEMS

1. As noted at the beginning of the chapter, an old Chinese proverb states that, "If we do not change direction, we are likely to end up where we are headed." Prepare a list with two columns, the first labeled "Direction headed" and the second "Where we will go." List several poorly sustainable directions in which we are headed along with the corresponding likely consequences of heading in each direction.

2. Earth appears to be entering a new epoch, the anthropocene. Look up the meaning of epoch. In which epoch has humankind been living? What are some of the past epochs on Earth? What have been the consequences of transitions between epochs?

3. Choose a familiar pollutant based upon your knowledge of pollution phenomena. Using Internet resources, look up the "pollution control" regulations designed to control this pollutant. Then attempt to find a "pollution prevention" alternative for this potential pollutant. Is there any legislation that has promoted pollution prevention? Finally, see if you can find a "design for environment" or a "sustainable development" approach pertaining to the pollutant you chose.

4. At the beginning of Section 1.3 is a description of problems that arose from overuse of shared pastureland, a phenomenon called "The Tragedy of the Commons." Suggest a time frame that describes the decline of the commons. Do you think it was a gradual continuous decline, or do you think there might have been a major discontinuity, sometimes referred to as a "tipping point"? Explain.

5. Consider the potential effects of modern transgenic (genetic engineering) techniques on loss of biodiversity in food crops and livestock that provide food. Why might such loss be a problem? Suggest ways that it might be prevented.

6. In light of Figure 1.4, outline a life-cycle analysis of personal computers. Suggest ways in which computers can be designed and manufactured in ways more amenable to minimizing their environmental impact.

7. One of the essential aspects of eco-efficiency is service extension by shifting from a supply-driven to a demand-driven economy. What do you think this means? Suggest a specific example, such as provision of transportation. Suggest how the demand-driven segment of the economy may be influenced by the supply-driven segment (does advertising play a role?).

8. Consider the durability of green products. In which respects is high durability desirable? Why may it be detrimental in some cases?

9. Consider the first of the Twelve Principles of Green Science and Technology which asserts that "humans will deplete Earth's resources and damage Earth's environment to an extent that conditions for human existence on the planet will be seriously compromised or even become impossible." Either check out the book by Jared Diamond, *"How Societies Choose to Fail or Succeed,"* or look up a detailed review of it on the internet. List some of the ways that societies in the past have failed because of environmental degradation. Suggest how current societies may fail because of destruction of the environment and natural capital.

10. The fourth principle of green science and technology asserts that it is essential to recognize the anthrosphere as one of five basic spheres of the environment. Look up material pertaining to environmental science, such as some of the more popular books on the subject or course curricula dealing with environmental science. How do these sources treat the anthrosphere, if it is mentioned at all? Is this treatment fair and realistic?

11. The fifth principle of green science and technology recognizes the key importance of energy and asserts that provision of sufficient energy will require hard decisions and compromise. Suggest what these decisions may be and the compromises that they may entail.

12. The sixth principle of green science and technology deals with the importance of maintaining climate conducive to life on Earth and the seventh addresses Earth's capacity for biological and food productivity. Suggest ways in which these two principles are strongly interrelated.

13. The eighth principle of green science and technology states that materials should come from sustainable sources and be recyclable to the maximum extent possible. Compile a list of materials that do not come from sustainable sources and possible substitutes that are sustainable.

14. The ninth principle of green science and technology advises minimization of the use of toxic, dangerous, persistent substances. How might this principle conflict with the eleventh, which asserts that there are risks in not taking risks? Suggest how these conflicts may be resolved.

15. The tenth principle of green science and technology addresses the need to consider environment and sustainability in systems of economics and government and in creeds and personal lifestyles. Suggest ways in which these are in conflict with environment and sustainability and ways in which they may be made more compatible.

2. THE FIVE ENVIRONMENTAL SPHERES

2.1. INTRODUCTION

The discussion of green science and technology in this book is organized around five major, interacting environmental spheres: (1) hydrosphere, (2) atmosphere, (3) geosphere, (4) biosphere, and (5) anthrosphere, as shown in Figure 2.1. All of these spheres are introduced in this chapter because it is important to have a basic understanding of what each entails to discuss the remainder of the material in the book. Later in the book, three chapters are devoted to each of the environmental spheres.

Each of the environmental spheres interacts with every other in some respect and in some locations on Earth. For example, fish in water, organisms that are part of the biosphere living in a segment of the hydrosphere, require dissolved oxygen for their respiration. This oxygen comes from the atmosphere. Minerals dissolved in the water come from the geosphere. If the body of water is a reservoir resulting from damming a stream, the dam, its spillway, and its other features are part of the anthrosphere. Countless other examples like this can be given.

The interaction of the various environmental spheres and the exchanges of materials and energy between them are described by cycles known as **biogeochemical cycles**. These cycles are often profoundly affected by human activities. Section 2.7 discusses biogeochemical cycles and some of their aspects appear in later sections of the book.

Often, the relatively thin boundaries between environmental spheres are particularly important. One such interface is that at a soil surface comprising a boundary between the geosphere and the atmosphere. It is here where most of the plants that support life on Earth grow. This region typically extends only a meter or so into the atmosphere and a similar distance into soil. Within this layer, itself, are boundaries between the plant roots and soil, a biosphere/geosphere boundary, and plant leaf surfaces and the atmosphere, and a biosphere/atmosphere boundary.

2.2. THE HYDROSPHERE

Water (chemical formula H_2O) comprises the hydrosphere (Figure 2.2). As discussed in Chapter 4, although it has a simple chemical formula, water is actually a

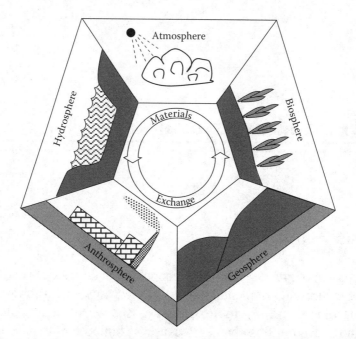

Figure 2.1.   There are five major spheres of the environment. Strong interactions, especially exchanges of materials and energy, occur among them and they are very much involved with bio-geochemical cycles.

very complex substance. It participates in one of the great natural cycles of matter, the hydrologic cycle discussed in some detail in Chapter 4, Section 4.3, and illustrated in Figure 4.2. Basically, the hydrologic cycle is powered by solar energy that evaporates water as atmospheric water vapor from the oceans and bodies of freshwater from where it may be carried by wind currents through the atmosphere to fall as rain, snow, or other forms of precipitation in areas far from the source. In addition to carrying water, the hydrologic cycle conveys energy absorbed as latent heat when water is evaporated by solar energy and released as heat when the water condenses to form precipitation.

Figure 2.2 illustrates major aspects of the hydrosphere and its relationship to the other four environmental spheres. About 97.5% of the water in the hydrosphere is contained in the oceans leaving only 2.5% as freshwater. Furthermore, about 1.7% of Earth's total water is held immobilized in ice caps in polar regions and in Greenland. This leaves only 0.77% of Earth's water — commonly designated as freshwater — which is regarded as potentially accessible for human use.

As shown in Figure 2.2, freshwater occurs in several places. Surface water is found on land in natural lakes, rivers, impoundments or reservoirs produced by damming rivers, and underground as groundwater. Water occurs on and beneath the surface of the geosphere. Groundwater is an especially important resource and one susceptible to contamination by human activities. Through erosion processes, moving water shapes the geosphere and produces sediments that eventually become sedimentary deposits, such as vast deposits of limestone. Water is essential to all

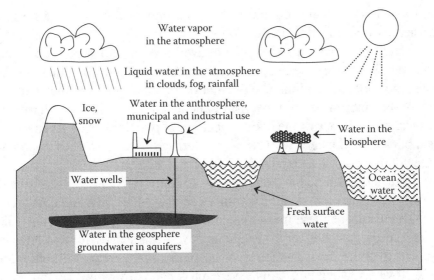

Figure 2.2.   The main aspects of the hydrosphere and its relationship to the other environmental spheres. Water is driven through the hydrosphere by the solar-energy-powered hydrologic cycle, which uses solar energy to evaporate water from the oceans and wind currents to carry water vapor through the atmosphere to other locations where it falls as precipitation.

organisms in the biosphere. Rooted plants draw water from soil and carry it to the plant leaves where it evaporates through a process called *transpiration*. This enables movement of nutrients from plant roots to leaf tissues.

Water is the most widely used substance by organisms in the anthrosphere. Humans use water in their households for drinking, food preparation, cleaning, and disposal of wastes. Moving water is perhaps the oldest form of power harnessed by humans, through waterwheels dating back to more than 2000 years, and is the leading current source of renewable energy through hydroelectric power production. Water converted to steam (hot water vapor) is widely used for heat transfer in industry and in buildings and is the largest means of electrical power generation by running through steam turbines coupled to electrical generators. Humans draw water from rivers, from impoundments made by damming rivers, and by pumping from underground aquifers.

Water is a key resource in the maintenance of sustainability. Shortages of water from climate-induced droughts have been responsible for the declines of major civilizations. Variations in water supply cause severe problems for humans and other life-forms throughout the world. Devastating floods displace and even kill large numbers of people throughout the world and destroy homes and other structures. Severe droughts curtail plant productivity resulting in food shortages for humans and animals in natural ecosystems and often necessitating slaughter of farm animals during catastrophic droughts. It is feared that both drought and the severity of occasional flooding will become much worse as the result of global warming brought on by rising carbon dioxide levels in the atmosphere (see Chapter 9, Section 9.2).

Both water quality and quantity are important factors in sustaining healthy and prosperous human populations. Water is an important vector for disease. In the past, waterborne cholera and typhoid have killed millions of people and these and other waterborne diseases, especially dysentery, are problems in less developed areas lacking proper sanitation. The prevention of water pollution has been a major objective of the environmental movement, and avoiding discharge of harmful water pollutant chemicals is one of the main objectives of the practice of green technology. Water supplies are a concern with respect to terrorism because of their potential for deliberate contamination with biological or chemical agents.

For centuries, humans have endeavored to manage water by measures such as building reservoirs to store water for future use and with dikes and dams to control flooding. The results of these measures have been mixed. Typically, construction of reservoirs to provide water for arid regions have been successful and have enabled development in these areas, which may occur for many years without significant problems. However, when unusually severe, prolonged droughts do occur, the effects are exacerbated by the fact that control of water supplies has enabled excessive growth in water-deficient areas. The Las Vegas metropolitan region of the U.S. and Mexico City in Mexico are examples of metropolitan regions that have outgrown the natural water capital available to them. Also, construction of river dikes has enabled agricultural and other development in flood-prone areas. But when a record flooding event occurs, such as happened with a 500-year flood along the Missouri river in 1993 or Hurricane Katrina in New Orleans in 2005, failure of the protective systems causes much greater devastation than would otherwise be the case. Following the 1993 Missouri River flood, sensible actions were taken in some areas where farm property along the Missouri River was purchased by government agencies and allowed to revert to wildlife habitat in its natural state, which included periodic flooding. It would be most sensible for the future of New Orleans to relocate areas located below sea level and flooded by levee failure in 2005 to higher ground and to avoid trying to thwart the natural tendency of water to seek lower levels where humans may try to live.

Problems with water supply are discussed in Chapter 4. Figure 4.5 in Chapter 4 shows rainfall patterns in the continental U.S. It is seen that the eastern U.S. has generally adequate rainfall. However, the western U.S. is water deficient. Furthermore, some of the fastest growing areas of the U.S. are in the most water-deficient areas including southern California, Arizona, Nevada, and Colorado. Severe water supply problems exist in other parts of the world, such as sections of Africa and the area of Palestine and Israel.

The world abounds with examples of groundwater depletion, one of the most obvious manifestations of water overuse. Water pumped from below the ground in Mexico City, which is built on an old lake bed, has caused much of the city to sink, damaging many of the structures in it. In the U.S., wasteful use of groundwater is illustrated by the depletion of the High Plains Aquifer commonly called the Ogallala Aquifer. Shown in Figure 2.3, this formation lies beneath much of Nebraska,

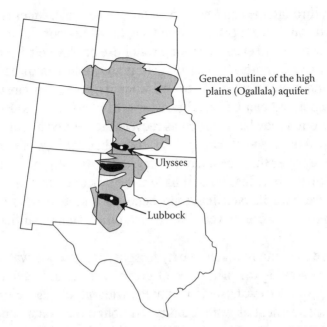

Figure 2.3.   General outline of the Ogallala aquifer, the largest source of fresh groundwater in the world. Although the entire aquifer has suffered significant depletion since the mid-1900s, the problem is especially severe in areas around Ulysses, Kansas, Lubbock, Texas, and the northern Texas panhandle outlined in black.

western Kansas, the Oklahoma and Texas panhandles, and small sections of eastern Wyoming, Colorado, and New Mexico. Although it is recharged from surface water in parts of Nebraska, it is largely composed of fossil water from the last Ice Age. It contains an astounding amount of water, enough to cover the entire U.S. to a depth of around 1.5 m!

Since the 1940s, water has been pumped from the Ogallala aquifer in enormous amounts to support the growth of corn and other crops not normally adapted to the High Plains region. The result has been a rapidly dropping water table (level reached by water in a well drilled into an aquifer), exceeding 50 m in some areas. In the middle of Kansas' irrigated corn belt around Ulysses, the water level dropped by 6 m in just the decade from 1995 to 2005. Clearly, such water exploitation is unsustainable and will force a shift from thirsty crops, such as corn, to those that require less water, such as milo.

The depletion of water supplies has numerous implications for sustainability. Clearly, the U.S. can meet its food demands without exploiting the Ogallala aquifer and other diminishing sources of water. However, as of 2006, a major thrust of the U.S. energy plan — such as it is — has been increased production of ethanol from corn and biodiesel fuel from soybeans. Both of these crops require large amounts of water, and irrigation would have to be practiced on a large scale to produce enough to have an impact on fuel supplies. Some underground water supplies should be regarded as depletable resources and reserved for municipal water supplies and manufacturing, which require only a fraction as much water as does irrigation.

Fortunately, through the solar-powered hydrologic cycle, water is one of nature's most renewable resources. As discussed in Chapter 4, Section 4.4 and illustrated in Figure 4.4, the trend toward ever-increasing water use has leveled off in the U.S. since about 1980 because of more efficient irrigation and industrial processes. It should be kept in mind that water is never destroyed when it is used and substantial amounts that have not evaporated can be reclaimed for other applications. Even water infiltrating into the ground may be regarded as recycled water because it renews groundwater sources. Although the thought of so doing has largely prevented efforts to completely recycle water that has been through domestic sewage systems, recycling of this water following purification will have to be practiced in some water-deficient areas in the future and, in fact, has long been the practice where municipalities take their water supplies from rivers into which other municipalities have dumped sewage.

Humans have long manipulated the hydrosphere to provide water, often with ill effect. However, following the principles of green science and technology, it is now possible to supply water to water-deficient areas without damaging the environment and even with enhancement of water quality. Although the major southwestern U.S. rivers, the Colorado and Rio Grande, are overutilized, the Mississippi river maintains an enormous flow that could be tapped for water-deficient areas. One possibility is to divert a fraction of this flow near the mouth of the Mississippi and pump the water using abundant wind power to arid regions of the U.S. Southwest and northern Mexico. By constructing wetlands near the point of the diversion, sediment could be collected from the river water, and aquatic plants growing in the wetlands could remove nutrients that are now harmful to water quality in the Gulf of Mexico.

Groundwater recharge is another key to water sustainability that is being practiced in parts of the world. Most groundwater recharge occurs naturally, although it has been reduced by paving and surface modifications in the anthrosphere. Anthrospheric constructs on the surface can be designed to maximize recharge. For example, some paving surfaces in China have been made of porous materials that allow water to penetrate into the ground below. Two of the more active approaches to groundwater recharge are shown in Figure 2.4. One of these is water pumped into a shaft that extends underground and even to the aquifer itself. Surfaces of these conduits can become clogged with silt, bacterial growths, and other materials suspended in the recharge water and may have to be cleaned periodically. A spreading basin consists of a reservoir of water excavated into porous geospheric material from which water flows into the aquifer. An advantage is the purification of water that occurs through contact with mineral matter, but the process does not work well if aquifers are overlain by poorly pervious layers.

2.3. THE ATMOSPHERE

The atmosphere is a layer of gases with a total mass of about 5.15×10^{15} tons that surrounds the solid Earth. The density of atmospheric air rapidly becomes less

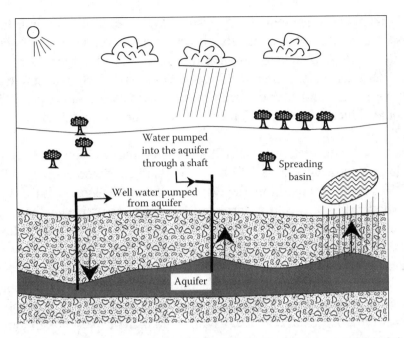

Figure 2.4.   Artificial groundwater recharge. The two methods shown here are a shaft for pumping water directly into the aquifer and a spreading basin from which water percolates into the aquifer. The arrows above the aquifer show that the water level is lowered by pumping from a well and raised by recharge through a shaft or spreading basin.

with increasing altitude. People who travel to the tops of higher mountains readily notice the lower density of the air in reaching altitudes of only 3000 or 4000 m. The air is so thin at the approximately 13,000 m altitude at which commercial jet air-craft fly, that a person exposed to air at this altitude could remain conscious for only about 15 sec without supplementary oxygen. More than 99% of the atmosphere's mass is within 40 km of Earth's surface, with the majority of the air below 10-km altitude (compared to Earth's diameter of almost 13,000 km). Although there is no clearly defined upper limit to the atmosphere, at altitudes exceeding 1000 km, air molecules are lost to space and this may be considered as a practical upper limit to the atmosphere.

Humans think first of the atmosphere as a source of oxygen essential to their existence; deprived of oxygen, they expire within a few minutes. In addition to pro-viding oxygen, air provides many other support functions essential for life. One of these is to act as a blanket to keep Earth's surface at an average temperature of about 15°C at sea level and within a temperature range that enables life to exist. As well as maintaining temperature levels conducive to life, Earth's atmosphere absorbs very short wavelength ultraviolet radiation from the sun, which, if it reached organisms on Earth's surface, would tear apart the complex biomolecules essential for life. In addition to providing oxygen required by animals, the atmosphere is the source of carbon dioxide, which green plants absorb from the atmosphere and convert into bio-mass. The atmosphere's low percentage of carbon dioxide (only about 0.038%) is the source of carbon for the biomass in all food and biomass-based raw materials. The chemically bound nitrogen that all organisms require for their protein is taken from the atmosphere, which consists mostly of elemental nitrogen gas. Bacteria growing on the roots of some plants accomplish this task as do nitrogen-fixing industrial processes operating under rather severe conditions. It should also be noted that the atmosphere contains and carries water vapor evaporated from oceans that forms rain and other kinds of precipitation over land in the hydrologic cycle (Chapter 4, Figure 4.2).

Dry air (excluding water vapor) is 78.1% by volume nitrogen gas (N_2), 21.0% O_2, 0.9% noble gas argon, and 0.038% CO_2, a level that keeps increasing by a little more than 0.001% per year. In addition, there are numerous trace gases in the atmosphere at levels below 0.002% including ammonia, carbon monoxide, helium, hydrogen, krypton, methane, neon, nitrogen dioxide, nitrous oxide, ozone, sulfur dioxide, and xenon. The atmosphere is 1 to 3% by volume water vapor, though the level is somewhat higher in tropical regions and lower in some areas. As air becomes cooler with increasing altitude, water vapor condenses into droplets and the air becomes drier.

Figure 2.5 shows some of the main aspects of the atmosphere and its relationship to other environmental spheres. Living organisms experience the lowest layer called the **troposphere** characterized with decreasing temperature and density with increasing altitude. The troposphere extends from surface level, where the average temperature is 15°C, to about 11 km (the approximate cruising altitude of commercial jet aircraft) where the average temperature is –56°C. The very cold region at the upper limits of the troposphere is called the *tropopause*. It is especially significant for life on Earth because water vapor reaching have forms ice crystals which tend to settle. Were this not the case, over the eons the intense ultraviolet radiation above the tropopause would long ago have broken apart the water molecules in the atmosphere and the very light elemental hydrogen (H_2) produced would have escaped to outer space leadng to a dry planet that could not support life.

Above the troposphere is the **stratosphere** in which the average temperature increases from about –56°C at its lower boundary to about –2°C at its upper limit. The heating effect in the stratosphere is because of energy absorbed by air molecules from the intense solar radiation that impinges on it. Because the ultraviolet radiation from the sun can break apart molecules of elemental oxygen (O_2) in the stratosphere, it contains significant levels of oxygen atoms (O) and a form of elemental oxygen consisting of molecules of three O atoms (O_3) called ozone. Although humans do not venture into the stratosphere, unless they are inside pressurized suits or aircraft compartments, this layer of the atmosphere is critical for sustaining the lives of humans and virtually all other organisms. That is because the very dilute ozone layer, which, if it were in a pure form at sea level, would be only 3 mm thick, absorbs a portion of ultraviolet radiation, which otherwise would penetrate to Earth's surface and destroy essential life biomolecules in exposed plants and animals. Some classes of chemical species, especially the chlorofluorocarbons or freons used as refrigerants, are known to react in ways that destroy stratospheric ozone, and their elimination from commerce has been one of the major objectives of efforts to sustain life on

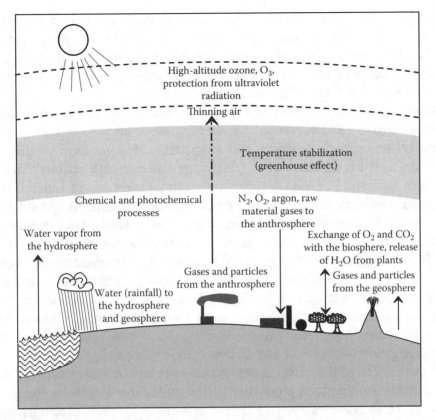

Figure 2.5.   The atmosphere in relation to the other environmental spheres. In addition to being a source and sink of materials, including gases essential for life, the atmosphere provides vital protective temperature-stabilizing and ultraviolet radiation-absorbing functions, is the site of numerous chemical and photochemical reactions (those involving ultraviolet radiation and light), and is the conduit for water transport in the hydrologic cycle.

Earth. Above the stratosphere are the atmospheric mesosphere and thermosphere, which are relatively less important in the discussion of the atmosphere.

Earth's atmosphere is crucial in absorbing, distributing, and radiating the enormous amount of energy that comes from the sun. The flux of solar energy is 1340 W/m^2, which means that a square meter of surface directly exposed to sunlight unfiltered by air would receive energy from the sun at a power level of 1,340 W, ample to power an electric iron or thirteen 100-W light bulbs plus a 40-W bulb! The energy from the sun is in the form of electromagnetic radiation, which has a wavelike character in which shorter wavelengths are more energetic. The incoming radiation is mostly in the form of visible light and more energetic forms with a maximum intensity at a wavelength of 500 nm (1 nm = 10^{-9} m). This energy is largely absorbed and converted to heat in the atmosphere and at Earth's surface. Heat energy is radiated back out into space as infrared radiation between about 2 µm and 40 µm (1 µm = 10^{-6} m) with a maximum intensity at about 10 µm). As this energy is radiated back out through the atmosphere, it is delayed by being reabsorbed by water molecules, carbon dioxide, methane, and other minor species in the atmosphere. The result is that Earth's atmosphere is kept warmer than it would be without these energy-absorbing gases, a greenhouse effect that is very important in sustaining life on Earth. As discussed in Chapter 9, anthrospheric discharges of greenhouse gases, especially carbon dioxide and methane, are likely causing an excessive greenhouse effect, which will have harmful effects on global climate.

Meteorology is the science of the atmosphere familiar to most people from (sometimes erroneous) forecasts of weather including such factors as rain, wind, cloud cover, atmospheric pressure, and temperature. Longer term weather conditions are described as **climate**, such as the warm, low-humidity weather that prevails in southern California or the cool, generally rainy conditions of Ireland. Weather and climate conditions are driven by the fact that the incoming flux of solar energy is very intense in regions around the equator and very low in polar regions due to the angles at which the solar flux impinges Earth in these regions. Heated equatorial air tends to expand and flow away from the equatorial regions, creating winds and carrying large quantities of energy and water vapor evaporated from the oceans with it. As the air cools, water vapor condenses forming precipitation and warming the air from heat released when water goes from a vapor to a liquid. When this occurs on a massive scale, huge, even catastrophic weather events with torrential rainfalls and damaging winds can result as hurricanes and typhoons.

Climate is one of the most important factors in sustainability, especially for its key role in maintaining food productivity. If warming due to greenhouse gas emissions to the atmosphere causes widespread drought in formerly productive agricultural regions, the ability to support human populations with sufficient food will be severely curtailed. Favorable climate conditions and a relatively unpolluted atmosphere are important parts of natural capital. Meteorological phenomena have an important influence on air pollution. Winds may carry pollutants for some distance, such as occurs when air polluted with sulfur dioxide from coal-burning power plants in the Ohio Valley is carried to New England or parts of Canada where it forms acid rain. Temperature inversions are atmospheric phenomena in which a warmer layer of air overlays a cooler one, thereby limiting the normal vertical circulation of air. This phenomenon can enable stagnant masses of polluted air to remain in place such that photochemical smog formed by the action of sunlight on atmospheric regions contaminated with hydrocarbons and nitrogen oxides gets "cooked up" in the stationary body of air.

2.4. THE GEOSPHERE

The **geosphere** is the solid Earth. It is the medium on which most food is grown and is a source of metals, plant fertilizers, construction materials, and fossil fuels. Large quantities of consumer and industrial wastes have been discarded to the geosphere, a practice that is ultimately unsustainable. As shown in Figure 2.6, the geosphere interacts strongly with the hydrosphere, atmosphere, biosphere, and

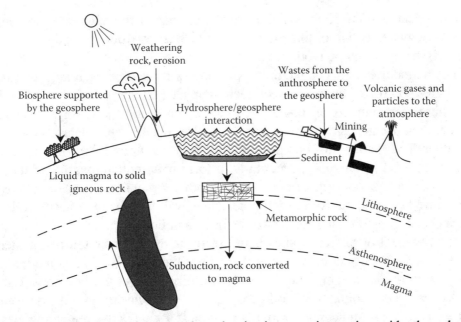

Figure 2.6.   Representation of the geosphere showing its strong interactions with other spheres of the environment.

anthrosphere. The geosphere is obviously a huge part of natural capital, and managing and preserving it is of utmost importance to achieve sustainability.

The geosphere consists of an outer solid layer of Earth composed of rocks, the minerals that make up rocks, soil, and sediments; a hot liquid layer of molten rock constituting the outer core; and the highly compressed iron-rich inner core which, though extremely hot, is solid because of the extreme pressures to which it is subjected. The outermost solid layer of Earth is the lithosphere composed of relatively strong solid rock, varying in thickness from just a few to as much as 400 km, averaging about 100 km. Earth's crust is the outer layer of the lithosphere, which is only 5 to 40 km thick. Below the lithosphere — to a depth of approximately 250 km and resting on the viscous liquid rock mantle — is the asthenosphere composed of hot rock that is relatively weak and plastic.

The shape of solid Earth is that of a geoid defined by the levels of the oceans and a hypothetical sea level beneath the continents. Rather than being a perfect sphere, the Earth geoid is somewhat pear-shaped because of differences in gravitational attraction in different parts of Earth. Although humans have flown hundreds of thousands of kilometers into space, they have been unable to penetrate more than a few kilometers into Earth's crust.

Whereas the asthenosphere is largely continuous, the lithosphere is divided into huge lithospheric plates that behave as units. The continents and the Pacific ocean rest upon their own plates. The lithospheric plates move relative to each other by, typically, a few centimeters per year. Earthquakes arise from abrupt movement of plates relative to each other, and volcanoes that emit hot and molten rock, ash, and gases are located along plate boundaries. The science of the relative movements of lithospheric plates is that of plate tectonics, which dates from about the mid-1900s and is still being developed today.

The boundaries between tectonic plates are of three types. Divergent boundaries are those on ocean floors between tectonic plates that are moving apart and are where hot magma undergoes upwelling and cooling to form new solid lithospheric rock, creating ocean ridges. Convergent boundaries occur where plates move toward each other. The matter thus compressed may be forced downward into the asthenosphere in subduction zones, eventually to form new molten magma. Matter that is forced upward at convergent boundaries produces mountain ranges. Where two plates move laterally relative to each other, transform fault boundaries exist. Earthquakes occur along the fault lines of transform fault boundaries.

A key aspect of plate tectonics is the **tectonic cycle** illustrated in Chapter 10, Figure 10.3. In this cycle, there is upwelling of molten rock magma at the boundaries of divergent plates. This magma cools and forms new solid lithospheric material. At convergent boundaries, solid rock is forced downward and melts from the enormous pressures and contact with hot magma at great depths, reforming magma. This cycle and the science of plate tectonics explain once-puzzling observations of geospheric phenomena including the opening and spreading of ocean floors that create and enlarge oceans, the movement of continents relative to each other, formation of mountain ranges, volcanic activity, and earthquakes.

There are about 2000 known minerals composing Earth and characterized by definite chemical composition and crystal structure. Most rocks in the geosphere are composed of only about 25 common minerals. The crust is 49.5% oxygen and 25.7% silicon, and this is reflected in the chemical composition of rocks, which are mostly various chemical combinations of oxygen and silicon with smaller amounts of aluminum, iron, carbon, sulfur, and other elements. Only about 1.6% of Earth's crust consists of the kinds of rock that must serve as important resources of metals (other than aluminum and iron), phosphorus required for plant growth, and other essential minerals. Careful management of this resource of essential minerals is one of the primary requirements for sustainability.

Molten rock that penetrates to near the top of Earth's crust cools and solidifies to form igneous rock. Exposed to water and the atmosphere, igneous rock undergoes physical and chemical changes in a process called *weathering*. Weathered rock material carried by water and deposited as sediment layers may be compressed to produce secondary minerals, of which clays are an important example. Molded and heated to high temperatures to make pottery, brick, and other materials, clays were one of the first raw materials used by humans and are still widely used today.

A crucial part of the crust is the thin layer of weathered rock, partially decayed organic matter, air spaces, and water composing soil (see Chapter 11), obviously of great importance because it supports plant life that provides food on which humans and other animals depend for their existence. The top layer of soil that is most productive of plants is topsoil, which is often only a few centimeters thick in many locations, or even nonexistent where poor cultivation practices and adverse climatic conditions have led to its loss by wind and water erosion. The conservation of soil and enhancement of soil productivity are key aspects of sustainability (see Chapter 11).

All other environmental spheres are tied strongly with the geosphere. With perhaps the exception of a few communications satellites orbiting around Earth in space, practically all other things and creatures commonly regarded as parts of Earth's environment are located on, in, or just above the geosphere. The atmosphere exchanges gases with the geosphere. For example, organic carbon made by photosynthetic plants from carbon dioxide from the atmosphere may end up as soil organic matter in the geosphere, and the photosynthetic processes of plants growing on the geosphere put elemental oxygen back into the atmosphere. Major segments of the hydrosphere including the oceans, rivers, and lakes rest on the geosphere, and groundwater exists in aquifers underground. The majority of biomass of organisms in the biosphere is located on or just below the surface of the geosphere. Most structures that are parts of the anthrosphere are located on the geosphere, and a variety of wastes from human activities are discarded to the geosphere. Modifications and alterations of the geosphere can have important effects on other environmental spheres and vice versa.

2.5. THE BIOSPHERE

The **biosphere** includes living organisms and the materials and structures that they produce. Here is a brief introduction to the biosphere as a distinct sphere of the environment. Biology and its chemistry are discussed in Chapter 3 and the role of the biosphere in green science and green technology is covered in Chapter 13 to Chapter 15.

Biology is the science of the living organisms in the biosphere. Living organisms are characterized by particular classes of biochemical substances (biomolecules). These include proteins that are the basic building blocks of organisms, carbohydrates made by photosynthesis and metabolized by organisms for energy, lipids (fats and oils), and the all-important nucleic acids (DNA, RNA), genetic material which defines the essence of each individual organism and acts as codes to direct protein biosynthesis and reproduction. In the biosphere a hierarchical organization prevails, beginning at the molecular level and progressing through living cells, organs, organisms, and the biosphere itself. Organisms are capable of carrying out metabolic processes in which they chemically alter substances. Organisms reproduce and their young undergo various stages of development. And organisms demonstrate heredity and may be altered by mutation processes resulting from modifications of DNA.

Organisms comprising the biosphere may be as simple as individual cells of bacteria and photosynthetic algae so small that a microscope is required to view them or as complex as human beings consisting of a variety of organs and with brains that can carry out thought and reasoning processes. Biologists recognize six kingdoms of organisms. Three of these may exist as single cells. Archaebacteria and eubacteria do not have defined cell nuclei. Protists are single-celled but do have defined nuclei in which cellular DNA is located and may have distinct structures, such as mouthlike structures for ingesting food and light-sensitive structures that act as primitive "eyes." At more complex levels are generally multicelled plantae (plants) and animalia (animals), as well as fungi including yeasts, molds, and mushrooms.

A group consisting of the same species of organisms makes up a population. Several populations coexisting in the same location constitute a community. An ecosystem is made up of interacting communities in a defined environment with which they interact. The entire biosphere is composed of all the ecosystems on Earth. A key aspect of a biological ecosystem is its productivity, that is, its ability to produce biomass by photosynthesis. The organisms that photosynthetically remove carbon dioxide from the atmosphere and fix it in the form of organic matter that is further converted by biochemical processes to proteins, fats, DNA, and other life molecules constitute the basis of the whole ecosystem food chain on which the remainder of the organisms in the food chain depend for their existence. The major photosynthetic organisms in terrestrial ecosystems are plants growing in soil. In water, the major contribution is made by algae, plants growing suspended in water. Some protozoa and some bacteria also have photosynthetic capabilities.

Figure 2.7 illustrates some of the major aspects of the biosphere. As shown in Figure 2.7, dominant plant species anchor the community as major producers of biomass. In addition to providing most of the food through photosynthesis that the rest of the ecosystem uses, the dominant plant species often act to modify the physical environment of the ecosystem in ways that enable the other species to exist in it. For

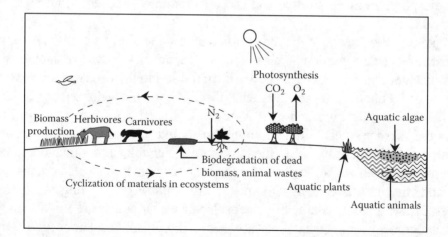

Figure 2.7.   The biosphere. Organisms in the biosphere occupy ecosystems in which materials undergo complete recycle. These systems are based upon photosynthetic plants and (in water) photosynthetic algae that form the basis of the food chain in different ecosystems. Photosynthesis removes carbon dioxide from the atmosphere and returns elemental oxygen. Elemental nitrogen from the atmosphere is chemically combined into biomolecules by the action of nitrogen-fixing bacteria, such as those that grow on legume plant roots. Dead biomass is broken down to simple inorganic materials by fungi and bacteria.

example, the trees in a forest ecosystem provide the physical habitat in which birds can nest, relatively safe from predators. In addition, the trees provide shade that significantly modifies the habitat at ground level, producing a hospitable microclimate and preventing the growth of most kinds of low-growing plants.

Fossil evidence shows the enormous changes that the biosphere has undergone over millions of years of evolution and climatic change. Recent human influences on the biosphere have also caused dramatic, much more rapid changes. The single most important event was the colonization of the Western Hemisphere following Columbus' 1492 voyage of discovery. Ecosystems that had evolved separately in the two hemispheres over millions of years were suddenly united with sometimes striking results. In a phenomenon called *ecological release*, populations of some species introduced into new ecosystems without their natural predators exploded. Clover and bluegrass introduced from Europe into North America grew explosively and became firmly established in what is now the state of Kentucky before the first settlers reached that region. Peach trees introduced into the Carolinas and Georgia grew so well that fears were expressed over the potential development of a "peach tree wilderness." Andean valleys were inundated with mint introduced into Peru. In a reverse exchange, corn from the Western Hemisphere became rapidly established in Europe and Africa. It is possible that the exceptionally high food productivity of corn enabled a population explosion in Africa so great that slave traders were able to take millions of people from the continent without seriously depleting its population. Smallpox introduced to North America by Europeans decimated Native American populations, reducing populations of some tribes by 90%.

The biosphere is strongly connected to the other environmental spheres. It has strong effects on these spheres and, in turn, is strongly influenced by them. Some of the major effects are the following:

- Organisms in the biosphere generate a variety of materials that humans use.

- Conditions in the geosphere, especially of soil, are crucial in determining biospheric productivity.

- The availability of water in the hydrosphere has a strong influence on the types and growth of organisms.

- The biosphere is strongly influenced by large quantities of pesticides and fertilizers produced by anthrospheric processes to enhance the health and productivity of organisms, particularly farm crops, in the biosphere.

- Most of the major efforts to control pollutants, wastes, and toxic substances from the anthrosphere are designed to protect organisms, including humans, in the biosphere.

- The nature of the anthrosphere and the ways in which it is operated strongly influence the nature of the biosphere and its productivity.

As well as being strongly influenced by the other environmental spheres, the biosphere has a strong influence upon the other spheres. For example, the oxygen that all humans breathe was put into the atmosphere by the actions of photosynthetic bacteria eons ago. Lichens, mixed communities of algae and fungi growing synergistically (to mutual advantage), act upon geospheric rock to break it down to soil that is hospitable to plant life. In less industrialized societies, the nature of the anthrosphere made and operated by humans depends strongly upon the surrounding biosphere. Abundant herds of bison provided the sturdy hides used by Native Americans for their mobile-home teepees. The availability and kinds of wood have largely determined the nature of dwellings constructed in many societies. More than 2000 years ago, domesticated animals harnessed to carts, wagons, and plows provided humans with mobility and the means to cultivate soil, and many societies, including the Amish farmers in the U.S., still use horses, donkeys, mules, oxen, and water buffalo for such purposes. As petroleum dwindles as a source of raw materials, such as plastics, it is likely that materials produced in the biosphere will play a stronger future role in determining the kinds of materials used in the anthrosphere.

Over eons of evolution, the biosphere has developed as an exemplary survivor that can teach humans important lessons in achieving sustainability. One important factor is that organisms carry out their activities under very mild conditions, which is one of the goals of green technology. Even the conditions of boiling water in which some thermophilic bacteria grow in locations such as the thermal hot springs of Yellowstone National Park are mild compared to the high-temperature, high-pressure conditions required in many chemical syntheses. Furthermore, the intolerance of living organisms to toxic substances demonstrates ways in which such substances can be avoided in performing chemical processes in the anthrosphere.

Over hundreds of millions of years of evolution, organisms have had to evolve sustainable ecosystems for their own survival, completely recycling materials and enhancing their environment. In contrast, humans have behaved in ways that are unsustainable with respect to their own existence, exploiting nonrenewable resources and polluting the environment on which they depend for their own survival. The complex, sustainable ecosystems in which organisms live sustainably in relationships with each other and their surroundings serve as models for anthropogenic systems. By taking lessons from the biosphere and its long-established ecosystems, humans can develop much more sustainable industrial systems as discussed under the topic of industrial ecology in Chapter 17.

The biosphere is a key to achieving sustainability, especially with respect to its ability to perform photosynthesis. Biogenic materials have the potential to be produced in a much greener, more sustainable manner than do materials produced by humans, because plants use carbon dioxide from the atmosphere as a carbon source and are solar powered. Organisms perform syntheses under inherently mild — therefore safe — conditions. Furthermore, organisms are particularly well adapted to make a variety of complex and specialized materials that are very difficult or impossible to make by purely chemical means.

2.6. THE ANTHROSPHERE

The **anthrosphere** may be defined as that part of the environment made or modified by humans and used for their activities. Although some environmentalists object to recognizing the anthrosphere as an environmental sphere, examination of any area in which humans dwell reveals that it is definitely a part of the surroundings. Buildings, roads, airports, factories, power lines, and numerous other things constructed and operated by humans provide visible evidence of the existence of the anthrosphere on Earth. The anthrosphere is such a distinct part of the environment and its potential effects so profound that many scientists agree with the contention of Nobel Prize-winning atmospheric chemist Paul Crutzen that Earth is undergoing a transition from the holocene geological epoch to a new one, the anthropocene. This is occurring because human activities are now quite significant compared to Nature's in their impact on Earth's environment and are changing Earth's fundamental physics, chemistry, and biology. Some scientists fear that, especially through changes in global climate, activity in the anthrosphere will detrimentally alter Earth's relatively stable, nurturing environment and produce one that is much more challenging to human existence.

The commonly held belief that pre-industrial humans had little influence on the biosphere and ecosystems is not completely accurate. For example, some pre-Columbian native South Americans were expert landscape managers who constructed terraces, drained wetlands, and constructed raised agricultural fields to grow a variety of food crops, thereby creating their own primitive, but still significant anthrosphere. In areas now occupied by the present day U.S., Native Americans employed fire to destroy forests and create grasslands that supported game animals. Their pyromaniacal tendencies created as much as one third of the Midwestern prairies first encountered by early settlers and, indeed, forestation, largely with European tree varieties, often followed in the paths of these settlers. Compared to the modern industrial era, however, the impact of humans until just a few centuries ago was comparatively benign. Particularly with the development of fossil-fuel resources and the large machines powered by these fuels, during the last approximately 200 years, humans have built a pervasive anthrosphere that is having massive environmental impact.

The anthrosphere is tied intimately with the other environmental spheres, and the boundaries between them are sometimes blurred. Most of the anthrosphere is anchored to the geosphere. But anthrospheric ships move over ocean waters in the hydrosphere, and airplanes fly through the atmosphere. Farm fields are modifications of the geosphere, but the crops raised on them are part of the biosphere. Many other such examples could be given.

There are many distinct segments of the anthrosphere as determined by factors such as the following:

- Where humans dwell

- How humans and their goods move

- How commercial goods and services are provided

- How renewable food, fiber, and wood are provided

- How energy is obtained, converted to other forms, and distributed

- How wastes are collected, treated, and disposed

Considering these aspects, it is possible to list a number of specific things that are parts of the anthrosphere as shown in Figure 2.8. These include dwellings as well as other structures used for manufacturing, commerce, education, and government functions. Utilities include facilities for the distribution of water, electricity, and fuel, systems for the collection and disposal of municipal wastes and wastewater (sewers), and — of particular importance to sustainability — systems for materials recycle. Transportation systems include roads, railroads, airports, and waterways constructed or modified for transport on water. The anthrospheric segments used in food production include cultivated fields for growing crops and water systems for irrigation. A variety of machines, including automobiles, trains, construction machinery, and airplanes are part of the anthrosphere. The communications sector of the anthrosphere includes radio transmitter towers, satellite dishes, and fiber-optic

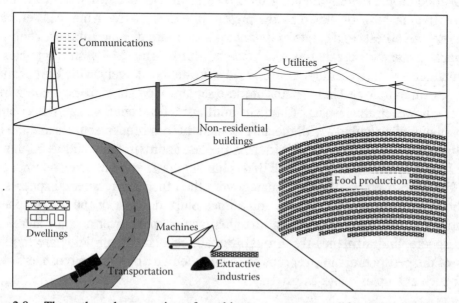

Figure 2.8.   The anthrosphere consists of machines, structures, modifications to the geosphere, and other things made and operated by humans.

networks. Oil and gas wells are employed for extracting fuels from the geosphere and mines for removing coal and minerals.

Much of the anthrosphere may be classified as **infrastructure** made up of utilities, facilities and systems that large numbers of people must use in common and that are essential for a society to operate properly. A large portion of the infrastructure consists of physical components including electrical power generating facilities and distribution grids, communications systems, roads, railroads, air transport systems, airports, buildings, water supply and distribution systems, and waste collection and disposal systems. Another essential part of the infrastructure consists of laws, regulations, instructions, and operational procedures. Components of the infrastructure may be in the public sector, such as U.S. highways or some European railroads, or privately owned and operated, such as the trucks that use the highways or airlines that use publicly owned airports.

Infrastructure has been compared to a computer operating system, which enables a computer to operate programs for word processing, record keeping, calculation, drawing, communication, and other common computer operations, and to properly record, store, correlate, and output the products of such programs. The infrastructure of the anthrosphere enables acquisition and processing of materials, the conversion of materials to manufactured items, the distribution of such items, and most other activities that occur in the anthrosphere. Many computer users are all too familiar with the loss of productivity that can occur with outdated and poorly designed operating systems that have a tendency to crash. Similarly, an outdated, cumbersome, poorly designed, worn-out anthrospheric infrastructure causes economic systems and societies to operate in a very inefficient manner that is inconsistent with sustainability. Catastrophic failure can result, such as has occurred with several massive failures of electrical power grids.

Parts of the infrastructure tend to deteriorate with age. One of the greatest problems is corrosion, a chemical process in which metals, such as the steel that composes bridge girders, tend to revert to the state in which they occur in nature (in the case of steel, rust). Human negligence and deliberate damage — misuse, neglect, vandalism, even terrorism — can cause premature loss of infrastructure function. Often the problem starts with improper design of elements of the infrastructure. Sustainability requires that elements of the infrastructure be properly designed, maintained, and protected to avoid the expense and material and energy use entailed having to rebuild the infrastructure prematurely.

To date, much of the resources put into infrastructure have been dedicated to "conquering," or at least temporarily subduing nature. For example, publicly funded levees have been constructed along major rivers to protect surrounding areas from flooding — which may succeed for decades until a "100-year" event such as the devastating 2005 hurricane Katrina that struck New Orleans causes the levee system to fail. Now the emphasis with respect to infrastucture must be directed toward sustainability and the maintenance of environmental quality. Such elements as highly effective waste treatment systems with recovery of materials and energy from wastes, high-speed rail systems to replace inefficient movement of people and freight by private carriers, and electrical systems that use wind power to the maximum extent possible are examples of sustainable infrastructure.

Consideration must be given to the fundamental purposes of the infrastructure in human activities. For example, one of the most important parts of the infrastructure is the transportation system. In more developed countries, transportation tends to emphasize the private automobile in which individuals commute to work, a very inefficient means of transport. However, in present times, much employment involves handling of information, now greatly facilitated by technologies for instantaneously transmitting large quantities of information anywhere in the world. This has enabled substantial numbers of people to work from their homes with a commute that involves simply walking to a home office wired to the internet. This capability has led to outsourcing of substantial numbers of jobs to workers in countries thousands of miles distant leading to concerns about employment opportunities for people within a country.

A term that is sometimes used in connection with sustainability of the anthrosphere is *exergy*. Exergy is expressed in energy units and is the maximum work potential in energy units of a system with respect to its surrounding environment. Exergy is based upon the first and second laws of thermodynamics and is amenable to mathematical analysis. For example, the energy content of a gallon of gasoline can be the same as that of a mass of warm, thermally polluted water discharged from a cooling system. But the exergy of the former is much greater because it can be used for beneficial applications, whereas the energy in the warm water has little potential use.

The Sociosphere

A crucial part of the anthrosphere and its infrastructure is the **sociosphere**, the societal organization of people. The sociosphere includes governments, laws, cultures, religions, families, and social traditions. A fundamental unit of the sociosphere is the community consisting of a group of people living largely within defined boundaries, interacting on a personal basis, and with a sense of belonging to a distinct human population. A well-functioning sociosphere is crucial to human welfare, prosperity, and sustainability. People in countries with dictatorial, corrupt governments that do not maintain human rights will not usually fare well, even if the countries have substantial natural resources. (The quality of life in several petroleum-rich nations is very poor despite income from the oil.) Fanatical creeds that oppress segments of the population or that are exploitive of natural capital without recognizing the importance of sustainability are counterproductive to the maintenance of a good sociosphere.

An important discipline dealing with the sociosphere is that of **economics**, the science that describes the production, distribution, and use of income, wealth, and materials (commodities). This science has developed with a rather limited scope

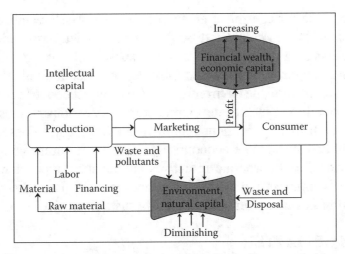

Figure 2.9.   Conventional view of economics in which the environment and its natural capital are viewed as a part of the economic system to be exploited and become diminished by removal of resources and disposal of pollutants.

that is inconsistent with the sustainability on which economic systems ultimately depend. The emphasis has been placed upon growth, which is almost universally regarded by economists as good. Capital and net worth have been measured in terms of financial and material possessions. The environment has been regarded as a part of the economic system from which materials may be extracted, which is to be "developed" with structures and other artifacts of the anthrosphere, and into which wastes are to be discarded.

There is now much evidence that the prevalent world economic system is putting an unacceptable strain on environmental support systems. Examples of such adverse effects include the virtual elimination of the once abundant Newfoundland fishery, once productive grasslands in northwestern China that are turning into deserts, falling water tables, steadily increasing atmospheric carbon dioxide levels, record high temperatures, and extreme weather events.

It is useful to regard segments of natural capital as endowments. Endowments invested in stock and interest-bearing instruments can produce income indefinitely. So long as only the income or, preferably, only part of it is spent, an endowment can keep on producing income indefinitely. But when the body of the endowment itself is spent, the endowment disappears and can never again produce income. So it is with endowments of natural capital. Grasslands properly managed and grazed by animals at or below their carrying capacity will maintain their productivity indefinitely. However, putting too many animals on pasture land, though briefly raising yield, will damage the endowment of productive grassland and, in the worst-case scenario, cause it to become a totally unproductive desert that has lost essential topsoil and can never again be restored to its former productivity.

It is essential that economics be viewed as a part of Earth's greater environmental system, rather than viewing the environment as a subsection of a world economic system. Taking such a view is essential for achieving sustainability. This imperative would seem to argue for controlled economic systems. The greatest experiment in controlled economics was conducted by the former Soviet Union and other Communist countries during the latter 1900s. That experiment was a dismal failure that led to low living standards, poverty, and lifestyles in which individual initiative was discouraged. Furthermore, the environment in Communist countries suffered severe deterioration from things such as development of heavy industries, especially steel and cement manufacturing, without regard to environmental consequences. The most successful and sustainable economic systems in the future will be those that allow for individual and corporate initiatives and accumulation of wealth defined more in terms of happiness and well-being rather than just material possessions, and operating within rules that promote and require sustainability.

2.7. CYCLES OF MATTER

There is a constant exchange of matter among the five major environmental spheres as described by cycles of matter. It is convenient to classify the various cycles by the chemical species that they involve, most commonly elements such as carbon and nitrogen. One particularly important cycle is that of a chemical compound, water, shown in the hydrologic cycle in Figure 4.2. Another important cycle is the rock cycle in which molten rock solidifies, undergoes weathering, may be carried by water and deposited as sedimentary rock, is converted to metamorphic rock by heat and pressure, and is eventually buried at great depths and melted to produce molten rock again. Since natural cycles of matter usually involve geochemical and biological processes, they are frequently called **biogeochemical cycles**, a term that particularly describes plant and animal nutrients including carbon and nitrogen. It should be kept in mind that anthrospheric processes are also very much involved in the important cycles of matter.

Cycles of matter involve various reservoirs of matter, such as the ocean, the atmosphere, and parts of the geosphere, as well as conduits through which matter moves between these reservoirs. Two main classifications of cycles of matter are endogenic cycles that occur below or directly on the surface of the geosphere and exogenic cycles, which involve surface phenomena and an atmospheric component. The rock cycle is an endogenic cycle and the hydrologic cycle is an exogenic cycle.

Figure 2.10 illustrates one of the key biogeochemical cycles, the carbon cycle. An important reservoir of carbon is in the atmosphere as carbon dioxide. Photosynthetic processes by plants extract significant amounts of carbon from the atmosphere and fix it as biological carbon in the biosphere. In turn, animals and other organisms in the biosphere release carbon dioxide back to the atmosphere through the respiration processes by which they utilize oxygen and food for energy production. More carbon dioxide is released to the atmosphere by the combustion of biological materials such as wood and fossil fuels including coal and petroleum. Carbon dioxide from the atmosphere dissolves in water to produce dissolved carbon dioxide and inorganic carbonates. Solid carbonates, particularly limestone, dissolve in bodies of

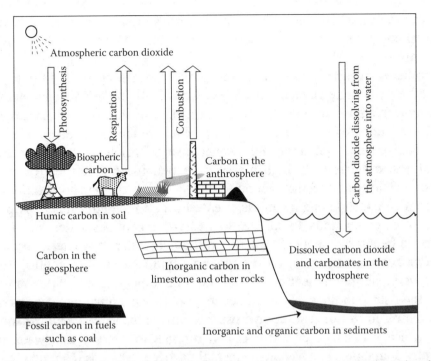

Figure 2.10.   The carbon cycle describes locations in various environmental spheres in which carbon is found and the interchanges of carbon in various chemical forms among these spheres. Consideration of the carbon cycle is of utmost importance in maintaining sustainability.

water to also produce dissolved inorganic carbon. Respiration of organic matter by organisms in water and sediments also produces dissolved inorganic carbon species in water. Large amounts of carbon are held in the geosphere. The major forms of this carbon are fossil carbon of ancient plant origin held by fossil fuels, such as coal and petroleum, and inorganic carbon in carbonate rocks, especially limestone. Organic carbon from partially degraded plant material is present as humic material in soil, a source that is added to by organisms breaking down plant biomass from the biosphere.

The carbon cycle is extraordinarily important in maintaining sustainability because a major part of it is the fixation of carbon from highly dilute atmospheric carbon dioxide into biomass by photosynthesis carried out by green plants. Biomass is a source of food, chemical energy, and raw materials, and the carbon cycle contains the main pathway by which solar energy is captured and converted to a form of energy that can be utilized by organisms and as fuel.

Other important cycles of matter are linked to the carbon cycle. The oxygen cycle describes movement of oxygen in various chemical forms through the five environmental spheres. At 21% elemental oxygen by volume, the atmosphere is a vast reservoir of this element. This oxygen becomes chemically bound as carbon dioxide by respiration processes of organisms and by combustion. The reservoir of atmospheric oxygen is added to by photosynthesis. Oxygen is a component of biomass in the biosphere and most rocks in Earth's crust are composed of oxygen-containing compounds. With its chemical formula of H_2O, water in the hydrosphere is predominantly oxygen.

Another important life-element cycle is the nitrogen cycle. The atmosphere is almost 80% by volume elemental nitrogen. Chemically bound nitrogen is essential for life-molecule proteins, but elemental nitrogen is such a stable species that binding atmospheric nitrogen into forms that can be utilized in biomolecules is a difficult thing to do. Humans do it in the anthrosphere using precious-metal catalysts at high temperatures and extremely high pressures, expending a great deal of energy in the process. Nitrogen becomes bound to oxygen under the harsh conditions inside an internal combustion engine and is released back to the atmosphere as pollutant nitrogen oxide compounds. In contrast to the harsh conditions under which elemental nitrogen becomes chemically bound in the anthrosphere, bacteria, such as the *Rhizobium* bacteria that grow on the roots of legume plants accomplish the same feat in a very energy-efficient manner at the mild temperature and pressure conditions of their surroundings. Nitrogen occurs in chemical compounds (ammonia and nitrate) dissolved in water in the hydrosphere and as the same chemical species in soil, to which it is added as fertilizer. Microorganisms in soil and water convert chemically bound nitrogen back to elemental nitrogen, which is released back to the atmosphere.

Two other biogeochemical cycles involving important life elements are the phosphorous cycle and the sulfur cycle. Sulfur occurs abundantly in the geosphere and is released to the atmosphere as organic sulfur by organisms in the oceans and as pollutant sulfur oxides by anthrospheric processes, such as combustion of sulfur-containing coal. Phosphorus is an essential life-element in DNA and other nucleic acids and in species designated ATP and ADP required for energy transfer in living systems. Unlike the other cycles mentioned here, the phosphorus cycle is exclusively endogenous without a significant atmospheric reservoir.

SUPPLEMENTARY REFERENCES

Barry, Roger G. and Richard J. Chorley, *Atmosphere, Weather, and Climate*, 8th ed., Routledge, New York, 2004.

Bengtsson, Lennart O. and Claus U. Hammer, Eds., *Geosphere-Biosphere Interactions and Climate*, Cambridge University Press, New York, 2001.

Brutsaert, Wilfried, *Hydrology: An Introduction*, Cambridge University Press, New York, 2005.

Chapin, F. Stuart, Pamela A. Matson, and Harold A. Mooney, *Principles of Terrestrial Ecosystem Ecology*, Springer-Verlag, New York, 2002.

Hayes, Brian, *Infrastructure: A Field Guide to the Industrial Landscape*, W. W. Norton, New York, 2005.

Hunt, Elizabeth, *Thirsty Planet: Strategies for Sustainable Water Management*, Zed Books, New York, 2004.

Kaufman, Donald G., and Cecilia M. Franz, *The Biosphere: Protecting Our Global Environment*, 4th ed., Kendall/Hunt Publishing Co., Dubuque, IA, 2005.

Sarmiento Jorge L. and Nicolas Gruber, *Ocean Biogeochemical Dynamics*, Princeton University Press, Princeton, NJ, 2006.

Sigel, Astrid, Helmut Sigel, and Roland K.O. Sigel, Eds., *Biogeochemical Cycles of Elements*, Taylor & Francis, London, 2005.

van der Pluijm, Ben A. and Stephen Marshak, *Earth Structure: An Introduction to Structural Geology and Tectonics*, 2nd ed., W. W. Norton, New York, 2004.

QUESTIONS AND PROBLEMS

1. The boundaries between the environmental spheres are very important. Discuss the special significance of the geosphere/atmosphere boundary.

2. Name two respects in which the hydrologic cycle is solar powered.

3. Only a very small fraction of Earth's water is freshwater. Suggest how global warming may reduce this fraction even further.

4. Depletion of atmospheric oxygen is not a problem. Suggest how the atmospheric oxygen levels might be reduced and what would be a much greater problem before O_2 depletion could become a consideration.

5. It is generally believed that the atmosphere of Mars once contained significant amounts of water. Suggest how this water was lost.

6. Why does the temperature of the stratosphere increase with increasing altitude?

7. Suggest two reasons why the solar energy input to Earth's surface is less at higher latitudes than it is at the Equator.

8. What is the special significance of tectonic plate boundaries with respect to natural geospheric disasters?

9. Which part of the hydrosphere is most intimately associated with the geosphere. Explain.

10. What are the major classes of biochemical substances that are synthesized and utilized by organisms?

11. Suggest some of the ways in which organisms modify their physical environment.

12. Suggest how humans fit in with the classification of living organisms shown in Figure 2.7.

13. What is ecological release and how may it affect the biosphere?

14. Suggest how dwindling petroleum supplies may affect the biosphere and make it even more important for sustainability.

15. Why is it not totally accurate to suggest that preindustrial humans did not affect biospheric ecosystems?

16. Explain how infrastructure is an important part of the anthrosphere.

17. What is the sociosphere? How are you connected with it?

18. How are conventional views of economics inconsistent with the achievement of sustainability?

3. GREEN CHEMISTRY, BIOLOGY, AND BIOCHEMISTRY

3.1. INTRODUCTION

To understand green science and technology, it is essential to have a basic knowledge of chemistry, biology, and the interface of these two sciences, biochemistry. Therefore, this chapter briefly introduces these topics.

Chemistry is the science of matter whereas **biology** is the science of living organisms. Because everything, including our own bodies, consists of matter, chemistry is obviously an essential discipline. Biology must be considered in this book because much of green science and technology involves living organisms. One of the basic tenets of green science and technology is to avoid harm to living organisms, including ourselves, which requires an understanding of biology. Furthermore, the sustainable production of materials required for modern civilization utilizes living organisms, particularly through biological photosynthesis. **Biochemistry** is the chemical science of living organisms, based upon fundamental chemical principles, but often much more complex than chemistry carried out in the laboratory.

3.2. INTRODUCTION TO CHEMISTRY: ATOMS AND ELEMENTS

Matter is anything that has mass and occupies space. It may consist of solids, liquids, or gases. Matter is broadly divided into the two general categories of **organic substances**, many of which are of biological origin, consisting of most of those that contain carbon, and **inorganic matter** composed of all substances that do not contain carbon plus a few, such as limestone, that do. The most fundamental kind of matter consists of the **elements**. Of these, **metals** such as copper or silver are generally solid, shiny in appearance, electrically conducting, and malleable. **Nonmetals** often have a dull appearance, are not at all malleable, and may exist as gases (atmospheric oxygen), liquids (bromine), or solids (sulfur). **Organic substances** consist of virtually all materials, such as wood, that contain carbon. All other substances are **inorganic substances**.

Atoms

All matter is composed of only about 100 elements, which combine in countless ways to produce the multimillion kinds of known matter. Each of the elements is composed of extremely small, chemically identical particles called **atoms**. Atoms, in turn, are made up of **subatomic particles**. Two of these kinds of subatomic particles occur in the very small **nucleus** of each atom located in the center of the atom and composing virtually all of its mass. The first of these is the **proton**, designated p, p^+, or +, having an electrical charge of +1 and a mass of 1.007 u, where the atomic mass unit is designated u and is defined as exactly 1/12 the mass of the most common kind of atom of carbon, carbon-12. The number of protons in the nucleus of each atom of an element is unique to that element and is called the **atomic number**. The other kind of subatomic particle in the atom nucleus is the **neutron**, which is uncharged, has a mass of 1.009 u, and is designated n. Surrounding the nucleus is a cloud composed of rapidly moving, negatively charged **electrons**, designated e, e^-, or −, with a charge −1 and a mass 0.0005 u. With their masses of essentially 1 u, the proton and neutron are said to each have a **mass number** of 1, whereas the electron, with a negligible mass compared to the proton and neutron, is assigned a mass number of 0. The charges and mass numbers of the three basic subatomic particles are denoted by the following symbols:

$$\, _1^1p, \, _0^1n, \, _{-1}^0e$$

The simplest atom that has all three subatomic particles is that of deuterium, a form of the lightest element, hydrogen, as represented on the left in Figure 3.1. On the right in Figure 3.1 is a representation of an atom of carbon-12, which, as noted

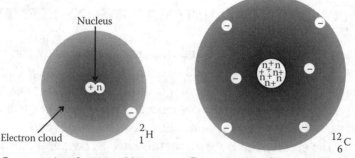

Representation of an atom of deuterium,   Representation of an atom of carbon-12,
a form (isotope) of hydrogen   the most common form of the element

Figure 3.1.   On the left is represented a deuterium (hydrogen-2) atom containing one proton (+) and one neutron (n) in its nucleus surrounded by a cloud of negative charge from an electron (−) moving rapidly around the nucleus. On the right is shown an atom of carbon-12 having six protons and six neutrons in its nucleus surrounded by a cloud of negative charge formed by six rapidly moving electrons. The nuclei, which contain virtually all the mass of each atom, are proportionately much smaller than shown in the figure.

above, is the form of elemental carbon on which the atomic mass unit is based. These atoms are **isotopes** of their elements. Atoms of a particular isotope of an element vary from other atoms (isotopes) of the same element only in the number of neutrons in the nucleus of the atom, but behave chemically the same as all other isotopes of the element. Deuterium, the isotope of hydrogen shown in Figure 3.1, composes only 0.0156% of naturally occurring hydrogen; the rest consists of atoms of hydrogen-1, the nuclei of which are composed of a single proton and no neutrons. (A very rare form of hydrogen made artificially is radioactive tritium the atoms of which have two neutrons and one proton in their nuclei.) Most carbon consists of carbon-12 shown in Figure 3.1; a small percentage is composed of carbon-13, each atom of which has a nucleus with six protons and seven neutrons. A very rare form of carbon is radioactive carbon-14, which has six protons and eight neutrons in its nucleus. Over thousands of years, carbon-14 is transmuted to stable nitrogen, emitting radiation in the form of high-energy electrons (beta particles) in the process, a phenomenon used in dating artifacts.

Each of the elements has its own **chemical symbol** consisting of one or two letters. As shown in Figure 3.1, the atomic symbols for hydrogen and carbon are H and C, respectively. Because C is already used to denote carbon, the chemical symbol for the element calcium is Ca. In some cases, the chemical symbols are based on Latin names for the elements. For example, K is the chemical symbol for potassium (Latin name kalium).

Chemical symbols with subscript and superscript numbers can be used to designate specific isotopes as shown by the symbols in Figure 3.1. In the case of deuterium, the subscript 1 shows that the isotope has 1 proton in its nucleus and the superscript 2 shows that the mass number of the isotope is 2 with a contribution of 1 from the proton and 1 from the neutron in the nucleus. In the case of the isotope symbol for carbon-12 shown in Figure 3.1, the subscript 6 denotes 6 protons in the nucleus and the superscript 12 indicates a mass number of 12 from 6 protons plus 6 neutrons.

In addition to an atomic number equal to the number of protons in the nucleus of its atoms, each element has an **atomic mass** equal to the average mass of all the element's atoms. Atomic mass is not an integer primarily because the atoms of an element generally consist of 2 or more isotopes. For example, the atomic mass of chlorine, Cl, is 35.453 because it consists of a mixture of isotopes with a mass number of 35 (17 protons plus 18 neutrons in the nucleus) and those with a mass number of 37 (17 protons and 20 neutrons).

3.3. ELEMENTS AND THE PERIODIC TABLE

One of the most useful tools for understanding chemistry is the **periodic table**, a systematic listing of the elements in order of increasing atomic number arranged to reflect the periodic chemical behavior of the elements with increasing atomic number. In the periodic table, each element is shown in a box with its chemical symbol, atomic number, and atomic mass. In general, chemical properties vary systematically across rows or **periods** in the periodic table, whereas elements in the same vertical columns or **groups** have generally similar chemical properties. Complete periodic tables showing the more than 100 known elements can be found in any standard beginning chemistry textbook. In this section, an abbreviated periodic table of the first 20 elements is developed to illustrate their periodic properties.

Atoms of the first element in the periodic table, hydrogen, atomic number 1, were described above. Most hydrogen is composed of the isotope with a mass number of 1, so the average mass of hydrogen atoms is very close to 1 u (atomic mass units, see preceding section). It is in fact 1.0079 u meaning that the atomic mass of H is 1.0079. All of this information is shown in the first box of the abbreviated periodic table in Figure 3.4.

The chemical behavior of atoms is determined by the number and distribution of electrons in the atoms, especially the **outer electrons** or **valence electrons**. Such electrons can be shown as dots by **electron-dot symbols** or **Lewis symbols**. With only one electron, the Lewis symbol of hydrogen is simply

H•

The energy levels and arrangement of the continually moving electrons in the cloud of negative charge surrounding the nuclei of atoms determine their chemical behavior and are called their **electron configurations**. Electrons occupy distinct **energy levels** and are contained in **electron shells**, each of which can hold a certain maximum number of electrons. An atom with a **filled electron shell** has essentially no tendency to lose, gain, or share electrons, required to bond with other atoms to produce chemical compounds. Such chemically unreactive atoms exist as gas-phase atoms and are called **noble gases**.

The element with atomic number two is **helium**, all of the atoms of which contain two protons and two electrons. Most helium atom nuclei also contain two neutrons and the atomic mass of helium is 4.00260. The two electrons in the helium atoms, shown by the Lewis symbol in Figure 3.2, constitute a filled electron shell so that it is the first element that is a noble gas.

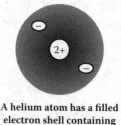

A helium atom has a filled   It can be represented by the
electron shell containing   Lewis symbol above.
2 electrons.

He:

Figure 3.2.   Two representations of the helium atom, the lightest element having a filled electron shell.

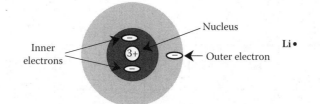

Figure 3.3.   An atom of lithium (Li) has two inner electrons and one outer electron that it can lose to produce Li⁺ ions in chemical compounds of lithium.

Hydrogen and helium are the only two elements in the first period of the periodic table. Each period ends with a noble gas.

After helium in the periodic table comes the metal lithium (Li), atomic number 3, atomic mass 6.941. Most helium atom nuclei contain four neutrons; a smaller percentage contain only three. As shown in Figure 3.3, lithium has two **inner electrons** electrons in an **inner shell**, as in the immediately preceding noble gas helium, and an **outer electron** in an **outer shell** that is farther from, and less strongly attracted to, the nucleus. The inner electrons are virtually impossible to remove and the electron shell in which they are located behaves like that in the noble gas helium. However, the lithium atom readily loses its outer shell electron leaving a charged atom or **ion**, Li⁺. This ion is produced when lithium bonds with other kinds of atoms to form chemical compounds as described in Section 3.4. Lewis symbols normally show only outer shell electrons as denoted in the single dot for the Lewis symbol of lithium on the right of Figure 3.3.

Lithium is the first element in the second period of the periodic table. The other elements in this period are, in order of increasing atomic number, the following: 4, beryllium, Be; 5, boron, B; 6, carbon, C; 7, nitrogen, N; 8, oxygen, O; 9, fluorine, F; and 10, neon, Ne. The Lewis symbol of **neon**, the last element in the second period,

$$: \overset{\cdot\,\cdot}{\underset{\cdot\,\cdot}{Ne}} :$$

shows that the neon atom has eight outer electrons, which gives neon a filled electron shell. Like the preceding noble gas, helium, which has a filled shell of two electrons, neon has no tendency to lose, gain or share any electrons, is chemically unreactive, and is likewise a noble gas. The elements in the second period of the periodic table are shown in Figure 3.4 along with their atomic numbers, atomic masses, and Lewis symbols.

Like neon, all other atoms with eight outer electrons are chemically unreactive noble gases. In addition to neon, these are argon (atomic number 18, shown in Figure 3.4), krypton (atomic number 36), xenon (atomic number 54), and radon (atomic number 86). Each of the noble gases beyond helium may be represented by the Lewis symbol consisting of the symbol for the element surrounded by eight dots. The eight electrons constitute an **octet**. In forming chemical compounds, atoms tend

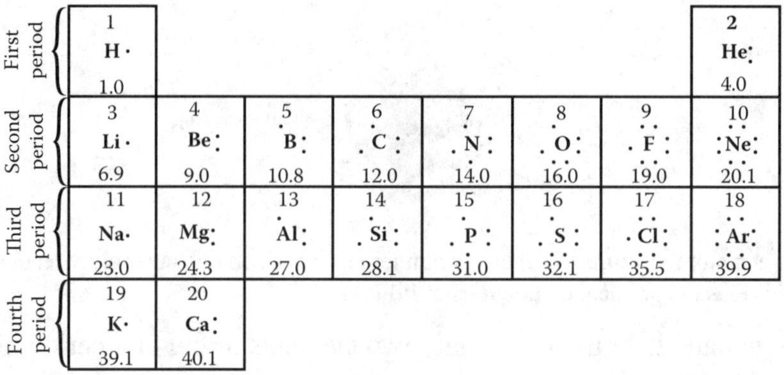

Figure 3.4.   Abbreviated 20-element version of the periodic table showing Lewis symbols of the elements. The bottom number in each entry is the atomic mass of the element.

to acquire the **noble gas outer electron configuration** octet by losing, gaining, or sharing electrons. This is the basis of the **octet rule** of chemical bonding, which is very useful in predicting the formation of chemical compounds.

The abbreviated periodic table is completed by adding a third period and the first two elements of the fourth period, the atomic numbers, names, and symbols of which are the following: 11, sodium, Na; 12, magnesium, Mg; 13, aluminum, Al; 14, silicon, Si; 15, phosphorus, P; 16, sulfur, S; 17, chlorine, Cl; 18, argon, Ar; 19, potassium, K; and 20, calcium, Ca. As seen in Figure 3.4, the Lewis symbols of the elements in vertical columns, called **groups**, show the same outer electron configurations with the exception of helium, which is a noble gas with only two outer shell-electrons rather than eight. Other than hydrogen, H, which has unique chemical behavior, the chemical properties of elements in the same group are generally similar.

3.4. CHEMICAL COMPOUNDS AND CHEMICAL BONDS

With the exception of the noble gases, most atoms are joined by **chemical bonds** to other atoms. This is the case even with pure elements such as elemental hydrogen, which exists as **molecules**, each consisting of two H atoms (Figure 3.5). The hydrogen molecule is denoted by the **chemical formula**, H_2, in which the subscript 2 shows that there are two H atoms per molecule. These are held together by a **covalent bond** made up of two shared electrons, each contributed by one of the H atoms. This bond can be shown by **electron dot formulas** or **Lewis formulas** for the H_2 molecule illustrated on the bottom of Figure 3.5.

Chemical Compounds with Covalent Bonds

Most substances are **chemical compounds** consisting of two or more elements, the atoms of which are joined by chemical bonds. The most common example of a chemical compound is that of water consisting of molecules each composed of an

The H atoms in elemental   Are held together by chemical   That have the chemical
hydrogen   bonds in molecules   formula H_2.

The formation of H_2 can also be represented with Lewis symbols for atoms and a Lewis formula for the molecule as shown below:

$$H \overset{\cdot}{\underset{\cdot}{\rightleftharpoons}} H \longrightarrow H : H$$

Figure 3.5. Formation of a molecule of H_2 showing covalent bonds and Lewis formula of the hydrogen molecule.

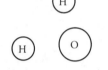

Two hydrogen atoms and an   To form a molecule in which 2 H   The chemical formula of
oxygen atom bond together   atoms are attached to the O atom   the water product is H_2O.

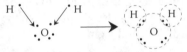

The formation of H_2O can also be represented with
Lewis symbols for atoms and a
Lewis formula for the molecule as shown below:

Figure 3.6. The water molecule. The dashed circles around each H atom show that by sharing electrons the atom attains the helium noble gas electron configuration, and the dashed circle around the O atom shows that by sharing electrons with H it attains the stable octet of outer shell electrons.

oxygen (O) atom bonded to two H atoms by covalent chemical bonds each consisting of two shared electrons as illustrated in Figure 3.6. The chemical formula of water is H_2O in which the subscript 2 denotes two H atoms per molecule and the absence of a subscript following O implies one O atom. The Lewis formula of H_2O at the bottom of Figure 3.6 shows two important aspects of chemical bonding. The first of these is that the most common type of covalent bond consisting of (two) shared electrons is illustrated by the two H–O bonds. Secondly, in forming chemical compounds, atoms tend to attain the outer electron shell configuration of the nearest noble gases in the periodic table. By sharing electrons with O, each of the H atoms attains the outer electron configuration of the two electrons in helium as illustrated by the dashed circles around each of the H atoms. In the case of O, the O atom attains the outer electron configuration of eight electrons as in neon — the stable octet — by sharing electrons with H atoms.

Ionic Bonds

In addition to covalent bonds in which electrons are shared, there exist **ionic bonds** in which oppositely charged entities called *ions* are held together by

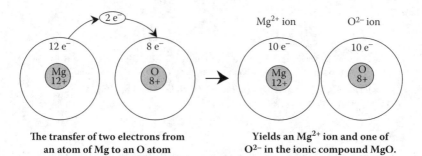

The transfer of two electrons from
an atom of Mg to an O atom

Yields an Mg^{2+} ion and one of
O^{2-} in the ionic compound MgO.

Figure 3.7.   Formation of an ionic compound of MgO consisting of positively charged cations and negatively charged anions by the transfer of electrons between atoms.

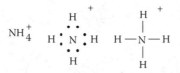

Figure 3.8.   The ammonium ion is composed of four H atoms bonded to a central N atom by covalent bonds represented as pairs of dots or by straight line. Note the octet of electrons around the N atom and the two electrons shared by each H atom.

electrostatic attractions. Positively charged **cations** and negatively charged **anions** are bonded together within a **crystalline lattice**. The formation of ions bonded by ionic bonds is illustrated for magnesium oxide in Figure 3.7.

The ions shown in Figure 3.7 are simply atoms that have acquired charges by losing or gaining electrons. Many ions consist of groups of atoms held together by covalent bonds but having an overall positive or negative charge. A very common example is the ammonium ion, NH_4^+, shown in Figure 3.8.

The composition of chemical compounds is denoted by chemical formulas, such as that of water, H_2O. Figure 3.9 illustrates a chemical formula of a typical compound, ammonium sulfate. Each **formula unit** of a compound is the smallest entity of the compound. In the case of a compound with only covalent bonds, such as H_2O,

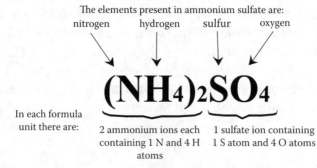

The elements present in ammonium sulfate are:
nitrogen   hydrogen   sulfur   oxygen

$$(NH_4)_2SO_4$$

In each formula
unit there are: 2 ammonium ions each   1 sulfate ion containing
                     containing 1 N and 4 H   1 S atom and 4 O atoms
                     atoms

Each formula unit contains 2 N atoms, 8 H atoms, 1 S atom, 4 O atoms

Figure 3.9.   The chemical formula of ammonium sulfate showing its elemental composition.

the formula unit is simply the molecule. But ammonium sulfate consists of ammonium ions, NH_4^+, and sulfate ions, SO_4^{2-}, bonded together with ionic bonds and each held together internally by covalent bonds, such as those shown in Figure 3.8 for the ammonium ion. Because specific cations in ionic compounds are not bound to specific anions and *vice versa*, it is more correct to refer to formula units rather than molecules of ionic compounds. A chemical formula contains a lot of information as summarized for ammonium sulfate in Figure 3.9.

3.5. DEALING WITH MATTER QUANTITATIVELY: THE MOLE

An important property of a compound is its **formula (molecular) mass**, which is the mass in atomic mass units of a formula unit of the compound, often expressed without units. In the case of H_2O, for example, the molecular mass is 2×1.0 u for the two H atoms plus 16.0 u for the O atom, a sum of 18.0 u or simply 18.0. For the more complicated case of ammonium sulfate (Figure 3.9), the formula mass is given by the following:

$$2 \text{ N atoms} \times 14.0 \text{ u/N atom} = 28.0 \text{ u}$$
$$8 \text{ H atoms} \times 1.0 \text{ u/H atom} = 8.0 \text{ u}$$
$$1 \text{ S atom} \times 32.1 \text{ u/H atom} = 32.1 \text{ u}$$
$$4 \text{ O atoms} \times 16.0 \text{ u/H atom} = \underline{64.0 \text{ u}}$$
$$\text{Formula mass} = 132.1 \text{ u or } 132.1$$

For quantitative calculations, it is convenient for the chemist to work in masses of grams (abbreviated g, where a cubic container of water the size of a sugar cube has a mass of approximately 1 g). This has led to the use of the mole (abbreviated mol) for expressing quantities of matter where a **mole** of a substance contains a mass in grams equal numerically to its formula mass, a quantity called the **molar mass**. Therefore, a mole of water, formula mass 18.0, has a mass of 18.0 g and a mole of ammonium sulfate, formula mass 132.1, has a mass of 132.1 g. The *mole* is defined as the quantity of substance that contains the same number of specified entities as there are atoms of C in exactly 0.012 kg (12 g) of carbon-12, the most common carbon isotope that contains 6 protons and 6 neutrons in its nucleus. To specify the mass of a mole of a substance, simply state the atomic mass (of an element) or the formula mass (of a compound) and affix "grams." A mole of a substance contains an enormous number of formula units of the substance, specifically 6.02×10^{23}. This number is **Avogadro's number**.

3.6. CHEMICAL REACTIONS AND EQUATIONS

The process that occurs when chemical substances are changed to other substances is a **chemical reaction**. Chemical reactions involve breaking of chemical bonds and formation of new bonds. Using chemical formulas to denote elements and compounds, a chemical reaction can be stated as a **chemical equation**. An example is the equation for the reaction of elemental hydrogen with elemental nitrogen to produce ammonia:

$$3H_2 + N_2 \rightarrow 2NH_3 \tag{3.6.1}$$

This equation states that hydrogen (H_2) reacts with nitrogen (N_2) to yield ammonia (NH_3). The H_2 and N_2 on the left of the equation are the **reactants** and the NH_3 on the right is the **product**. The arrow separating the reactants and products is read "yields."

All correctly written chemical equations are **balanced** showing the same number of each kind of atom on both sides of the equation. In the example above there are six H atoms, two in each of the three H_2 molecules on the left (reactants) and six H atoms, three in each of the two NH_3 molecules on the right. Also, on the left there are two N atoms in the N_2 reactant molecule on the left and two N atoms, one in each of the two NH_3 molecules on the right. Therefore, the chemical equation is balanced as written. The process of balancing chemical equations consists of changing the numbers in front of the formula but does not allow changing chemical formulas.

The reaction for the synthesis of ammonia occurs over a material called a **catalyst**, which enables a chemical reaction to occur faster without itself being consumed in the process. Catalysts are very important in green technology because they enable efficient utilization of raw materials under safe and mild conditions. The ultimate catalysts are the enzymes that act in living organisms. Whereas the industrial chemical process for the synthesis of ammonia requires elevated temperatures and very high pressures, *Rhizobium* bacteria growing in nodules on the roots of green plants, such as soybeans, accomplish the same thing under the mild conditions just below the soil surface.

Chemical equations can also be used to show whether substances involved in a chemical reaction are gases, liquids, solids, or dissolved in water. For example, the chemical equation

$$2Na(s) + 2H_2O(l) \rightarrow H_2(g) + 2NaOH(aq) \tag{3.6.2}$$

shows that solid (*s*) sodium reacts with liquid (*l*) water to produce gaseous hydrogen (*g*) and sodium hydroxide, NaOH, dissolved in aqueous solution (*aq*).

3.7. PHYSICAL PROPERTIES AND STATES OF MATTER

Physical properties of matter are those that can be measured without altering the chemical composition of the matter. Three physical properties important in describing and identifying particular kinds of matter are density, color, and solubility. **Density** (d) is defined as mass per unit volume and is expressed by the formula

$$d = \frac{\text{mass}}{\text{volume}} \qquad (3.7.1)$$

The densities of liquids and solids are commonly expressed in units of grams per cubic centimeter (g/cm^3). (A U.S. penny is almost 2 cm in diameter and a cubic centimeter of water at 4 degrees Celsius temperature has a mass of exactly 1 g. The cm^3 is a volume equal to 1 milliliter, abbreviated ml, which is 1/1000 of a liter. A liter is a volume slightly greater than 1 qt.) The densities of gases are much less than those of liquids and solids and are commonly expressed in units of grams per liter, g/l). **Color** is a readily observed physical property. **Solubility**, the degree to which a substance dissolves in water or other liquids, is another commonly measured physical property.

Gases, liquids, and solids are **states of matter**. A quantity of **gas**, such as the air in an automobile tire, assumes the volume and shape of its container and can be compressed to smaller volumes by applying pressure. **Liquids**, such as water in a pitcher, take on the shape of their container but have an essentially constant volume. A quantity of a **solid** has both a defined shape and volume.

Changes in states of matter — particularly of water — are very important in environmental processes. For example, water in atmospheric water vapor condenses from the gas state to the liquid state forming clouds and precipitation, releasing heat in the process. Pure water can be obtained from seawater by heating the water to produce the pure vapor, leaving salt behind, and condensing the water vapor to produce salt-free liquid water.

Gas is the "loosest" state of matter because gas molecules bounce around in otherwise empty space. As temperature is raised, the rate of movement of gas molecules increases, and the gas either expands (under constant pressure) or exerts a higher pressure (if the volume is maintained constant). The relationships among gas pressure, volume, and quantity (in moles) are described by the **gas laws**, which are discussed in more detail in Chapter 7.

The molecules of liquids occupy essentially all the space in the liquid, so it is not easily compressed. However, liquid molecules move readily relative to each other, so the shape of a quantity of liquid varies and is the same as the container within which the liquid is held. Liquids enter the gas state by the process of **evaporation** and form liquid from the gas state by **condensation**. Equilibrium between these two processes results in a steady-state level of vapor above a liquid, which exerts a pressure called **vapor pressure**. When water boils, the pressure of the water vapor above the liquid water is equal to atmospheric pressure.

Liquids that dissolve solids, gases, or other liquids are said to be acting as **solvents**, the material that dissolves in the liquid is called a **solute**, and the mixture of solvent and solute is a **solution**. A solution that is at equilibrium with excess solute so that it contains the maximum amount of solute that it can dissolve is called a **saturated solution**. One that can still dissolve more solute is called an **unsaturated**

solution. The maximum degree to which a solute dissolves in a liquid is the solute's **solubility**.

The amount of solute dissolved in a specific quantity of solvent or solution is the **solution concentration**. A convenient way for chemists to express solution concentration is in units of moles solute dissolved per liter of solution (m/l), *M*:

$$M = \frac{\text{moles of solute}}{\text{number of liters of solution}} \qquad (3.7.2)$$

As an example of molar concentration, suppose that a quantity of 51.0 g of ammonia gas (NH_3), is dissolved in enough water to make a final solution with a volume of 2.00 l. What is the molar concentration, *M*, of the resulting solution? Given atomic masses of 1.0 and 14.0 for H and N, respectively, the formula mass of ammonia is $3 \times 1.0 + 14.0 = 17.0$ and the molar mass is 17.0 g/mol. The number of moles of NH_3 in 51.0 g of this compound is given by:

$$\text{moles} = \frac{\text{mass}}{\text{molar mass}} = \frac{51.0 \text{ g}}{17.0 \text{ g/mol}} = 3.0 \text{ mol} \qquad (3.7.3)$$

The molar concentration of the solution is calculated by substituting into Equation 3.7.2:

$$M = \frac{3.0 \text{ mol}}{2.00 \text{ l}} = 1.50 \text{ mol/l} \qquad (3.7.4)$$

The molar concentration of a solute, "X", is conventionally expressed as [X].

The atoms, molecules, and ions in the solid state fill the space in the solid and are in fixed positions relative to each other, giving a quantity of solid a fixed volume and shape. Molecules of solids do not enter the vapor state readily and, to the extent that they do so, the solid is said to **sublime**.

3.8. THERMAL PROPERTIES OF MATTER

Thermal properties relating how matter behaves with respect to heat and temperature are important properties of matter. In the sciences, temperature is commonly measured in Celsius units (°C), a scale under which water freezes at 0°C and boils at 100°C, and heat is expressed in units of joules (J). The **melting point** of a pure substance is the temperature at which the substance changes from a solid to a liquid. At the melting temperature, pure solid and pure liquid composed of the substance may be present together in a state of equilibrium. Boiling occurs when a liquid is heated to a temperature such that bubbles of vapor of the substance are evolved. When the surface of a pure liquid substance is in contact with the pure vapor of the substance at 1 atmosphere (atm, where 1 atm is the pressure of the atmosphere at sea level) pressure, boiling occurs at a temperature called the **normal boiling point**.

As the temperature of a substance is raised, energy must be put into it to enable the molecules of the substance to move more rapidly relative to each other and to overcome the attractive forces between them. The **specific heat** of a substance is defined as the amount of heat energy required to raise the temperature of a gram of substance by 1 degree Celsius. The specific heat of liquid water is 4.18 J/g-°C. The **heat of vaporization** required to convert liquid water to vapor is 2260 J/g (2.26 kJ/g) for water boiling at 100°C at 1 atm pressure. This amount of heat energy is about 540 times that required to raise the temperature of 1 g of liquid water by 1°C. When water vapor condenses, large amounts of heat energy called **heat of condensation** are released. **Heat of fusion** is the quantity of heat taken up in converting a unit mass of solid entirely to liquid at a constant temperature. The heat of fusion of water is 330 J/g for ice melting at 0° C and is 80 times the specific heat of water.

3.9. ACIDS, BASES, AND SALTS

Acids

Acids, bases, and salts, which are discussed in this section, constitute most of the inorganic compounds and many of the organic compounds that are known. These compounds are very important in life processes, in the environment, and as industrial chemicals.

An **acid** is a substance that produces hydrogen ions (H^+) in water. (Actually, in water, H^+ ion is associated with water molecules in clusters such as the **hydronium ion**, H_3O^+, but for simplicity in this book, hydrogen ion in water will be shown simply as H^+.) For example, when HCl gas is dissolved in water,

$$HCl \xrightarrow{H_2O} H^+(aq) + Cl^-(aq) \tag{3.9.1}$$

It is entirely in the form of H^+ ions and Cl^- ions (a solution called hydrochloric acid and usually written simply as HCl).

Bases

A **base** is a substance that produces hydroxide ion (OH^-) and/or accepts H^+. Many bases consist of metal ions and hydroxide ions. For example, solid sodium hydroxide dissolves in water

$$NaOH(s) \rightarrow Na^+(aq) + OH^-(aq) \tag{3.9.2}$$

to yield a solution containing OH^- ions. When ammonia gas is bubbled into water, a limited number of the molecules of NH_3 remove hydrogen ion from water and produce ammonium ion (NH_4^+) and hydroxide ion as shown by the following reaction:

$$NH_3 + H_2O \leftrightarrow NH_4^+ + OH^- \tag{3.9.3}$$

The double arrows denote that the reaction is reversible. In fact, the equilibrium of this reaction lies to the left so that most of the ammonia in water is in the form of dissolved molecular NH_3 rather than the NH_4^+ ion.

Salts

Whenever an acid and a base are brought together, water is always a product, leaving a negative ion from the acid and a positive ion from the base:

$$\underset{\text{hydrochloric acid}}{H^+ + Cl^-} + \underset{\text{sodium hydroxide}}{Na^+ + OH^-} \rightarrow \underset{\text{sodium chloride}}{Na^+ + Cl^-} + \underset{\text{water}}{H_2O} \tag{3.9.4}$$

Sodium chloride is a **salt**, a compound composed of a positively charged cation other than H^+ and a negatively charged anion other than OH^-.

Dissociation of Acids and Bases in Water

The reactions discussed above have shown that acids and bases dissociate, or ionize, in water to produce ions. There is a great difference in how much various acids and bases ionize. Some, like HCl or NaOH, are completely dissociated in water. Because of this, hydrochloric acid is called a **strong acid**. Sodium hydroxide is a **strong base**. Toxic hydrocyanic acid (HCN) acts as an acid to produce hydrogen ions in water:

$$HCN \leftrightarrow H^+ + CN^- \tag{3.9.5}$$

When the HCN molecule comes apart, it is said to **ionize**, and the process is called **ionization**. At all but extremely low HCN concentrations, only a small percentage of the acid molecules release H^+ ion. Its tendency to remain as undissociated HCN is shown by the reverse arrow in Equation 3.9.5. Partially ionized acids and bases, such as hydrocyanic acid and ammonia, mentioned above are **weak acids** and **weak bases**. Some common strong acids that ionize completely in water are sulfuric acid (H_2SO_4) and nitric acid (HNO_3); weak acids include acetic acid ($HC_2H_3O_2$, of which only one of the four Hs can form H^+ ion) and hypochlorous acid (HClO). The percentage of weak-acid molecules that are ionized depends upon the concentration of the acid. The lower the concentration, the higher the percentage of ionized molecules.

Hydrogen Ion Concentration and pH

Hydrogen ion concentration, commonly denoted as $[H^+]$ and expressed in mol/l (M) is a very important characteristic of some solutions. For example, the value of

$[H^+]$ in human blood must stay within relatively narrow ranges, or the person will become ill or even die. Fortunately, there are mixtures of chemicals that keep the H^+ concentration of a solution relatively constant. Reasonable quantities of acid or base added to such solutions do not cause large changes in H^+ concentration. Solutions that resist changes in $[H^+]$ are called **buffers**.

Because of the fact that water, itself, produces both H^+ and hydroxide ion,

$$H_2O \leftrightarrow H^+ + OH^- \tag{3.9.6}$$

there is always some H^+ and some OH^- in any solution. (The reverse arrow shows that H^+ and OH^- ions recombine to give H_2O molecules.) In an acid solution, the concentration of OH^- is always very low and, in a solution of base, the concentration of H^+ is very low. If the value of either $[H^+]$ or $[OH^-]$, concentrations in moles per liter (M), is known, the value of the other can be calculated from the following relationship:

$$[H^+][OH^-] = 1.00 \times 10^{-14} = K_w \text{ (at 25°C)} \tag{3.9.7}$$

For example, in a solution of 0.100 M HCl in which $[H^+] = 0.100\ M$,

$$[OH^-] = \frac{K_W}{[H^+]} = \frac{1.00 \times 10^{-14}}{0.100} = 1.00 \times 10^{-13}\ M \tag{3.9.8}$$

Molar concentrations of hydrogen ion, $[H^+]$, range over many orders of magnitude and are conveniently expressed by pH defined as

$$pH = -\log[H^+] \tag{3.9.9}$$

In absolutely pure water $[H^+] = [OH^-]$, the value of $[H^+]$ is exactly 1×10^{-7} mol/L at 25° C, the pH is 7.00, and the solution is **neutral** (neither acidic nor basic). **Acidic** solutions have pH values of less than 7, and **basic** solutions have pH values of greater than 7. When the H^+ ion concentration is 1 times 10 to a power (the superscript number, such as –2, –7, etc.), the pH is simply the negative value of that power. Thus, when $[H^+]$ is 1×10^{-3}, the pH is 3; when $[H^+]$ is 1×10^{-4}, the pH is 4. For a solution with a hydrogen ion concentration between 1×10^{-4} and 1×10^{-3}, such as 3.16 x 10^{-4}, the pH is obviously going to be between 3 and 4. The pH is calculated very easily on an electronic calculator by having it compute the negative log of 3.16×10^{-4}, which is equal to 3.50.

3.10. ORGANIC CHEMISTRY

Most carbon-containing compounds are **organic chemicals** and are addressed by the subject of **organic chemistry**. Organic chemistry is a vast, diverse, discipline because of the enormous number of organic compounds that exist as a consequence of the versatile bonding capabilities of carbon. Such diversity is because of the ability of carbon atoms to bond to each other through single (two shared electrons) bonds, double (four shared electrons) bonds, and triple (six shared electrons) bonds, in a limitless variety of straight chains, branched chains, and rings. All organic compounds, of course, contain carbon. Virtually all also contain hydrogen and have at least one C–H bond.

Among organic chemicals are included the majority of important industrial compounds, synthetic polymers, agricultural chemicals, biological materials, and most substances that are of concern because of their toxicities and other hazards. Pollution of the water, air, and soil environments by organic chemicals is an area of significant concern.

Molecular Geometry in Organic Chemistry

The three-dimensional shape of a molecule, called its **molecular geometry**, is particularly important in organic chemistry. This is because its molecular geometry determines, in part, the properties of an organic molecule, particularly its interactions with biological systems. Shapes of molecules are represented in drawings by lines of normal, uniform thickness for bonds in the plane of the paper, and with broken lines for bonds extending away from, and heavy lines for bonds extending toward, the viewer. These conventions are shown by the example of dichloromethane, CH_2Cl_2, an important organochloride solvent and extractant, illustrated in Figure 3.10.

3.11. HYDROCARBONS

The simplest and most easily understood organic compounds are **hydrocarbons**, which contain only hydrogen and carbon. As shown in Figure 3.11, the major types of hydrocarbons are alkanes, alkenes, alkynes, and aromatic (aryl) compounds. In the structures shown, C=C represents a **double bond** in which four electrons are shared, and C≡C is a **triple bond** in which six electrons are shared.

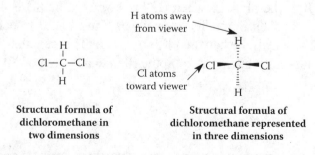

Figure 3.10.   Structural formulas of dichloromethane, CH_2Cl_2; the formula on the right provides a three-dimensional representation.

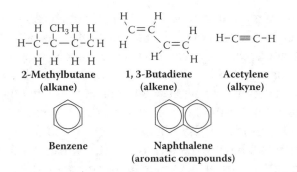

Figure 3.11.   Examples of major types of hydrocarbons.

Alkanes

Alkanes, also called **aliphatic hydrocarbons**, are hydrocarbons in which the C atoms are joined by single covalent bonds consisting of two shared electrons. Some examples of alkanes are shown in Figure 3.12. The three major kinds of alkanes are **straight-chain alkanes**, **branched-chain alkanes**, and **cycloalkanes**. In one of the molecules shown in Figure 3.12, all of the carbon atoms are in a straight chain, and in two they are in branched chains, whereas in a fourth molecule six of the carbon atoms are in a ring.

Formulas of Alkanes

Formulas of organic compounds present information at several different levels of sophistication. **Molecular formulas**, such as that of octane (C_8H_{18}), give the

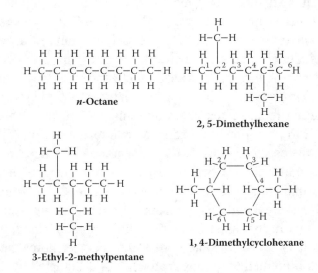

Figure 3.12.   Structural formulas of four hydrocarbons, each containing eight carbon atoms, that illustrate the structural diversity possible with organic compounds. Numbers used to denote locations of atoms for purposes of naming are shown on two of the compounds.

number of each kind of atom in a molecule of a compound. As shown in Figure 3.12, however, the molecular formula of C_8H_{18} may apply to several alkanes, each one of which has unique chemical, physical, and toxicological properties. These different compounds are designated by **structural formulas** showing the order in which the atoms in a molecule are arranged. Compounds that have the same molecular, but different structural, formulas are called **structural isomers**. Of the compounds shown in Figure 3.12, n-octane, 2,5-dimethylhexane, and 3-ethyl-2-methylpentane are structural isomers, all having the formula C_8H_{18}, whereas 1,4-dimethylcyclohexane is not a structural isomer of the other three compounds because its molecular formula is C_8H_{16}.

Most organic compounds can be derived from alkanes, and many important parts of organic molecules contain one or more alkane groups minus a hydrogen atom bonded as substituents onto the basic organic molecule. As a consequence, the names of many organic compounds are based upon alkanes. Two important substituent groups derived from alkanes are the methyl group, $-CH_3$ (derived from methane, CH_4) and the ethyl group, $-C_2H_5$ (derived from ethane, C_2H_6).

Naming Organic Compounds

Systematic names, from which the structures of organic molecules can be deduced, have been assigned to all known organic compounds. The more common organic compounds, including many toxic and hazardous organic substances, likewise have **common names** with no structural implications. To provide some idea of how organic compounds are named, consider the alkanes shown in Figure 3.12. The fact that n-octane has no side chains is denoted by "n", that it has eight carbon atoms by "oct," and that it is an alkane by "ane." The names of compounds with branched chains or atoms other than H or C attached make use of numbers that stand for positions on the longest continuous chain of carbon atoms in the molecule. This convention is illustrated by the second compound in Figure 3.12. It gets the hexane part of the name from the fact that it is an alkane with six carbon atoms in its longest continuous chain ("hex" stands for 6). However, it has a methyl group (CH_3) attached on the second carbon atom of the chain and another on the fifth. Hence the full systematic name of the compound is 2,5-dimethylhexane, where "di" indicates two methyl groups. In the case of 3-ethyl-2-methylpentane, the longest continuous chain of carbon atoms contains five carbon atoms, denoted by pentane; a methyl group is attached to the second carbon atom, and an ethyl group, C_2H_5, on the third carbon atom. The substituent ethyl and methyl groups are listed in alphabetical order. The last compound shown in the figure has six carbon atoms in a ring, indicated by the prefix "cyclo," so it is a cyclohexane compound. Furthermore, the carbon in the ring to which one of the methyl groups is attached is designated by "1", and another methyl group is attached to the fourth carbon atom around the ring. Therefore, the full name of the compound is 1,4-dimethylcyclohexane.

Reactivity of Alkanes

Alkanes are relatively unreactive. At elevated temperatures they readily burn with molecular oxygen in air as shown by the following reaction of propane:

$$C_3H_8 + 5O_2 \rightarrow 3CO_2 + 4H_2O + heat \qquad (3.11.1)$$

Common alkanes are highly flammable, and the more volatile lower molecular mass alkanes form explosive mixtures with air.

In addition to combustion, alkanes undergo **substitution reactions** in which one or more H atoms on an alkane are replaced by atoms of another element. The most common such reaction is the replacement of H by chlorine, to yield **organochlorine** compounds. For example, methane reacts with chlorine to give chloromethane, as shown below:

$$Cl_2 + CH_4 \rightarrow CH_3Cl + HCl \qquad (3.11.2)$$

Unsaturated Hydrocarbons

Alkenes (olefins) are hydrocarbons that have double bonds consisting of four shared electrons. The simplest and most widely manufactured alkene is ethene (ethylene),

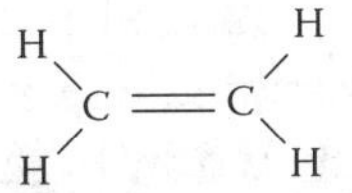

Ethene (ethylene)

used for the production of polyethylene polymer. Another example of an important alkene is 1,3-butadiene (Figure 3.11), widely used in the manufacture of polymers, particularly synthetic rubber.

Acetylene (Figure 3.11) is an **alkyne**, a class of hydrocarbons characterized by carbon–carbon triple bonds consisting of six shared electrons. Highly flammable, dangerously explosive acetylene is used in large quantities as a chemical raw material and fuel for oxyacetylene torches. The double and triple bonds in alkenes and alkynes have "extra" electrons capable of forming additional bonds. Therefore, the carbon atoms attached to these bonds can add atoms without losing any atoms already bonded to them, and the multiple bonds are said to be unsaturated. Therefore, alkenes and alkynes both undergo addition reactions in which pairs of atoms are added across unsaturated bonds as shown in the hydrogenation reaction of ethylene with hydrogen to give ethane

$$
\underset{H}{\overset{H}{\diagdown}}C=C\underset{H}{\overset{H}{\diagup}} + H-H \longrightarrow H-\underset{\underset{H}{|}}{\overset{\overset{H}{|}}{C}}-\underset{\underset{H}{|}}{\overset{\overset{H}{|}}{C}}-H \qquad (3.11.3)
$$

Addition reactions, which are not possible with alkanes, add to the chemical and metabolic versatility of compounds containing unsaturated bonds and constitute a factor contributing to their generally higher toxicities. Addition reactions make unsaturated compounds much more chemically reactive, more hazardous to handle in industrial processes, and more active in atmospheric chemical processes, such as smog formation (see Chapter 8). However, because addition reactions do not generate by-products that may require disposal, they are favored in the green synthesis of chemicals.

Benzene (Figure 3.11) is the simplest of a large class of **aromatic (aryl) hydrocarbons**. Many important aryl compounds have substituent groups containing atoms of elements other than hydrogen and carbon and are called **aromatic compounds** or **aryl compounds**. Most aromatic compounds discussed in this book contain 6-carbon-atom benzene rings as shown for benzene, C_6H_6, in Figure 3.13. The atoms in aromatic compounds are held together in part by particularly stable bonds that contain delocalized clouds of so-called π (pi, pronounced "pie") electrons. In a simplified sense, the structure of benzene can be visualized as resonating between the two equivalent structures shown on the left in Figure 3.13 by the shifting of electrons in chemical bonds. This structure can be shown more simply and accurately by a hexagon with a circle in it.

Many toxic substances, environmental pollutants, and hazardous waste compounds, such as benzene, toluene, naphthalene, and chlorinated phenols, are aromatic compounds (see Figure 3.14). As shown in Figure 3.14, some aromatic compounds, such as naphthalene and the polycyclic aromatic compound, benzo[a]pyrene, contain fused rings.

Benzo[a]pyrene is the most studied of the polycyclic aryl hydrocarbons (PAHs), which are characterized by condensed ring systems ("chicken wire" structures). These compounds are formed by the incomplete combustion of other hydrocarbons. Some PAH compounds, including benzo[a]pyrene, are of toxicological concern because they are precursors to cancer-causing metabolites.

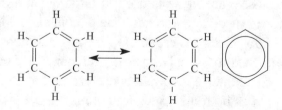

Figure 3.13.   Representation of the aryl benzene molecule with two resonance structures (left) and, more accurately, as a hexagon with a circle in it (right). Unless shown by symbols of other atoms, it is understood that a C atom is at each corner and that one H atom is bonded to each C atom.

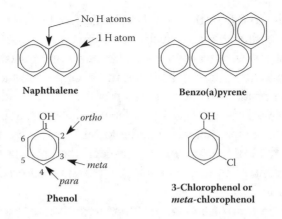

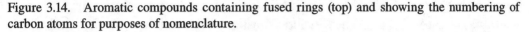

Figure 3.14.   Aromatic compounds containing fused rings (top) and showing the numbering of carbon atoms for purposes of nomenclature.

3.12. GREEN CHEMISTRY

Of all the "green sciences" green chemistry is arguably the most well developed. **Green chemistry** is the practice of chemical science and manufacturing within a framework of industrial ecology in a manner that is sustainable, safe, and nonpolluting and that consumes minimum amounts of materials and energy while producing little or no waste material.[1] Figure 3.15 illustrates this definition. There are certain basic principles of green chemistry. Some publications recognize "The Twelve Principles of Green Chemistry."[2] This section addresses the main ones of these.

As anyone who has ever spilled the contents of a food container onto the floor well knows, it is better to not make a mess than to clean it up once made. As applied to green chemistry, this basic rule means that waste prevention is much better than

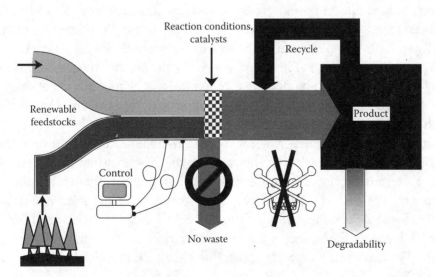

Figure 3.15.   Illustration of the definition of green chemistry.

waste cleanup. Failure to follow this simple rule has resulted in most of the troublesome hazardous waste sites that are causing problems throughout the world today.

One of the most effective ways to prevent generation of wastes is to make sure that insofar as possible all materials involved in making a product should be incorporated into the final product. Therefore, the practice of green chemistry is largely about incorporation of all raw materials into the product, if at all possible. We would not likely favor a food recipe that generated a lot of inedible by-product. The same idea applies to chemical processes. In that respect, the concept of atom economy, expressed as the percentage of reagents that get into the final product, is a key component of green chemistry.

The use or generation of substances that pose hazards to humans and the environment should be avoided. Such substances include toxic chemicals that pose health hazards to workers. They include substances that are likely to become air or water pollutants and harm the environment or organisms in the environment. Here the connection between green chemistry and environmental chemistry is especially strong.

Chemical products should be as effective as possible for their designated purpose, but with minimum toxicity. The practice of green chemistry is making substantial progress in designing chemicals and new approaches to the use of chemicals such that effectiveness is retained and even enhanced while toxicity is reduced.

Chemical syntheses as well as many manufacturing operations make use of auxiliary substances that are not part of the final product. In chemical synthesis, such a substance consists of solvents in which chemical reactions are carried out. Another example consists of separating agents that enable separation of product from other materials. Because these kinds of materials may end up as wastes or (in the case of some toxic solvents) pose health hazards, *the use of auxiliary substances should be minimized and preferably totally avoided.*

Energy consumption poses economic and environmental costs in virtually all synthesis and manufacturing processes. In a broader sense, the extraction of energy, such as fossil fuels pumped from or dug out of the ground, has significant potential to damage the environment. Therefore, *energy requirements should be minimized.* One way in which this can be done is through the use of processes that occur near ambient conditions, rather than at elevated temperature or pressure. One successful approach to this has been the use of biological processes, which, because of the conditions under which organisms grow, must occur at moderate temperatures and in the absence of toxic substances. Such processes are discussed further in Chapter 12.

Raw materials extracted from Earth are depleting in that there is a finite supply that cannot be replenished after they are used. So, wherever possible, *renewable raw materials should be used instead of depletable feedstocks.* As discussed further in Chapter 12, biomass feedstocks are highly favored in those applications for which they work. For depleting feedstocks, recycling should be practiced to the maximum extent possible.

In the synthesis of an organic compound, it is often necessary to modify or protect groups on the organic molecule during the course of the synthesis. This often results in the generation of by-products not incorporated into the final product, such as occurs when a protecting group is bonded to a specific location on a molecule, then removed when protection of the group is no longer needed. Because these processes generate by-products that may require disposal, *the use of protecting groups in synthesizing chemicals should be avoided insofar as possible.*

Reagents should be as selective as possible for their specific function. In chemical language, this is sometimes expressed as a preference for selective catalytic reagents over nonselective stoichiometric reagents.

Products that must be dispersed into the environment should be designed to break down rapidly into innocuous products. One of the oldest, but still one of the best, examples of this is the modification of the surfactant in household detergents 15 or 20 years after they were introduced for widespread consumption to yield a product that is biodegradable. The poorly biodegradable surfactant initially used caused severe problems of foaming in wastewater treatment plants and contamination of water supplies. Chemical modification to yield a biodegradable substitute solved the problem.

Exacting "real-time" control of chemical processes is essential for efficient, safe operation with minimum production of wastes. This goal has been made much more attainable by modern computerized controls. However, it requires accurate knowledge of the concentrations of materials in the system measured on a continuous basis. Therefore, *the successful practice of green chemistry requires real-time, in-process monitoring techniques coupled with process control.*

Accidents, such as spills, explosions, and fires, are a major hazard in the chemical industry. Not only are these incidents potentially dangerous in their own right, but they also tend to spread toxic substances into the environment and increase exposure of humans and other organisms to these substances. For this reason, it is best to *avoid the use or generation of substances that are likely to react violently, burn, build up excessive pressures, or otherwise cause unforeseen incidents in the manufacturing process.*

The principles outlined above are developed to a greater degree in the remainder of the book. They should be kept in mind in covering later sections.

3.13. BIOLOGY

Biology is the science of life and the organisms that comprise life. Living organisms are defined with respect to the following: (1) constitution by particular classes of life molecules, (2) hierarchical organization, (3) capability to carry out metabolic processes, (4) ability to reproduce, (5) development, and (6) heredity.

The kinds of molecules that comprise living organisms are discussed in later sections of this chapter. They are proteins, carbohydrates, lipids (fats), and nucleic acids. Along with salts and water, these materials compose most of living matter.

Hierarchical organization applies to living organisms from the level of atoms all the way to the biosphere as a whole. Proteins, carbohydrates, lipids, and nucleic acids in living organisms are organized into distinct microscopic bodies contained in cells and called **organelles**. Cells are bodies of several micrometers (μm) in size that are the basic building blocks of organisms. Cells with similar functions comprise **tissues** and tissues in turn make up **organs**, which may be organized into whole systems of organs. An **organism** is a collection of organs and organ systems. A group of organisms from the same species comprises a **population** and several different populations existing in the same locale make up a **community**. Numerous communities living in a particular environmental area, interacting with each other and with their environment, constitute an **ecosystem**. Finally, all Earth's ecosystems comprise the entire **biosphere**.

The process of **metabolism** is what occurs when organisms mediate chemical (biochemical) processes to get energy, make raw materials required for tissues in organisms or modify raw materials for this purpose, and reproduce. Though there are thousands of different metabolic reactions, two stand out. The first of these is photosynthesis in which plants use light energy to convert inorganic CO_2 and H_2O to glucose sugar, $C_6H_{12}O_6$. The second major type of metabolic reaction is the mirror image of photosynthesis, **cellular respiration** in which glucose sugar is reacted with oxygen ("burned") to CO_2 and H_2O, yielding energy that is used by the organism.

All organisms undergo **reproduction** to produce offspring to continue the species. In addition to continuing a species, reproduction enables evolution to occur that results in adaptation of species to their environment and development of new species.

Development is the process that occurs as an organism progresses from a fertilized egg to a juvenile and on to adulthood. Even single-celled bacteria that reproduce by cell division undergo development as the cells grow and produce additional organelles prior to further division. As anyone who has observed an infant grow into young adulthood knows, humans undergo development as well.

Heredity refers to the process by which traits characteristic of a species of organism are passed on to later generations. Heredity occurs through the action of DNA. Heredity is the mechanism by which organisms have undergone evolution and adaptation to their environment.

Organisms that comprise living beings in the biosphere range in size and complexity from individual bacterial cells less than a micrometer in dimensions up to giant whales and human beings capable of thought and reasoning. Organisms comprising the biosphere belong to six kingdoms. **Archaebacteria** and **Eubacteria** are generally single-celled organisms without distinct, defined nuclei. **Protists** (protozoans) are generally single-celled organisms that have cell nuclei and may exhibit rather intricate structures. The three other kingdoms are **Plantae** (plants), **Animalia** (animals), and **Fungi** typified by molds and mushrooms.

Organisms are classified according to their food and energy sources and their utilization of oxygen. **Autotrophs** synthesize their food and biomass from simple inorganic substances, usually using solar energy to perform photosynthesis. **Chemautotrophs** mediate inorganic chemical reactions for their energy. **Heterotrophs**, including humans, derive their energy and biomass from the metabolism of organic matter, usually biomass from plants. **Oxic** (aerobic) organisms require oxygen, whereas **anoxic** (anaerobic) organisms use alternate sources of oxidants. **Facultative** organisms can use oxygen or other oxidants depending upon conditions.

3.14. CELLS: BASIC UNITS OF LIFE

As a fundamental unit of the biosphere, it is appropriate to choose the living **cell**. A single one of these very small entities visible only under a microscope may perform all the functions required for a single-celled organism to process nutrients and energy and to reproduce. Or cells in multicelled organisms may be highly specialized entities, such as human liver, brain, and red blood cells. There are two general classes of cells. **Prokaryotic cells** are those that make up bacteria and simple single-celled organisms that composed all of life on Earth for the first approximately two billion years of life on the planet. These cells are only about 1 to 2 μm in size, have only limited external appendages, and possess little differentiated internal structure. **Eukaryotic cells** compose all organisms other than bacteria, are typically 10 μm or more in size, sometimes have external appendages, and generally show well-differentiated internal structures with numerous distinct parts. These cells appeared only about 1.5 billion years ago in the estimated 3.5 billion years that life has existed on Earth. Figure 3.16 represents these two kinds of cells.

Prokaryotic Cells

Prokaryotic cells characteristic of bacteria are enclosed by strong **cell walls** composed largely of carbohydrates that hold the cells together. The plasma membrane controls passage of materials into and out of the cell and is the site of

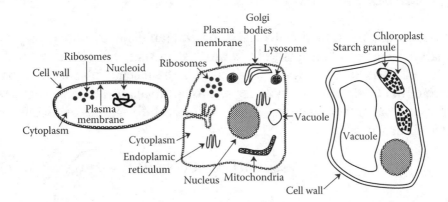

Figure 3.16. Representations of prokaryotic cells (left), eukaryotic animal cells (center), and eukaryotic plant cells (right). Eukaryotic cells are very complex and not all of the many organelles are shown.

photosynthesis in photosynthetic bacteria. Gelatinous **cytoplasm** composed largely of protein and water fills the cell. There is not a defined nucleus, but the cell has a mass of genetic material (DNA) that composes a **nucleoid**. The DNA directs cell metabolism and reproduction. Proteins are made in the cell in **ribosomes** that are distributed around the cell interior. Ribosomes and other bodies in the prokaryotic cell are not enclosed by separate defined membranes as is the case with more complex eukaryotic cells.

Eukaryotic Cells

Eukaryotic cells all are enclosed by **plasma membranes** (cell membranes), are filled with cytoplasm, and contain a variety of organelles that are enclosed by membranes. Figure 3.16 shows only a few of the most important organelles contained in eukaryotic cells. The genetic material in eukaryotic cells is contained in a **nucleus** that is enclosed by a membrane. This DNA is associated with proteins and RNA forming **chromosomes**. **Mitochondria** in eukaryotic cells are bodies in which oxidative metabolism, the process by which the cells use oxygen to gain energy from "burning" food nutrients, is carried out. **Lysosomes** are bodies that contain enzymes capable of breaking down cellular macromolecules (proteins, carbohydrates, nucleic acids, and lipids). This very important process destroys waste macromolecules that otherwise would accumulate and stop the cells from performing their necessary functions and in so doing recycles the small molecules that cells need for their metabolic processes. In a sense, therefore, lysosomes do green chemistry at the cellular level. The system of internal membranes composing the **endoplasmic reticulum**, one of the major features that distinguish eukaryotic cells from prokaryotic cells, contains surface-bound enzymes that synthesize proteins, such as proteinaceous enzymes, that are exported from the cell. **Golgi bodies** act to expel materials from the cell.

3.15. METABOLISM AND CONTROL IN ORGANISMS

Living organisms continually carry out **metabolism** by which they process materials and energy. Photosynthesis, in which solar energy is used to convert atmospheric carbon dioxide and water to glucose, is the metabolic action that provides the base of the food chain for most organisms. Animals break down complex food materials to smaller molecules through the process of **digestion. Respiration** occurs as nutrients are metabolized to yield energy:

$$C_6H_{12}O_6 \text{ (glucose)} + 6O_2 \rightarrow 6CO_2 + 6H_2O + \text{energy} \qquad (3.15.1.)$$

Organisms assemble small molecules to produce biomolecules, such as proteins, by a **synthesis** process.

In addition to considering metabolism as a phenomenon within an individual organism, it can be viewed as occurring within groups of organisms living in an ecosystem. Consider, for example, the metabolism of nitrogen within an ecosystem. Elemental nitrogen from the atmosphere may be fixed as organic nitrogen by bacteria living symbiotically on the roots of leguminous plants, then converted to nitrate when the nitrogen-containing biomass decays. The nitrate may be taken up by other plants and incorporated into protein. The protein may be ingested by animals and the nitrogen excreted as urea in their urine to undergo biological decay and return to the atmosphere as elemental nitrogen. Carbon from carbon dioxide in the atmosphere may be incorporated into biomass by plant photosynthesis, then eventually returned to the atmosphere as carbon dioxide as the biomass is used as a food source by animals.

In Section 3.6, catalysts were defined as materials that enable a reaction to occur without themselves being consumed. Living organisms have catalysts, special proteins that enable biochemical reactions to take place called **enzymes**. In addition to making reactions go much more rapidly, enzymes are often highly specific in the reactions that they catalyze. The reason for the specificity of enzymes is that they have defined structures that fit with the substances on which they act.

Enzymes in Metabolism

Figure 3.17 illustrates the action of enzymes. The first step is the reversible formation of an enzyme/substrate complex that forms because of the complementary shapes of the enzyme (more specifically the active site on the enzyme) and the substrate. The second step is the formation of products accompanied by release of the unchanged enzyme molecule. This reaction implies that the substrate is split apart by enzyme action, a very common enzymatic process called **hydrolysis** when it is accompanied by the addition of water with an H atom going to one of the products and an OH group to the other. Other types of enzyme-catalyzed reactions occur, including the joining of two molecules, modifications of functional groups

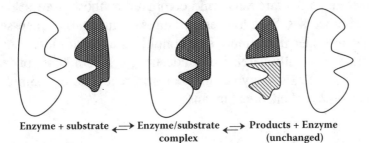

Enzyme + substrate ⇌ Enzyme/substrate ⇌ Products + Enzyme
                        complex   (unchanged)

Figure 3.17.   Enzyme action. The enzyme recognizes the substrate on which it acts because of the complementary shapes of the enzyme and the substrate. The double arrows indicate that the processes are reversible.

consisting of particular groupings of atoms on organic molecules, and rearranging the structures of molecules.

The names of enzymes, usually ending in "-ase" often reflect their functions and may also indicate where they operate. An example is gastric proteinase, a name that indicates the enzyme acts in the stomach (gastric) and hydrolyzes proteins (proteinase). The enzyme released by the pancreas that hydrolyzes fats is called pancreatic lipase.

A number of factors can affect enzyme action. One important factor is temperature. Organisms without temperature-regulating mechanisms have enzymes that increase in activity as temperature increases up to the point where the heat damages the enzyme, after which the activity declines precipitously with increasing temperature. Enzymes in mammals function optimally at body temperature (37°C for humans) and are permanently destroyed by about 60°C. There is particular interest in enzymes that function in bacteria that live in hot springs and other thermal areas where the water is at or near boiling. These enzymes may turn out to be very useful in commercial biosynthesis operations where the higher temperature enables reactions to occur faster.

A significant concern with potentially toxic substances is their adverse effects upon enzymes. As an example, organophosphate compounds, such as insecticidal parathion and military poison sarin "nerve gas" bind with acetylcholinesterase required for nerve function, causing it not to act in its normal function of stopping nerve action once its proper action has been accomplished. Some substances cause the intricately wound protein structures of enzymes to come apart, a process called *denaturation*, which stops enzyme action. Heavy metals, such as lead and cadmium, have a strong affinity for –SH groups in enzymes and may bind at enzyme active sites thus destroying the function of the enzymes.

Enzymes are of significant concern in the practice of green technology. One obvious relationship is that between enzymes and chemicals that damage them. In carrying out green chemical processes, such chemicals should be avoided wherever possible. The moderate temperatures under which enzymes must operate are of particular interest in green technology because, enzyme-catalyzed processes must occur at temperatures that are very mild compared to those used in most conventional industrial processes. Such low temperatures mean that the processes catalyzed by enzymes occur under much more safe conditions and with less consumption of energy than those normally carried out industrially. In syntheses of biochemicals and some pharmaceuticals, enzymes can perform reactions that simply cannot be achieved by conventional chemical means.

Nutrients

The raw materials that organisms require for their metabolism are **nutrients**. Those required in larger quantities include oxygen, hydrogen, carbon, nitrogen, phosphorus, sulfur, potassium, calcium, and magnesium and are called **macronutrients**.

Plants and other autotrophic organisms use these nutrients in the form of simple inorganic species, such as H_2O and CO_2, which they obtain from soil, water, and the atmosphere. Heterotrophic organisms obtain much of the macronutrients that they need as carbohydrates, proteins, and lipids (see Section 3.16) from organic food material.

An important consideration in plant nutrition is the provision of **fertilizers** consisting of sources of nutrient nitrogen, phosphorus, and potassium. A large segment of the chemical manufacturing industry is involved with fixing nitrogen from the atmosphere as ammonia (NH_3) and converting it to nitrate (NO_3^-), urea (CON_2H_4), or other compounds that are applied to the soil as nitrogen fertilizer. Phosphorus is mined as mineral phosphate that is converted to biologically available phosphate ($H_2PO_4^-$ and HPO_4^{2-} ions) by treatment with sulfuric or phosphoric acid. Potassium is mined as potassium salts and applied directly as fertilizer. The ongoing depletion of sources of phosphorus and potassium fertilizer is a sustainability issue of significant concern.

Organisms also require very low levels of a number of **micronutrients**, which are usually used by essential enzymes that enable metabolic reactions to occur. For plants, essential micronutrients include the elements boron, chlorine, copper, iron, manganese, sodium, vanadium, and zinc. The bacteria that fix atmospheric nitrogen required by plants require trace levels of molybdenum. Animals require in their diet elemental micronutrients including iron and selenium as well as micronutrient vitamins consisting of small organic molecules.

Regulation of Metabolism

Organisms must be carefully regulated and controlled in order to function properly. A major function of these regulatory functions is the maintenance of the organism's **homeostasis**, its crucial internal environment. The most obvious means of control in animals is through the **nervous system** in which messages are conducted very rapidly to various parts of the animal as **nerve impulses**. More advanced animals have a brain and spinal cord that function as a **central nervous system** (CNS). This sophisticated system receives, processes, and sends nerve impulses that regulate the behavior and function of the animal. Effects on the nervous system are always a concern with toxic substances. For example, exposure to organic solvents that dissolve some of the protective lipids around nerve fibers can lead to a condition (in which limbs do not function properly) called **peripheral neuropathy**. Therefore, a major objective of green chemistry is to limit the use of and human exposure to such solvents.

Both animals and plants employ **molecular messengers** that move from one part of the organism to another to carry messages by which regulation occurs. Messages sent by these means are much slower than those conveyed by nerve impulses. Molecular messengers are often **hormones** that are carried by a fluid medium in the organism, such as the bloodstream, to cells where they bind to **receptor proteins**

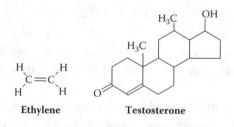

Ethylene   Testosterone

Figure 3.18.   A simple molecule that acts as a plant hormone to promote maturation processes (ethylene) and a common animal hormone, testosterone, the male sex hormone.

causing some sort of desired response. For example, the process may cause the cell to synthesize a protein to counteract an imbalance in homeostasis. Some hormones called **pheromones** carry messages from one organism to another. They commonly serve as sex attractants. Some biological means of pest control use sex pheromones to cause sexual confusion in insects, thus preventing their reproduction. Figure 3.18 shows a common plant hormone and a common animal hormone.

In animals, regulatory hormones are commonly released by **endocrine glands**, such as the anterior pituitary gland that releases human growth hormone, the parathyroid gland that releases a hormone to stimulate uptake of calcium into the blood from bones and the digestive tract, and the pancreas that releases insulin to stimulate glucose uptake from blood. These hormones are carried to target cells in fluids external to the cells. A significant concern with toxic substances is their potential to interfere with the function of endocrine glands. Another concern is that some toxic substances may mimic the action of hormones. For example, evidence exists to suggest that premature sexual development in some young female children can be caused by ingestion of synthetic chemicals that mimic the action of the female sex hormone estrogen.

3.16. BIOCHEMICALS AND BIOCHEMISTRY

The science of the chemical processes that occur in living organisms is **biochemistry**. Chemical species produced by organisms are called **biochemicals**. Some of these, such as enzymes and hormones, have been mentioned in this chapter. Biochemistry is very important in green science and technology because of the importance of biochemicals as "green" materials and biochemical processes as green processes that can be very environmentally friendly. It is important to recognize four major classes of biochemicals. These are carbohydrates, proteins, lipids, and nucleic acids. They are discussed briefly in this section.

Carbohydrates

Carbohydrates are biomolecules consisting of carbon, hydrogen, and oxygen having the approximate simple formula CH_2O. One of the most common carbohydrates is the simple sugar glucose (Figure 3.19). Units of glucose and other simple

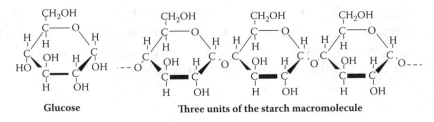

Glucose   Three units of the starch macromolecule

Figure 3.19.   Glucose, a monosaccharide or simple sugar, and a segment of the starch molecule, which is formed when glucose molecules join together with the elimination of one H_2O molecule per glucose monomer.

sugars called **monosaccharides** join together in chains with the loss of a water molecule for each linkage to produce macromolecular **polysaccharides**. The most common polysaccharides are **starch** and **cellulose** in plants and starchlike glycogen in animals.

Glucose carbohydrate is the biological material generated from water and carbon dioxide when solar energy in sunlight is utilized in photosynthesis. The overall reaction is

$$6CO_2 + 6H_2O \rightarrow C_6H_{12}O_6 + 6O_2 \qquad (3.16.1)$$

This is obviously an extremely important reaction because it is the one by which inorganic molecules are used to produce high-energy carbohydrate molecules that are, in turn, converted to the vast number of biomolecules that comprise living systems. There are other simple sugars, including fructose, mannose, and galactose that have the same simple formula as glucose ($C_6H_{12}O_6$), but which must be converted to glucose before being utilized by organisms for energy. Common table sugar, sucrose ($C_{12}H_{22}O_{11}$) consists of a molecule of glucose and one of fructose linked together (with the loss of a water molecule) and, because of these two simple sugars of which it is composed, is called a **disaccharide**.

Starch molecules, which may consist of several hundred glucose units joined together, are readily broken down by organisms to produce simple sugars used for energy and to produce biomass. For example, humans readily digest starch in potatoes or bread to produce glucose used for energy (or to make fat tissue). Another chemically very similar polysaccharide consisting of even more glucose units is cellulose, which comprises much of the biomass of plant cells. Humans and other animals cannot digest cellulose directly to use as a food source but some bacteria and fungi do so readily. Such bacteria living in the stomachs of termites and ruminant animals (cattle, sheep, moose) break down cellulose to small molecules that are converted to molecules that can be absorbed through the digestive systems of animals and utilized as food.

Carbohydrates are potentially very important in green chemistry. For one thing, they are a concentrated form of organic energy that results from the capture of solar energy by photosynthetic processes. Carbohydrates can be utilized directly for energy or fermented to produce ethanol, C_2H_6O, a combustible alcohol that is added to gasoline or can even be used in place of gasoline. Secondly, carbohydrates represent a rich source of organic raw material that can be converted to other organic molecules to make plastics and other useful materials.

Proteins

Proteins are macromolecules that are composed of nitrogen, carbon, hydrogen, and oxygen along with smaller quantities of sulfur. The small molecules from which proteins are made are composed of 20 naturally occurring **amino acids**. The simplest of these, glycine is shown in the first structure in Figure 3.20 along with two other amino acids. As shown in Figure 3.20, amino acids join together with the loss of a molecule of H_2O for each linkage formed. The three amino acids in Figure 3.20 are shown linked together as they would be in a protein in the bottom structure in the figure. Many hundreds of amino acid units may be present in a protein molecule.

The three-dimensional structures of protein molecules are of the utmost importance and largely determine what the proteins do in living systems and how they are recognized by other biomolecules. Enzymes, special proteins that act as catalysts to enable biochemical reactions to occur, recognize the substrates on which they act by the complementary shapes of the enzyme molecules and substrate molecule. The loss of protein structure, called **denaturation**, can be very damaging to proteins and to the organism in which they are contained.

Two major kinds of proteins are tough **fibrous proteins** that compose hair, tendons, muscles, feathers, and silk, and spherical or oblong-shaped **globular proteins**, such as hemoglobin in blood or the proteins that comprise enzymes. Proteins serve

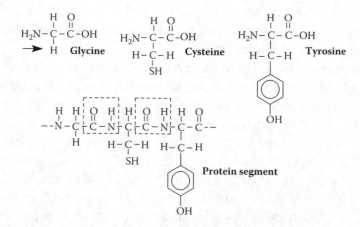

Figure 3.20.   Three amino acids. Glycine is the simplest amino acid. All others have the basic glycine structure except that different groups are substituted for the H designated in glycine by an arrow. The lower structure shows these three amino acids linked together in a segment of a macromolecular protein chain. For each linkage, one molecule of H_2O is lost. The peptide linkages holding amino acids together in proteins are outlined by dashed rectangles.

many functions. These include **nutrient proteins**, such as casein in milk, **structural proteins**, such as collagen in tendons, **contractile proteins**, such as those in muscle, and **regulatory proteins**, such as insulin, that regulate biochemical processes.

Lipids: Fats, Oils, and Hormones

Lipids differ from most other kinds of biomolecules in that they are repelled by water. Lipids can be extracted from biological matter by organic solvents, such as diethyl ether or toluene. Recall that proteins and carbohydrates are distinguished largely by chemical structures (chemically similar amino acids for proteins and chemically similar simple sugar units for carbohydrates). However, lipids have a variety of chemical structures that share the common physical characteristic of solubility in organic solvents. Many of the commonly encountered lipid fats and oils are esters of glycerol alcohol, $CH_2(OH)CH(OH)-CH_2(OH)$, and long-chain carboxylic acids (fatty acids), such as stearic acid, $CH_3(CH_2)_{16}CO_2H$. The glycerol molecule has three $-OH$ groups to each of which a fatty acid molecule may be joined through the carboxylic acid group with the loss of a water molecule for each linkage that is formed. Figure 3.20 shows a fat molecule formed from three stearic acid molecules and a glycerol molecule. Such a molecule is one of many possible **triglycerides**. Also shown in this figure is cetyl palmitate, the major ingredient of spermaceti wax extracted from sperm whale blubber and used in some cosmetics and pharmaceutical preparations. Cholesterol shown in Figure 3.21 is one of several important lipid **steroids**, which share the ring structure composed of rings of five and six carbon atoms shown in the figure for cholesterol. Steroids act as **hormones** (Section 3.15).

Although the structures shown in Figure 3.20 are diverse, they all share a common characteristic. This is the preponderance of hydrocarbon chains and rings, so that lipid molecules largely resemble hydrocarbons. This is the characteristic that makes lipids soluble in organic solvents.

Lipids are important in green chemistry for several reasons. Lipids are very much involved with toxic substances, the generation and use of which are always important in green chemistry. Poorly biodegradable substances, particularly organochlorine compounds, that are always an essential consideration in green chemistry, tend to accumulate in lipids in living organisms, a process called **bioaccumulation**. Lipids can be valuable raw materials and fuels. Therefore, the development and cultivation of plants that produce oils and other lipids is a major possible route to the production of renewable resources.

Nucleic Acids

Nucleic acids are biological macromolecules that store and pass on the genetic information that organisms need to reproduce and synthesize proteins. The two major kinds of nucleic acids are **deoxyribonucleic acid**, **DNA**, which basically stays in

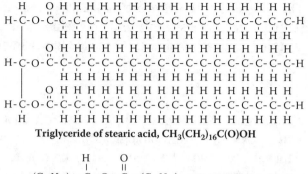

Triglyceride of stearic acid, CH₃(CH₂)₁₆C(O)OH

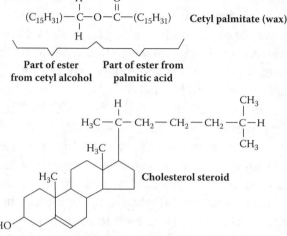

Figure 3.21.   Three examples of lipids formed in biological systems. Note that a line structure is used to show the ring structure of cholesterol. The hydrocarbon-like nature of these compounds which makes them soluble in organic compounds is obvious. The vertical dashed box on the left outlines the part of the triglyceride derived from glycerol and the horizontal dashed box represents the part derived from stearic acid.

place in the cell nucleus of an organism and **ribonucleic acid, RNA,** which is spun off from DNA and functions throughout a cell. Molecules of nucleic acids contain three basic kinds of materials. The first of these is a simple sugar, 2-deoxy-β-D-ribo-furanose (deoxyribose) contained in DNA and β-D-ribofuranose (ribose) contained in RNA. The second major kind of ingredient consists of nitrogen-containing bases: cytosine, adenine, and guanine, which occur in both DNA and RNA, thymine, which occurs only in DNA, and uracil, which occurs only in RNA. The third constituent of both DNA and RNA is inorganic phosphate, PO_4^{3-}. These three kinds of substances occur as repeating units called **nucleotides** joined together in astoundingly long chains in the nucleic acid polymer as shown in Figure 3.22.

The remarkable way in which DNA operates to pass on genetic information and perform other functions essential for life is the result of the structure of the DNA molecule. In 1953, James D. Watson and Francis Crick deduced that DNA consisted of two strands of material counterwound around each other in a structure known as an α-helix, an amazing bit of insight that earned Watson and Crick the Nobel

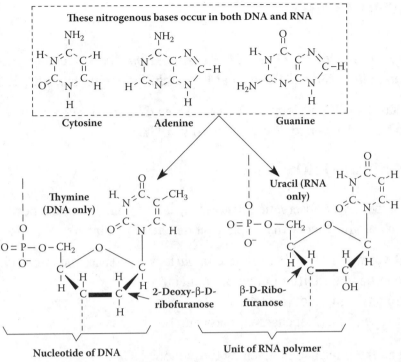

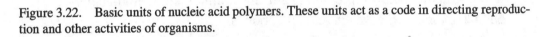

Figure 3.22.   Basic units of nucleic acid polymers. These units act as a code in directing reproduction and other activities of organisms.

Prize in 1962. These strands are held together by hydrogen bonds between complementary nitrogenous bases. Taken apart, the two strands resynthesize complementary strands, a process that occurs during reproduction of cells in living organisms. In directing protein synthesis, DNA becomes partially unravelled and generates a complementary strand of material in the form of RNA, which in turn directs protein synthesis in the cell.

Nucleic acids have an enormous, as of yet largely unrealized, potential in the development of green chemistry. Much of the hazard of many chemical substances results from potential effects of these substances upon DNA. Of most concern is the ability of some substances to alter DNA and cause uncontrolled cell replication characteristic of cancer.

In recent years, humans have developed the ability to alter DNA so that organisms synthesize proteins and perform other metabolic feats that would otherwise be impossible. Such alteration of DNA is commonly known as **genetic engineering** and **recombinant DNA** technology. Organisms produced by recombinant DNA techniques that contain DNA from other organisms are called **transgenic organisms**. The potential of this technology to produce crops with unique characteristics, to synthesize pharmaceuticals, and to make a variety of useful raw materials as renewable feedstocks is discussed in later chapters.

LITERATURE CITED

1. Manahan, Stanley E., *Green Chemistry and the Ten Commandments of Sustainability*, 2nd ed., ChemChar Research, Inc, Columbia, MO, 2006.

2. Anastas, Paul T. and John C. Warner, *Green Chemistry Theory and Practice*, Oxford University Press, New York, 1998.

QUESTIONS AND PROBLEMS

1. Match the law or observation denoted by letters below with the portion of Dalton's atomic theory that explains it denoted by numbers:

   a. Electron   i. Number of these in nucleus equals atomic number
   b. Neutron   ii. Differ in number of neutrons
   c. Proton   iii. Not in nucleus
   d. Isotopes   iv. Not charged, mass number 1

2. State all the information given by the symbol $^{37}_{17}Cl$

3. In the structure below representing a formula unit of ammonium chloride, NH_4Cl, indicate all covalent and ionic bonds:

$$H-\overset{\overset{\displaystyle H}{|}}{\underset{\underset{\displaystyle H}{|}}{N}}-H^{+-}\ Cl$$

4. Summarize the information given in the formula $Al_2(SO_4)_3$.

5. Balance the following chemical equations remembering that chemical formulas must not be altered in balancing equations: (a) $CH_4 + Cl_2 \rightarrow CHCl_3 + HCl$, (b) $CH_4 + O_2 \rightarrow CO_2 + H_2O$, (c) $AlCl_3 + H_2O \rightarrow Al(OH)_3 + HCl$, (d) $S + O_2 \rightarrow SO_3$.

6. Calculate the number of grams of NaCl that must be dissolved and diluted to a volume of 2.5 L to give a 0.110 M solution of NaCl.

7. Starting with 7.50 g of ice at 0°C, calculate the amount of heat required to (1) melt the ice, (2) raise the temperature of the liquid water to 100°C, and (3) convert the water to steam at 100°C.

8. Starting with the Lewis formula of molecular hydrogen, show the formation of hydrogen ions from this molecule.

9. The sulfate ion, SO_4^{2-}, is produced by the dissociation of sulfuric acid, H_2SO_4. Show with a chemical equation what happens when sulfuric acid reacts with sodium hydroxide and name the products.

10. Using Equations 3.9.7 and 3.9.9, if necessary, calculate the molar concentration of H^+ ion, $[H^+]$, and pH for solutions that are (a) 0.100 mol of HCl in exactly 1 L of solution, (b) 0.002 mol of HCl in exactly 1 L of solution, (c) 1 mole of NaOH in exactly 1 L of solution,

11. Where the hydrocarbon with 7 carbon atoms in a chain is heptane, the methyl group is $-CH_3$, the ethyl group is $-C_2H_5$, and the Cl atom bonded to a carbon atom is called "chloro," name the compounds below:

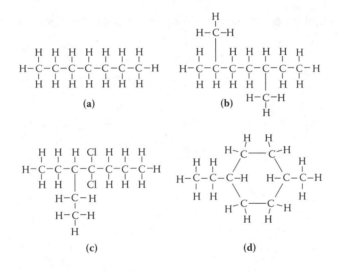

12. Describe the significance of the discovery made by Watson and Crick.

13. Distinguish among oxic, anoxic, and facultative organisms.

14. Look up exoenzymes in a biochemistry book or on the Internet. Suggest how fungi, organisms that cannot chew and ingest wood, are able to use cellulose in wood as a food source.

15. What are the two major means of regulation and control in organisms?

16. What mediates virtually all aspects of metabolism, that is, what kind of substance enables metabolic reactions to occur?

17. Distinguish between macronutrients and micronutrients.

18. Distinguish between eukaryotic cells and prokaryotic cells.

19. The formula of common simple sugars, such as glucose, is $C_6H_{12}O_6$. The simple formula of higher carbohydrates is $C_6H_{10}O_5$. Of course, many simple

sugar units are required to make a molecule of starch or cellulose. If higher carbohydrates are formed by joining together molecules of simple sugars, why is there a difference in the ratios of C, H, and O atoms in the higher carbohydrates as compared to the simple sugars?

20. From the formulas shown in Figure 3.20, give the structural formula of the largest group of atoms common to all amino acid molecules.

21. Look up the structures of ribose and deoxyribose. Explain where the "deoxy" came from in the name, deoxyribose.

22. In what respect is an enzyme and its substrate like two opposite strands of DNA?

23. Define metabolic processes and the two major categories into which they may be divided.

4. WATER: A UNIQUE SUBSTANCE ESSENTIAL FOR LIFE

4.1. A FANTASTIC MOLECULE

All of matter and of life are all about the vast variety of molecules composing matter and living organisms. We are rightfully highly impressed by the remarkably complex molecules of DNA composed of billions of atoms that make up the genetic code of organisms. Chemists are constantly trying to synthesize new molecules, such as those in pharmaceuticals that may be extremely complex and fiendishly difficult to make. But among all molecules, one stands out for its diversity, its occurrence throughout the environment, its vast variety of uses, and its role as a medium for life. This is the molecule of water, which has the very simple formula of H_2O. Essential for life, useful for a vast variety of applications, and totally recyclable, water is the ultimate green chemical compound.

Although the chemical formula of water is simple and the H_2O molecule is small, the behavior of this unique substance is unusual and complex. The special properties of water are because of its molecular structure and the interaction of water molecules with each other. These aspects of water are discussed along with its chemical properties in Chapter 5. But at this point, several things need to be emphasized regarding the water molecule. The formation of the water molecule from atoms of hydrogen and oxygen were shown in Figure 3.6. Rather than showing the water molecule as a linear structure of H–O–H, it was illustrated with the two H atoms forming an angle on the same "side" of the O atom. The reason for this structure is that the outer shell of eight electrons in the O atom of H_2O is composed of four pairs of electrons, two pairs of which are in the covalent bonds between the H and O atoms and two pairs of which are not involved in bonds. It turns out that such pairs of electrons tend to be arranged in the imaginary sphere around the O atom to be as far from each other as possible. This is illustrated in Figure 4.1.

The structure of the water molecule shown in Figure 4.1 does two things that determine water's properties. The first of these is that the end of the water molecule with the two unshared pairs of electrons is relatively negative compared to the end with the two H atoms, so that the water molecule is electrically **polar**. Secondly, an unshared pair of electrons on the oxygen atom of one molecule can bond to an

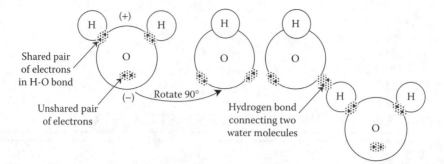

Figure 4.1.   Because of the arrangements of the four pair of electrons making up the outer electron shell of the O atom in the water molecule, the molecule is electrically polar and can form special bonds called hydrogen bonds with other water molecules. These characteristics determine the chemical and physical diversity of water.

H atom on another water molecule in a special kind of bond called the **hydrogen bond**. These two characteristics, especially hydrogen bonding, mean that water molecules are strongly attracted to each other. So, when water is heated, which means that the water molecules move more rapidly, a large amount of heat energy must be put into a mass of the substance to raise its temperature. Furthermore, when solid ice is melted, a very large amount of heat energy is required (to enable the molecules of H_2O to move from their relatively fixed positions in the solid to their mobile state in the liquid) compared to that needed to melt other substances, and an even larger amount of heat energy is required to convert liquid water to vapor (steam). Similarly large amounts of heat energy are released when steam condenses to liquid water, as liquid water cools, and when liquid water freezes.

The heat-storing and heat-transfer capabilities of water are very important in its practical use and influence on the environment. Steam produced by heating water in a boiler can be moved in steam lines to buildings where it is condensed in radiators releasing heat to the buildings. Some modern "green" buildings have cooling systems in which water is frozen at night by heat pumps when cooler temperatures and lower electricity demand make it relatively efficient to do so; the ice is then melted during hot daytime periods absorbing heat and cooling the buildings. Europe owes its relatively mild winters, despite its more northern latitudes, to the Gulf Stream, consisting of water warmed in the Gulf of Mexico region that flows near the surface of the Atlantic Ocean to European shores, cools as it releases its heat to the air in European regions, then sinks as colder, denser water and flows back to the region from which it came.

Water has several important properties that are very important in its environmental influence and practical uses. These are listed in Table 4.1 and discussed further in this chapter.

4.2. WATER AS AN ESSENTIAL RESOURCE

Throughout history, the quality and quantity of water available to humans have been vital factors in determining their well-being. Whole civilizations have

Table 4.1.   Important Properties of Water

Property	Effect or Significance
Excellent solvent, especially for ionic substances	Transport of nutrients and wastes in biological systems; makes nutrients available to plant roots; dissolves, transports, and deposits minerals
High surface tension	Controlling factor in physiology; governs drop and surface phenomena such as in rainfall formation
Transparent	Enables light to penetrate water to some depth enabling algae and other plants in water to carry out photosynthesis
Maximum density as a liquid at 4°C	Ice floats, water in bodies of water is stratified into layers during summer months
High heat of evaporation	Large quantities of heat are absorbed when water evaporates and are released when water vapor condenses enabling use of steam for heat transfer and strongly influencing weather phenomena
High latent heat of fusion	Temperature stabilized at the freezing point of water
High heat capacity	Stabilization of temperatures of organisms and geographical areas

disappeared because their supplies of essential water became unsustainable because of changes in climate, abuse of soil, deterioration of the watershed land on which water was collected, and exhaustion of underground water aquifers. Even in temperate climates, fluctuations in precipitation cause problems. Devastating droughts and destructive floods are problems in many areas of the world.

Waterborne diseases such as cholera, typhoid, and dysentery killed millions of people in the past and still cause great misery in less developed countries. Ambitious programs of dam and dike construction have reduced flood damage, but they have had a number of undesirable side effects in some areas, such as inundation of farmland by reservoirs, failure of unsafe dams, and destruction of fisheries. Globally, problems with quantity and quality of water supply remain and in some respects are becoming more serious. These problems include increased water use because of population growth, contamination of drinking water by improperly discarded hazardous wastes and destruction of wildlife by water pollution.

Water chemistry and biology are discussed in more detail in Chapter 5 and require some understanding of the sources, transport, characteristics, and composition of water. The chemical reactions that occur in water and the chemical species found in it are strongly influenced by the environment in which the water is found. The chemistry of water exposed to the atmosphere is quite different from that of water at the bottom of a lake. Groundwater characteristics are strongly affected by the geochemical characteristics of the underground aquifers that contain the water. Microorganisms play an essential role in determining the chemical composition of water. Thus, in discussing water chemistry, it is necessary to consider the many general factors that influence this chemistry.

The study of water is known as hydrology, which has several subcategories. Limnology is the branch of the science dealing with the characteristics of freshwater including biological properties, as well as chemical and physical properties. Oceanography is the science of the ocean and its physical and chemical characteristics. The chemistry and biology of the Earth's vast oceans are unique because of the oceans' high salt content, great depth, and other factors.

4.3. OCCURRENCE OF WATER

Water circulates throughout Earth's environment by means of the solar-powered **hydrologic cycle** (Figure 4.2). Excluding chemically bound water, the total amount of water on Earth is about 1.4 billion cubic kilometers (1.4×10^9 km^3). Of this amount, about 97.6% is present as salt water in Earth's oceans. This leaves about 33 million km^3 to be distributed elsewhere on Earth as shown in Table 4.2. Even of this amount about 87% is present in solid form, predominantly as polar snowcap and another 12% as groundwater. Therefore, just slightly over 1% of all Earth's fresh water is distributed among surface water, atmospheric water, and biospheric water. This very small fraction comprises water in lakes, including water in the Great Lakes of North America, water in all the Earth's vast rivers (Mississippi, Congo, and Amazon), groundwater to a depth of 1 km, water in the atmosphere, and water in the biosphere.

Earth's water can be considered in several **compartments**. The amounts of water and the **residence times** of water in these compartments vary greatly. The

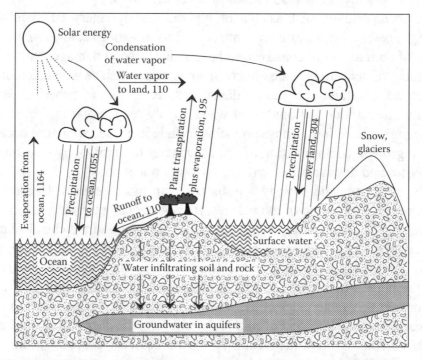

Figure 4.2.   The hydrologic cycle, quantities of water in trillions of liters per day.

Table 4.2.   Distribution of Earth's Water Other Than Ocean Water

Location	Quantity (l)	Percent Total Freshwater
Snow, snowpack, ice, glaciers	2.90×10^{19}	86.9
Accessible groundwater	4.00×10^{18}	12.0
Lakes, reservoirs, ponds	1.25×10^{17}	0.37
Saline lakes	1.04×10^{17}	0.31
Soil moisture	6.50×10^{16}	0.19
Moisture in living organisms	6.50×10^{16}	0.19
Atmosphere	1.30×10^{16}	0.039
Wetlands	3.60×10^{15}	0.011
Rivers, streams, canals	1.70×10^{15}	0.0051

largest of these compartments consists of the **oceans**, containing more than 97% of all Earth's chemically unbound water with a residence time of about 3000 years. Oceans serve as a huge reservoir for water and as the source of most water vapor that enters the hydrologic cycle. As vast heat sinks, oceans have a tremendous moderating effect on climate.

The majority of Earth's water not held in the oceans is bound as snow and ice. Antarctica contains about 85% of all the ice in the world. Most of the rest is contained in the permanent ice pack in the Arctic ocean and in the Greenland ice pack; a small fraction is present in mountain glaciers and snowpack. As discussed in Chapter 9, a major concern with respect to global warming is the melting of ice pack around the world and the accompanying rise in sea levels that may flood low-lying areas of the world.

Groundwater is water held below the surface in porous rock formations called **aquifers**. It influences, and is strongly influenced by the mineral matter with which it is in contact. It dissolves minerals from mineral formations and deposits them on rock surfaces with which it is in contact. Groundwater is replenished by water flowing in from the surface, and it discharges into bodies of water that are below its level.

Bodies of freshwater include **lakes**, **ponds**, and **reservoirs**. Water flows from higher elevations back to the ocean through **rivers** and **streams**. The rate at which water flows in a stream is called its **discharge**. For the Mississippi River, the average discharge is 50 billion liters per hour. Collectively, water in lakes, ponds, reservoirs, rivers, and streams is called **surface water**. Groundwater and surface water have appreciably different characteristics. Many substances either dissolve in surface water or become suspended in it on its way to the ocean. Surface water in a lake or reservoir that contains the mineral nutrients essential for algal growth may support a heavy growth of algae. Surface water with a high level of biodegradable organic material, used as food by bacteria, normally contains a large population of bacteria. All these factors have a profound effect upon the quality of surface water.

There is a strong connection between the hydrosphere, where water is found, and the lithosphere, which is that part of the geosphere accessible to water. Human activities affect both. For example, disturbance of land by conversion of grasslands or forests to agricultural land or intensification of agricultural production may reduce vegetation cover, decreasing **transpiration** (loss of water vapor by plants) and affecting the microclimate. The result is increased rain runoff, erosion, and accumulation of silt in bodies of water. The nutrient cycles may be accelerated, leading to nutrient enrichment of surface waters. This, in turn, can profoundly affect the chemical and biological characteristics of bodies of water.

An environmentally important compartment of water consists of **wetlands**, in which the water table is essentially at surface level. Wetlands consist of marshes, meadows, bogs, and swamps that usually support lush plant life and a high population of animals as well. Wetlands serve as a reservoir and stabilized supply of water. They are crucial nurseries for numerous forms of wildlife.

The **atmosphere** is the smallest compartment of water and the one with the shortest residence time of about 10 d. Atmospheric water is of utmost importance in the movement of water from the oceans to inland in the hydrologic cycle. The atmosphere provides the crucial precipitation that gives water required by all land organisms and to sustain river flow, fill lakes, and replenish groundwater. Furthermore, latent heat contained in atmospheric water, and released when water vapor condenses to form rain, is a major energy transport medium and one of the main ways that solar energy is moved from the equator toward Earth's poles.

Water enters the atmosphere by **evaporation** from liquid water, **transpiration** from plants, and **sublimation** from snow and ice. Water in the atmosphere is present as water vapor and as suspended droplets of liquid and ice. Water vapor in the atmosphere is called **humidity,** and the percentage of water vapor compared to the maximum percentage that can be held at a particular temperature is called the **relative humidity. Condensation** occurs at a temperature called the **dew point** when water molecules leave the vapor state and form liquid or ice particles. This process is aided by the presence of **condensation nuclei** consisting of small particles of sea salt (produced by the evaporation of water from ocean spray), bacterial cells, ash, spores, and other matter on which water vapor condenses. Condensation, alone, does not guarantee precipitation in the form of rain or snow because the condensed water vapor may remain suspended in **clouds**; precipitation occurs when the conditions are right for the cloud particles to coalesce into particles large enough to fall.

Topographical conditions can strongly influence the degree and distribution of precipitation. A striking example of this is provided by the effects of coastal mountain ranges upon precipitation. Moisture-laden air flowing in from the ocean is forced up the sides of coastal mountain ranges, cooling as it does so, and releasing rain as it becomes supersaturated. On the other side of the range, the air warms

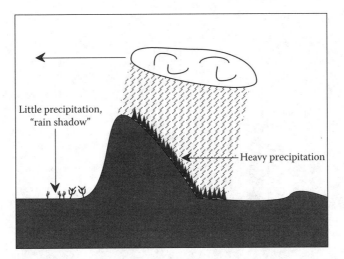

Figure 4.3.   Illustration of a rain shadow.

so that the level of water vapor becomes much less than the saturation concentration, the water stays in the vapor form, clouds disappear, and rain does not fall. The area of low rainfall on the leeward side of a coastal mountain range is called a **rain shadow** (see Figure 4.3).

4.4. WATER UTILIZATION

In the continental U.S., an average of approximately 1.48×10^{13} l of water fall as precipitation each day, which translates to 76 cm/year. Of that amount, approximately 1.02×10^{13} l/d, or 53 cm/year, are lost by evaporation and transpiration. Thus, the water theoretically available for use is approximately 4.6×10^{12} l/d, or only 23 cm/year. At present, the U.S. uses 1.6×10^{12} l/d, or 8 cm of the average annual precipitation. This amounts to an almost tenfold increase from a usage of 1.66×10^{11} l/d in 1900. Since 1900, per capita use has increased from about 40 l/d in 1900 to around 600 l/d now. Much of this increase is accounted for by high agricultural and industrial use, which each account for approximately 46% of total consumption. Municipal use consumes the remaining 8%.

Although water use in the U.S. increased steadily throughout most of the 1900s, suprisingly, around 1980, growth in water use in the U.S. leveled out, despite increasing population (Figure 4.4). This encouraging trend has been attributed to the success of efforts to conserve water, especially in the industrial (including power generation) and agricultural sectors. Conservation and recycling have accounted for much of the decreased use in the industrial sector. Irrigation water has been used much more efficiently by replacing spray irrigators, which lose large quantities of water to the action of wind and to evaporation, with irrigation systems that apply water directly to soil. Trickle irrigation systems that apply just the amount of water needed directly to plant roots are especially efficient. These aspects of water use and conservation are discussed in more detail in Chapter 6.

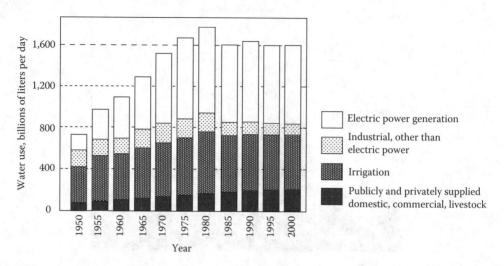

Figure 4.4.   Trends in Water Use in The United States (data from Lumia, Deborah S., Kristin S. Linsey, and Nancy L. Barber, Estimated Use of Water in the United States in 2000, U.S. Geological Survey Circular 1268, U.S. Department of the Interior, U.S. Geological Survey, Reston, Virginia: 2004).

A major problem with water supply is its nonuniform distribution with location and time. As shown in Figure 4.5, precipitation falls unevenly in the continental U.S. This causes difficulties because people in areas with low precipitation often consume more water than people in regions with more rainfall. Rapid population growth in the more arid southwestern states of the U.S. during recent decades has further aggravated the problem and, as of 2005, much of the western U.S. was suffering the effects of a severe drought classified by some experts as a 500-year event. Water shortages are becoming more acute in this region, which contains six of the nation's eleven largest cities (Los Angeles, Houston, Dallas, San Diego, Phoenix,

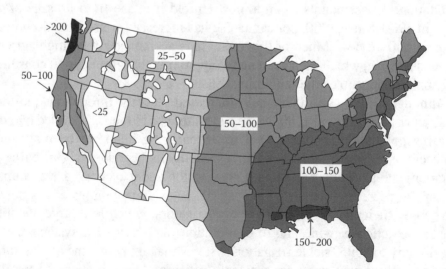

Figure 4.5.   Distribution of precipitation in the continental U.S. showing average annual rainfall in centimeters.

and San Antonio). Other problem areas include Florida, where overdevelopment of coastal areas threatens Lake Okeechobee; the Northeast, plagued by deteriorating water systems; and the High Plains, ranging from the Texas panhandle to Nebraska, where irrigation demands on groundwater are dropping the water table steadily with no hope of recharge. These problems are minor, however, in comparison to those in some parts of Africa where water shortages are contributing to real famine conditions.

Available, renewable water supplies are largely determined by **runoff** equal to total precipitation minus that lost by evaporation/transpiration and infiltration. Average annual runoff does not give a complete picture of water availability, however. One reason is that some of the highest runoff occurs at times of the year when water demand is relatively low. Another reason is the highly seasonal nature of precipitation in many regions that receive large amounts of precipitation such that there are rainy seasons with too much water and flooding alternating with dry seasons in which drought conditions prevail. Therefore, a more meaningful measure of available water is **stable runoff**, which refers basically to available water. Periodic droughts, which in the continental U.S. seem to occur in approximately 30-year cycles, also greatly complicate water supply.

The water that humans use is primarily fresh surface water and groundwater. In arid regions, a small fraction of the water supply comes from desalinated ocean water, a source that is likely to become more important as the world's supply of freshwater dwindles relative to demand. Saline or brackish groundwaters may also be utilized in some areas.

A particularly striking example of water use — and misuse — is the extraordinarily heavy pumping of water from the Ogallala aquifer, which underlies much of the U.S. High Plains, ranging from the Texas panhandle to Nebraska. The water in this aquifer is "fossil water" left from the melting of glaciers after the last Ice Age. The quantity is enormous, several times that of all Earth's surface freshwater. However, intense pumping of water from this resource for irrigation has dropped the water table greatly — in some areas up to about 3 m/year — with no hope of recharge. In less than a century, this vast resource, which could have served the water needs of municipalities and to irrigate high-value crops for centuries, has been depleted to grow corn, alfalfa, and other crops not well adapted to the High Plains region, areas where challenging climate and poor soil make it difficult to live and to utilize the available water.

4.5. STANDING BODIES OF WATER

The physical condition of a body of water strongly influences the chemical and biological processes that occur in water. **Surface water** occurs primarily in streams, lakes, and reservoirs. Lakes may be classified as oligotrophic, eutrophic, or dystrophic, an order that often parallels the life cycle of the lake. **Oligotrophic** lakes are deep, generally clear, deficient in nutrients, and without much biological activity.

Eutrophic lakes have more nutrients, support more life, and are more turbid. **Dystrophic** lakes are shallow, clogged with plant life, and normally contain colored water with a low pH. **Wetlands** are flooded areas in which the water is shallow enough to enable growth of bottom-rooted plants.

Some constructed reservoirs are very similar to lakes, whereas others differ a great deal from them. Reservoirs with a large volume relative to their inflow and outflow are called **storage reservoirs.** Reservoirs with a large rate of flow through compared to their volume are called **run-of-the-river reservoirs.** The physical, chemical, and biological properties of water in the two types of reservoirs may vary appreciably. Water in storage reservoirs more closely resembles lake water, whereas water in run-of-the-river reservoirs is much like river water. Impounding water may have profound effects on its quality.

Estuaries constitute another type of body of water, consisting of arms of the ocean into which streams flow. The mixing of fresh and salt water gives estuaries unique chemical and biological properties. Estuaries are the breeding grounds of much marine life, which makes their preservation very important.

Water's unique temperature–density relationship results in the formation of distinct layers within nonflowing bodies of water, as shown in Figure 4.6. During the summer a surface layer (**epilimnion**) is heated by solar radiation and, because of its lower density, floats upon the bottom layer, or **hypolimnion.** This phenomenon is called **thermal stratification.** When an appreciable temperature difference exists between the two layers, they do not mix, but behave independently and have very different chemical and biological properties. The epilimnion, which is exposed to light, may have a heavy growth of algae. As a result of exposure to the atmosphere and (during daylight hours) because of the photosynthetic activity of algae, the epilimnion contains relatively higher levels of dissolved oxygen, and is said to be aerobic or oxic. In the hypolimnion, bacterial action on biodegradable organic material consumes oxygen and may cause the water to become anaerobic or anoxic, that is, essentially free of oxygen. As a consequence, chemical species in a relatively reduced form tend to predominate in the hypolimnion.

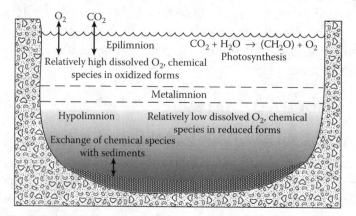

Figure 4.6.   Stratification of a lake.

The shear-plane, or layer between epilimnion and hypolimnion, is called the **metalimnion**. During the autumn, when the epilimnion cools, a point is reached at which the temperatures of the epilimnion and hypolimnion are equal. This disappearance of thermal stratification causes the entire body of water to behave as a hydrological unit, and the resultant mixing is known as **overturn**. An overturn also generally occurs in the spring. During the overturn, the chemical and physical characteristics of the body of water become much more uniform, and a number of chemical, physical, and biological changes may result. Biological activity may increase from the mixing of nutrients. Changes in water composition during overturn may cause disruption in water-treatment processes.

The chemistry and biology of the Earth's vast oceans are unique because of the ocean's high salt content, great depth, and other factors. Oceanographic chemistry is a discipline in its own right. The environmental problems of the oceans have increased greatly in recent years because of release of pollutants to oceans, oil spills, and increased utilization of natural resources from the oceans.

4.6. FLOWING WATER

Surface water that flows in streams and rivers originates from precipitation that falls initially on areas of land called the **watershed**. Watershed protection has become one of the most important aspects of water conservation and management. To a large extent, the quantity and quality of available water depends upon the nature of the watershed. An important characteristic of a good watershed is the ability to retain water for a significant length of time. This reduces flooding, allows for a steady flow of runoff water, and maximizes recharge of water into groundwater reservoirs (aquifer recharge). Runoff is slowed and stabilized by several means. One is to minimize cultivation and forest cutting on steeply sloping portions of the watershed, another is to use terraces and grass-planted waterways on cultivated land. The preservation of wetlands maximizes aquifer recharge, stabilizes runoff, and reduces turbidity of the runoff water. Small impoundments in the feeder streams of the watershed have similar beneficial effects.

Sedimentation by Flowing Water

The action of flowing water in streams cuts away stream banks and carries sedimentary materials for great distances. Sedimentary materials may be carried by flowing water in streams as the following:

- **Dissolved load** from sediment-forming minerals in solution

- **Suspended load** from solid sedimentary materials carried along in suspension

- **Bed load** dragged along the bottom of the stream channel

The transport of calcium carbonate as dissolved calcium bicarbonate provides a straightforward example of dissolved load. Water with high dissolved carbon dioxide content (usually present as the result of bacterial action) in contact with calcium carbonate formations contains Ca^{2+} and HCO_3^- ions. Flowing water containing calcium in this form may become more basic by loss of CO_2 to the atmosphere, consumption of CO_2 by algal growth, or contact with dissolved base, resulting in the deposition of insoluble $CaCO_3$:

$$Ca^{2+} + 2HCO_3^- \rightarrow CaCO_3(s) + CO_2(g) + H_2O \qquad (4.6.1)$$

Most flowing water that contains dissolved load originates underground, where it dissolves minerals from the rock strata that it flows through.

Most sediments are transported by streams as suspended load, obvious from the appearance of "mud" in the flowing water of rivers draining agricultural areas or finely divided rock in Alpine streams fed by melting glaciers. Under normal conditions, finely divided silt, clay, or sand make up most of the suspended load, although larger particles are transported in rapidly flowing water. The degree and rate of movement of suspended sedimentary material in streams are functions of the velocity of water flow and the settling velocity of the particles in suspension.

Bed load is moved along the bottom of a stream by the action of water "pushing" particles along. Particles carried as bed load do not move continuously. The grinding action of such particles is an important factor in stream erosion.

Typically, about 2/3 of the sediment carried by a stream is transported in suspension, about 1/4 in solution, and the remaining relatively small fraction as bed load. The ability of a stream to carry water increases with both the overall rate of flow of the water (mass per unit time) and the velocity of the water. Both of these are higher under flood conditions, so floods are particularly important in the transport of sediments.

Streams mobilize sedimentary materials through **erosion**, **transport** materials along with stream flow, and release them in a solid form during **deposition**. Deposits of stream-borne sediments are called **alluvium**. As conditions such as lowered stream velocity begin to favor deposition, larger, more settleable particles are released first. This results in **sorting** such that particles of a similar size and type tend to occur together in alluvium deposits. Much sediment is deposited in floodplains where streams overflow their banks.

Free-Flowing Rivers

Rivers in their natural state are free flowing. Unfortunately, the free-flowing characteristics of some of the world's finest rivers have been lost to development for power generation, water supply, and other purposes. Many beautiful river valleys have been flooded by reservoirs, and other rivers have been largely spoiled by straightening channels and other measures designed to improve navigation. One of the greater losses from dam construction has consisted of highly productive farmland in river floodplains. Esthetically, an unfortunate case was the flooding, early in the 1900s, of the Hetch Hetchy Valley in Yosemite National Park by a dam designed to produce hydroelectric power and water for San Francisco. More recently proposals have been made to drain the valley in an attempt to restore it to some of its original beauty.

4.7. GROUNDWATER

Most **groundwater** originates as **meteoric** water from precipitation in the form of rain or snow and enters underground aquifers through **infiltration** (Figure 4.7). The rock and soil layer in which all pores are filled with liquid water is called the **zone of saturation**, the top of which is defined as the **water table**. Water infiltrates into aquifers in areas called **recharge zones**. Groundwater may dissolve minerals from the formations through which it passes. Most microorganisms originally present in groundwater are filtered out as it seeps through mineral formations. Occasionally, the content of undesirable salts may become excessively high in groundwater, although it is generally superior to surface water as a domestic water source. Groundwater is a vital resource in its own right; it plays a crucial role in geochemical processes, such as the formation of secondary minerals. The nature, quality, and mobility of groundwater are all strongly dependent on the rock formations in which the water is held. Physically, an important characteristic of such formations is their **porosity**, which determines the percentage of rock volume available to contain water. A second important physical characteristic is **permeability**, which describes the ease of flow of the water through the rock. High permeability is usually associated with high porosity. However, clays, which are common secondary mineral constituents

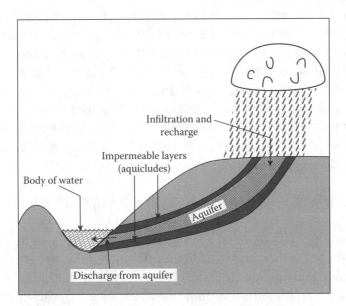

Figure 4.7.   Groundwater in an aquifer.

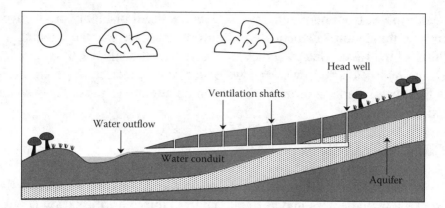

Figure 4.8.   Qanat used to drain water from aquifers by gravity.

of soil, tend to have low permeability even when a large percentage of the volume is filled with water.

Groundwater that is used is usually taken from **water wells**. Poor design and mismanagement of water wells can result in problems of water pollution, land subsidence where the water is pumped out, and severely decreased production. As an example, when soluble iron or manganese are present in groundwater, exposure to air at the well wall can result in the formation of solid deposits of insoluble iron and manganese oxides produced by bacterially catalyzed oxidation processes that result from contact with oxygen in the air. The deposits fill the spaces that water must traverse to enter the well. As a result, they can seriously impede the flow of water into the well from the water-bearing aquifer. This creates major water source problems for municipalities using groundwater for water supply. As a result of this problem, chemical or mechanical cleaning, drilling of new wells, or even acquisition of new water sources may be required.

An interesting ancient technology for utilizing groundwater is the **qanat** (Figure 4.8). Basically, the qanat consists of a tunnel drilled into an aquifer and leading to an outflow located below the elevation of the aquifer. Typical lengths of the conduits in qanats are 10 to 16 km, although they have reached almost 30 km in length. Iran is the world's center of qanat technology dating back to ancient Persia some 3000 years ago. There are about 22,000 qanat units containing over 270,000 km of conduits in Iran that supply about 75% of the nation's water for irrigation and domestic use. Now the system in Iran is threatened by tube wells drilled into the aquifers that are lowering water tables and draining water from the qanats.

4.8. IMPOUNDMENT AND TRANSFER OF WATER

Efforts to cope with demand for water have resulted in significant alterations in the pattern of water flow and distribution. Vast irrigation systems were known to antiquity, notably in irrigation from the Nile River in ancient Egypt and by the Incas in pre-Columbian South America. In a feat of engineering, the Romans built large aqueducts, some of which are still in use more than 20 centuries later. Typically, water is impounded in a reservoir produced by damming a river. From the reservoir the water may be moved through pipes, tunnels, or canals. One of the most massive projects utilizing all such measures is the one providing the water supply for Los Angeles and southern California. This project started with the transfer of water from the Owens Valley in northern California, a distance of about 250 mi. It now involves transfer of water from the Feather and Sacramento Rivers, as well as from the Colorado River. The huge Aswan High Dam forming Lake Nasser has stabilized the flow of Egypt's Nile River. China has undertaken a huge project to move water over 600 mi from the Yangtze River to the area around Bejing. Another huge project designed largely to provide hydroelectric power is located in Quebec.

Impounding water in reservoirs may have some profound effects upon water quality resulting from factors such as different velocities, changed detention times, and altered surface-to-volume ratios relative to the streams that were impounded. Beneficial changes due to impoundment include decreases in organic matter, turbidity, and hardness (calcium and magnesium content). Some detrimental changes are lower oxygen levels due to decreased reaeration, decreased mixing, accumulation of pollutants, lack of bottom scour produced by flowing stream water, and increased growth of algae. Algal growth may be enhanced when suspended solids settle from impounded water, causing increased exposure of the algae to sunlight. Stagnant water in the bottom of a reservoir may be of low quality. Oxygen levels frequently go to almost zero near the bottom, and odorous hydrogen sulfide is produced by the biochemical reduction of sulfur compounds in the low-oxygen environment. Insoluble oxidized iron and manganese species are reduced to soluble ions, which must be removed prior to using the water. A major detrimental effect of water impoundment is simply loss of water by evaporation and by infiltration into surrounding rock strata. In some cases, this can amount to 10% or more of the flow of the river impounded. Evaporative losses can cause increased water salinity. Impounded water drops its load of suspended silt, which will eventually fill a reservoir. Unforeseen detrimental effects of impoundment are illustrated vividly by Lake Nasser. The silt deposited in the lake is no longer available to provide nutrients for the fields irrigated by the river water downstream. Esthetically, irrigation canals are totally inadequate substitutes for wild, free-flowing rivers.

One of the greater environmental disasters resulting from water diversion has been the deterioration of the Aral Sea, a landlocked body of saline water in the former Soviet Union republics of Uzbekistan and Kazakhstan in the south-central part of Asia. Beginning with diversion of water flowing into the sea from the Amu Dar'ya and Syr Dar'ya rivers to irrigate cotton in 1918, the water flow into the Aral Sea was severely curtailed. By 1990 the sea had lost 2/3 of its volume, the salinity of the water had increased proportionally, and the once thriving body of water had essentially "died." Salty dust blown from its shore area has contaminated surrounding farmland. This dead body of water serves as a stark reminder of the consequences of thoughtless technology that does not adequately consider environmental consequences.

4.9. WATER: A VERY USEFUL GREEN SUBSTANCE

Water is the ultimate green substance, essentially indestructible and totally renewable, required by all living organisms and nontoxic, except in cases of gross abuse, such as from drinking enormous quantities of it, and the ultimate green solvent with a vast variety of applications. Water as a green substance is discussed in detail in Chapter 6.

As illustrated in Figure 4.4, the major uses of water are in electrical power generation (for cooling water) and for irrigation. Withdrawal of water for industrial cooling applications is relatively sustainable because usually no more than 5% is lost to evaporation leaving the remainder available for other applications. Some manufacturing operations do require very large amounts of water — as examples, about 8000 l to produce a kilogram of aluminum and an astounding 350,000 to 400,000 l to manufacture an automobile.

Most agricultural use of water is for irrigation. Sometimes, as little as 10% of the water withdrawn for irrigation actually reaches the crops, with the rest lost to evaporation and infiltration. Furthermore, irrigation water may become seriously contaminated, such as by salts leached from the soil and by fertilizer. However, more efficient irrigation and cropping practices have a high potential for water conservation, while still serving irrigation needs effectively. One such technique is "trickle irrigation," in which just enough water drips on the plant roots to keep them moist. Buildup of salts in soil can be a problem with low-consumption irrigation techniques.

Water is not destroyed, but it can be lost for practical use. The three ways in which this may occur are the following:

- **Evaporative losses**, such as occur during spray irrigation and when water is used for evaporative cooling

- **Infiltration** of water into the ground, often in places and ways that preclude its later uses as groundwater

- **Degradation** from pollutants, such as salts picked up by water used for irrigation

The total of the three factors above is called water **consumption**. In many cases it is only a fraction of **withdrawal** consisting of the amount of water that is taken and run through a water system for some purpose. An example in which consumption is only a small fraction of withdrawal is when water is used for cooling, in which all but a small amount lost to evaporation is returned to a stream or body of water slightly warmer, but undamaged.

Some very marked changes in the pattern of water use are inevitable. The impact may be particularly severe upon agriculture, as illustrated, for example, by drastic curtailment of irrigation water to California agriculture in 1991. In the southwestern U.S., for example, agriculture accounts for the bulk of total water usage — 85% in California, 90% in New Mexico and Arizona, and 68% in Texas. In some areas, industries and municipalities are willing to buy their water at prices up to ten times that paid for irrigation water. The increased cost of water could have marked effects on food prices and availability in the U.S.

Water continues to be the subject of heated disputes among land owners and governmental agencies. The state of South Dakota has protested the release of water from reservoirs in the state to maintain barge traffic on the lower Missouri River. Suggestions to transfer water from Washington, Oregon, or northern California to meet growing demand in southern California generate heated discussion. Numerous international disputes have arisen over water supplies, and these can be expected to intensify even to the point of warfare as water becomes scarcer relative to demand.

As with other scarce resources, **conservation** provides much of the answer for water shortages. Conservation can begin at the point where precipitation falls by proper management of the watershed. Proper agricultural practices that minimize and slow runoff, retain soil moisture, and maximize aquifer recharge are very helpful. Use of low tillage agriculture that retains water in plant residues on the surface is very helpful as is the older practice of summer fallowing in which fields are allowed to stand without crop growth in order to build up soil moisture. Construction of terraces, small ponds, and similar water-retaining structures are helpful conservation practices. Other water conservation measures may be applied to minimize domestic consumption. These include use of shower heads that consume minimum amounts of water, low-consumption toilets, appliances (washing machines, dishwashers) designed for low water consumption, and lawns that are planted with native grass species that require minimal water.

A limiting factor in water conservation and reuse is buildup of dissolved salt content. Furthermore, much of the water available for use is saltwater from the ocean or from brackish groundwater. Therefore, salt removal, **desalination**, can be an effective way of augmenting, recycling, or extending water supplies. This topic is addressed further in Chapter 6.

4.10. WATER TECHNOLOGY AND INDUSTRIAL USE

Next to air, water is the cheapest and most universally available raw material. As a consequence, it has found a vast number of uses. Some of these are mentioned here, and water technology is briefly addressed.

Water is used for some of its unique physical properties. The most common of these is the high heat of vaporization/condensation of water. Because of this characteristic, enormous amounts of energy can be put into water by evaporating it to steam, enabling it to be transferred over long distances by pipe to remote locations where the heat can be released by allowing the water to condense. Therefore, water in the form of steam has become a favorite means of transferring heat. The high heat of fusion for ice enables the use of solid water to absorb heat.

The heat energy imparted to steam may be converted to mechanical energy very efficiently in a turbine or steam piston engine (the latter largely of historical interest). In modern times, steam is largely generated in **water-tube** boilers in which liquid water is introduced in small (2 to 8 cm inside diameter) tubes and converted to steam by hot combustion gases circulating outside the tubes. Larger components of the system, particularly the water and steam drums, are isolated from the fire, so that if they fail, their contents will not blast into the firebox and cause an explosion. The failure of a water tube is a relatively less serious event.

A modern water-tube boiler does more than just convert liquid water to steam. Furnace walls are cooled with water, which is converted to steam. Steam is heated after it is generated by passing hot flue gas over a **superheater** consisting of tubes containing steam to further increase efficiency. To further increase efficiency, combustion air is preheated, and flue gas is cooled with an **air heater**. The aspects of water and steam used to generate steam power are outlined in Figure 4.9.

Boiler feed water used to raise steam must meet some stringent requirements to prevent scaling, fouling, and corrosion. Often this is accomplished to a large extent by steam condensate that is recirculated through the system so that the only additonal water required is **makeup water**. Makeup water is treated to remove insoluble contaminants and those that become insoluble when the water is heated. Hardness in the form of dissolved calcium and magnesium is removed by treatment with phosphate salts:

$$5Ca^{2+} + 3PO_4^{3-} + OH^- \rightarrow Ca_5OH(PO_4)_3(s) \qquad (4.10.1)$$

Silicon must be kept to low levels because it can carry over with steam and cause damaging deposits of SiO_2 on turbine blades. Corrosive, dissolved oxygen and

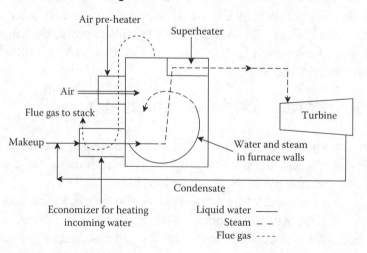

Figure 4.9.   Aspects of water and steam in a boiler.

carbon dioxide can be removed by heating the feed water in an open boiler system. Traces of residual oxygen can be removed by reaction with added hydrazine:

$$N_2H_4 + O_2 \rightarrow 2H_2O + N_2 \qquad\qquad (4.10.2)$$

Anticorrosive agents, such as cyclohexylamine are added to boiler makeup water.

Water is an effective mechanical agent. It can be used to mine and process some kinds of minerals. Examples are sand and gravel harvested from river beds by pumped water. Water under ultra-high pressure can be used to cut materials very effectively and precisely. These applications produce water with high levels of suspended materials that must be removed before the water is released.

One of the greatest uses of water is for its solvent properties. Especially with its surface tension reduced by adding a surface-active agent, water becomes a powerful solvent that is used in a large number of cleaning operations. Water has even begun to replace organic solvents, which pose pollution and disposal problems, in washing small parts, such as electronic constituents, and substitution of water for petroleum-based solvents is an important aspect of green technology. Mixed with suspended lubricants, water is very useful as a lubricant and cooling agent, such as in metal-stamping operations. As a solvent for chemical reactants, water serves as the solvent medium for a number of important chemical synthesis and processing applications. Substances can be purified by dissolving them in water, then evaporating some of the water off to leave a quantity of the substance in a purified form.

Water is a chemical ingredient for a number of industrial chemical reactions. It is required for the hardening of Portland cement to make concrete. It can be used as the reagent in treating some kinds of hazardous wastes by hydrolysis.

Waterpower is the use of moving water for the generation of energy. The water so used may be water descending by the force of gravity from a dammed stream, or it may be rising and falling water from tides. It is the oldest source of nonanimal energy, and has been in use for many centuries. Prior to invention of the steam engine, waterwheels used falling water to provide energy, often for flour mills, and for many decades after steam power was developed, waterpower was competitive in some applications.

At the current time, waterpower is most widely used to power electricity generators, an application called **hydroelectric power**. By 1993, total hydroelectric capacity in the U.S. stood at about 100,000 megawatts representing 13% of the total power-generating capacity (a large coal-fired or nuclear power plant typically has a capacity of 1000 MW). If all available sites in the U.S., including Alaska, were utilized for hydroelectric power, the capacity could be doubled. However, for environmental and economic reasons this will not happen. A useful adaptation of hydroelectric power is **pumped storage** in which water is used to run turbines attached to generators to generate power at times of high demand; the process is reversed under low-demand conditions such that the generator acts as a motor and the turbine as a pump in which it acts to pump water to an elevated storage reservoir.

The economic and environmental advantages of waterpower are obvious. Not the least of these is that the "fuel" is free. Environmentally, there are no emissions or ash with which to deal. There are numerous disadvantages. Under conditions of extreme drought, water may become unavailable, so a system dependent upon waterpower is vulnerable, though much less so than wind-powered or solar-powered systems. Unfortunately, the vast reservoirs required for most waterpower developments destroy free-flowing rivers and are detrimental to fish migration, such as that required for salmon reproduction. Most remaining available sites are in remote regions from which the transfer of electricity requires massive power lines, which also present environmental problems.

Much of what may be regarded as water technology deals with water purification and treatment. These topics are addressed in Chapter 6.

SUPPLEMENTARY REFERENCES

Calhoun, Yael, Ed., *Water Pollution*, Chelsea House Publishers, Philadelphia, 2005.

Gray, Nick, *Water Technology: An Introduction for Environmental Scientists and Engineers*, 2nd ed., Elsevier Butterworth-Heinemann, Oxford, U.K., 2005.

Harman, Rebecca, *The Water Cycle*, Heinemann, Chicago, IL, 2005.

Lazarova, Valentina and Akic a Bahri, *Water Reuse for Irrigation: Agriculture, Landscapes, and Turf Grass*, CRC Press, Boca Raton, FL, 2005.

Lehr, Jay, Jack Keeley, Janet Lehr, and Thomas B. Kingery III, *Water Encyclopedia*, John Wiley & Sons, Hoboken, NJ, 2005.

Mays, Larry W., *Water Resources Engineering*, John Wiley & Sons, Hoboken, NJ, 2005.

Schmitz, P. Michael, *Water and Sustainable Development,* Peter Lang, New York, 2005.

QUESTIONS AND PROBLEMS

1. Compare the physical conditions of water in a deep lake in summer with those in a free-flowing stream, and explain how these affect the environmental chemistry and biology of water.

2. Some evidence suggests that rainfall is increased downwind from sources of pollution, such as large industrial cities. Offer a possible explanation in terms of what this chapter says about the process by which precipitation is formed in the atmosphere.

3. Modify Figure 4.2, "The Hydrologic Cycle," to show involvement of the anthrosphere.

4. Western areas of the U.S. states of Oregon and Washington are quite wet, whereas eastern regions of these states beyond the Cascade Mountains are generally very dry. Give a possible explanation. What phenomenon does this observation illustrate?

5. Try to relate some of the "Important Properties of Water" in Table 4.1 to the nature of the water molecule and to hydrogen bonds.

6. Distinguish between dissolved load, suspended load, and bed load in streams.

7. What are deposits formed from suspended solids in streams called? Why do particles of a similar size and type tend to occur together in such deposits?

8. Under what circumstances does water in a stream or lake tend to enter an aquifer? What are the circumstances under which the opposite occurs?

9. Water infiltrating through soil may pick up high concentrations of dissolved CO_2 from bacterial decay before flowing into rock. If the bedrock is limestone, $CaCO_3$, a sinkhole may form. Explain.

10. Above a reservoir constructed by damming a free-flowing mountain stream, the stream water contains a large concentration of O_2, some NO_3^-, and some suspended solid $Fe(OH)_3$. Water from the bottom of the reservoir is fed through turbines, and the water exiting from the turbines contains little O_2, some NH_4^+, and some dissolved Fe^{2+}. Explain.

11. What are the advantages of "trickle irrigation"?

12. How is use made of the high heat of evaporation/condensation of water used industrially?

13. List as many industrial uses of water as you can based on such things as its heat, solvent, and mechanical properties.

14. Explain how pumped storage of water can lead to more efficient electricity production and utilization.

15. Using the model of the water molecule shown on the left of Figure 4.1, suggest how H_2O molecules might be oriented around (1) a dissolved positively charged cation and (2) a dissolved negatively charged anion. What does this have to do with water's solvent properties for salts?

16. Describe an important respect in which the chemistry of water exposed to the atmosphere differs from that at the bottom of a lake.

17. Some paving in China is now composed of porous materials that allow water infiltration. What beneficial effect might this have on water resources?

18. Suggest a major use of run-of-the-river reservoirs as compared to storage reservoirs.

19. What kinds of conditions might be expected in areas where the geosphere surface dips below the level of the water table?

20. Explain why "water consumption" is a somewhat misleading term.

21. Significant amounts of fertile topsoil and nutrients are eroded from soil by water and carried away by rivers. Suggest structures that humans might make and operate to salvage some of this resource.

5. AQUATIC BIOLOGY, MICROBIOLOGY, AND CHEMISTRY

5.1. ORGANISMS IN WATER

Much of Earth's biosphere exists in the hydrosphere and, because life requires water, all organisms must be near sources of water. Water-dwelling organisms serve as sources of food for many organisms that exist on land, including humans. Life in water is crucial to sustainability. Millions of people derive much of their food, including virtually all protein, from fish and other aquatic organisms. The photosynthetic productivity of aquatic plants and algae provides the basis of the food chains for key ecosystems and is a significant factor in removing excess carbon dioxide from the atmosphere. One of the major concerns of sustainability is the reduced biomass output of aquatic life, particularly in the oceans. Overfishing has depleted Earth's ocean fisheries. Increased atmospheric carbon dioxide levels can raise the dissolved CO_2 levels in oceans, slightly, but enough to alter the ability of key organisms to thrive in the oceanic environment. Pollution has significantly affected fish and other food organisms in water, and consumption of some fish in significant quantities is not recommended because of contaminant mercury. Obviously, aquatic life is a key factor in green science, and its preservation and enhancement is a key goal of the practice of green science and technology.

Prior to discussing the living organisms (**biota**) in water, it is useful to define some terms that apply to them. **Plankton** are small plants, animals, and single-celled organisms that float, drift, or move weakly under their own power near the surface of water. At the bottom of the food chain are photosynthetic plankton or **phytoplankton**. Small animals that float freely or propel themselves weakly are called **zooplankton**. **Invertebrate planktivores** (minute crustaceans, insect larvae) and **vertebrate planktivores** (small fish, minnows) feed on zooplankton. **Water plants** anchored to the bottom with roots inhabit shallow regions of bodies of water. Among these kinds of plants are spike rushes that grow in very shallow water, reeds that grow in somewhat deeper water, pond lilies that float in water, and, at still greater depths, plants such as pondweed that are actually submerged. **Periphyton**, more picturesquely named **aufwuchs**, are small organisms, such as diatoms, algae, and water moss, that grow attached to twigs, rocks, and debris in water. Water supports a large variety of animal life, including fish, sponges, snails, hydras, and insects. **Arthropods**, animals with hard, segmented exoskeletons and jointed legs, are normally abundant in water. Especially prominent among arthropods in water are **crustaceans** ("water insects" in a sense), such as crayfish, lobsters, crabs, shrimps, and water fleas. Animals that can move about entirely on their own power, such as fish, are called **nekton**.

Organisms that live in water may be classified as either autotrophic or heterotrophic. **Autotrophic** biota utilize solar or chemical energy to fix elements from simple, nonliving inorganic material into complex life molecules that compose living organisms. Algae are typical autotrophic aquatic organisms. Generally, CO_2, NO_3^-, and $H_2PO_4^-/HPO_4^{2-}$ are sources of C, N, and P, respectively, for autotrophic organisms. Organisms that utilize solar energy to synthesize organic matter from inorganic materials are called **producers**.

Heterotrophic organisms utilize the organic substances produced by autotrophic organisms as energy sources and as the raw materials for the synthesis of their own biomass. **Grazers** feed on living organic matter. **Detritovores** are organisms that feed on nonliving plant and animal matter called **detritus**. **Decomposers** (or **reducers**) are a subclass of detritovores consisting chiefly of bacteria and fungi, which ultimately break down material of biological origin to the simple compounds originally fixed by the autotrophic organisms.

The ability of a body of water to produce living material is known as its **productivity**. Productivity results from a combination of physical and chemical factors. Water of low productivity generally is desirable for water supply or for swimming.

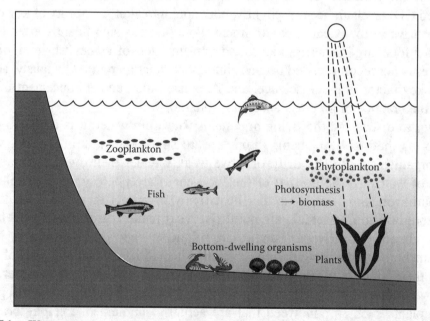

Figure 5.1.   Water supports many forms of life. Photosynthetic organisms in water produce biomass that forms the basis of food chains that support a variety of life including large segments of the human population.

Relatively high productivity is required for the support of fish. Excessive productivity can result in choking by weeds and can cause odor problems. The growth of algae may become quite high in very productive waters, with the result that the concurrent decomposition of dead algae reduces oxygen levels in the water to very low values. This set of conditions is commonly called **eutrophication**.

Physical and Chemical Factors in Aquatic Life

Aquatic organisms are strongly influenced by the physical and chemical properties of the body of water in which they live. Temperature, transparency, and turbulence are the three main physical properties affecting aquatic life. Very low water temperatures result in very slow biological processes, whereas very high temperatures are fatal to most organisms. The solubility of oxygen from air decreases with increasing temperature, which can cause stress for fish living in water. A difference of only a few degrees can produce large differences in the kinds of organisms present in water. Thermal discharges of hot, spent, cooling water from power plants frequently kill heat-sensitive fish while increasing the growth of fish and other species that are adapted to higher temperatures. The transparency of water is particularly important in determining the growth of algae. Thus, turbid water may not be very productive of biomass, even though it has the nutrients, optimum temperature, and other necessary conditions. Turbulence is an important factor in mixing and transport processes in water. Some small plankton depend upon water currents for their own mobility. Water turbulence is largely responsible for the transport of nutrients to living organisms and of waste products away from them. It plays a role in the transport of oxygen, carbon dioxide, and other gases through a body of water and in the exchange of these gases at the water–atmosphere interface. Moderate turbulence is generally beneficial to aquatic life.

Dissolved oxygen (DO) frequently determines the extent and kinds of life in a body of water. Oxygen deficiency is fatal to many aquatic animals such as fish. The presence of oxygen can be equally fatal to many kinds of anaerobic bacteria.

Biochemical oxygen demand (BOD), the amount of oxygen utilized when the organic matter in a given volume of water is degraded biologically, is another important water-quality parameter. A body of water with a high BOD and no means of rapidly replenishing the oxygen, obviously cannot sustain organisms that require oxygen.

Carbon dioxide is produced by respiratory processes in waters and sediments and can also enter water from the atmosphere. Carbon dioxide is required for the photosynthetic production of biomass by algae and in some cases is a limiting factor. High levels of carbon dioxide produced by the degradation of organic matter in water can cause excessive algal growth and productivity.

The levels of nutrients in water frequently determine its productivity. Aquatic plant life requires an adequate supply of carbon (CO_2), nitrogen (nitrate, NO_3^-), phosphorus ($H_2PO_4^-$, HPO_4^{2-}), and trace elements such as iron. In many cases, phosphorus is the limiting nutrient and is generally controlled in attempts to limit excess productivity.

The salinity of water also determines the kinds of life forms present. Irrigation waters may pick up harmful levels of salt. Marine life obviously requires or tolerates saltwater, whereas many freshwater organisms are killed by it.

5.2. LIFE IN THE OCEAN

As a medium for life, the marine environment differs substantially from freshwater environments. The most obvious such difference is the salt content of the ocean, consisting mostly of sodium chloride (NaCl). This affects both the organisms in the ocean and the oceanic environment. Because of osmotic phenomena, freshwater organisms cannot live in the sea, and *vice versa*. Because of its high dissolved salt content, the temperature/density relationships of ocean water are unlike those of freshwater. Other factors that influence life in the ocean include tides, waves, its huge volume, and its great depth in some places.

The most shallow of the three life zones in oceans is the **photic zone** in which there is sufficient light to enable a high level of photosynthesis. Organic matter from the photic zone falls to lower zones as **detrital matter**, which serves as a food source for the organisms below the photic zone. The next layer down is the dimly illuminated **mesopelagic zone** inhabited by various sea creatures, such as sharks. Lower still is the **bathypelagic zone** characterized by enormously high pressures and a total absence of sunlight. This zone is inhabited by unique organisms, some of which generate light by bioluminescence. The **benthic region** on the sea bottom is populated by a variety of worms, crustaceans, crabs, and clams. Life can be especially abundant in the vicinity of **hydrothermal vents** ("black smokers"), which warm the surrounding water and emit hot water rich in metal sulfides and H_2S. Unique bacteria in the vicinity of the vents synthesize biomass chemosynthetically, gaining energy by mediating the oxidation of sulfide to sulfate (SO_4^{2-}), and providing the food needed by other organisms. (See Section 5.9 for a discussion of oxidation and reduction reactions.)

Compared to land and freshwater, the biomass productivity of the ocean is relatively low, largely because of a shortage of nutrients. Not surprisingly, productivity is highest in areas where upwelling currents provide nutrients, and in coastal areas.

With the exception of the unique life forms around hydrothermal vents noted above, primary productivity in the oceans is from *phytoplankton*, such as dinoflagellates and diatoms, consisting of small floating photosynthetic organisms. The next kind of organism up the food chain consists of herbivorous zooplankton, largely planktonic arthropods that feed directly on phytoplankton. These in turn are fed upon by carnivorous zooplankton, a diverse group consisting largely of larvae of larger organisms. The herbivorous and carnivorous zooplankton are fed upon by larger nekton — fish, sharks, whales. In general, larger creatures of this type feed on organisms that are the next step down in size, although there are some exceptions —

the huge baleen whale feeds predominantly on very small organisms, which it filters from water.

5.3. LIFE AT THE INTERFACE OF SEAWATER WITH FRESHWATER AND WITH LAND

Areas in which seawater meets the shore or freshwater flows into the sea are especially active locations for life. These two areas are addressed briefly here. These regions are particularly vulnerable to environmental disruption, such as from oil spills from tankers, or by human modifications of the physical nature of the interface.

The ocean meets with land on sandy or rocky coastlines. A particularly prominent phenomenon that occurs at this interface is alternate exposure of shoreline rock, sand, and mud to seawater and to air, the consequence of wave and tidal action. This results in a striking zonation of life forms.

On rocky shores, the zone between the high-tide and low-tide marks, alternately covered with seawater and exposed to the atmosphere by tidal action, is called the **littoral zone,** further subdivided into subzones. The most common types of plants that grow in the littoral zone are seaweeds of the genus *Fucus*, and related types characterized by their olive-brown color, branched fronds, and air bladders, which enable them to float. Prominent among animal life in this region are periwinkles (single-shelled organisms that move by means of a wide, muscular foot), mussels (bivalve mollusks), and barnacles (marine crustaceans that cling strongly to rocks and other objects).

Sandy and muddy (mudflat) coastal areas offer the opportunity for the growth of a large number of burrowing animals. Among these in sand are various kinds of crabs, shrimp, clams, and sand dollars. Many of the animals that live in sand and mud are very small, ranging from about 60 μm to slightly less than 1 mm. These include nematodes (unsegmented roundworms), copepods (a type of crustacean), ostracods (seed shrimp), and gastrotrichs (small bottle-shaped organisms with cilia on their undersides). Spending most of their lives in, and deriving their food from sedimentary materials, these organisms are especially susceptible to the effects of pollutants, such as organochlorine compounds, heavy metals, and organometallic compounds that accumulate in sediments.

A specialized type of seashore structure supporting unique marine ecosystems consists of **coral reefs** built up by calcium carbonate deposits produced by coral and other marine organisms. Coral reefs form in tropical regions where a firm geological formation is available at shallow depths, conditions that often exist around volcanic islands. These structures provide habitats for the coral itself, associated algae, crustaceans, echinoderms, mollusks, sponges, and fish.

Estuaries are the locations where freshwater from rivers mixes with seawater from the ocean. These regions show gradations of salinity, which vary with the ebb and flow of tides. The food chain in estuaries is based on both detrital food sources

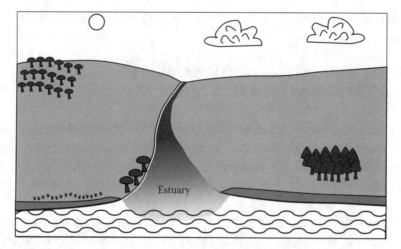

Figure 5.2.   Estuaries where freshwater mixes with ocean water are important nurseries for a number of aquatic species.

and phytoplankton. They are especially important as nurseries for marine fish and shellfish, in part because potential predators from the ocean are intolerant of the lower salinity of estuarine waters.

The alluvial plains of estuary regions that are alternatively covered with seawater, and then drained as the tides rise and fall, may support a heavy growth of grass and other salt-tolerant plants and are called **salt marshes**. These marshes are the sites of tidal creeks and tidal pools. Among the kinds of plants that they support are salt marsh cordgrass, salt grass, marsh hay cordgrass, black grass, and marsh elder. Among the animals that grow in salt marshes are mussels, fiddler crabs (highly adaptable because they can breathe with either lungs or gills), marsh periwinkles, and sand hoppers. Exposed by low tide, these tasty animals attract predators, such as egrets, gulls, and herons.

5.4. FRESHWATER LIFE

A variety of life inhabits bodies of freshwater, such as ponds, lakes, and reservoirs. The types of organisms are strongly influenced by physical characteristics of the body of water. One of the most important of these is the stratification into the upper epilimnion and hypolimnion layers (see Section 4.5 and Figure 4.6). The mixing of these layers that occurs in the spring and fall — the **overturn** — is very important for aquatic life because it brings nutrients to the surface, thereby increasing fertility. Several zones of a body of water are particularly important to its biota. Rooted plants grow in the **littoral zone** in which sunlight penetrates to the bottom. Further from shore, the region in which light penetrates sufficiently to support significant photosynthesis, but not to the bottom, is called the **limnetic zone**, populated by phytoplankton, zooplankton, and fish. Open water below the limnetic zone is called the **profundal zone**. Biological activity in the profundal zone is variable, but often relatively low because it is too deep for photosynthesis, and detrital matter raining down from higher levels does not stay long enough to support much life. However, biological activity is usually high in the **benthic zone** at the bottom of a body of water because of the large amount of detrital biomass that it receives. Organisms that predominate in this region are decomposers, detritovores, and, in benthic waters below the profundal zone, anaerobic bacteria.

The nature and abundance of life in a standing body of water depend significantly on whether the body of water is oligotrophic, eutrophic, or dystrophic (see Section 4.5). Because they are nutrient-poor, oligotrophic lakes lack the inorganic phosphate and nitrate nutrients required to support a high level of photosynthesis and the input of organic matter to support large populations of detritovores. Nutrient-rich eutrophic lakes enable high levels of photosynthetic activity, and the biomass produced also supports a large population of organisms that feed on it and its decomposition products. Dystrophic lakes are so rich in organic matter washed in from surrounding areas that they support a high level of biological activity in their littoral zones. They have relatively low planktonic activity, however.

Life in flowing-water streams and rivers is dependent upon the input of detrital food from the land, although relatively clear streams may generate significant amounts of biomass photosynthetically. Another major factor affecting such life is the influence of often strong currents in flowing water. Organisms have to adapt to currents so that they are not swept downstream, which they do by having streamlined or flat shapes, or by being attached to rocks or to the stream bed. Insect larvae frequently stay under stones where the current is weak and where they can cling to the stone surface. Aquatic water moss and algae may have strong "holdfasts" with which they are attached to stones.

Organic food material in streams may be in the form of large or small particles (detritus) or as dissolved organic matter. Various organisms have developed mechanisms to deal with these food sources. **Shredders** tear apart larger particles of organic matter; their feces and the shredded particles that they don't ingest become part of the supply of fine-particle organic matter. The finely divided organic matter, in turn, is ingested by **collectors** ("fine particulate detritovores") that collect the food by filtering and gathering mechanisms. Aquatic **grazers** feed on algae that grow on

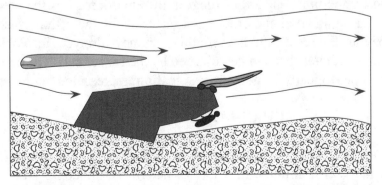

Figure 5.3.   Water-dwelling organisms adapt to strong currents in various ways, such as streamlining, holding tight to surfaces, or staying under rocks.

solids in the streams, and **gougers** burrow into larger particles of wood. **Piercers** suck the juices from bottom-rooted plants. Aquatic bacteria and fungi grow on particulate organic matter, such as dead leaves. At the top of the food chain, **predators** feed on other organisms.

5.5. MICROORGANISMS IN WATER

Microorganisms in water may consist of bacteria, fungi, protozoans, and algae (see Section 3.12). Microorganisms have a strong influence on water. The algae are the primary producers that generate biomass, $\{CH_2O\}$, photosynthetically. Fungi and bacteria are responsible for breaking down and mineralizing organic matter and are essential in nutrient recycling. Bacteria are responsible for the major elemental oxidation-reduction conversions that occur in water. Various protozoans exhibit characteristics of bacteria, fungi, and algae.

Algae in Water

Algae in water may be considered as generally microscopic organisms that subsist on inorganic nutrients and produce organic matter from carbon dioxide by photosynthesis. The general nutrient requirements of algae are carbon (from CO_2 or HCO_3^-), nitrogen (generally as NO_3^-), phosphorus (as $H_2PO_4^-$ or HPO_4^{2-}), sulfur (as SO_4^{2-}), and some trace elements. The crucial role played by algae in water is their production of organic matter by algal photosynthesis as described by the reaction

$$CO_2 + H_2O + h\nu \rightarrow \{CH_2O\} + O_2(g) \tag{5.5.1}$$

where $\{CH_2O\}$ represents a unit of carbohydrate and $h\nu$ stands for the energy of a quantum of light.

Fungi in Water

The most important function of fungi in the environment is the breakdown of insoluble cellulose in wood and other plant materials by the action of extracellular cellulase enzyme. Excreted from the cell, this enzyme hydrolyzes insoluble cellulose to soluble carbohydrates that can be absorbed by the fungal cell. The carbohydrates are oxidized biochemically by **aerobic respiration**, represented by the reaction,

$$\{CH_2O\} + O_2(g) \rightarrow CO_2 + H_2O \tag{5.5.2}$$

which is, in a sense, the opposite of photosynthesis. Although the role of fungi growing directly in water is limited, they largely determine the composition of natural waters and wastewaters because of the large amount of their decomposition products that enter water. An example of such a product is humic material, which interacts with hydrogen ions and metals (see Section 5.11).

Bacteria in Water

Bacteria obtain the energy and raw materials needed for their metabolic processes and reproduction by mediating chemical reactions. Bacterial species have evolved that utilize many of these reactions that are possible in water. As a consequence of their participation in such reactions, bacteria are involved in many biogeochemical processes in water and soil. Bacteria are essential participants in the important elemental cycles in nature, including those of nitrogen, carbon, and sulfur. They are responsible for the formation of many mineral deposits, including some of iron and manganese. On a smaller scale, some of these deposits form through bacterial action in natural water systems and even in pipes used to transport water.

Factors Affecting Bacterial Growth

Because of their large surface-to-volume ratios that enable very rapid exchange of nutrients and waste products with their surroundings, and because of their short reproduction times (less than 1 h in many cases), bacterial populations may increase very rapidly. This is illustrated by the **population curve** for bacterial growth shown in Figure 5.4, illustrating numbers of bacteria resulting from an initially small number of bacteria present at time zero. Following a **lag phase** of little or no growth, is a **log phase**, or exponential phase, during which the population doubles over a regular time interval called the **generation time**. It is during the log phase that very large amounts of chemical species may be transformed in water during a short time period, such as occurs when dissolved oxygen in water is rapidly used up by aerobic respiration (Reaction 5.5.2). The log phase terminates and the **stationary phase** begins when a limiting factor, such as exhaustion of nutrients occurs. The stationary phase is followed by the **death phase**.

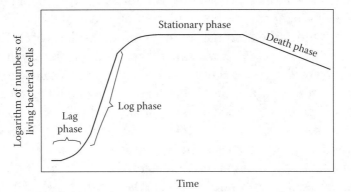

Figure 5.4.   Population curve for a bacterial culture.

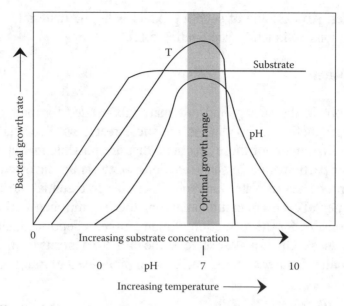

Figure 5.5.   Effects of increasing substrate concentration, temperature, and pH on bacterial growth.

Figure 5.5 illustrates the effects of three factors affecting growth of bacteria such as those that live in water. The first of these is **substrate concentration**, that is, the availability of food to the organisms. It is seen that growth increases in a linear fashion up to a saturation value, beyond which increasing substrate levels do not cause increased growth.

Temperature has a strong effect upon bacterial activity, growth, and metabolism. As shown in Figure 5.5, a curve of growth rate as a function of temperature is skewed toward the high temperature end of the curve and exhibits an abrupt drop-off beyond the temperature maximum. This occurs because enzymes are destroyed by being denatured at temperatures not far above the optimum. The temperature/growth curves of bacteria vary with the species. The temperature range for optimum growth of bacteria is remarkably wide, with some bacteria being able to grow at 0°C, and others existing at temperatures as high as 80°C. **Psychrophilic bacteria** are bacteria having temperature optima below approximately 20°C. The temperature optima of **mesophilic bacteria** lie between 20°C and 45°C. Bacteria having temperature optima above 45°C are called **thermophilic bacteria**. Thermophilic bacteria are of interest in green technology because bacteria can be used to carry out some chemical transitions, and those that thrive at higher temperatures can enable such biochemical processes to occur more rapidly and efficiently. Bacteria used for studies of such processes have been taken from colonies of "extremophiles" that grow in boiling hot springs in Yellowstone National Park.

Other factors such as pH can affect bacterial growth rate. Figure 5.5 shows a plot of pH vs. bacterial growth rate. Although the optimum pH will vary somewhat, enzymes that govern bacterial activity typically have a pH optimum around neutrality. Enzymes tend to become denatured at pH extremes. This behavior likewise is reflected in plots of bacterial metabolism as a function of pH. For some bacteria, such as those that generate sulfuric acid by the oxidation of sulfide (production of pollutant acid mine water), the pH optimum may be quite acidic. Like bacteria that thrive at high temperatures, those that can function at relatively high or low pH values may have some uses in green technology processes, such as biological waste treatment processes, where pH values may vary significantly from neutrality (ph 7).

5.6. MICROORGANISMS AND ELEMENTAL TRANSITIONS

As a consequence of their participation in oxidation/reduction reactions, microorganisms, particularly bacteria, are involved in many biogeochemical processes in water. There is not room in this section to cover these in detail. However, some of the more important oxidation/reduction reactions mediated by microorganisms in water are briefly discussed here.

As discussed in more detail in Section 5.9, a substance is oxidized when it loses electrons or reacts with molecular oxygen (O_2) and it is reduced when it gains electrons. Whenever an oxidation process occurs, a reduction process occurs as well. Oxidation/reduction is mentioned here because so many of the processes carried out by microorganisms involve oxidation/reduction. As an example, when bacteria or other organisms including humans metabolize carbohydrates (sugar, starch), represented by the simplified chemical formula $\{CH_2O\}$,

$$\{CH_2O\} + O_2(g) \rightarrow CO_2 + H_2O \qquad (5.5.2)$$

it is seen that oxygen from molecular oxygen is added to carbon in $\{CH_2O\}$ to give the more oxidized species CO_2. In this reaction, O_2 is reduced. In photosynthesis,

$$CO_2 + H_2O + h\nu \rightarrow \{CH_2O\} + O_2(g) \qquad (5.5.1)$$

carbon in CO_2 is reduced to carbon in $\{CH_2O\}$.

Bacterially mediated reactions of carbon include photosynthesis by algae and photosynthetic bacteria (Reaction 5.5.1) and aerobic respiration (Reaction 5.5.2). Anaerobic respiration of organic matter produces methane (in which the carbon from half of the $\{CH_2O\}$ is reduced) and carbon dioxide (in which half of the carbon from $\{CH_2O\}$ has been oxidized):

$$2\{CH_2O\} \rightarrow CH_4 + CO_2 \qquad (5.6.1)$$

Biodegradation of carbonaceous organic matter by bacteria is addressed later in this section.

The most important bacterially mediated oxidation/reduction reactions of nitrogen species in water are (1) nitrogen fixation, whereby molecular nitrogen is fixed as organic nitrogen; (2) nitrification, the process of oxidizing ammonia to nitrate;

(3) nitrate reduction, the process by which nitrogen in nitrate ion is reduced to form compounds having nitrogen in a lower oxidation state; and (4) denitrification, the reduction of nitrate and nitrite to N_2, with a resultant net loss of nitrogen gas to the atmosphere. The overall microbial process for **nitrogen fixation**, the binding of atmospheric nitrogen in a chemically combined form, is the following:

$$3\{CH_2O\} + 2N_2 + 3H_2O + 4H^+ \rightarrow 3CO_2 + 4NH_4^+ \qquad (5.6.2)$$

Among the few aquatic bacteria that can fix atmospheric nitrogen are *Azotobacter*, several species of *Clostridium*, and *Cyanobacteria*. Bacterial fixation of atmospheric nitrogen is an important green process in that it accomplishes, under very mild conditions, the same reaction that requires extreme conditions of temperature and pressure as well as consumption of energy and natural gas to manufacture ammonia in a chemical plant. **Nitrification**, the conversion of NH_4^+ to NO_3^-, occurs by the following reaction:

$$2O_2 + NH_4^+ \rightarrow NO_3^- + 2H^+ + H_2O \qquad (5.6.3)$$

This reaction is important in nature because it provides nitrogen in the nitrate form that most plants utilize. **Denitrification** is the process by which nitrate is reduced by bacteria to gaseous nitrogen and returned to the atmosphere:

$$4NO_3^- + 5\{CH_2O\} + 4H^+ \rightarrow 2N_2 + 5CO_2 + 7H_2O \qquad (5.6.4)$$

This reaction is important in wastewater treatment because it removes fixed nitrogen from the wastewater effluent which, released to the aquatic environment, would cause excessive algal growth and eutrophication.

Some bacteria metabolize inorganic sulfur species in water. Acting in conjunction with other bacteria, *Desulfovibrio* can reduce sulfate ion to H_2S:

$$SO_4^{2-} + 2\{CH_2O\} + 2H^+ \rightarrow H_2S + 2CO_2 + 2H_2O \qquad (5.6.5)$$

Because of the high concentration of sulfate ion in seawater, bacterially mediated formation of H_2S causes pollution problems in some coastal areas and is a major source of atmospheric sulfur. In waters where sulfide formation occurs, the sediment is often black in color because of the formation of FeS.

Some bacteria can oxidize hydrogen sulfide to higher oxidation states, particularly to sulfate ion:

$$H_2S + 2O_2 \rightarrow 2H^+ + SO_4^{2-} \qquad (5.6.6)$$

This process produces strong sulfuric acid, and one of the colorless sulfur bacteria, *Thiobacillus thiooxidans*, is tolerant of solutions containing up to 1 mol of H^+ per liter, a remarkable acid tolerance.

Of all the metals, iron is the one most commonly acted upon by bacteria in water. A variety of bacteria, including *Ferrobacillus*, *Gallionella*, and some forms of *Sphaerotilus*, utilize iron compounds in obtaining energy for their metabolic needs. These bacteria catalyze the oxidation of iron(II) to iron(III) by molecular oxygen:

$$4FeCO_3(s) + O_2 + 6H_2O \rightarrow 4Fe(OH)_3(s) + 4CO_2 \qquad (5.6.7)$$

One consequence of bacterial action on iron compounds is acid mine drainage from coal mines. Acid mine water results from the presence of sulfuric acid produced from pyrite, FeS_2, a sulfur-containing mineral associated with coal. The first of these bacterially mediated reactions is the oxidation of pyrite:

$$2FeS_2(s) + 2H_2O + 7O_2 \rightarrow 2Fe^{2+} + 4H^+ + 4SO_4^{2-} \qquad (5.6.8)$$

The next step is the bacterially catalyzed oxidation of iron(II) ion to iron(III) ion,

$$4Fe^{2+} + O_2 + 4H^+ \rightarrow 4Fe^{3+} + 2H_2O \qquad (5.6.9)$$

The Fe^{3+} ion reacts chemically with pyrite to dissolve it,

$$FeS_2(s) + 14Fe^{3+} + 8H_2O \rightarrow 15Fe^{2+} + 2SO_4^{2-} + 16H^+ \qquad (5.6.10)$$

and this reaction in conjunction with Reaction 5.6.8 constitutes a cycle for the dissolution of pyrite. Fe^{3+} is an acidic ion, and at pH values much above 3, the iron(III) precipitates as the hydrated iron(III) oxide:

$$Fe^{3+} + 3H_2O \leftrightarrow Fe(OH)_3(s) + 3H^+ \qquad (5.6.11)$$

The beds of streams afflicted with acid mine drainage often are covered with "yellowboy," an unsightly deposit of amorphous, semigelatinous $Fe(OH)_3$. The most damaging component of acid mine water, however, is sulfuric acid, H_2SO_4. It is directly toxic because of its strong acidity and has other undesirable effects.

Microbial Degradation of Organic Matter

The biodegradation of organic matter in the aquatic and terrestrial environments is a crucial environmental process. The biodegradation of organic matter by microorganisms occurs by way of a number of stepwise, microbially catalyzed reactions. Often, several microbial species are involved in sequence to completely degrade an organic chemical.

The degradation of hydrocarbons by microbial oxidation is the primary means by which organic compounds, such as herbicides and petroleum wastes, are eliminated from water and soil. Bacteria capable of degrading hydrocarbons include *Micrococcus*, *Pseudmonas*, *Mycobacterium*, and *Nocardia*. The complete biodegradation of hydrocarbons converts them to carbon dioxide and water as shown below for octane, an ingredient of gasoline:

$$2C_8H_{18} + 25O_2 \xrightarrow{\text{Bacterial oxidation}} 16CO_2 + 18H_2O \qquad (5.6.12)$$

Complete oxidation of a compound to the most stable simple inorganic molecules, in this case CO_2 and H_2O, is called **mineralization**. Biodegradation of organic compounds may also occur partially to give intermediate organic compounds.

In addition to their being oxidized, organic compounds can undergo other chemical processes carried out by microorganisms. One of the most common of these is **hydrolysis** in which a molecule is split apart with the addition of a water molecule. In the absence of oxygen, organic compounds can be **reduced** by the action of anoxic bacteria that function without oxygen.

5.7. ACID-BASE PHENOMENA IN AQUATIC CHEMISTRY

Alkalinity

The capacity of water to accept H^+ ions (protons) is called **alkalinity**. Alkalinity is important in water treatment and in the chemistry and biology of natural waters. Frequently, the alkalinity of water must be known to calculate the quantities of chemicals to be added in treating the water. Highly alkaline water often has a high pH and generally contains elevated levels of dissolved solids. These characteristics may be detrimental for water to be used in boilers, food processing, and municipal water systems. Alkalinity serves as a pH buffer and reservoir for inorganic carbon, thus helping to determine the ability of water to support algal growth and other aquatic life. It is used by limnologists, who study freshwater systems, as a measure of water fertility because the high levels of HCO_3^- ion normally associated with high alkalinity serve as a source of carbon for the photosynthetic production of biomass. Generally, the basic species responsible for alkalinity in water are bicarbonate ion (usually the major contributor), carbonate ion, and hydroxide ion:

$$HCO_3^- + H^+ \rightarrow CO_2 + H_2O \qquad (5.7.1)$$

$$CO_3^{2-} + H^+ \rightarrow HCO_3^- \qquad (5.7.2)$$

$$OH^- + H^+ \rightarrow H_2O \qquad (5.7.3)$$

Acidity

Acidity as a term applied to natural water systems is the capacity of the water to neutralize OH^-. Most natural waters are alkaline, and acidic water usually indicates severe pollution. Acidity generally results from the presence of weak acids such as $H_2PO_4^-$, CO_2 (see Reaction 5.7.4), H_2S, proteins, fatty acids, and acidic metal ions, particularly Fe^{3+} (see Reaction 5.6.11). Free mineral acid is applied to strong acids such as H_2SO_4 and HCl in water. Pollutant acid mine water contains an appreciable concentration of free mineral acid.

Carbon Dioxide in Water

Carbon dioxide, CO_2, is a weak acid in water. Because of the presence of carbon dioxide in air and its production from microbial decay of organic matter, dissolved CO_2 is present in virtually all natural waters and wastewaters. Carbon dioxide is a weak acid so that rainfall from even an absolutely unpolluted atmosphere is slightly acidic due to the presence of dissolved CO_2.

The concentration of gaseous CO_2 in the atmosphere varies with location and season; it is increasing by about 1 part per million (ppm) by volume per year. For purposes of calculation here, the concentration of atmospheric CO_2 will be taken as 380 ppm (0.0380%) in dry air. At 25°C water in equilibrium with unpolluted air containing 380 ppm carbon dioxide has a $CO_2(aq)$ concentration of 1.244×10^{-5} mol/l (M, see the discussion of molar concentration in Section 3.7).

Although CO_2 in water is often represented as H_2CO_3, most carbon dioxide dissolved in water is present simply as molecular CO_2 (aq). The CO_2–HCO_3^-–CO_3^{2-} system in water may be described by the reactions and equations that follow, which describe the interchange among these three species. Dissolved CO_2 in water acts as an acid and ionizes with water as follows:

$$CO_2 + H_2O \leftrightarrow H^+ + HCO_3^- \tag{5.7.4}$$

As denoted by the double arrows, this reaction is reversible and its equilibrium is described by the following equilibrium constant expression where the formulas in brackets represent molar concentrations of the species shown:

$$\frac{\left[H^+\right]\left[HCO_3^-\right]}{\left[CO_2\right]} = K_{a1} = 4.45 \times 10^{-7} \quad pK_{a1} = 6.35 \text{ (negative log } K_{a1}) \tag{5.7.5}$$

A second ionization may occur of HCO_3^-,

$$HCO_3^- \leftrightarrow H^+ + CO_3^{2-} \tag{5.7.6}$$

as described by the following equilibrium constant expression:

$$\frac{\left[H^+\right]\left[CO_3^{2-}\right]}{\left[HCO_3^-\right]} = K_{a2} = 4.69 \times 10^{-11} \quad pK_{a2} = 10.33 \tag{5.7.7}$$

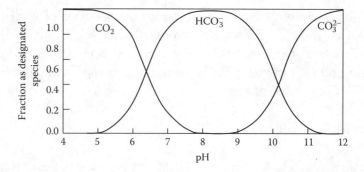

Figure 5.6. Distribution of species diagram for the CO_2–HCO_3^-–CO_3^{2-} system in water.

The predominance of the species in the CO_2–HCO_3^-–CO_3^{2-} system depends upon the pH of the water. At low pH, dissolved CO_2 predominates as is the case with carbonated beverages or some mineral water "with gas." At relatively basic pHs CO_3^{2-} may predominate. Throughout the normal pH range of natural waters and drinking water, HCO_3^- is the predominant species and is the species responsible for most of the water alkalinity. The pH ranges in which the various species predominate are shown by a **distribution of species diagram** with pH as a master variable as illustrated in Figure 5.6. The plots in this diagram may be calculated from the K_a expressions given in the preceding equations.

5.8. PHASE INTERACTIONS AND SOLUBILITY

Homogeneous chemical reactions occurring entirely in aqueous solution are rather rare in natural waters and wastewaters. Instead, most significant chemical and biochemical phenomena in water involve interactions between species in water and another phase, as shown by the examples illustrated in Figure 5.7. A typical such process is production of solid biomass through the photosynthetic activity of algae occurring within a suspended algal cell and involving exchange of dissolved

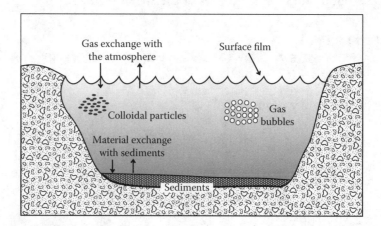

Figure 5.7. Most important environmental chemical processes in water involve interactions between water and another phase.

solids and gases between the surrounding water and the cell. Similar exchanges occur when bacteria degrade organic matter (often in the form of small particles) in water. Chemical reactions occur that produce solids or gases in water. Iron and many important trace-level elements are transported through aquatic systems as chemical compounds in the form of very small colloidal particles or are adsorbed onto solid particles. Pollutant hydrocarbons and some pesticides may be present on the water surface as an immiscible liquid film. Sediment can be washed physically into a body of water.

In addition to water, the phases in a body of water may be divided between sediments (bulk solids) and suspended colloidal material. An important aspect of phase interactions involves solubilities of gases and solids in water. Colloidal material, consisting of very fine particles of solids, gases, or immiscible liquids suspended in water, is involved with many significant aquatic chemical phenomena. It is very reactive because of its high surface area-to-volume ratio.

Gases

The solubilities of gases in water are calculated with **Henry's Law**, which states that the solubility of a gas in a liquid is proportional to the partial pressure of that gas in contact with the liquid. Dissolved gases — O_2 for fish and CO_2 for photosynthetic algae — are crucial to the welfare of living species in water. Some gases in water can also cause problems; for example, fish may die from bubbles of nitrogen formed in their blood after they have been exposed to water supersaturated with N_2. Volcanic carbon dioxide evolved from volcanic Lake Nyos in the African country of Cameroon asphyxiated 1700 people in 1986. (Since that incident, pipes have been inserted into the lake bed to gradually release carbon dioxide.)

Without enough dissolved oxygen, many kinds of aquatic organisms cannot exist in water. Dissolved oxygen is consumed by the degradation of organic matter in water. Many fish kills are caused not from the direct toxicity of pollutants but by a deficiency of oxygen because of its consumption in the biodegradation of pollutants.

Most elemental O_2 comes from the atmosphere, which is 20.95% oxygen by volume of dry air. The solubility of O_2 decreases with increasing water temperature. The concentration of O_2 in water at 25°C in equilibrium with air at atmospheric pressure is only 8.32 mg/l. Thus, water in equilibrium with air cannot contain a high level of dissolved oxygen compared to many other solute species. Oxygen-consuming processes in the water may cause the dissolved oxygen level to rapidly approach zero unless some efficient mechanism for the reaeration of water is operative, such as turbulent flow in a shallow stream or air pumped into the aeration tank of an activated sludge secondary-sewage treatment facility. At higher temperatures, the decreased solubility of O_2, combined with the increased respiration rate of aquatic organisms, frequently causes a condition in which a higher demand for oxygen accompanied by its lower solubility in water results in severe oxygen depletion.

Carbon Dioxide and Carbonate Species in Water

An important aquatic system involving interchanges among gaseous, aquatic, and solid mineral phases is the one that relates gaseous carbon dioxide, carbon dioxide and related species dissolved in water, and solid mineral carbonates, particularly limestone ($CaCO_3$) and dolomite ($CaCO_3 \cdot MgCO_3$). Carbon dioxide (CO_2), bicarbonate ion (HCO_3^-), and carbonate ion (CO_3^{2-}) have an extremely important influence upon the chemistry of water. Many minerals are deposited as salts of the carbonate ion, CO_3^{2-}. Algae in water utilize dissolved CO_2 and HCO_3^- in the synthesis of biomass. The equilibrium of dissolved CO_2 with the atmosphere,

$$CO_2(atmosphere) + H_2O \leftrightarrow CO_2(aq) \leftrightarrow H^+ + HCO_3^- \tag{5.8.1}$$

and equilibrium of CO_3^{2-} ion between aquatic solution and solid carbonate minerals,

$$MCO_3(\text{slightly soluble carbonate salt}) \leftrightarrow M^{2+} + CO_3^{2-} \tag{5.8.2}$$

have a strong buffering (stabilizing) effect upon the pH of water.

As a consequence of the low level of atmospheric CO_2, water totally lacking in alkalinity in equilibrium with the atmosphere contains only a very low level of carbon dioxide. A large share of the carbon dioxide found in water is a product of the breakdown of organic matter by bacteria (see Reaction 5.5.2). The formation of HCO_3^- and CO_3^{2-} greatly increases the solubility of carbon dioxide. High concentrations of free carbon dioxide in water may adversely affect respiration and gas exchange of aquatic animals and may even be fatal.

As water seeps through layers of decaying organic matter while infiltrating the ground, it may dissolve a great deal of CO_2 produced by the respiration of organisms in the soil. Later, as water goes through limestone formations, it dissolves calcium carbonate because of the presence of the dissolved CO_2, the process by which limestone caves are formed:

$$CaCO_3(s) + CO_2(aq) + H_2O \leftrightarrow Ca^{2+} + 2HCO_3^- \tag{5.8.3}$$

As shown in Figure 5.8, the equilibrium of the above reaction is important in determining water hardness (dissolved calcium) and alkalinity.

Sediments

Sediments, which typically consist of mixtures of clay, silt, sand, organic matter, and various minerals, may vary in composition from pure mineral matter to predominantly organic matter. Physical, chemical, and biological processes may all result in the deposition of sediments in the bottom regions of bodies of water. Sedimentary material may be simply carried into a body of water by erosion or through

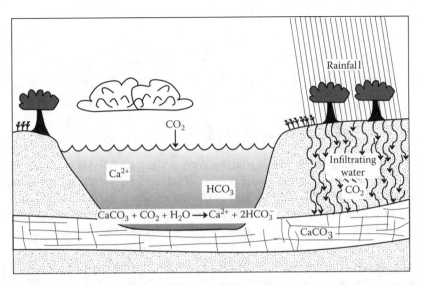

Figure 5.8.   Water containing dissolved carbon dioxide from the atmosphere or from bacterial degradation of organic matter underground reacts with limestone, $CaCO_3$, to produce dissolved Ca^{2+} (water hardness) and dissolved HCO_3^{2-} (water alkalinity).

sloughing (caving in) of the shore. Thus, clay, sand, organic matter, and other materials may be washed into a lake and settle out as layers of sediment.

Sediments may be formed by simple precipitation reactions. A common example is the formation of calcium carbonate sediment when water rich in carbon dioxide and containing a high level of calcium along with HCO_3^- anions loses carbon dioxide to the atmosphere,

$$Ca^{2+} + 2HCO_3^- \leftrightarrow CaCO_3(s) + CO_2(g) + H_2O \qquad (5.8.4)$$

or when the pH is raised by a photosynthetic reaction:

$$Ca^{2+} + 2HCO_3^- + h\nu \leftrightarrow \{CH_2O\} + CaCO_3(s) + O_2(g) \qquad (5.8.5)$$

The latter reaction is an example of the influence of aquatic life on sediment formation. Another example is the bacterially mediated oxidation of iron(II) to iron(III) to produce a precipitate of insoluble iron(III) hydroxide:

$$4Fe^{2+} + 10H_2O + O_2 \rightarrow 4Fe(OH)_3(s) + 8H^+ \qquad (5.8.6)$$

Colloids in Water

Many minerals, some organic pollutants, proteinaceous materials, some algae, and some bacteria are suspended in water as very small **colloidal particles** (Figure 5.9). Such particles have some characteristics of both species in solution and larger

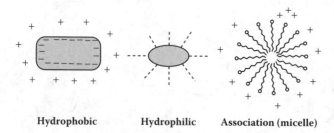

Hydrophobic   Hydrophilic   Association (micelle)

Figure 5.9.   Representations of hydrophobic, hydrophilic, and association colloidal particles. The negatively charged hydrophobic and association colloidal particles are shown surrounded by positively charged counter ions. The dashed lines in the hydrophilic colloidal particles show bonds to water. In the micelle of the association colloid, the jagged lines represent organophilic (usually hydrocarbon) "tails" and the circles represent anionic "heads" attracted to water.

particles in suspension, range in diameter from about 0.001 μm to about 1 μm, and scatter white light as a light blue hue observed at right angles to the incident light. The characteristic light-scattering phenomenon of colloids results from their being the same order of size as the wavelength of light and is called the **Tyndall effect**. The unique properties and behavior of colloidal particles are strongly influenced by their physical-chemical characteristics, including high surface area relative to their volume, high interfacial energy, and high surface/charge density ratio.

There are three classes of colloidal particles as illustrated in Figure 5.9. **Hydrophilic colloids** generally consist of macromolecules (such as proteins and synthetic polymers) that are characterized by strong interaction with water. **Hydrophobic colloids** interact to a lesser extent with water and are stable because of their positive or negative electrical charges. Examples of hydrophobic colloids are clay particles, petroleum droplets, and very small gold particles. **Association colloids** consist of special aggregates of ions and molecules called **micelles**. To understand how micelle formation occurs, consider sodium stearate, a typical soap with the following formula:

$$CH_3CH_2CH_2CH_2CH_2CH_2CH_2CH_2CH_2CH_2CH_2CH_2CH_2CH_2CH_2CH_2CH_2CO_2^-Na^+$$

This molecule (ion) has a long organic "tail" consisting of the chain of CH_2 groups and a charged ionic "head," $-CO_2^-Na^+$. As a result of this structure, stearate anions in water tend to form clusters consisting of as many as 100 anions grouped together with their hydrocarbon "tails" on the inside of a spherical colloidal particle and their ionic "heads" on the surface in contact with water and with Na^+ counterions. This results in the formation of colloidal particles called **micelles** (Figure 5.9).

The stability of colloids is a prime consideration in determining their behavior. It is involved in important aquatic chemical phenomena, including the formation of sediments, dispersion and agglomeration of bacterial cells, and dispersion and removal of pollutants (such as crude oil from an oil spill).

5.9. OXIDATION-REDUCTION

Oxidation-reduction (**redox**) reactions are those involving changes of oxidation states of reactants. The term oxidation comes from chemical reactions with molecular oxygen in which oxygen atoms acquire electrons from another species. For example, when calcium metal reacts with O_2 to produce calcium oxide, CaO,

$$2Ca + O_2 \rightarrow 2\{Ca^{2+}O^{2-}\} \tag{5.9.1}$$

each calcium atom loses 2 electrons to an oxygen atom. As denoted by the charged symbols shown in the formula of the CaO product in the above reaction, each calcium atom is *oxidized* by losing 2 electrons to become a Ca^{2+} *cation* and each oxygen atom is *reduced* to become an O^{2-} *anion* (charged chemical species are called *ions*).

Oxidation-reduction phenomena are highly significant in the environmental chemistry and microbiology of natural waters and wastewaters. The reduction of oxygen by organic matter in a lake,

$$\{CH_2O\} \text{ (becomes oxidized)} + O_2 \text{ (becomes reduced)} \rightarrow CO_2 + H_2O \quad (5.9.4)$$

results in oxygen depletion, which can be fatal to fish. The rate at which sewage is oxidized is crucial to the operation of a waste-treatment plant. Reduction of insoluble iron(III) to soluble iron(II),

$$Fe(OH)_3(s) + 3H^+ + e^- \rightarrow Fe^{2+} + 3H_2O \tag{5.9.5}$$

in a reservoir contaminates the water with iron, which is hard to remove in the water-treatment plant. Oxidation of NH_4^+ to NO_3^- in water,

$$NH_4^+ + 2O_2 \rightarrow NO_3^- + 2H^+ + H_2O \tag{5.9.6}$$

is essential for getting the ammonium nitrogen into a form assimilable by algae in the water. Many other examples can be cited of the ways in which the types, rates, and equilibria of redox reactions largely determine the nature of important solute species in water.

The relative tendency for a water solution to be oxidizing or reducing is because of the activity of electrons. A high electron activity denotes reducing, and a low electron activity indicates oxidizing conditions. Because electron activity in water varies over many orders of magnitude, it is convenient to discuss oxidizing and reducing tendencies in terms of pE, a parameter analogous to pH and defined conceptually as the negative log of the electron activity. Water containing a relatively high concentration of dissolved O_2, such as water in contact with the atmosphere, has a low

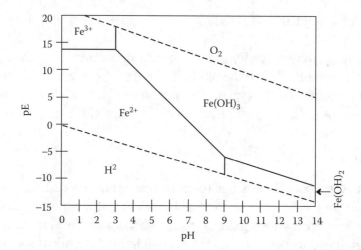

Figure 5.10.   A simplified pE–pH diagram for iron in water (maximum total soluble iron concentration 1.0×10^{-5} mol/l).

electron activity because of the tendency of oxidant O_2 to take electrons and it has a high pE. Water with a low pE and high electron activity contains essentially no dissolved O_2, is reducing, and is characterized by the presence of chemical species with a tendency to donate electrons (reducing agents).

The nature of chemical species in water is usually a function of both pE and pH. A good example of this may be illustrated by a simplified pE–pH diagram for iron in water, assuming that iron is in one of the four forms of Fe^{2+} ion, Fe^{3+} ion, solid $Fe(OH)_3$, or solid $Fe(OH)_2$ as shown in Figure 5.10. Water in which the pE is higher than that shown by the upper dashed line is thermodynamically unstable toward oxidation (with a tendency to decompose to yield O_2) and, below the lower dashed line, water is thermodynamically unstable toward reduction (with a tendency to decompose to yield H_2). It is seen that Fe^{3+} ion is stable only in a very oxidizing, acidic medium such as that encountered in acid mine water, whereas Fe^{2+} ion is stable over a relatively large region as reflected by the common occurrence of soluble iron(II) in oxygen-deficient groundwaters. Highly insoluble $Fe(OH)_3$ is the predominant iron species over a very wide pE–pH range.

The pE–pH diagram for the iron system can be used to illustrate an interesting phenomenon that can cause problems in water supply. Consider groundwater at a neutral pH of around 7. Being underground and not in contact with molecular oxygen, it may have a low pE due to its reducing environment. In contact with iron minerals, the water may develop a relatively high concentration of soluble Fe^{2+} ion, the reduced form of iron. When the water is taken to the surface and exposed to molecular O_2, the pE is high and insoluble $Fe(OH)_3$ is the stable iron species so that the following reaction occurs:

$$4Fe^{2+} + O_2 + 10H_2O \rightarrow 4Fe(OH)_3(s) + 8H^+ \tag{5.9.7}$$

The $Fe(OH)_3$ product of this reaction is a slimy gelatinous orange precipitate that eventually turns to a form of rust. In water supplies it can form ugly, intractable deposits on clothing and bathroom fixtures.

5.10. METAL IONS AND CALCIUM IN WATER

The formula of a metal ion in aqueous solution usually is written M^{n+}, which signifies the simple hydrated metal cation $M(H_2O)_x^{n+}$. A bare metal ion, such as Mg^{2+}, cannot exist as a separate entity in water. In order to secure the highest stability of their outer electron shells, metal ions in water are bonded, or coordinated, to water molecules or other stronger bases (electron-donor partners) that might be present.

Metal ions in aqueous solution seek to reach a state of maximum stability through chemical reactions, including acid-base,

$$Fe(H_2O)_6^{3+} \longleftrightarrow FeOH(H_2O)_5^{2+} + H^+ \tag{5.10.1}$$

precipitation,

$$Fe(H_2O)_6^{3+} \longleftrightarrow Fe(OH)_3(s) + 3H_2O + 3H^+ \tag{5.10.2}$$

and oxidation-reduction reactions:

$$Fe(H_2O)_6^{2+} \longleftrightarrow Fe(OH)_3(s) + 3H_2O + e^- + 3H^+ \tag{5.10.3}$$

Calcium and Hardness in Water

Of the cations found in most freshwater systems, calcium generally has the highest concentration. The chemistry of calcium, although complicated enough, is simpler than that of the transition metal ions, such as iron or chromium, found in water. Calcium is a key element in many geochemical processes, and minerals constitute the primary sources of calcium ion in water. Among the primary contributing minerals are gypsum, $CaSO_4 \cdot 2H_2O$; anhydrite, $CaSO_4$; dolomite, $CaCO_3 \cdot MgCO_3$; and calcite and aragonite, which are different mineral forms of $CaCO_3$.

As mentioned in Section 5.8 and illustrated in Figure 5.8, water containing a high level of carbon dioxide readily dissolves calcium from its carbonate minerals:

$$CaCO_3(s) + CO_2(aq) + H_2O \longleftrightarrow Ca^{2+} + 2HCO_3^- \tag{5.8.3}$$

When the above equation is reversed and CO_2 is lost from the water, calcium carbonate deposits are formed. The concentration of CO_2 in water determines the extent of dissolution of calcium carbonate.

Calcium ion, along with magnesium and sometimes iron(II) ion, accounts for **water hardness**. The most common manifestation of water hardness is the curdy precipitate formed by soap in hard water. Temporary hardness is because of the presence of calcium and bicarbonate ions in water and may be eliminated by boiling the water, thus causing the reversal of Reaction 5.8.3:

$$Ca^{2+} + 2HCO_3^- \leftrightarrow CaCO_3(s) + CO_2(g) + H_2O \qquad (5.10.4)$$

Increased temperature in water having temporary hardness may force the above reaction to the right by evolving CO_2 gas. As a result, a white precipitate of calcium carbonate may form in boiling water having temporary hardness. The equilibrium between dissolved carbon dioxide and calcium carbonate minerals is important in determining several natural water chemistry parameters such as alkalinity, pH, and dissolved calcium concentration (Figure 5.8).

5.11. COMPLEXATION AND SPECIATION OF METALS

The properties of metals dissolved in water depend largely upon the nature of metal species dissolved in the water. Therefore, **speciation** of metals plays a crucial role in their environmental chemistry in natural waters and wastewaters. In addition to the hydrated metal ions, for example, $Fe(H_2O)_6^{3+}$ and hydroxo species such as $FeOH(H_2O)_5^{2+}$ (see Reaction 5.10.1), metals may exist in water reversibly bound to inorganic anions or to organic compounds as **metal complexes**, or they may be present as **organometallic** compounds containing carbon-to-metal bonds. The solubilities, transport properties, and biological effects of such species are often vastly different from those of the metal ions themselves.

A metal ion in water may combine with an electron donor (Lewis base) to form a **complex** or **coordination compound** (or ion). Thus, cadmium ion in water combines with a ligand, cyanide ion, to form a complex ion as shown below:

$$Cd^{2+} + CN^- \leftrightarrow CdCN^+ \qquad (5.11.1)$$

Additional cyanide ligands may be added to form the progressively weaker (more easily dissociated) complexes: $Cd(CN)_2$, $Cd(CN)_3^-$, and $Cd(CN)_4^{2-}$.

In this example, the cyanide ion is a **unidentate ligand**, meaning that it has only one site that bonds to the cadmium metal ion. Of considerably more importance than complexes of unidentate ligands in natural waters are complexes with **chelating agents**. A chelating agent has more than one atom that may be bonded to a central metal ion at one time to form a ring structure. Chelates of metals with chelating groups from humic substances are shown in Figure 5.11. In general, because a chelating agent may bond to a metal ion in more than one place simultaneously, chelates are more stable than complexes involving unidentate ligands. Metal chelate stability tends to increase with the number of chelating sites available on the ligand.

Humic substances are the most important class of naturally occurring complexing agents. These are degradation-resistant materials formed during the decomposition of vegetation that occur as deposits in soil, marsh sediments, peat, coal, lignite, or other locations where large quantities of vegetation have decayed. The types of humic substances are classified on the basis of their solubilities: **Humin** is insoluble, **humic acid** is soluble only in base, and **fulvic acid** is soluble in both acid and base. Because of their acid-base, sorptive, and complexing properties, both the soluble and insoluble humic substances have a strong effect upon the properties of water. In general, fulvic acid dissolves in water and exerts its effects as the soluble species. Humin and humic acid remain insoluble and affect water quality through exchange of species, such as cations or organic materials, with water.

Some feeling for the nature of humic substances may be obtained by considering the structure of a hypothetical molecule of fulvic acid below:

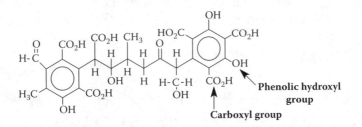

The binding of metal ions by humic substances is one of the most important environmental qualities of humic substances. Iron and aluminum are very strongly bound to humic substances, whereas magnesium is rather weakly bound. This binding can occur as chelation between a carboxyl group and a phenolic hydroxyl group, as chelation between two carboxyl groups, or as complexation with a carboxyl group (see Figure 5.11).

Fulvic-acid complexes of metals in natural waters have a number of effects. Among these is that they keep some of the biologically important transition-metal ions in solution and are particularly involved in iron solubilization and transport. Fulvic acid-type compounds are associated with color in water. These yellow materials, called **Gelbstoffe**, frequently are encountered along with soluble iron.

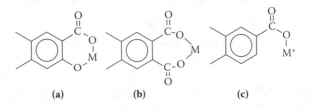

Figure 5.11.   Binding of a metal ion, M^{2+}, by humic substances: (a) by chelation between carboxyl and phenolic hydroxyl, (b) by chelation between two carboxyl groups, and (c) by complexation with a carboxyl group.

Organometallic Compounds

Another type of environmentally important metal species consists of **organo-metallic compounds**, differing from complexes and chelates in that the organic portion is bonded to the metal by a carbon-metal bond and the organic ligand is frequently not capable of existing as a stable separate species. Example organometallic compound species are monomethylmercury ion and dimethylmercury:

$$Hg^{2+} \qquad HgCH_3^+ \qquad Hg(CH_3)_2$$

Organometallic compounds may enter the environment directly as pollutant industrial chemicals and some, including organometallic mercury, tin, selenium, and arsenic compounds, are synthesized biologically by bacteria. Some of these compounds are particularly toxic because of their mobilities in living systems and abilities to cross cell membranes.

SUPPLEMENTARY REFERENCES

Benjamin, Mark M., *Water Chemistry*, McGraw-Hill, Boston, MA, 2002.

Berk, Zeki, Ed., *Water Science for Food, Health, Agriculture and Environment*, Technomic Publishing Co., Lancaster, PA, 2001.

Dasch, E. Julius, Ed., *Water: Science and Issues*, Macmillan Reference, New York, 2003.

Ford, Timothy Edgcumbe, Ed., *Aquatic Microbiology: An Ecological Approach*, Blackwell Scientific Publications, Boston, MA, 1993.

Giller, Paul S. and Bjorn Malmqvist, *The Biology of Streams and Rivers*, Oxford University Press, New York, 1999.

Poleyoy, S.L., *Water Science and Engineering*, Krieger Publishing Co., Malabar, FL, 2003.

Sigee, David, *Freshwater Microbiology: Biodiversity and Dynamic Interactions of Microorganisms in the Aquatic Environment*, John Wiley & Sons, New York, 2005.

Smirnov, N.N. and A.N. Smirnov, *Elsevier's Dictionary of Aquatic Biology*, Elsevier Science, New York, 1999.

Stober, Ingrid and Kurt Bucher, Eds., *Water-Rock Interaction*, Bucher, Kluwer Academic Publishers, Boston, MA, 2002.

Sullivan, Patrick, Franklin J. Agardy, and James J.J. Clark, *The Environmental Science of Drinking Water*, Butterworth-Heinemann, Burlington, MA, 2005.

QUESTIONS AND PROBLEMS

1. What are plankton? In which sense are phytoplankton at the bottom of the food chain?

2. Define productivity of water. Name and describe the undesirable pollution condition resulting from excess productivity.

3. Distinguish between DO and BOD in water.

4. Explain why the plot of bacterial growth as a function of temperature is skewed?

5. Because of changes in water density, the water in stratified lakes or reservoirs turns over, a phenomenon known as overturn. Suggest how overturn may affect the water and aquatic life.

6. With regard to animal life in streams, distinguish among shredders, collectors, grazers, and gougers.

7. What is the most important function of fungi insofar as water is concerned?

8. What are humic substances, and why are they significant in water?

9. How do organometallic compounds differ from metal complexes?

10. In what sense is the interface of seawater and freshwater of particular importance for aquatic life?

11. What are oxidation/reduction reactions? Give some examples of such reactions in water.

12. The numbers of a specific kind of bacteria in water as a function of time were found to be the following:

Elapsed Time (h)	Millions of Bacteria (per milliliter)	Elapsed Time (h)	Millions of Bacteria (per milliliter)
0.0	0.103	2.0	1.59
0.5	0.199	2.5	2.13
1.0	0.400	3.0	2.49
1.5	0.815	3.5	2.52

What do these numbers indicate about the phase or phases of the population curve of the bacteria?

13. What is a metal complex? Give an example of one.

14. Explain using chemical reactions how bacteria are involved with the nitrogen cycle.

15. What are epoxidation and β-oxidation, and how are they involved with biodegradation?

16. Deposits of iron oxides, some used as iron ore, are taken as evidence of the action of photosynthetic cyanobacteria in producing the atmosphere's elemental O_2 in eons past. Using chemical equations to illustrate, suggest how this may have taken place.

17. What is water alkalinity? What substance dissolved in solution is most responsible for water alkalinity? What are the benefits of alkalinity?

18. What is the significance of other phases, such as mineral matter or suspended bacteria, in water?

19. How is Henry's Law used?

20. Water hardness is because of calcium ions. Show a chemical reaction by which hardness is added to water that also explains the formation of caves in limestone formations.

21. What are colloidal particles? What are their major effects? What are the types of colloidal particles?

22. Association colloidal particles, such as those formed by soap ions, will hold miniscule droplets of oil within the particles (micelles). Offer an explanation for this phenomenon.

23. Consider water containing iron at a low pE of −6 and a pH of 7. What would be the stable iron species? What iron species would form if the water were exposed to atmospheric oxygen?

6. KEEPING WATER GREEN

6.1. WATER POLLUTION: GREEN WATER MAY NOT BE SO GREEN

During periods of low flow, the San Marcos/Cazones River that goes by the city of Poza Rico, Mexico, appears an intense, beautiful green color. But this appearance masks a major problem because what makes the river so green is the rich growth of algae and plants in it. This plant life thrives so well in the tropical climate because of nutrient-rich wastes that enter the river from untreated sewage. It illustrates a major water pollution problem called eutrophication discussed later in this chapter.

This chapter discusses keeping water green, not in color but in terms of water quality. Four major topics are addressed. The first of these is water pollution, which degrades water quality and can make it unfit for use or to support life. The second major area addressed is how water quality can be maintained and enhanced, largely through various water treatment processes. A third major area is water pollution prevention and a fourth is recycling of the water resource, which is arguably nature's most recyclable material.

It should be kept in mind that the best way to treat water is to not mess it up in the first place, a concept that certainly is in keeping with the best practice of green science and sustainability. It is a bit ironic that water in a municipal system is used as a medium to carry away human wastes, converting the clean drinking water that enters the system to a rather unpleasant mix of feces, urine, macerated food wastes, microorganisms, and all the other things that get flushed down the drain. This wastewater normally is treated in a wastewater treatment plant and discharged to a waterway from which, in many cases, it is pumped to be further purified for use in another municipal water system. In less developed countries, the same thing happens minus the wastewater treatment step.

6.2. NATURE AND TYPES OF WATER POLLUTANTS

Water pollutants can be divided among some general classifications, as summarized in Table 6.1. Most of these categories of pollutants, and several subcategories, are discussed in this chapter. An enormous amount of material is published on this

Table 6.1.   General Classes of Water Pollutants

Class of Pollutant	Significance
Toxic elements, heavy metals	Health, aquatic life
Organically bound metals	Metal transport
Inorganic pollutants	Toxicity, aquatic biota
Algal nutrients	Eutrophication
Radionuclides	Toxicity
Acidity, alkalinity, salinity (in excess)	Water quality, aquatic life
Sewage	Water quality, oxygen levels
Biochemical oxygen demand	Water quality, oxygen levels
Trace organic pollutants	Toxicity
Pesticides	Toxicity, aquatic biota, wildlife
Polychlorinated biphenyls	Possible biological effects
Chemical carcinogens	Incidence of cancer
Petroleum wastes	Effect on wildlife, esthetics
Pathogens (bacteria, virus, protozoa)	Health effects
Detergents	Eutrophication, wildlife, esthetics
Sediments	Water quality, aquatic biota, wildlife
Taste, odor, and color	Esthetics

subject each year, and it is impossible to cover it all in one chapter. To be up to date on this subject, the reader may want to survey journals and books dealing with water pollution, such as those listed in the Supplementary References section at the end of this chapter.

6.3. HEAVY METALS

The **heavy metals** are those metals of relatively higher atomic numbers. Some heavy metals are considered among the most troublesome and toxic water pollutants. The heavy metals of most concern in water are addressed briefly here.

Cadmium (Cd) is widely used in metal plating and in making small batteries, such as those used in some cameras. Cadmium is very toxic, damaging red blood cells, kidney tissue, and testicular tissue. Cadmium can enter water from industrial pollution sources.

Lead (Pb — from its Latin name of plumbum) is arguably the most common heavy metal pollutant because of its widespread use in industry, in the manufacture of lead storage batteries, and formerly as a leaded additive to gasoline, as a pigment in white house paint, and as an anticorrosive primer applied prior to painting steel.

Exposure to lead causes a number of adverse health effects and is suspected of causing mental retardation in exposed children. Lead was widely used in plumbing, and its use in solder to join together copper water pipe is now banned.

The most tragic modern incident of poisoning from **mercury** (Hg) occurred in the Minamata Bay area of Japan from 1953 through 1960 when people consumed seafood from the bay, which had been polluted by drainage of mercury wastes from a chemical plant. The total number of cases of mercury poisoning reported was 111, and there were 43 deaths and 19 cases of congenital birth defects in babies whose mothers had eaten the contaminated seafood. Hazardous methylated forms of mercury are discussed under organometallic compounds below.

Arsenic (As) is a metalloid (among the elements bordering metals and nonmetals in the periodic table), but its environmental and toxicological effects are much like those of heavy metals. Arsenic has been employed in hundreds of dastardly murder plots over the years, and can cause both chronic poisoning over a long period of time and acute poisoning from the ingestion of as little as 100 mg of the element. Before the advent of more modern pesticides, arsenic compounds were used in huge quantities to kill pests in orchards and on crops. Among the deadly compounds used for this purpose were lead arsenate, $Pb_3(AsO_4)_2$; sodium arsenite, Na_3AsO_3; and Paris Green, $Cu_3(AsO_3)_2$.

In modern times, the most tragic occurrence of water pollution by arsenic has taken place in Bangladesh, an impoverished nation on the Indian subcontinent. Around 1980, a program sponsored by the United Nations to drill wells in this country brought abundant, pathogen-free drinking water to many areas of Bangladesh. However, about 20 years later, symptoms of arsenic poisoning appeared among the people using this water source, leading to debilitating illnesses and death. These otherwise wholesome water sources were contaminated with dangerous levels of arsenic.

Organically Bound Metal Water Pollutants

In a number of instances, simple hydrated metal ions are not soluble enough in water to cause pollution problems. However, organically bound metals may be considerably more mobile and in some cases more toxic. Binding of metals as metal chelates was discussed in Section 5.11. Another form of metal binding occurs when metals are bonded directly to carbon in hydrocarbon groups such as the methyl group ($-CH_3$) to produce **organometallic compounds**.

A nasty surprise related to the formation of organometallic compounds in water was revealed in 1970 when it was found that fish in some areas, such as in Lake Saint Clair located between Michigan and Ontario, Canada, had dangerously high levels of mercury. It was known that the electrically driven chloralkali method of producing sodium hydroxide and elemental chlorine — both important industrial chemicals — used mercury electrodes in electrolyzing sodium chloride solutions, and that each unit in this process was releasing up to 14 kg of mercury per day. However, it was known that the inorganic forms of mercury released formed very insoluble precipitates in water and were thought to be safely buried with lake and river sediments. Subsequent investigation showed that anoxic bacteria (those that live in the absence of O_2) growing in the sediments were attaching methyl groups, $-CH_3$, to mercury:

$$HgCl_2 \xrightarrow[\text{bacteria}]{\text{Anoxic}} CH_3HgCl + Cl^- \tag{6.3.1}$$

The monomethylmercury ion in this compound (CH_3Hg^+) is soluble and mobile in water and the dimethylmercury — $(CH_3)_2Hg$ — also produced is volatile as well. These mobile species were released from the sediments and became concentrated in fish tissue.

More recently, the organometallic compounds of most concern in water have been the organotin compounds. Up to 40,000 t/year of organotin compounds, such as tetra-*n*-butyltin,

$$(C_4H_9) - \underset{\underset{(C_4H_9)}{|}}{\overset{\overset{(C_4H_9)}{|}}{Sn}} - (C_4H_9) \quad \textbf{Tetra-\textit{n}-butyltin}$$

used to be produced each year as industrial biocides to prevent biological growths on surfaces. Boat and ship hulls were painted with organotin-containing paints to prevent the growth of "aufwuchs," organisms that attach themselves to such surfaces and greatly increase the friction, hence the fuel costs, of propelling these vessels through water. Because of water pollution concerns, in 1998, the International Maritime Organization agreed to ban organotin antifouling paints on all ships by 2003.

Heavy Metal Pollutants and Green Technology

From the discussion above, it is seen that several heavy metals are among the more troublesome water pollutants, with a number of others in more isolated cases. Obviously, it is important to prevent such elements from getting into water. Here the practice of green technology plays an important role. One approach is to strictly forbid the release of heavy metals into water. This has worked reasonably well but, as with all command and control measures, it is subject to human oversight, accident, and even deliberate releases to try to avoid disposal costs.

A much better approach, where possible, is to use the principles of green chemistry and technology to avoid any possibility of pollutant release. For example, a command-and-control approach to preventing the release of cadmium in electroplating operations would be to strictly control any releases. But, a green technology approach is to come up with safer substitutes for cadmium in metal treatment so that there is never any cadmium around to be released. Similar approaches can be tried with any other water pollutant.

6.4. INORGANIC WATER POLLUTANTS

Cyanide is deadly as volatile hydrogen cyanide, HCN, or as cyanide ion, CN^-; as little as 60 mg of cyanide can be fatal to a human. Cyanide is produced from coke ovens and is widely used in the metals industry for metal extraction from ores and for metal cleaning and electroplating. Cyanide is sometimes released to water, especially from metal extraction operations. One such incident occurred in 1995 when cyanide-containing water mixed with red clay from mine tailings was released from a gold mining operation in the South American country of Guyana. A breached dam allowed release of approximately 2.7 billion liters of cyanide-contaminated wastes. The cyanide present in the water at a level of approximately 25 ppm killed all the fish in the small Omai Creek leading from the site of release to the Essequibo River, where the dilution from the river flow reduced levels of cyanide to below fatal concentrations. In 1992, cyanide and heavy metals spilling from the Summitville mine in southern Colorado killed all life in a 17-mi section of the Alamosa River. The state of Colorado agreed to settle for damages totaling $30 million in late 2000.

Although no human fatalities resulted from these incidents, the scope of the spill and the extreme toxicity of cyanide point to the dangers of using large quantities of a reagent as toxic as cyanide. Regulations forbidding release of cyanide were not helpful in this case — it happened. A green chemistry approach to this problem would be to find safer alternatives to cyanide so that there is no possibility of its release.

Excessive levels of **ammoniacal nitrogen** in the form of ammonium ion (NH_4^+) or molecular ammonia (NH_3) cause water-quality problems and may be harmful to aquatic life. However, ammoniacal nitrogen at lower levels is a normal constituent of water and is even added deliberately to drinking water so it can react with chlorine used for disinfection to provide for residual disinfection in water distribution systems.

Hydrogen sulfide (H_2S) is a toxic gas with a foul odor that is produced by anoxic (growing in the absence of oxygen) bacteria acting upon inorganic sulfate (see Section 5.6 and Reaction 5.6.5), from geothermal sources (hot springs) and as a pollutant from chemical plants, paper mills, textile mills, and tanneries. Because of its bad odor and toxicity, it is an undesirable pollutant in water.

Microbial degradation under ground may generate **carbon dioxide** (CO_2) that exists as free carbon dioxide in water (see Section 5.7). Excessive levels can be toxic to aquatic organisms and can make water corrosive because of its acidity and tendency to dissolve protective $CaCO_3$ coatings on pipe.

Nitrite ion (NO_2^-) can be generated by the action of bacteria on inorganic nitrogen species and is added to water as a corrosion inhibitor in some industrial processes. Normally, levels are low because of the narrow range of conditions under which nitrite is stable. Nitrite is quite toxic, causing methemoglobinemia by converting the hemoglobin in blood to methemoglobin, a form useless for transporting oxygen. **Nitrate ion** (NO_3^-) is a more common water contaminant, but is of less concern because of the high tolerance of adult humans for it. However, infants and multistomached ruminant animals (cattle, sheep, goats, deer) have conditions in their stomachs that can result in the chemical reduction of nitrate to toxic nitrite. This has killed a large number of animals and has resulted in fatal cases of methemoglobinemia in infants.

Acidity

Strong acid pollutants that cause water to have a low pH are very damaging to organisms living in water. Although spills of acids can pollute water, the most common acid pollutant comes from the bacterial mediated oxidation of iron pyrite (FeS_2) to produce sulfuric acid. The overall process is represented by the reaction,

$$4FeS_2(s) + 2H_2O + 15O_2 \rightarrow 2H_2SO_4 + 2Fe_2(SO_4)_3 \qquad (6.4.1)$$

which produces sulfuric acid (H_2SO_4) and $Fe_2(SO_4)_3$, which also acts as an acid. The bacteria that carry out these reactions include *Thiobacillus ferrooxidans*, *Thiobacillus thiooxidans*, and *Ferrobacillus ferrooxidans*.

Another source of water pollutant acid is from acid rain. Hydrogen chloride (HCl) emitted to the atmosphere forms hydrochloric acid, whereas nitric oxide (NO) and sulfur dioxide (SO_2) emitted to the atmosphere can be oxidized in the presence of atmospheric water vapor to produce nitric acid (HNO_3) and sulfuric acid, respectively. Falling from the atmosphere as acid rain, these acids are especially damaging to life in lake water that does not have contact with the kinds of minerals that can neutralize the acid and keep the pH from falling low enough to damage aquatic life. Lakes in some parts of Canada and in New England in the U.S. are especially susceptible to this kind of damage.

There are not many direct pollutant sources of excessively high **alkalinity**, which is because of salts such as sodium carbonate (Na_2CO_3) that tend to raise the pH to levels high enough to be harmful to aquatic life. Some soils and rocks associated with mining have high alkalinity, and human activities can cause this alkalinity to be leached into water.

Water **salinity** is because of dissolved salts, such as sodium chloride and calcium chloride. Each pass of water through a municipal water system adds salinity, especially from NaCl flushed into water from recharging water softeners with sodium chloride. Irrigation also adds salinity and is responsible for the high levels of salt in California's Salton Sea, an artificial body of water with no outlet to the ocean created artificially by runoff from irrigated lands. Fertilizers are salts, and they get into runoff water during irrigation.

Water Pollutants That Are Just Too Nutritious

Some inorganic species are pollutants, not because they are toxic, but because they are very nutritious for algae in water. Algae and other plants require a number of different inorganic nutrients. Those required in the greatest quantity are inorganic phosphorus ($H_2PO_4^-$, HPO_4^{2-}), nitrogen (NH_4^+, NO_3^-), and potassium (K^+).

The plant fertilizers described above can get into water from a number of sources, including fertilizers put on soil to enhance crop growth, from some industrial pollutants, and — especially in the cases of phosphates and nitrates — from the degradation of sewage in wastewater. So what is wrong with having nutrient-rich water? If the levels of nutrients are too high, algae grow too well and generate too much biomass. This material eventually dies and decays, which uses up all the oxygen in the water and clogs a body of water with dead plant matter. This unhealthy condition of excessive plant growth is called **eutrophication**, derived from the Greek word meaning "well-nourished."

Eutrophication is usually curtailed by limiting phosphate input into bodies of water and streams. The reason for doing this is that phosphorus is usually the limiting nutrient, much like the limiting reactant in a chemical reaction. So if phosphate levels are cut down, algal growth and resulting eutrophication are curtailed. Around 1970, the most common source of pollutant phosphorus was from phosphates added as builders to household detergents and discharged with sewage. Treated wastewater discharged to streams and other natural waters added phosphate that caused eutrophication. By the application of green chemistry (though not known as such then) other chemicals were found that could substitute for phosphate in detergents without causing eutrophication.

6.5. ORGANIC WATER POLLUTANTS

A whole host of water pollutants are organic compounds, which include virtually all carbon-containing compounds. These may be nontoxic, highly biodegradable materials, such as waste food in sewage, that are nevertheless bad for water because of the dissolved oxygen consumed when they degrade (see next subsection). At the other extreme are very poorly degradable substances, such as polychlorinated biphenyls (PCBs), that tend to accumulate in sediments and in the lipid (fat) tissues of fish and birds that eat fish. Organic water pollutants are addressed briefly here.

Oxygen-Demanding Substances

One of the most common water pollutants consists of substances that are not toxic, but serve as excellent food sources for bacteria in water. When such substances, represented here as $\{CH_2O\}$, undergo biodegradation,

$$\{CH_2O\} + O_2 \xrightarrow{\text{Microorganisms}} CO_2 + H_2O \qquad (6.5.1)$$

oxygen is consumed. The resulting low oxygen levels are detrimental to fish and some other forms of aquatic life and also include the likelihood that odorous reduced species, particularly H_2S, will be evolved. High levels of oxygen-demanding substances are associated with the water pollution phenomenon of eutrophication discussed above.

The oxygen-consuming potential of biodegradable materials in water is called **biochemical oxygen demand, BOD**. BOD is commonly expressed as the amount of oxygen consumed in biodegrading the organics in a liter of water.

Sewage

One of the most common sources of BOD is sewage from domestic, commercial, and industrial sources. In addition to BOD from fecal matter and food wastes, sewage contains oil, grease, grit, sand, salt, soap, detergents, degradation-resistant organic compounds, and an incredible variety of objects that get flushed down the drain. Sewage used to be simply discarded to the nearest handy stream or body of water, an unfortunate practice that still continues in many parts of the world. There are several reasons that this should not be done, the most obvious of which is that sewage stinks. But it also carries disease-causing (pathogenic) microorganisms (bacteria, virus, protozoa) and it exerts a high BOD in the water to which it is discarded. Sewage treatment to reduce BOD is addressed in Section 6.11.

Detergents in Sewage

A major problem with sewage that was solved by the application of green chemistry (before it was known as such) was the presence of detergents in the sewage. After detergents came into widespread use in the 1940s, massive layers of foam that sometimes covered most of the treatment plant developed at sewage treatment plants in which air is blown through sewage to promote biodegradation and at the outfalls where treated sewage was discharged into streams. There were even cases of treatment plant workers who walked into the foam, fell, and were asphyxiated by the gases entrained by the foam. The phenomenon even extended to surface water and groundwater contaminated with treated sewage, which developed startling heads of foam on being drained from a water faucet. Furthermore, fish fingerlings were killed by whatever was getting through the sewage treatment plant.

The culprit in the troublesome incidents of foam associated with sewage was found to be the surfactant (the active ingredient that makes water "wetter") in waste detergents, a material called **alkyl benzene sulfonate (ABS)** for which the structural formula is shown as follows:

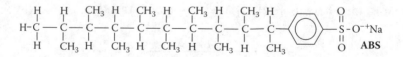

(Recall from Chapter 3 that the hexagon with a circle in it stands for an aromatic benzene ring.) It is seen in this structure that there is a chain of 10 C atoms (counting the one that is part of the end –CH$_3$ group), and attached to alternate C atoms on the chain are –CH$_3$ groups. This molecule and similar ones with different C-atom chain lengths were found to be very effective stable surfactants. There was only one problem, which is that bacteria hate branched chains and degrade this compound only slowly. Hence it stayed around forming foam in sewage treatment plants and significant amounts remained even after the treated water was discharged.

The solution to the ABS detergent problem was found in the 1960s. Although it predates the concepts of green chemistry and industrial ecology, it is still cited as one of the best applications of green chemistry. The solution to the problem created by chemistry was to use chemistry to devise another kind of molecule, equally effective as a surfactant, but much more to the liking of bacteria. The substance developed is still used today and is known as **linear alkyl sulfonate (LAS)**, shown as follows:

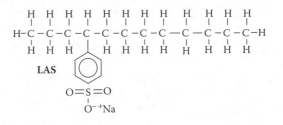

Similar to ABS, LAS has a chain of carbon atoms. But it is a straight chain, not a branched one. Bacteria like straight chains, the LAS degraded readily in the sewage treatment plant, and the problem with surfactants in sewage went away.

6.6. PESTICIDES IN WATER

Food producers and gardeners constantly struggle with pests of various kinds that consume food, ruin crops, compete for space, water, and nutrients, and otherwise make life difficult and expensive. In addition to **insecticides** used to kill insects and **herbicides** to control weeds, there are many other kinds of pesticides, including **bactericides** to control bacteria, **slimicides** to control slime-causing organisms in water, and **algicides** used against algae, all potential water pollutants. When the major water pollution control laws were enacted around 1970, a great concern was DDT and other very persistent insecticides that polluted water and sediments. In rural areas the greater problem is from herbicides which, because they must be applied directly to cropland, have an inherent tendency to get into runoff and into water sources. Fortunately, pesticides have become generally safer and less persistent in the environment. Although pesticide use has leveled off, it is still enormous, around 350 million kilograms per year for U.S. agriculture as well as large amounts for nonagricultural uses, including forestry, landscaping, gardening, food distribution, and home pest control.

Insecticides

Figure 6.1 shows some insecticides that either are potential water pollutants or have been water pollutants in the past. The insecticides that caused the greatest water pollution problems, and that do so even now in developing countries where their use is still allowed, are the **organochlorine** insecticides. There were many of these, probably best exemplified by DDT and including others, such as chlordane. DDT was widely sprayed in the environment and was instrumental in killing mosquitos that carry malaria. Highly persistent chlordane was the most effective insecticide against termites and was buried around buildings to prevent termite infestation.

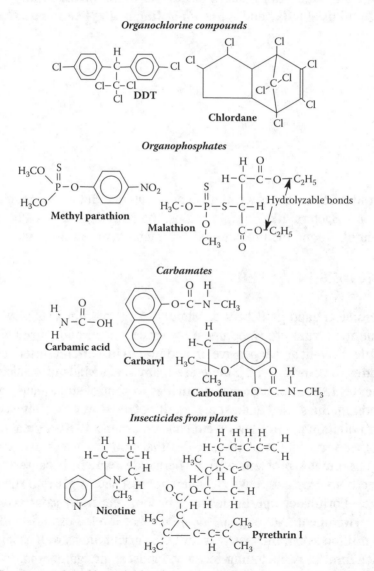

Figure 6.1.   Some examples of insecticides that are, or have been, potential water pollutants.

The organochlorine insecticides were dominant from the 1940s until the 1960s. In general, they are not particularly toxic to humans and other animals. However, DDT and related compounds have an even more detrimental characteristic because of their tendency to undergo **bioaccumulation** in fish and other organisms, concentrating in fat tissue. Furthermore, as organisms that have accumulated these compounds are eaten, and these organisms, in turn, are eaten by larger animals, the organochlorine compounds become progressively more concentrated in fat tissue, a process called **biomagnification**. It was this phenomenon that caused the birds of prey — eagles, falcons, hawks — to become so contaminated with DDT that they produced soft egg shells that broke before young could hatch. This threatened a number of species of birds, including the Bald Eagle that is the U.S. symbol, leading to the banning of DDT and most other organochlorine insecticides.

For a time, as DDT and other organochlorine insecticides lost favor, **organophosphate insecticides**, which are organic derivatives of phosphoric acid (H_3PO_4) came into common use. These insecticides had a big advantage in being biodegradable with no tendency to undergo bioaccumulation. Two examples are methyl parathion and malathion. Parathion, once the most widely used organophosphate insecticide, is a very effective insecticide and, because of its rather rapid biodegradability, it was not usually a significant water pollutant. It acts by inhibiting the action of acetylcholinesterase, an enzyme essential to nerve function. This is the same mode of action of military poison "nerve gases," such as sarin, and a significant number of fatal poisonings occurred because of parathion exposure. Although parathion is now banned, malathion remains on the market and is only about 1/100 as toxic to mammals as is parathion. This is because, as shown in its structural formula in Figure 6.1, malathion can be cleaved with addition of water by enzymes possessed by humans and other mammals, but not by insects.

Carbamates, in which various hydrocarbon groups are substituted for H on carbamic acid, largely replaced phosphates, and several carbamates are still widely used. **Carbaryl** (Sevin) is used to kill insects on lawns or gardens and, because of its low toxicity to mammals, can be sprinkled on pets in (often futile) attempts to rid them of flea infestations. Highly water-soluble **carbofuran** is a plant **systemic insecticide** that is taken up by roots and leaves and distributed through the plants, killing insects that feed on the plants. Like organophosphates, carbamates are acetylcholinesterase inhibitors, but are not highly toxic to animals other than insects. They are very biodegradable and are not generally serious water pollutants.

Some important insecticides have been derived from plants. These include **nicotine** from tobacco and **rotenone** extracted from some plant roots. **Pyrethrins** are excellent, biodegradable insecticides extracted from dried chrysanthemum or pyrethrum flowers. Noted for their low toxicities to mammals and excellent abilities to "knock down" flying insects, pyrethrins were probably used to kill insects in China 2000 years ago. The current major sources of these environmentally-friendly insecticides are chrysanthemum varieties grown in Kenya. Because of the excellent insecticidal properties of pyrethrins, their synthetic analogs, **pyrethroids**, have been synthesized and widely produced. These substances are now insecticides of choice for household applications. Because of their biodegradability, pyrethrins and pyrethroids are not serious water pollutants.

Herbicides

Herbicides are applied to vast areas of farmland to control weeds that crowd out corn, soybeans, cotton, and other economically important crops. The manner in which herbicides must be applied ensures that they are susceptible to being washed off fields by rainfall giving herbicides a high potential to contaminate water. Herbicides are commonly found in drinking water supplies and some municipalities are required to use activated carbon filtration to remove herbicides from municipal drinking water.

Herbicides come in an enormous variety of chemical compounds. One of the most widely produced types consists of the **triazines**, which have 6-membered rings in which C atoms alternate with N atoms ("triazine" denotes 3 nitrogen atoms). Three of these widely used to control weeds on corn and soybeans are atrazine, simazine, and cyanazine:

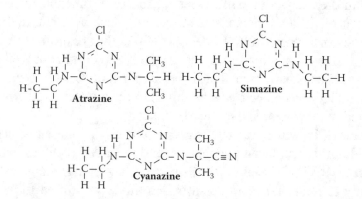

Glyphosate (see following structure) kills weeds by interfering with the synthesis of some kinds of amino acids essential for plant proteins. It is a postemergence herbicide, which means that it is applied directly to weeds after they have started to grow. It is effective against broadleaf weeds, grasses, and perennial plants. Glyphosate is the active ingredient in Monsanto's Roundup herbicide. Monsanto has developed genetically modified soybeans and other crops that are "Roundup ready," meaning that they are not harmed by the direct application of this herbicide, which kills competing weeds.

The Infamous Dioxin

Some of the more severe pollution problems associated with pesticides have come from their manufacture. One of the more notorious of these was **dioxin**, known chemically as 2,3,7,8-tetrachlorodibenzo-*p*-dioxin (TCDD), the structure of which is shown as follows:

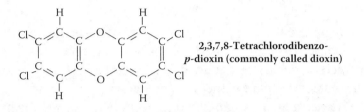

2,3,7,8-Tetrachlorodibenzo-*p*-dioxin (commonly called dioxin)

Dioxin was produced as a by-product of the manufacture of 2,4,5-trichlorophenoxyacetic acid, 2,4,5-T, the infamous "Agent Orange" used to defoliate jungles in the Vietnam war. Dioxin is essentially insoluble in water, melts at 305°C (very high for an organic compound), is chemically stable up to 700°C, and does not undergo biodegradation well. Although it is remarkably toxic to some animals (especially guinea pigs), it is not extremely toxic to humans. It has no uses, but is generated as a by-product of the manufacture of some chemicals and during the incineration of chlorine-containing plastics, such as polyvinyl chloride. It was a highly undesirable impurity from the synthesis of 2,4,5-T herbicide mentioned above.

Because of its extremely low water solubility, dioxin is not a common water pollutant. However, it can accumulate in sediments. It gained notoriety as a hazardous waste in the early 1970s when dioxin-contaminated wastes were mixed with waste oil and sprayed for dust abatement on roads, horse arenas, and other areas in the state of Missouri. Horses and birds were killed, and the entire town of Times Beach, Missouri, was contaminated with the contaminated oil. The U.S. Environmental Protection Agency bought out the whole town in March 1983, at a cost of $33 million. Soil from the town along with other dioxin-contaminated soil was subsequently incinerated at a total estimated cost of about $80 million.

6.7. POLYCHLORINATED BIPHENYLS (PCBS)

Polychlorinated biphenyls (PCBs) consist of a class of 209 compounds made from substituting from 1 to 10 chlorine atoms for H atoms on biphenyl as shown by the examples in Figure 6.2. PCBs are notable for their extreme chemical and thermal stability, resistance to biodegradation, low vapor pressure, and high dielectric constants. They even survive ordinary combustion processes and are dispersed to the atmosphere with stack gases. Until their manufacture and use were banned by the Toxic Substances Control Act of 1976, they were widely used as coolant-insulation fluids in transformers and capacitors, as plasticizers to make plastics more

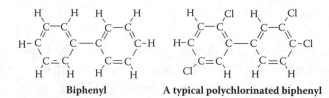

Biphenyl   A typical polychlorinated biphenyl

Figure 6.2.   Biphenyl and one of a possible 209 polychlorinated biphenyls (PCBs) produced by substituting from 1 to 10 Cls for the Hs on biphenyl.

flexible, to impregnate cotton and asbestos, and as additives to some epoxy paints. The extreme stability that led to these uses also contributed to their widespread dispersion and persistence in the environment. With the best practice of green technology, PCBs would never have come into general use.

PCBs are denser than water and tend to accumulate in sediments. One of the worst incidents of sediment pollution by PCBs occurred during 1950 to 1976 as thousands of kilograms of PCBs were dumped — legally at the time — into New York's Hudson River from electrical equipment manufacturing operations. In late 2005, the U.S. Environmental Protection Agency and General Electric reached an agreement that called for General Electric to begin dredging PCBs from the river starting with the 2007 dredging season. Dredging and disposal of the wastes may eventually cost as much as $500 million.

6.8. RADIOACTIVE SUBSTANCES IN WATER

Radioactive isotopes, or **radionuclides** can get into water from either natural sources or from the fission of uranium or plutonium in nuclear power reactors or (formerly) as fallout from above-ground weapons testing. Radionuclides have unstable nuclei that change spontaneously to nuclei of atoms of different elements by the emission of **ionizing radiation** in the form of alpha particles, beta particles, and gamma rays. These emissions are called *ionizing radiation* because they produce reactive ions in materials including flesh. These ions and other reactive species can damage DNA, impede the body's ability to make hemoglobin, and cause a number of biological effects, including severe anemia, mutations, cancer, and death.

There are three major kinds of ionizing radiation that are most commonly given off by radionuclides. Some of the heavier elements, such as uranium and radium emit **alpha particles**, a helium atom nuclei composed of two neutrons and two protons, denoted as $_2^4\alpha$. The penetrating ability of these very heavy particles is low, so that they do not pose a hazard outside the body. However, if they enter the body, such as by ingestion of contaminated drinking water, they cause enormous damage to exposed tissue. **Beta radiation** is in the form of high-energy electrons designated $_{-1}^0\beta$ or, less commonly, positive electrons, called *positrons*. These particles are more penetrating than alpha particles. **Gamma rays** are not particles, but are electromagnetic radiations that behaves like very short wavelength, high-energy X-rays. The energies of gamma rays are highly specific for the isotope nuclei that produce them and can be measured by sophisticated gamma ray spectrometers as a means of identifying the kinds and quantities of radionuclides emitting the radiation. Radionuclides decay with specific half-lives that can range from fractions of a second to millions of years. After the passage of each half-life, the radioactivity of a specific radionuclide is half of what it was at the beginning of the half-life. This means that all radionuclides are eventually converted to nonradioactive forms, although this may take a very long time.

Aside from some special circumstances involving production of radionuclides in military production facilities, the greatest concern with respect to these materials in water arises from natural geological sources. Specifically, the alpha particle emitter radium-226 (^{226}Ra — half-life 1620 years) is a particular concern in drinking water. A number of municipal water supplies contaminated with this radionuclide by leaching into groundwater from rock formations underground have been shut down because of the hazard presented.

Immediately following World War II, aboveground testing of nuclear weapons was a significant source of environmental pollution of radionuclides. Nuclear bombs, typically detonated on towers or dropped from aircraft in remote regions of New Mexico, Nevada, the South Pacific, and Russia, generated large quantities of uranium fission products and produced more radionuclides by the absorption of neutrons from the bomb by dirt and sand. A large mass of radioactive dust and debris would be entrained in a rapidly ascending column of hot gases and dispersed throughout the world. "Fallout" from these tests got onto land, causing particular concern about strontium-90 falling on pastureland and getting into cow's milk. Radionuclides were also scavenged from the atmosphere by rainfall, which could get into water supplies. The radioactive products of most concern from these tests were of elements that the body recognizes as material to be incorporated into tissue. These were strontium-90 (^{90}Sr — half-life 28 years) which is in the same chemical group as calcium and is incorporated into bone; cesium-137, (^{137}Cs), an alkali metal that the body handles much like sodium and potassium ions; and iodine-131 (^{131}I) — half-life 8 days — that is attracted to the thyroid and can impair its function and even cause thyroid cancer. The banning of aboveground nuclear weapons testing has largely stopped release of these elements to the environment. The last and largest such release was the catastrophic explosion and fire at the Soviet power reactor at Chernobyl in 1986 (see Chapter 17).

Fortunately, radionuclides are easily and sensitively detected in water. The most common such contaminant, radium-226, is readily removed by water softening processes that involve treatment of water with lime (see Section 6.9).

All nations seem to be abiding well with international treaties forbidding aboveground testing of nuclear weapons, so that should not be a source of radionuclide contamination. We can hope that there will not be any hostile exchanges of nuclear weapons in the future, although as of 2006 concern was increasing because several countries not noted for their reliability were pursuing nuclear weapons development programs. Improperly disposed radionuclides remain a threat to some water supplies. One example is a reported plume of radioactive groundwater flowing from the Hanford atomic energy installation to the Columbia River in Washington State.

6.9. MUNICIPAL WATER TREATMENT AND GREEN OZONE FOR WATER DISINFECTION

Water requires varying degrees of treatment before it is used for households or for industrial applications. Furthermore, wastewater from sewage or from industrial applications normally must be treated before it is released. Some of the basic treatment processes are similar, regardless of the intended use or mode of disposal of the water. They are discussed briefly here for municipal water, and some of the specialized treatment processes applied to wastewater are summarized at the end of the chapter.

Water being purified for municipal use is usually **aerated** by contacting it with air. This treatment blows out odor-causing impurities. When, as is often the case, dissolved iron is present as soluble iron(II) ion (Fe^{2+}), oxygen from the air oxidizes it,

$$4Fe^{2+} + O_2 + 10H_2O \rightarrow 4Fe(OH)_3(s) + 8H^+ \qquad (6.9.1)$$

forming a gelatinous solid of $Fe(OH)_3$. As it settles, this solid entrains and carries with it very small colloidal solid impurities suspended in the water, such as mud, a process called **coagulation**.

Water often contains excessive levels of dissolved calcium along with bicarbonate ion, HCO_3^-. This temporary hardness can be removed from water by adding **lime**, $Ca(OH)_2$:

$$Ca^{2+} + 2HCO_3^- + Ca(OH)_2 \rightarrow 2CaCO_3(s) + 2H_2O \qquad (6.9.2)$$

The $CaCO_3$ solid formed settles in a settling basin. The process can be somewhat slow and incomplete. So to avoid deposition of solid $CaCO_3$ precipitation in the water distribution system, which can clog pipes, CO_2 gas is added to the water.

If residual organic matter, such as humic substances (Section 5.11) is present in the water, it may be treated by running it over granular activated carbon. This is a form of carbon that has been reacted with steam at a high temperature to cause the formation of an enormous number of microscopic pores in the carbon, giving it a huge internal surface area of up to 2000 m^2/g! Carbon so treated is very effective in removing dissolved organic matter from water.

As a final step in water purification, water is disinfected. This may be done by adding elemental chlorine to the water, which reacts with water as follows:

$$Cl_2 + H_2O \rightarrow H^+ + Cl^- + HOCl \qquad (6.9.3)$$

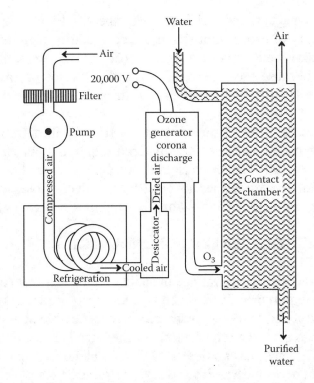

Figure 6.3. A schematic diagram of a system to generate ozone by electrical discharge through refrigerated, dried air followed by disinfection of water by the ozone product.

Hypochlorous acid (HOCl) is a good disinfecting agent that kills virus and bacteria in the water. Salts of HOCl including NaOCl and $Ca(OCl)_2$ can also be used for water disinfection.

When used as a disinfectant, chlorine may form undesirable organic compounds called *trihalomethanes*. To avoid this problem **chlorine dioxide** (ClO_2) an effective water disinfectant that does not produce trihalomethanes, is now commonly used. This compound is a dangerously reactive gas that is unsafe to move from a manufacturing site to where it is needed, so it is made by reacting sodium chlorite with elemental chlorine:

$$2NaClO_2(s) + Cl_2(g) \leftrightarrow 2ClO_2(g) + 2NaCl(s) \qquad (6.9.4)$$

This process of making potentially dangerous chlorine dioxide only in the quantities needed, where needed, when needed is in keeping with the best practice of green chemistry and technology.

Another disinfection agent that is superior to chlorine in some respects is **ozone**, O_3. An ozone-based water disinfection system is outlined in Figure 6.3. This reactive form of elemental oxygen is made by passing an electrical discharge at approximately 20,000 V through air that has been filtered, cooled, dried, and pressurized:

$$3O_2(g) \xrightarrow[\text{discharge}]{\text{Electrical}} 2O_3(g) \qquad (6.9.5)$$

The ozone contacts water in a contact chamber for 10 to 15 min. Although the low water solubility of ozone limits its disinfective ability, there is no formation of organochlorine disinfection by-products. And ozone is more effective than chlorine against viruses. Because ozone lasts for only a short period of time in water, some chlorine must be added to maintain residual disinfection in the water distribution system.

The use of ozone for water disinfection is a virtually ideal example of the practice of green chemistry. The only raw material required is air, which is available without cost everywhere. The ozone is produced only as needed, where it is needed, and it is not stored. The ozone does not persist, and it decomposes to benign molecular oxygen. The likelihood of forming harmful by-products is very low.

6.10. TREATMENT OF WATER FOR INDUSTRIAL USE

Water for industrial use may range in quality all the way from cooling water, for which the major requirement is that it be wet, to hyperpure water used in the semiconductor industry. For economic reasons, water for industrial applications is usually treated only to meet the requirements of the intended use. As examples, water used in food processing must be treated to destroy pathogens and boiler feed water must be treated to remove corrosive and scale-forming solutes.

There are a number of specific treatment operations to which industrial feed water may be subjected. Dissolved oxygen that may be corrosive can be removed with hydrazine or sulfite. Precipitants may be added to remove specific contaminants, the most common of which is calcium ion, Ca^{2+}. Calcium can form harmful deposits, especially when the water is heated. The addition of phosphate precipitates calcium, reducing the levels of soluble calcium to very low values. In some cases, the calcium can be tolerated in the water if chelating agents are added to prevent the precipitation of calcium solids. Scale can be inhibited by the addition of dispersants. It may be necessary to adjust pH by adding acid or base. Even water not destined for food use may have to be disinfected to prevent growth of harmful bacteria, such as in cooling systems.

An important consideration with industrial water is the possibility for **recycling** and **sequential use**. As the name implies, recycling refers to running water back through a system for essentially the same use. Sequential use recognizes that several applications may require water of successively lower quality. The water is first employed for the application requiring the best quality of water. The next use of the water requires a somewhat lower quality, and the last use before discharge requires the least quality. In some favorable cases, the water leaving the system can be applied to grass or fields for irrigation.

Both organic and inorganic (salt) solutes can make water unsuitable for recycling. Biodegradable organic materials can be degraded by biological waste treatment measures, which are discussed under the category of sewage treatment below. Filtration over activated carbon mentioned earlier in this section can remove harmful organic

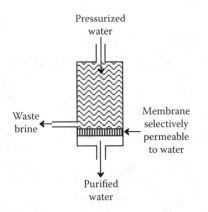

Figure 6.4.   Dissolved impurities can be removed from water by forcing pure water through a membrane selectively permeable to water in the reverse osmosis process.

solutes. The most straightforward way to remove dissolved salts is distillation in which water is evaporated leaving the salts behind, and then condensed to a pure H_2O product. Although used to get freshwater from seawater in some energy-rich arid Middle Eastern countries, this approach is too costly for water recycling in most parts of the world. A more cost-effective method of water purification is **reverse osmosis**, the basic principle of which is shown in Figure 6.4. With this method, water is forced under pressure through a semipermeable membrane that attracts pure water but rejects dissolved salts. The purified water is readily recycled for a variety of uses, but disposal is required of the concentrated salt brine left behind.

6.11. WASTEWATER TREATMENT

The primary objective of sewage treatment is to remove oxygen-demanding substances from wastewater. These are substances of mostly biological origin, abbreviated {CH_2O}, that undergo biodegradation and consumption of dissolved oxygen, thus exerting a BOD. Sewage treatment can be divided into the three main categories of (1) primary treatment to remove grit, grease, and solid objects, (2) secondary treatment to reduce BOD, and (3) tertiary treatment to further refine the quality of the effluent water. This section addresses primarily secondary treatment.

Secondary wastewater treatment uses a mass of microorganisms in contact with sewage and atmospheric oxygen to eliminate BOD by the reaction,

$$\{CH_2O\} + O_2 \rightarrow CO_2 + H_2O \tag{6.11.1}$$

a process that also generates additional biomass of organisms capable of degrading more BOD. The microorganisms can be on a support that alternately contacts liquid sewage and air. One system that uses that approach is the **trickling filter** in which sewage is sprayed over rocks coated with microorganisms. However, the most commonly used process at present is the **activated sludge process** shown in Figure 6.5. In this process, wastewater is pumped into one end of a large aeration tank through

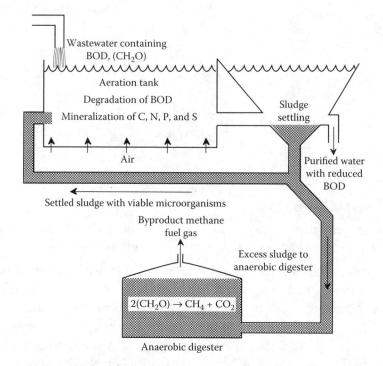

Figure 6.5.   Activated sludge process.

which air is bubbled. Viable microorganisms suspended in the tank metabolize the degradable materials composing the BOD in the sewage, converting some of the carbon to carbon dioxide and synthesizing additional microorganisms by the overall process represented in Reaction 6.11.1. In addition to converting organic C to CO_2, the bacteria in the aeration tank convert organic N, P, and S to simple inorganic ions, such as NH_4^+, $H_2PO_4^-$, and SO_4^{2-}, a process called **mineralization**. After an appropriate residence time in the tank, the treated water is taken to a settling tank where the microorganisms settle out as sludge. Most of this sludge is pumped back to the front of the aeration tank to contact additional BOD. Some excess sludge is taken to an anoxic (anaerobic) digester, where the sludge is acted upon by methane-forming bacteria in the absence of oxygen to generate methane gas, CH_4. The purified water from the treatment plant can be discharged directly or subjected to additional treatment. One option is to allow the water to flow slowly through a constructed wetland system where it is purified by natural processes.

Sewage treatment has a significant potential for the practice of green technology. One of the most obvious ways in which this can occur is by efficient utilization of the methane produced in the anoxic digestion of the spent sludge from the activated sludge process. Processes that rely primarily on anoxic digestion have the potential to produce even more methane because little of the biomass is consumed by oxic processes in the presence of air. Such anoxic processes are used in some large livestock feeding operations, providing significant amounts of useful methane. Even after anoxic digestion, the leftover sludge has value as fertilizer to provide nitrogen and phosphorus for plant growth and is commonly spread on soil in locations where sufficient land is available. Another alternative is to chemically gasify the leftover sludge with oxygen by high-temperature processes similar to those used to gasify coal. Properly designed and operated wetlands for advanced treatment of secondary sewage effluent are a means of purifying the effluent without consuming additional chemicals and energy while providing desirable areas for wildlife. It should be kept in mind that all of these applications require sewage that is relatively free of persistent pollutants, particularly poorly degradable organic compounds and heavy metals, so control of materials discharged with sewage is important.

6.12. WATER AS A GREEN RESOURCE

If one were to name the greenest substance, it would likely be water. Overall, water is extraordinarily abundant. Through the solar-powered hydrologic cycle (Figure 4.2), water is recycled naturally, and in those areas with suitable climate conditions, freshwater is regularly supplied by simply falling from the sky. Earth's oceans contain an inexhaustible supply of this essential material.

Despite its overall abundance, access to sufficient supplies of usable water is a major problem in many areas of the world, some of which have insufficient supplies for basic domestic needs, to say nothing of adequate amounts for growing food on irrigated land. Many arid regions border oceans in which the water is unusable because of its salt content. The lack of available water causes hardship and misery for millions of people. Water has long been a source of human conflict. Access to water for livestock and irrigation has set off range wars in arid regions of the U.S. West and water has the potential to start much deadlier conflicts in Africa and the Middle East.

A major challenge facing humankind is the provision of adequate supplies of usable water. This will require the intense applications of the principles of green science and technology. The ways in which this can be done are discussed in the remainder of this chapter. The following are the main aspects of green science and technology applied to water:

- Utilization of available water in a sustainable manner including water conservation

- Transfer of water from areas with abundant sources to water-deprived areas

- Upgrading water quality from otherwise unusable sources

- Transfer of water from areas with abundant sources to water-deprived areas

- Seawater desalination

- Sequential use of water

- Utilization of wastewater

- Purification of wastewater for reuse

Water recycling, sometimes called water reclamation or water reuse, refers to utilization of water that has previously been used by humans for some purpose. There are two general categories of water recycling. The first of these is **unplanned water recycling** in which water discharged into waterways is withdrawn, treated, and placed back into water distribution systems. This leads to the rather unappetizing thought that some of the drinking water withdrawn from a faucet may have been used to flush a toilet only days earlier. **Planned water recycling** involves using water that has been through some sort of application, with or without additional treatment. Usually planned recycling entails **sequential use** of water for applications with progressively lower quality needs. For example, water collected from washing clothing and from showers could be used to flush toilets, treated to remove its most objectionable characteristics, and then employed for irrigation. The Irvine Ranch Water District in arid southern California uses treated sewage water to irrigate parklands and has even provided this water to flush toilets in high-rise buildings in Irvine. It is estimated that the cost of providing dual systems of water supply for these buildings has added only about 9% to the cost of the structures. Water reuse and recycling is covered in a comprehensive book dealing with water utilization.[1]

Municipal wastewater in the U.S. and Western Europe normally must be treated to the secondary level (water outlet in Figure 6.5). This water can be employed for some applications, but normally additional treatment is required. The degree of treatment required obviously depends upon the application. For example, secondary effluent might be suitable for landscape irrigation, could be used with some additional treatment and disinfection for growing crops such as animal forage, and would not be recommended at all for growing vegetables. Use of recycled water is generally restricted to nonpotable (nondrinking) purposes. Acceptable uses include irrigation for some agricultural applications, golf courses, parklands, and landscape. A popular use is for cooling water for power plants, petroleum refineries, and manufacturing applications. Industrial applications include process water for paper mills, carpet dyeing operations, and metal fabrication. Recycled water can be used in construction, to mix concrete, and for dust control.

A major application of recycled water is to replenish surface and subsurface reservoirs of water. Some lakes and reservoirs that do not supply potable water are maintained with treated wastewater. A particularly common application is replenishment of underground aquifers, many of which have been severely depleted in arid regions. In coastal areas and islands, pumping down aquifers can lead to saltwater intrusion from the sea, which can ruin the water quality in the aquifers making them unsuitable as water sources. Pumping recycled water into such aquifers can replenish them (Figure 6.6), and contact of the water with subsurface rocks can help to

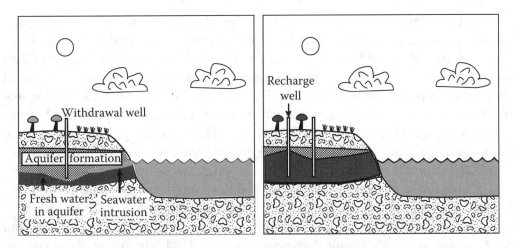

Figure 6.6.   Depletion of freshwater in an aquifer that is in contact with a body of saltwater can result in saltwater intrusion, ruining the aquifer. Pumping recycled water into the aquifer can remedy the problem, and contact of the recycled water with the aquifer mineral formation can further purify it.

purify it to a usable state. Underground aquifer augmentation with recycled water has been practiced in a number of areas for some time. The Water Factory 21 Direct Injection Project has been operating since 1976 in Orange County, CA to inject upgraded recycled water into an aquifer, both to prevent saltwater intrusion and to supplement supplies of potable water.

Augmentation of surface water reservoirs with recycled water has been less common than aquifer recharge. One such project has been carried out by the Occoquan Sewage Authority since 1978 by discharging recycled water to a stream that feeds the Occoquan Reservoir supplying potable water to Fairfax County, VA. One of the most advanced such projects was the proposed San Diego, CA, Water Repurification Project designed to convert reclaimed water, already upgraded for nonpotable applications, to water that could be placed in a surface reservoir from which the municipal water supply was withdrawn. The water was to be treated with multiple state-of-the-art technologies including advanced filtration, ultraviolet disinfection, ozonation, and reverse osmosis. Approved by the California State Department of Health Services in 1996 for operation to begin in 2001, the project was ultimately turned down by voters, some of whom expressed horror at the thought of "drinking potty water."

Water recycling provides a number of benefits. For water-deficient areas, arguably the greatest benefit is provision of a reliable water supply under local control. This is an especially important consideration in areas in which population pressures result in excess water demand over supply, such as parts of the southwestern U.S. shown in Figure 4.5. Water recycling decreases diversion of freshwater from sensitive ecosystems that require water for wildlife, fish, and plants. Throughout the Southwestern U.S. and other arid regions, such diversions have caused ecosystems to collapse because of loss of water and deterioration of water quality. Such damage can be prevented and even reversed with recycled water.

Discharge of wastewater to sensitive bodies of water — streams, estuaries, and the ocean — can cause significant damage that can be mitigated with water recycling. An example in which this was done is the $140 million San Jose/Santa Clara Water Pollution Control Plant that went into operation in 1997. Discharge of treated wastewater to San Francisco Bay had damaged the Bay's natural saltwater marsh converting it to a brackish marsh, threatening two endangered species and harming other organisms in the sensitive ecosystem. The recycling project upgraded the wastewater to 80 million gallons of water suitable for irrigation and industrial use, diverting it from the marsh and reducing dilution of the saltwater in the marsh.

In addition to mitigating damage to bodies of water and sensitive ecosystems, recycled water can be used to create and enhance stream (riparian) habitats and wetlands. Streams with water flows thus augmented can be prevented from drying out during dry seasons and provide enhanced aquatic and wildlife habitat. Wetlands can act to diminish floods, improve water quality, provide fisheries breeding areas, and enhance wildlife and wildfowl habitat. Recycled water can provide a steady, dependable source of water for existing wetlands and constructed wetlands.

Water recycling can be employed to reduce and prevent pollution. Processing water to make it suitable for reuse automatically removes pollutants that otherwise might be discharged. Wastewater from sewage contains nutrients (particularly nitrogen and phosphorus) that can cause excessive plant growth and eutrophication. When this water is used for irrigation, the nutrients are beneficial for plant growth and do not enter streams or bodies of water where they may cause a problem.

Water recycling faces several barriers. In cases where adequate supplies of freshwater are available, water recycling is relatively costly, largely because of the need to upgrade it for use. This is especially true if the recycled water is to be upgraded to potability. There is concern regarding the use of recycled water because of low levels of impurities that may escape treatment. These include pharmaceutical residues and metabolites, endocrine disruptors that can cause hormonal imbalance, and viruses. Salinity of recycled water can be a problem because of the salt inevitably picked up by water in going through a municipal water treatment system.

6.13. ALTERNATIVE GREEN USES OF WATER

Many pollution and contamination problems result from the widespread use of organic solvents. These include flammable hydrocarbons and chlorinated hydrocarbons, which tend to be toxic and to undergo bioaccumulation. The practice of green technology increasingly calls for substitution of water for organic solvents. One approach is to put additives in water so that it functions more like an organic solvent. For example, surfactants added to water enable it to suspend organic materials. For example, whereas the constituents of paint used to be dissolved and suspended in organic solvents, now most paints are water-based emulsions (latex paint), saving millions of barrels of petroleum formerly used as a paint vehicle and reducing hydrocarbon releases to the atmosphere substantially.

Although water is not a good solvent for organic substances, it becomes much more so at elevated temperatures. Applications of superheated pressurized water and even hotter supercritical water are discussed in the following two sections.

The outstanding solvent qualities of water also mean that it is difficult to get into a completely pure form. Even silica from glass containers can contaminate water and cause problems for some uses. For some applications it is necessary to have extremely pure water, commonly called **ultrapure water** and an even more pure grade called **hyperpure water**. The main applications of ultrapure water are in chemical analysis, where very low levels of contaminants are important, and in the semiconductor industry.

6.14. WATER: A MOST USEFUL SUBSTANCE

Some of the most important of the many useful aspects of water are discussed in Section 4.10. There are several reasons that water is "a most useful substance." Water offers many advantages in industrial and other applications. It is generally very inexpensive and readily available, although it can be scarce and expensive in some places. As the medium of life, water is the ultimate non-toxic material (although one can drown in it and some people have managed to drink enough water to kill themselves). It offers a stunning array of different physical and chemical properties that have made it extraordinarily useful. Furthermore, these properties can be varied and enhanced by additives dissolved or suspended in water, such as oil lubricants suspended in water used as a coolant and lubricant in parts machining, and, as discussed in the following two sections, water's solvent properties can be altered greatly by heating it under pressure or converting it to a supercritical fluid at very high temperatures and pressures.

Water's uses can be divided among those that (1) take advantage of water's physical properties, (2) depend upon its unique chemical properties, and (3) rely upon water's special role in living systems. Often all three categories combined are utilized in essential applications of water. For example, in the human body, water acts as a solvent; cools the body when heat is absorbed as perspiration evaporates; acts as a chemical agent in the hydrolysis of complex carbohydrates, proteins, and lipids; and generally provides a hospitable medium in which the body exists.

The most commonly utilized physical properties of water as liquid, steam, or ice are those that depend upon the compound's special behavior with respect to heat. Water has the highest heat of vaporization of any common material. An extraordinarily large amount of heat, 40.7 kilojoules/mole (kJ/mole) at 100°C absorbed to evaporate liquid water, means that a very large amount of heat energy is required to convert a quantity of liquid water to steam, and a correspondingly large amount of heat energy is released when the steam condenses to liquid water. The applications of this characteristic of water can be observed on most U.S. university campuses in which steam is used to transfer heat from a centralized power plant to buildings where it condenses in radiators and releases heat. Some cities in Europe use the same kind of system to heat residential buildings, a concept called **district heating**. District heating has a very great potential for the efficient utilization of energy from sources other than scarce natural gas and petroleum, and may even take advantage of the combustion of municipal refuse and methane generated biologically from sewage or animal manure. The high heat of fusion for ice enables the use of solid water to absorb heat. Some large buildings have been designed in which energy at night, when energy demand is relatively low, is used to freeze water and the heat absorbed by the water thawing during the daytime hours is used for cooling.

Section 4.10 discusses the processes by which water is converted to steam in boilers and Figure 4.9 shows aspects of this process. Steam so generated can be used for its heat content and as a working fluid to convert the heat energy in the steam to mechanical energy by means of a steam turbine or (formerly) a piston engine. It is noted that requirements for boiler water are rather stringent and it almost always must be treated to reduce scaling and corrosion. Even very pure water can be corrosive and may have to be treated to remove oxidant oxygen and to reduce its corrosive properties. The purification of water to standards such that it can be used to generate steam is the subject of a major industry. Green chemical processes for water purification are of interest and can help to avoid pollution and wastes. One way in which the expense and potential harm from such purification processes can be minimized is by collection of the steam condensate and recirculation through the system. The steam locomotives that once dominated land transportation simply discharged steam through their stacks to create a draft in the boiler firebox. These machines used enormous quantities of water that had little or no treatment, and corrosion and scaling of their boilers from minerals dissolved in water necessitated frequent trips to the "roundhouse" where repairs were performed.

An interesting application of steam is **steam explosion pulping** of wood. In this process, wood is heated with pressurized steam at 180°C to 210°C, followed by very rapid release of pressure, called explosive decompression. The wood literally explodes, releasing the fibers that can be used in making paper pulp, fiberboard and other applications. Cellulose released by steam explosion pulping can be treated to produce fermentable sugars for the production of alcohol and other fermentation products. Steam explosion is an important process in sustainable utilization of the renewable wood resource and an important green alternative to the chemical processes and harsh reagents that have been used for wood pulping.

Nature provides a huge "boiler" in the form of the hydrologic cycle (Chapter 4, Figure 4.2) in which water is evaporated from the ocean and from other sources of liquid water and condenses in the atmosphere to produce precipitation. The vast amount of heat released by the condensation cannot be utilized by humans to produce mechanical energy, but the energy of the water that has fallen on land can certainly be utilized. As discussed in Sections 4.10 and 18.16, this energy was tapped for centuries by waterwheels and now is harnessed in turbines to produce hydroelectric energy. Where this can be done without undue environmental harm from damming streams, hydroelectric energy is one of the most sustainable means of energy utilization.

The use of flowing liquid water to move the turbine blades in a hydroelectric generator illustrates the ability of water to act as a mechanical agent. Its applications to mining sand, gravel, and other minerals were mentioned in Section 4.10 as was the use of ultra-high pressure liquid water to cut materials. As discussed in Section 16.9, water is especially useful in the hydraulic mining of placer deposits of minerals originally deposited as sediments from running water. Beaches lost to erosion are being restored in some areas with sand pumped as a slurry with water.

The single greatest use of water is as a solvent, an application that takes advantage of both the chemical and physical properties of water. Water acts as a solvent in all the environmental spheres. Water permeates organisms; the human body is mostly water. In organisms, water carries nutrients to cells and carries dissolved waste products away. In the atmosphere, water droplets dissolve gaseous and particulate species and remove them from the atmosphere as precipitation. Water acts in the geosphere to dissolve minerals. As shown in Reaction 5.8.3 in Chapter 5, water is one of the reactants in the dissoluton of mineral calcium carbonate, $CaCO_3$. Dissolved minerals can be carried by dissolved water for long distances, then deposited as sediments when they come out of solution. Water's solvent properties vary significantly depending upon its temperature, and advantage may be taken of this characteristic in various applications of water as a solvent. As discussed in the following two sections, water becomes a much better solvent for organic materials and has many uses when it is heated under pressure or is placed under such high temperatures and pressures that it becomes a supercritical fluid. The utilization of supercritical water and subcritical hot water are being increasingly realized and provides the basis of some excellent developing green technologies.

Water's generally non-toxic nature and compatibility with life make it a critically important substance in green technology. Processes in which liquid water are present are largely those hospitable to life and tend to meet the benign criteria characteristic of green technological processes.

6.15. HOT WATER: PRESSURIZED SUBCRITICAL HOT WATER

By keeping it under pressure, liquid water can be heated from its normal boiling point of 100°C to its critical temperature of 374°C, above which the distinction between liquid and vapor vanishes and water becomes a supercritical fluid (see Section 6.16). As water is heated in this range, the hydrogen bonds that hold the molecules of liquid water together and make it such a unique liquid and outstanding solvent for polar materials and ions are broken to an increasing degree, and the dielectric constant of water becomes lowered from the value of 78 at 25°C to 33 (the same as methanol at 25°C) at 205°C. At temperatures between 100°C and 205°C, water behaves like a mixture of water and methanol and becomes a much more effective solvent for organic substances. Advantage is taken of this phenomenon to use superheated water as an extractant for organic materials from plant biomass, to extract pollutant organic materials from contaminated soil and sediments, and as a reaction medium for organic chemical reactions. The decrease in dielectric constant of water and the increase in solubility of substances at higher temperatures mean that the solubilities of some substances in superheated water may be orders of magnitude higher than in water at 25°C.

A potentially useful application of pressurized hot water is its application as a medium along with pressurized O_2 in which organic substances are oxidized. This procedure has been used to make oxygenated products from organics as refractory as coal. It has also been used as a means of destroying oxidizable pollutants in water.

6.16. SUPERCRITICAL WATER

Water is a particularly good solvent for organic materials as a supercritical fluid at a temperature exceeding 374.4°C and an extremely high pressure of at least 217.7 atm. In this region, water does not have separate gaseous and liquid phases. The properties of supercritical water can be tuned by varying temperature and pressure. Although the conditions are very severe and require special equipment, the ability to use water in place of organic solvents can be quite useful, especially as a medium for organic synthesis reactions.

The properties and uses of supercritical water have been summarized in an excellent review of the topic.[2] By varying the pressure on supercritical water, its density can be changed continuously from gaslike to liquidlike with a corresponding variation in the properties of the fluid. Even when compressed to a density equal to that of liquid water, supercritical water has a low viscosity. This is important for chemical processes because it increases mass transfer and rates of diffusion in diffusion-controlled processes. Varying the temperature and pressure of supercritical water enables varying its properties throughout a wide range, and conditions can be attained in which the water is completely miscible with nonpolar compounds while retaining its ability to solubilize ions and polar species.

Supercritical water has enormous potential as a solvent medium and catalytic material for chemical reactions. Its high temperature is conducive to pyrolysis, it promotes the hydrolysis of compounds in which molecules are cleaved with addition of H_2O, and in the presence of O_2 and other oxidants it is a strongly oxidizing medium. Careful tuning of temperature, pressure, and oxidant levels enables supercritical water to facilitate partial oxidation of organic materials. One such application is the partial oxidation of methane,

$$CH_4 + \frac{1}{2}O_2 \rightarrow CH_3OH \tag{6.16.1}$$

to methanol. With this reaction, the potential exists to convert hard-to-transport methane gas in remote regions to liquid methanol that can serve as a motor fuel or in fuel cells and that can be moved by large tankers. The dielectric constant of supercritical water increases with increasing pressure, which favors formation of polar reaction products, such as methanol.

A particularly attractive characteristic of supercritical water is its ability to serve as an oxidant medium for organic wastes. This application takes advantage of the fact that under certain conditions, the water becomes a good solvent for waste substances such as polychlorinated biphenyls (PCBs) which, with added oxygen, can be completely oxidized to carbon dioxide, water, and inorganic halides (chloride ion). Oxygen, like other common inorganic gases, is completely miscible with supercritical water under extreme pressures, which increases its capability to act as an oxidant.

Supercritical water may exist in the geosphere at the high temperatures and extreme pressures at depths of around 30 km. In the presence of supercritical water at these depths, minerals may behave much differently compared to their behaviors under normal conditions, especially with respect to their dissolution and precipitation. There is also evidence to suggest that chemical processes may form methane under these conditions leading to a nonbiological pathway for hydrocarbon formation.

LITERATURE CITED

1. Tchobanoglous, George, Franklin L. Burton, and H. David Stensel, *Wastewater Engineering: Treatment and Reuse*, 4th ed., McGraw-Hill, New York, 2002.

2. Weingaertner, Hermann and Ernst Ulrich Franck, Supercritical water as solvent, *Angewandte Chemie*, International Edition, **44**, 2672–2692, 2005.

SUPPLEMENTARY REFERENCES

Addis, Bill, *Building with Reclaimed Components and Materials: A Design Handbook for Reuse and Recycling*, Earthscan, Sterling, VA, 2006.

American Water Works Association, *Water Treatment*, 3rd ed., American Water Works Association, Denver, CO, 2003.

Crittenden, John C., *Water Treatment Principles and Design*, 2nd ed., John Wiley & Sons, Hoboken, NJ, 2005.

Eckenfelder, W. Wesley, *Industrial Water Pollution Control*, 3rd ed., McGraw-Hill, Boston, MA, 2000.

El-Dessouky, H.T. and H.M. Ettouney, *Fundamentals of Salt Water Desalination*, Elsevier Science, New York, 2002.

Judd, Simon and Bruce Jefferson, Eds., *Membranes for Industrial Wastewater Recovery and Re-use*, Elsevier Science, New York, 2003.

Lauer, William C., Ed., *Desalination of Seawater and Brackish Water*, American Waterworks Association, Denver, CO, 2006.

Lazarova, Valentina and Akica Bahri, Eds., *Water Reuse for Irrigation: Agriculture, Landscapes, and Turf Grass*, CRC Press, Boca Raton, FL, 2005.

Parsons, Simon A. and Bruce Jefferson, *Introduction to Potable Water Treatment Processes*, Blackwell Publishing, Malden, MA, 2006.

Rommelmann, David W., *Industrial Water Quality Requirements for Reclaimed Water*, AWWA Research Foundation, Denver, CO, 2004.

Tchobanoglous, George, Franklin L. Burton, and H. David Stensel, *Wastewater Engineering: Treatment and Reuse*, 4th ed., McGraw-Hill, Boston, MA, 2003.

Vickers, Amy, *Handbook of Water Use and Conservation*, Waterplow Press, Amherst, MA, 2001.

Viessman, Warren and Mark J. Hammer, *Water Supply and Pollution Control*, 7th ed., Prentice Hall, Upper Saddle River, NJ, 2004.

Wang, Lawrence K., Yung-Tse Hung, and Nazih K. Shammas, *Advanced Physico-chemical Treatment Processes*, Humana Press, Totowa, NJ, 2006.

Wolverton, B.C. and John D. Wolverton, *Growing Clean Water: Nature's Solution to Water Pollution*, Wolverton Environmental Services, Picayune, MS, 2001.

QUESTIONS AND PROBLEMS

1. What do mercury and arsenic have in common in regard to their interactions with bacteria in sediments?

2. What are some characteristics of radionuclides that make them especially hazardous to humans?

3. Give a specific example of each of the following general classes of water pollutants: (1) trace elements, (2) organically bound metal, and (3) pesticides.

4. A sample of water contaminated by the accidental discharge of a radionuclide used for medicinal purposes showed an activity of 23,956 counts per second at the time of sampling and 2,993 cps exactly 46 d later. Estimate the half-life of the radionuclide.

5. What are the two reasons that soap is environmentally less harmful than ABS surfactant used in detergents?

6. Why can one not give an exact chemical formula of the specific compound designated as PCB?

7. Describe a process that operates both in a body of water receiving pollutant sewage and in an activated sludge wastewater treatment operation and explain why it is harmful in the body of water and beneficial in the wastewater treatment plant.

8. Look up the origin of the word "plumbing" and explain how some older plumbing might be a source of a toxic water pollutant.

9. Chemically, what is the distinction between metal chelates and organometallic compounds? Suggest a "natural" or biological source of an example of each.

10. Suggest why carbon dioxide levels in groundwater often have dissolved carbon dioxide levels much higher than those from carbon dioxide in air. In an upscale restaurant or supermarket, what would you request to get, water laden with carbon dioxide from a groundwater source? Suggest a much cheaper source of such "fizzy" water.

11. Suggest two ways in which iron pyrite (FeS_2) may cause acidification of surface water. One of these mechanisms involves bacteria and the other is exclusively chemical.

12. What are two ways in which the application of green chemistry (not known as such at the time) eliminated two pollutants in detergents?

13. Explain why readily biodegradable pesticides are unlikely to undergo bioaccumulation or biomagnification.

14. What is meant by aeration of water? What are two pollutants removed from water when it is treated by aeration?

15. What is a less expensive option to distillation for obtaining freshwater from saline water? How does this option operate?

16. Sewage sludge circulates in an activated sludge wastewater treatment system and builds up to excess. What are two kinds of useful by-products that may be obtained from this sludge?

17. Consider a scheme for collecting outflow of the Mississippi River where it flows into the Gulf of Mexico and pumping it by wind power to water-deficient regions of the southwestern United States and northern Mexico. Make a list of advantages and disadvantages of such a system including ways in which it would be "green" and sustainable and ways in which it would not be.

18. How do pharmaceuticals and their metabolic products constitute a deterrent to water recycling?

19. What are two major uses of water other than its life-support (such as for drinking or irrigation) applications?

20. Water condensed from steam in heating and power applications is carefully reclaimed. Suggest why this is done. Water from spent steam in old steam locomotives used on railroads was simply discharged up the locomotive stack to create draft in the firebox. Why do you suppose the locomotives had to make very frequent trips to the "roundhouse" for maintenance?

21. Distinguish between hydroelectric power and pumped storage.

22. What is observed when water becomes supercritical? What are some applications of supercritical water? What class of "nongreen" substances may supercritical water replace?

7. THE ATMOSPHERE: A PROTECTIVE BLANKET AROUND US

7.1. LIVING AT THE BOTTOM OF A SEA OF GAS

We live and breathe in the **atmosphere**, a sea of gas consisting primarily of elemental O_2 and N_2, with significant amounts of water vapor, noble gas argon, carbon dioxide, and trace quantities of other gases. Unlike the ocean with its well-defined volume and surface, the atmosphere extends for many kilometers from Earth's surface, becoming less dense with increasing altitude. With increasing altitude the air simply gets thinner and thinner, but there is not a distinct altitude at which the atmosphere ends and outer space begins.

You would likely have little concern about traveling a short distance of 11 km or so — around 7 mi — on Earth's surface because at your destination the air available for breathing would be the same. But, suppose you could travel the same distance straight up to an altitude of around 11 km or 35,000 ft at which jet aircraft typically fly, a miniscule distance compared to Earth's diameter of almost 13,000 km? Without breathing equipment, you would not reach your destination alive. In fact, in the unlikely event of sudden, catastrophic loss of pressure in the pressurized cockpit of an aircraft cruising at such an altitude, the pilot has only about 15 seconds to grab an oxygen mask before losing consciousness (the passengers in the cabin have an equally short time, but it is more important for the pilot to stay conscious and dive to a lower altitude). The reason for this is that virtually all the air in the atmosphere is below the altitude of around 11 km. It is interesting, and a bit frightening, to contemplate, as one looks straight up into the sky, that it is just a very short distance to an environment totally inhospitable to human existence from which we are protected by a very thin layer of air.

More than 99% of the total mass of the atmosphere is found within approximately 30 km (about 20 mi) of the Earth's surface. Such an altitude is miniscule compared to the Earth's diameter, so it is not an exaggeration to characterize the atmosphere as a "tissue-thin" protective layer. Although the total mass of the global atmosphere is enormous, approximately 5.14×10^{15} t, it is still only approximately one millionth of the Earth's total mass.

The Protective Atmosphere

We know that we must have air's life-giving oxygen to breathe, but air is far more than just a source of oxygen. That is because it protects Earth's organisms in ways that are absolutely essential for their existence. One major protective function is to act as a blanket to keep us warm by reabsorbing the infrared radiation with which Earth radiates the energy that it receives from the sun. By delaying the exit of this energy into outer space, the average temperature of Earth's surface remains at about 15°C at sea level, though much colder at certain times and places and significantly warmer at others. Without this warming effect, plants could not grow and most other known organisms could not exist. The second protective function of the atmosphere is absorption of very short wavelength ultraviolet solar radiation. Were this radiation to reach our level, it would tear apart biomolecules, making it impossible for most life-forms to exist.

The altitude at which high-flying jet aircraft cruise marks the upper limit of the lowest of several layers of the atmosphere, the **troposphere**, which extends from sea level to about 11 km (Figure 7.1). As anyone who has gone to the peaks of high mountains knows, the troposphere gets cooler with increasing altitude, from an average temperature of 15°C at sea level to an average at 11 km of –56°C. Above the layer of the troposphere, however, atmospheric temperature increases to an average of –2°C at 50-km altitude. The layer above the troposphere is the **stratosphere**, which is heated by the absorption of intense ultraviolet radiation from the sun (Figure 7.2). There is virtually no water vapor in the stratosphere, and it contains ozone (O_3) and O atoms as the result of ultraviolet radiation acting upon stratospheric O_2. Beyond

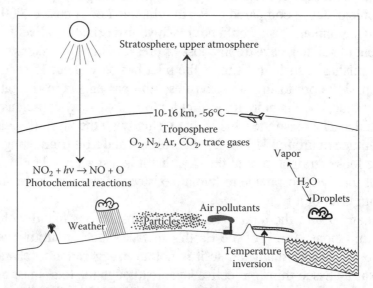

Figure 7.1.   The troposphere is the very thin layer of the atmosphere closest to Earth. It contains most of the atmosphere's air and water vapor. It is the source of oxygen, carbon dioxide, nitrogen, and water used by living organisms and as raw materials for manufacturing. With the important exception of stratospheric ozone destruction, it is where most air pollution phenomena occur.

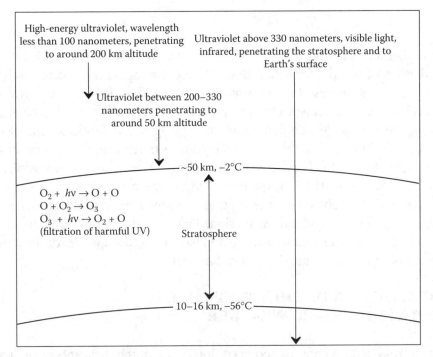

High-energy ultraviolet, wavelength less than 100 nanometers, penetrating to around 200 km altitude

Ultraviolet above 330 nanometers, visible light, infrared, penetrating the stratosphere and to Earth's surface

Ultraviolet between 200–330 nanometers penetrating to around 50 km altitude

~50 km, –2°C

$O_2 + h\nu \rightarrow O + O$
$O + O_2 \rightarrow O_3$
$O_3 + h\nu \rightarrow O_2 + O$
(filtration of harmful UV)

Stratosphere

10–16 km, –56°C

Figure 7.2.   The upper atmosphere including the stratosphere and regions beyond is particularly important in the absorption of radiation that would make life impossible if it reached Earth's surface. The layer of ozone (O_3) in the stratosphere is of the utmost importance and one that is subject to damage from anthropogenic species released into the atmosphere.

the stratosphere are layers called the mesosphere and thermosphere, but they are relatively less important in the discussion of the atmosphere.

Atmospheric Chemistry and Photochemistry

Chemical processes are extremely important in the atmosphere in maintaining its protective function and in pollution phenomena. Figure 7.2 shows some atmospheric chemical reactions. Consider the reaction $O_2 + h\nu \rightarrow O + O$. It is a **photochemical reaction** in which $h\nu$ represents the energy of a photon of ultraviolet radiation. Such photons have high energies that they can transfer to molecules absorbing them. In the case of the reaction shown, the energy is sufficient to split the O_2 molecule to produce O atoms, which in turn can react to produce molecules of ozone (O_3) which compose the critical protective ozone layer in the atmosphere. Photochemical and other atmospheric chemical reactions are discussed in detail in Chapter 8.

Formation of Ions in the Atmosphere

Absorption of extremely energetic ultraviolet radiation very high in the atmosphere can cause formation of ions as shown by the following reaction of molecular nitrogen:

$$N_2 + h\nu \rightarrow N_2^+ + e^- \qquad\qquad (7.1.1)$$

Although atmospheric ions are very reactive, at very high altitudes the air is so thin that ions can exist for some time before colliding with a species with which they can react. At altitudes of approximately 50 km and up, ions are so prevalent that the region is called the **ionosphere**. The presence of the ionosphere has been known since about 1901, when Marconi, attempting to bridge the Atlantic ocean with short-wave radio, discovered that radio waves could be transmitted over long distances where the curvature of the Earth makes line-of-sight transmission impossible. These radio waves bounce off the ionosphere. At night, there is less intense solar radiation to form ions, which recombine, raising the lower limit of the ionosphere and enabling longer-distance radio transmission. The emission of light from highly energized (excited) molecules and ions in the ionosphere is responsible for the spectacular displays of northern lights, the aurora borealis.

7.2. COMPOSITION OF THE ATMOSPHERE AND ATMOSPHERIC WEATHER

Aside from water vapor, the composition of air in the troposphere, which contains most of the mass of the atmosphere, is relatively uniform. Atmospheric air normally contains 1 to 3% water vapor by volume. On a dry basis, the two major constituents of air in the troposphere are elemental nitrogen (78.08% by volume of dry air) and elemental oxygen (20.95% by volume of dry air). Dry tropospheric air has two minor components, 0.934% noble gas argon and 0.038% carbon dioxide, a figure that is increasing by about 0.0001% per year because of input of carbon dioxide from fossil fuel combustion. In addition to argon, tropospheric air contains four other noble gases: Neon, 1.82×10^{-3}%; helium, 5.24×10^{-4}%; krypton, 1.14×10^{-4}%; and xenon, 8.7×10^{-6}%. In addition to these gases, atmospheric air contains a variety of trace gases, some of which are regarded as pollutants and whose levels may vary significantly. These are listed in Table 7.1.

The water vapor content of the troposphere is normally within a range of 1 to 3% by volume with a global average of about 1%. However, air can contain as little as 0.1% or as much as 5% water. The percentage of water in the atmosphere decreases rapidly with increasing altitude. Water circulates through the atmosphere in the hydrologic cycle as shown in Figure 4.2.

Water vapor absorbs infrared radiation even more strongly than does carbon dioxide, thus greatly influencing the Earth's heat balance. Clouds formed from water vapor and consisting of droplets of liquid water reflect light from the sun and have a temperature-lowering effect. On the other hand, water vapor in the atmosphere acts as a kind of "blanket" at night, retaining heat.

When ice particles in the atmosphere change to liquid droplets or when these droplets evaporate, heat is absorbed from the surrounding air. Reversal of these processes results in heat release to the air (as latent heat). This may occur many miles

Table 7.1. Atmospheric Trace Gases in Dry Tropospheric Air

Gas or Species	Volume Percentage[a]	Major Sources	Process for Removal from the Atmosphere
CH_4	1.6×10^{-4}	Biogenic[b]	Photochemical
CO	$\sim 1.2 \times 10^{-5}$	Photochemical, anthropogenic[c]	Photochemical
N_2O	3×10^{-5}	Biogenic	Photochemical
NO_x[d]	10^{-10}–10^{-6}	Photochemical, lightning, anthropogenic	Photochemical
HNO_3	10^{-9}–10^{-7}	Photochemical	Washed out by precipitation
NH_3	10^{-8}–10^{-7}	Biogenic	Photochemical, washed out by precipitation
H_2	5×10^{-5}	Biogenic, photochemical	Photochemical
H_2O_2	10^{-8}–10^{-6}	Photochemical	Washed out by precipitation
HO•[e]	10^{-13}–10^{-10}	Photochemical	Photochemical
HO_2•[e]	10^{-11}–10^{-9}	Photochemical	Photochemical
H_2CO	10^{-8}–10^{-7}	Photochemical	Photochemical
CS_2	10^{-9}–10^{-8}	Anthropogenic, biogenic	Photochemical
OCS	10^{-8}	Anthropogenic, biogenic, photochemical	Photochemical
SO_2	$\sim 2 \times 10^{-8}$	Anthropogenic, photochemical, volcanic	Photochemical
I_2	0–trace	—	—
CCl_2F_2[f]	2.8×10^{-5}	Anthropogenic	Photochemical
H_3CCCl_3[g]	$\sim 1 \times 10^{-8}$	Anthropogenic	Photochemical

[a] Levels in the absence of gross pollution.
[b] From biological sources.
[c] Sources arising from human activities.
[d] Sum of NO, NO_2, and NO_3, of which NO_3 is a major reactive species in the atmosphere at night.
[e] Reactive free radical species with one unpaired electron, transient species whose concentrations become much lower at night.
[f] A chlorofluorocarbon, Freon F-12.
[g] Methyl chloroform.

from the place where heat was absorbed and is a major mode of energy transport in the atmosphere. It is the predominant type of energy transition involved in thunderstorms, hurricanes, and tornadoes.

On a global basis, rivers drain only about one-third of the precipitation that falls on the Earth's continents. This means that two-thirds of the precipitation is lost as combined evaporation and transpiration, the process by which water enters plant roots and evaporates through leaves. During the summer, this **evapotranspiration** may exceed precipitation because of the large quantities of water stored in the root zone of the soil. To a degree, evapotranspiration furnishes atmospheric water vapor necessary for cloud formation and precipitation. Therefore, large-scale deforestation, soil damage (such as by plowing up grasslands in semiarid areas), and irrigation affect regional climate and rainfall. The very cold layer of air at the top of the troposphere (Figure 7.1) called the **tropopause** causes water to condense into ice crystals and serves as a barrier to the movement of water into the stratosphere.

7.3. THE PROPERTIES OF GASES AND THE GAS LAWS

To understand the atmosphere, it is important to know the properties of the gases from which it is made. The molecules and (in the case of noble gases, the atoms) that are contained in gases are relatively far apart and in constant, rapid motion. The distance between gas molecules make gases compressible into smaller volumes and their rapid motion causes gas to exert pressure. The motion of gas molecules is because of heat; the higher the temperature of gas, the greater the movement of the molecules.

Here are covered the fundamental gas laws that govern the behavior of gases in the atmosphere. In doing calculations with these laws, it is useful to keep in mind that the **quantity of gas** is most conveniently expressed as **moles** (see Section 3.5). Although pressure is expressed in several units, the most intuitively useful is the **atmosphere** in which the average pressure of the atmosphere at sea level is 1 atm. The **absolute temperature** is used for temperature in which each degree is the same as a Celsius degree and zero is absolute zero, the lowest attainable temperature at −273°C (273°C below the freezing point of water). Three important gas laws are the following:

Avogadro's law: At constant temperature and pressure the volume of a gas is directly proportional to the number of moles; doubling the number of moles doubles the volume.

Charles' law: At constant pressure the volume of a fixed number of moles of gas is directly proportional to the absolute temperature (degrees Celsius +273°) of the gas; doubling the absolute temperature at constant pressure doubles the volume.

Boyle's law: At constant temperature the volume of a fixed number of moles of gas is inversely proportional to the pressure; doubling the pressure halves the volume.

These three laws may be combined to give

$$PV = nRT \tag{7.3.1}$$

which is the **general gas law** relating volume (V), pressure (P), number of moles (n), and absolute temperature (T) in which R is a constant.

The general gas law can be used to do calculations relating n, P, T, and V in the following relationship in which R is a constant and the subscripts represent two different sets of conditions:

$$R = \frac{P_1 V_1}{n_1 T_1} = \frac{P_2 V_2}{n_2 T_2} \qquad (7.3.2)$$

This equation can be arranged in a form that can be solved for a new volume resulting from changes in P, n, or T:

$$V_2 = V_1 \times \frac{n_2 T_2 P_1}{n_1 T_1 P_2} \qquad (7.3.3)$$

As an example, calculate the volume of a fixed number of moles of gas initially occupying 15.0 l when the temperature is changed from 10°C to 90°C at constant pressure. In order to use these temperatures, they must be changed to absolute temperature by adding 273°. Therefore, $T_1 = 10° + 273° = 283°$, and $T_2 = 90° + 273° = 363°$. Because n and P remain constant, they cancel out of the equation yielding

$$V_2 = V_1 \times \frac{T_2}{T_1} = 15.0\ l \times \frac{363°}{283°} = 19.2\ l \qquad (7.3.4)$$

As another example, consider the effects of a change of pressure at a constant temperature and number of moles. Calculate the new volume of a quantity of gas occupying initially 16.0 l at a pressure of 1.20 atm when the pressure is changed to 1.60 atm. In this case, both n and T remain the same and cancel out of the equation giving the following relationship:

$$V_2 = V_1 \times \frac{P_1}{P_2} = 16.0\ l \times \frac{1.20\ atm}{1.60\ atm} = 12.0\ l \qquad (7.3.5)$$

Note that an increase in temperature increases the volume and an increase in pressure decreases the volume.

7.4. THE ATMOSPHERE AS A MEDIUM FOR THE TRANSFER OF MASS AND ENERGY

Earth receives an enormous amount of energy from the sun. As shown in Figure 7.3, the solar flux of energy from the sun is 1340 W/m². Some of this energy is scattered back to space, some is absorbed by Earth's atmosphere, and some reaches Earth's surface directly as illustrated in Figure 7.4. The maximum intensity of incoming radiation occurs at 0.5 μm (500 nm) in the visible region, with essentially none outside the range of 0.2 μm to 3 μm. This range encompasses the whole visible

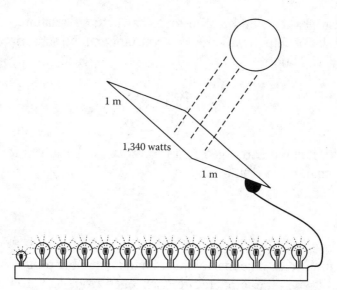

Figure 7.3.   A square meter of area perpendicular to incoming solar radiation at the top of the atmosphere receives energy from the sun at a power level of 1340 W, sufficient to power 13 100-W light bulbs plus a 40-W light bulb. This represents an enormous amount of energy, which is redistributed over Earth's surface and that must be radiated back into space.

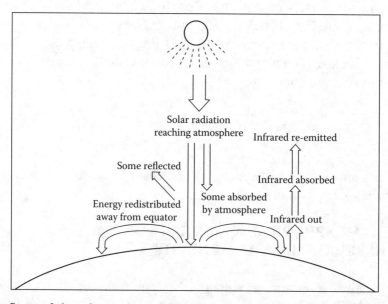

Figure 7.4.   Some of the solar energy reaching the top of the atmosphere penetrates to Earth's surface, some is absorbed in the atmosphere, which warms it, and some is scattered by the atmosphere and from Earth's surface. Outgoing energy is in the infrared region. Some of it is temporarily absorbed by the atmosphere before being radiated to space, causing a warming (greenhouse effect). The equatorial regions receive the most energy, part of which is redistributed by warm air masses and latent heat in water vapor away from the equator.

region and small parts of the ultraviolet and infrared adjacent to it. All of this energy must eventually be radiated back into space, which occurs primarily as infrared radiation at wavelengths between 2 μm and 40 μm with a maximum intensity around 10 μm. Thus the Earth loses energy by electromagnetic radiation of a much higher wavelength (lower energy per photon) than that of the radiation by which it receives energy. As the outbound radiation is transmitted through the atmosphere, some of it is reabsorbed by molecules of water, carbon dioxide, methane, and other minor species that have the ability to absorb infrared radiation. This temporary delay of the outbound energy has a warming effect on the atmosphere, commonly called the **greenhouse effect** and the carbon dioxide and other gases responsible for it are called **greenhouse gases**. This phenomenon is one of the crucial protective effects of the atmosphere that is responsible for keeping the average temperature of Earth's surface at about 15°C enabling life to exist on Earth. (Were this not the case, the surface temperature would average around –18°C.) However, as discussed in Chapter 9, excessive amounts of greenhouse gases carbon dioxide and methane released by human activities are almost certainly warming the global atmosphere leading to a major environmental and sustainability problem commonly called the "greenhouse effect" and "global warming."

The fraction of electromagnetic radiation from the sun that is reflected by Earth's surface is an important property called **albedo**. It is very important in determining how effective incoming radiation is in warming the surface. Freshly cultivated soil exposing a black surface to the sun has an albedo of only about 2.5% whereas that of fresh snow is about 90%. Anthrospheric effects on albedo can be substantial, such as occurs when urban areas are paved and covered with buildings creating "heat islands" that are several degrees warmer than surrounding land. Cultivation of soil in springtime can lead to warmer surface temperatures conducive to plant growth, whereas ground cover by green plants can cool the soil surface microclimate during summer's heat.

The maintenance of Earth's surface temperature is crucial to sustainability of the planet's life support systems and the mechanisms by which this critical balance is maintained are not completely understood. Geological and fossil records indicate that during long periods of time in the past Earth was significantly warmer than it is now, whereas in more recent times there were prolonged cold periods — ice ages — in which the temperature was lower. During the hotter periods, tropical jungles thrived in some areas and deserts covered large areas in other parts. During the ice ages, many of the areas closer to the Earth's poles were covered with ice a kilometer or more thick. It is distressing to contemplate that mean temperatures during these periods of climate extremes were only a few degrees lower than current temperatures. Convincing evidence also exists to suggest that past impacts by large asteroids blasted enormous amounts of material into the atmosphere leading to dark conditions in which photosynthetic production of plant biomass was minimal resulting in mass extinctions, such as those of the dinosaurs approximately 65 million years ago. Therefore, maintenance of climate conditions conducive to life on Earth is a key aspect of sustainability.

Weather and climate and related cycles, such as the hydrologic cycle, are driven by the redistribution of solar energy through the atmosphere. Solar energy impinges on Earth most directly at the equator, which is relatively warmer than regions farther north and south. The excess atmospheric energy in equatorial regions moves away from the equator, largely in moving masses of air, a process of **convection**. The heat in these air masses may be in the form of **sensible heat** due to the kinetic energy in rapidly moving air molecules or as **latent heat** in the form of water vapor. The heat absorbed when 1 gram of liquid water is converted to water vapor by the action of solar radiation is 2259 J, so the **heat of vaporization** of liquid water is 2259 J/g, the highest of any common liquid. When this water vapor condenses, in some cases hundreds or thousands of kilometers from where it was formed, an equal amount of heat is released. This can have a marked warming effect on the surrounding air. This kind of heat transfer accompanied by the rising and whirling motion of the heated air is the driving force behind tropical storms.

7.5. METEOROLOGY

Air moves in the atmosphere, picks up water vapor from oceans and other bodies of water, forms clouds, releases liquid water, undergoes pressure changes, expands, contracts, and is heated and cooled. The study of these phenomena occurring with atmospheric air constitutes the science of **meteorology**. Meteorological phenomena have significant effects on the atmosphere and on atmospheric chemistry as well as major influences on sustainability. Some examples of these effects are the following:

- Meteorology governs climate and determines if it is hospitable to life and conducive to biological productivity.

- Through the hydrologic cycle, meteorology determines rainfall amounts and distribution, crucial for crops and water availability.

- Air pollutants, such as sulfur and nitrogen oxides that produce acid rain, are transported in moving masses of air, whereas stationary masses of stagnant air that result during atmospheric temperature inversions can serve as vessels in which secondary pollutants, particularly photochemical smog, are formed.

- Wind, emerging as one of the most cost-effective renewable energy resources, is a meteorological phenomenon, and regions with strong, consistent winds will become significant energy-producing areas.

Atmospheric chemical processes can influence meteorological phenomena. The most obvious example of this is the formation of rain droplets around pollutant particles in the atmosphere.

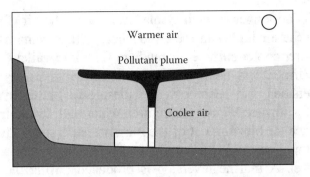

Figure 7.5.   Topographical features, such as a mountain range, may hinder movement of air. Temperature inversions may trap a body of air in which air contaminants may react to produce secondary pollutants.

Topographical Effects

Topography, the surface configuration and relief features of the Earth's surface may strongly affect winds and air currents. Differential heating and cooling of land surfaces and bodies of water can result in **local convective winds**, including land breezes and sea breezes at different times of the day along the seashore, as well as breezes associated with large bodies of water inland. Mountain topography causes complex and variable localized winds. The masses of air in mountain valleys heat up during the day causing upslope winds and cool off at night causing downslope winds. Upslope winds flow over ridge tops in mountainous regions. The blocking of wind and of masses of air by mountain formations some distance inland from seashores can trap bodies of air, particularly when temperature inversion conditions occur as illustrated in Figure 7.5.

Temperature Inversions

The complicated movement of air across the Earth's surface is a crucial factor in determining weather as well as the creation and dispersal of air pollution phenomena. When air movement ceases, air stagnation can occur with a resultant buildup of air pollutants in localized regions. Although the temperature of air relatively near the Earth's surface normally decreases with increasing altitude, certain atmospheric conditions can result in the opposite condition — increasing temperature with increasing altitude. Such conditions are characterized by high atmospheric stability and are known as **temperature inversions**. Because they limit the vertical circulation of air, temperature inversions result in air stagnation and the trapping of air pollutants in localized areas.

Inversions can occur in several ways. In a sense, the whole atmosphere is inverted by the warm stratosphere, which floats atop the troposphere, with relatively little mixing. An inversion can form when a warm air mass (warm front) overrides a cold air mass (cold front). Radiation inversions are likely to form in still air at night when the Earth is no longer receiving solar radiation. The air closest to the Earth cools faster than the air higher in the atmosphere, which remains warm, thus less dense. Furthermore, cooler surface air tends to flow into valleys at night, where it is overlain by warmer, less dense air. Subsidence inversions, often accompanied by radiation inversions, can become very widespread. These inversions can form in the vicinity of a surface high-pressure area when high-level air subsides to take the place of surface air blowing out of the high-pressure zone. The subsiding air is warmed as it compresses and can remain as a warm layer several hundred meters above ground level. A marine inversion is produced during the summer months when cool air laden with moisture from the ocean blows onshore and under warm, dry inland air.

As noted above, inversions contribute significantly to the effects of air pollution. This is because, as shown in Figure 7.5, inversions prevent mixing of air pollutants, thus keeping the pollutants in one area. This not only prevents the pollutants from escaping, but also acts like a container in which additional pollutants accumulate. Furthermore, in the case of secondary pollutants formed by atmospheric chemical processes, such as photochemical smog, the pollutants may be kept together such that they react with each other and with sunlight to produce even more noxious products.

Influence of Ocean Currents

The circulation of water in ocean currents significantly affects meteorology. The most prominent such effect is that of the **Gulf Stream**, which carries warm water from the Gulf of Mexico to northern Europe. Heat transferred from the Gulf Stream water to air in Europe serves to make much of Europe far more hospitable to human habitation than it would otherwise be. The cooled water sinks and moves westward in a current moving in the direction of Labrador before turning south to return to the Gulf of Mexico. Other massive oceanic water circulation patterns occur in the Pacific Ocean and affect weather along the Pacific coast. One manifestation of this pattern in the Pacific is a warm mass of surface water in the eastern tropical Pacific Ocean known as El Niño.

Oscillations in the circulation of oceanic water lasting two or three decades are associated with weather cycles on land. In Europe, a cool period in the 1950s and 1960s was followed by two decades of comparatively warm weather. In 1996, a distinct cooling of the water in the North Atlantic current resulted in a very cold winter in Europe. On a longer time span large fluctuations in the climate of land bordering the North Atlantic have probably been associated with temperature fluctuations in the circulating ocean water. Several centuries of unusually warm weather in medieval times enabled crops to thrive in England and allowed the Vikings to establish colonies in Greenland. These times were followed by about 300 years of unusually cold weather, the "Little Ice Age" that caused great hardship in Europe.

A major concern with sustainability is the potential effect of greenhouse warming and other effects of humans on ocean circulation patterns that might result in marked modifications in climate. A particular concern is that melting of polar and Greenland ice could change the salinity of Atlantic Ocean water resulting in a shifting of the Gulf Stream. This could cause significant cooling of the fragile European climate and conditions resembling those of the ice age.

Global Air Circulation

The factors discussed above that determine and describe the movement of air masses are involved in the massive movement of air, moisture, and energy that occurs globally. The central feature of global weather is the redistribution of solar energy that falls unequally on Earth at different latitudes (relative distances from the equator and poles). Consider Figure 7.6. Sunlight and its energy flux are most intense at the equator because, averaged over the seasons, solar radiation comes in perpendicular to Earth's surface at the equator. With increasing distance from the equator (higher latitudes), the angle is increasingly oblique and more of the energy-absorbing atmosphere must be traversed, so that progressively less energy is received per unit area of Earth's surface. The net result is that equatorial regions receive a much greater share of solar radiation, progressively less is received farther from the equator, and the poles receive a comparatively miniscule amount. The excess heat energy in the equatorial regions causes the air to rise. The air ceases to rise when it reaches the stratosphere because in the stratosphere the air becomes warmer with higher elevation. As the hot equatorial air rises in the troposphere, it cools by expansion and loss of water, then sinks again. The air circulation patterns in which this occurs are called **Hadley cells**. As shown in Figure 7.6, there are three major groupings of these

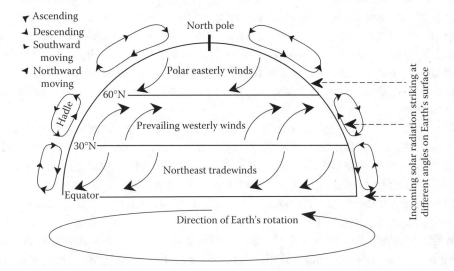

Figure 7.6.   Global circulation of air in the northern hemisphere.

cells, which result in very distinct climatic regions on Earth's surface. The air in the Hadley cells does not move straight north and south, but is deflected by Earth's rotation and by contact with the rotating Earth; this is the **Coriolis effect**, which results in spiral-shaped air circulation patterns, called cyclonic or anticyclonic, depending upon the direction of rotation. These give rise to different directions of prevailing winds, depending on latitude. The boundaries between the massive bodies of circulating air shift markedly over time and season, resulting in significant weather instability.

The movement of air in Hadley cells combined with other atmospheric phenomena results in the development of massive **jet streams** that are in a sense, shifting rivers of air that may be several kilometers deep and several tens of kilometers wide. Jet streams move through discontinuities in the tropopause, the cold layer at the top of the troposphere, generally from west to east at velocities around 200 km/h (well over 100 m/h); in so doing, they redistribute huge amounts of air and have a strong influence on weather patterns.

The air and wind circulation patterns described above redistribute massive amounts of energy. If it were not for this effect, the equatorial regions would be unbearably hot, and the regions closer to the poles intolerably cold. About half of the heat that is redistributed is carried as sensible heat by air circulation, almost 1/3 is carried by water vapor as latent heat, and the remaining approximately 20% by ocean currents.

7.6. WEATHER, WEATHER FRONTS, AND STORMS

Short-term variations in the meteorologic state of the atmosphere are described as **weather**. The weather is defined in terms of seven major factors: temperature, clouds, winds, humidity, horizontal visibility (as affected by fog, etc.), type and quantity of precipitation, and atmospheric pressure. All of these factors are closely interrelated. Longer-term variations and trends within a particular geographical region in those factors that compose weather are described as **climate** (Section 7.7). For example, the climate in desert regions of the world may be relatively hot and dry, but the weather in such regions may at times produce torrential rainfall or frigid temperatures.

Weather is driven by redistribution of energy in the atmosphere. A particularly important aspect of this redistribution is the energy released when precipitation forms. This energy can be enormous because of the high heat of vaporization of water. As an example, heat energy from sunlight and from hot masses of air is converted to latent heat by the evaporation of ocean water off the west coast of Africa. Prevailing winds drive masses of air laden with water vapor westward across the ocean. Rainfall forms, releasing the energy from the latent heat of water and warming the air mass. The hot mass of air that results rises, creating a region of low pressure into which air flows in a circular manner. This can result in the formation of a whirling mass of air in the form of a hurricane that may strike Puerto Rico, Cuba, Florida, or other areas thousands of miles from the area where the water was originally evaporated from the ocean.

A very obvious manifestation of weather consists of very small droplets of liquid water composing **clouds**. These droplets may coalesce under the appropriate conditions to form raindrops large enough to fall from the atmosphere. Condensation of water vapor, which forms clouds, must occur prior to the formation of precipitation in the form of rain or snow. For this condensation to happen, air must be cooled below a temperature called the **dew point**, and **condensation nuclei** must be present. These nuclei are hydroscopic substances such as salts, sulfuric acid droplets, and some organic materials including bacterial cells. Air pollution in some forms is a significant source of condensation nuclei and may thus influence weather and climate. As water condenses from atmospheric air, large quantities of heat are released. This is a particularly significant means for transferring energy from the ocean to land. Solar energy falling on the ocean is converted to latent heat by the evaporation of water, then the water vapor moves inland where it condenses. The latent heat released when the water condenses warms the surrounding land mass.

Clouds may absorb infrared radiation from Earth's surface, warming the atmosphere, but they also reflect visible light, which has a cooling effect. Because air pollutants such as H_2SO_4 are instrumental in forming clouds, air pollution may affect atmospheric temperature. It is believed that atmospheric H_2SO_4 pollution has a generally cooling effect on the atmosphere because it increases light-reflective cloud cover.

Air masses characterized by pressure, temperature, and moisture contents flow from regions of high atmospheric pressure to regions of low pressure. The boundaries between air masses are called **fronts**. The movement of air associated with horizontally moving air masses is **wind** and vertically moving air is an **air current**. Atmospheric air moves constantly, with behavior and effects that reflect the laws governing the behavior of gases. First of all, as shown in Figure 7.7, gases will move horizontally and/or vertically from regions of high atmospheric pressure to those of low atmospheric pressure. Furthermore, expansion of gases causes cooling, whereas compression causes warming. A mass of warm air tends to move from Earth's surface to higher altitudes, where the pressure is lower; in so doing, it expands adiabatically (that is, without exchanging energy with its surroundings) and becomes cooler. If there is no condensation of moisture from the air, the cooling effect is about 10°C per 1000 m of altitude, a figure known as the **dry adiabatic lapse rate**. A cold mass of air at a higher altitude does the opposite; it sinks and becomes warmer at about 10°C/1000 m. Often, however, when there is sufficient moisture in rising air, water condenses from it, releasing latent heat. This partially counteracts the cooling effect of the expanding air, giving a **moist adiabatic lapse rate** of about 6°C/1000 m. Parcels of air do not rise and fall, or even move horizontally, in a completely uniform way, but exhibit eddies, currents, and various degrees of turbulence.

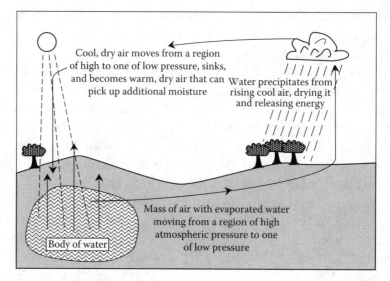

Figure 7.7.   Circulation patterns involved with movement of air masses and water and the uptake and release of solar energy as latent heat in water vapor.

Wind is involved in the movement of pollutants from one place to another and is also responsible for dispersing pollutants to harmless levels. Prevailing wind direction is an important factor in determining the areas most affected by an air pollution source. Wind plays an important role in the propagation of life by dispersing pollen, spores, seeds, and organisms, such as spiders. An enormous amount of energy is contained in wind, and it can be harnessed to generate electricity (see Figure 18.16 and the discussion of renewable energy resources in Section 18.16).

The interface between two masses of air that differ in temperature, density, and water content is called a **front**. A mass of cold air moving such that it displaces one of warm air is a **cold front**, and a mass of warm air displacing one of cold air is a **warm front**. Because cold air is denser than warm air, the air in a cold mass of air along a cold front pushes under warmer air. This causes the warm, moist air to rise, cooling it such that water condenses from it. The condensation of water releases energy, heating the air so that it rises farther. The net effect can be formation of massive cloud formations (thunderheads) that may reach stratospheric levels. These spectacular thunderheads may produce heavy rainfall and even hail, and sometimes violent storms with strong winds, including tornadoes. Warm fronts cause somewhat similar effects as warm, moist air pushes over colder air. However, the front is usually much broader, and the weather effects milder, typically resulting in widespread drizzles, rather than intense rainstorms.

Swirling **cyclonic storms**, such as typhoons, hurricanes, and tornadoes, are created in low-pressure areas by rising masses of warm, moist air. As such air cools, water vapor condenses, and the latent heat released warms the air more, sustaining and intensifying its movement upward in the atmosphere. Air rising from surface level creates a low-pressure zone into which surrounding air moves. The movement of the incoming air assumes a spiral pattern, thus causing a cyclonic storm.

7.7. CLIMATE

Perhaps the single most important influence on Earth's environment is **climate**, consisting of long-term weather patterns over large geographical areas. Climate is a particularly important aspect of sustainability and one that is susceptible to adverse influences by humans, such as by emissions of greenhouse warming gases. Although Earth's atmosphere is huge and has an enormous ability to resist and correct for detrimental change, it is apparent that human activities are reaching a point at which they are adversely affecting climate. Human effects on climate are addressed in Chapter 9, which discusses "Sustaining an Atmosphere for Life on Earth."

As a general rule, climatic conditions are characteristic of a particular region. This does not mean that climate remains the same throughout the year, of course, because it varies with season. One important example of such variation is the **monsoon**, seasonal variations in wind patterns between oceans and continents. The climates of Africa and the Indian subcontinent are particularly influenced by monsoons. In the latter, for example, summer heating of the Indian land mass causes air to rise, thereby creating a low-pressure area that attracts warm, moist air from the ocean. This air rises on the slopes of the Himalayan mountains, which also block the flow of colder air from the north, moisture from the air condenses, and monsoon rains carrying enormous amounts of precipitation fall. Thus, from May until into August, summer monsoon rains fall in India, Bangladesh, and Nepal. Reversal of the pattern of winds during the winter months causes these regions to have a dry season, but produces winter monsoon rains in the Philippine islands, Indonesia, New Guinea, and Australia.

Summer monsoon rains are responsible for tropical rain forests in central Africa. The interface between this region and the Sahara desert shifts over time. When the boundary is relatively far north, rain falls on the Sahel desert region at the interface, crops grow, and the people do relatively well. When the boundary is more to the south, a condition that may last for several years, devastating droughts and even starvation may occur.

It is known that there are fluctuations, cycles, and cycles imposed on cycles in climate. The causes of these variations are not completely understood, but they are known to be substantial, and even devastating to civilization. The last **Ice Age**, which ended only about 10,000 years ago and which was preceded by several similar ice ages, produced conditions under which much of the present land mass of the Northern Hemisphere was buried under thick layers of ice and uninhabitable. A "mini-ice age" occurred during the 1300s, causing crop failures and severe hardship in northern Europe. In modern times, the El Niño Southern Oscillation occurs over a period of several years when a large, semipermanent tropical low-pressure area shifts into the central Pacific region from its more common location in the vicinity of Indonesia. This shift modifies prevailing winds, changes the pattern of ocean currents, and affects upwelling of ocean nutrients with profound effects on weather, rainfall, and fish and bird life over a vast area of the Pacific from Australia to the west coasts of South and North America.

7.8. MICROCLIMATE

The preceding section described climate on a large scale, ranging up to global dimensions. The climate that organisms and objects on the surface are exposed to close to the ground, under rocks, and surrounded by vegetation, is often quite different from the surrounding macroclimate. Such highly localized climatic conditions are termed the **microclimate**. Microclimate effects are largely determined by the uptake and loss of solar energy very close to Earth's surface and by the fact that air circulation due to wind is much lower at the surface. During the day, solar energy absorbed by relatively bare soil heats the surface, but is lost only slowly because of very limited air circulation at the surface. This provides a warm blanket of surface air several centimeters thick, and a relatively thin layer of warm soil. At night, radiative loss of heat from the surface of soil and vegetation can result in surface temperatures several degrees colder than the air about 2 m above ground level. These lower temperatures result in condensation of **dew** on vegetation and the soil surface, thus providing a moister microclimate near ground level. Heat absorbed during early morning evaporation of the dew tends to prolong the period of cold experienced right at the surface.

Vegetation substantially affects microclimate. In denser growths, circulation may be virtually zero at the surface because vegetation severely limits convection and diffusion. The crown surface of the vegetation intercepts most of the solar energy, so that maximum solar heating may be a significant distance up from Earth's surface. The region below the crown surface of vegetation thus becomes one of relatively stable temperature. In addition, in a dense growth of vegetation, most of the moisture loss is not from evaporation from the soil surface, but rather from transpiration from plant leaves. The net result is the creation of temperature and humidity conditions that provide a favorable living environment for a number of organisms, such as insects and rodents.

Another factor influencing microclimate is the degree to which the slope of land faces north or south. South-facing slopes of land in the northern hemisphere receive greater solar energy. Advantage has been taken of this phenomenon in restoring land strip-mined for brown coal in Germany by terracing the land such that the terraces have broad south slopes and very narrow north slopes. On the south-sloping portions of the terrace, the net effect has been to extend the short summer growing season by several days, thereby significantly increasing crop productivity. In areas where the growing season is longer, better growing conditions may exist on a north slope because it is less subject to temperature extremes and to loss of water by evaporation and transpiration.

A particularly marked effect on microclimate is that induced by urbanization. In a rural setting, vegetation and bodies of water have a moderating effect, absorbing modest amounts of solar energy and releasing it slowly. The stone, concrete, and asphalt pavement of cities have an opposite effect, strongly absorbing solar energy, and reradiating heat back to the urban microclimate. Rainfall is not allowed to accumulate in ponds, but is drained away as rapidly and efficiently as possible. Human activities generate significant amounts of heat and produce large quantities of CO_2 and other greenhouse gases that retain heat. The net result of these effects is that a city is capped by a **heat dome** in which the temperature is as much as 5°C warmer than in the surrounding rural areas, such that large cities have been described as "heat islands." The rising warmer air over a city brings in a breeze from the surrounding area and causes a local greenhouse effect that probably is largely counterbalanced by reflection of incoming solar energy by particulate matter above cities. Overall, compared to climatic conditions in nearby rural surroundings, the city microclimate is warmer and foggier, overlain with more cloud cover a greater percentage of the time, subject to more precipitation, and generally less humid.

This discussion suggests that sustainability can be enhanced by measures that improve the microclimate. Several such measures, such as south-facing slopes of land in northern regions to extend the growing season, are mentioned above. Some agricultural practices can be used to enhance microclimate. For example, the microclimate in fields of perennial plants, for example, grasslands used to produce hay, is milder and conserves moisture better than fields used to grow monocultures of annual crops. In cities, some structures can be covered with sod that serves to moderate temperatures and conserve moisture. The microclimate produced by constructed wetlands can favor bioproductivity and esthetics.

7.9. ATMOSPHERIC PARTICLES

Particles are common significant components of the atmosphere, particularly the troposphere. Colloidal-sized particles in the atmosphere are called **aerosols**. Most aerosols from natural sources have a diameter of less than 0.1 μm. These particles originate in nature from sea sprays, smokes, dusts, and the evaporation of organic materials from vegetation. Other typical particles of natural origin in the atmosphere are bacteria, fog, pollen grains, and volcanic ash.

As shown in Figure 7.8, atmospheric particles undergo a number of processes in the atmosphere. Small colloidal particles are subject to diffusion processes. Smaller particles coagulate together to form larger particles. Sedimentation and scavenging by raindrops and other forms of precipitation are the major mechanisms by which particles are removed from the atmosphere. Particles also react with atmospheric gases.

Many important atmospheric phenomena involve aerosol particles, including electrification phenomena, cloud formation, and fog formation. Particles help determine the heat balance of the Earth's atmosphere by reflecting light. Probably the most important function of particles in the atmosphere is their action as nuclei for the formation of ice crystals and water droplets. Current efforts at rain-making are

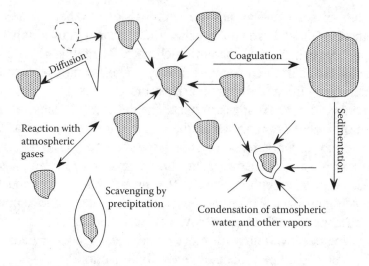

Figure 7.8.   Processes that particles undergo in the atmosphere.

centered around the addition of condensing particles to atmospheres supersaturated with water vapor. Dry ice was used in early attempts; later, silver iodide, which forms huge numbers of very small particles, has been used.

Particles are involved in many chemical reactions in the atmosphere. Neutralization reactions, which occur most readily in solution, may take place in water droplets suspended in the atmosphere. Small particles of metal oxides and carbon have a catalytic effect on oxidation reactions. Particles may also participate in oxidation reactions induced by light.

7.10. THE ATMOSPHERE AS A RESERVOIR OF NATURAL CAPITAL

As discussed in Chapter 1, **natural capital** consists of the resources that Earth and its ecosystems provide that are potentially useful to humans and that include not only materials and energy, but waste assimilative capacity and esthetics as well. The atmosphere is a huge part of natural capital. It provides a variety of materials, is an enormous reservoir for by-products and wastes, and largely determines the degree to which our surroundings are pleasant and conducive to our existence. These aspects of natural capital are discussed briefly in this section.

The first aspect of the atmosphere's contribution to natural capital is its protective and regulatory function, keeping Earth's surface at an acceptably warm temperature through the greenhouse effect discussed in this chapter and in more detail in Chapter 9, and protecting life on Earth from the destructive effects of solar and cosmic radiation. Without the greenhouse effect, the blue color of Earth due to its vast liquid oceans would be a sterile white because the water, including the oceans' surfaces, would be ice. Without the ability of the atmosphere to absorb incoming destructive radiation, exposed life would not be possible.

The second major aspect of Earth's natural capital is that it is a source of essential raw materials. Humans and most other nonphotosynthetic organisms use the atmosphere's inexhaustible pool of oxygen for their respiration. Plants get their carbon for photosynthesis from the relatively small pool of carbon dioxide in the atmosphere. Fixed nitrogen essential for life is extracted from the atmosphere by bacteria that convert the refractory elemental nitrogen molecule to nitrogen compounds that plants and ultimately humans can use in synthesizing protein. Atmospheric nitrogen is also fixed in the anthrosphere in the chemical synthesis of fertilizers, explosives, and other nitrogen compounds. Elemental nitrogen is extracted from the atmosphere and liquefied to use in cryogenics, the technology of very cold materials; people are alive today who were once stored as frozen embryos in a liquid nitrogen bath. Simultaneously, pure oxygen is extracted for industrial uses and for breathing by people with respiratory problems. Noble gas argon is also extracted from the atmosphere to serve as an inert atmosphere in specialized welding and other industrial applications.

Although oceans are the ultimate source of water, the atmosphere is the conduit that is mainly responsible for delivering this essential resource to organisms and humans that depend on it for their existence. Conditions in the atmosphere largely determine the quantity, quality, and distribution of water as it is carried through the hydrologic cycle. Hot, dry climatic conditions result in droughts that periodically cause great hardship and even starvation in some areas, particularly in Africa. Contamination with acid from acid-forming gases containing sulfur or nitrogen can degrade the quality of water from rainfall and has even caused loss of trees and fish fingerlings in some areas. Although the total amount of precipitation falling from the atmosphere is sufficient for overall global needs, its distribution in time and space is often such that water is in short supply and, in some cases, in excess.

A large part of the atmosphere's natural capital is its ability to assimilate materials. This is part of nature's crucial cycles. Water carried from soil to plant leaf surfaces and discharged to the atmosphere is part of the transpiration process by which water and the nutrients carried by it are transported to plants. Humans and other organisms that carry out aerobic respiration discharge by-product carbon dioxide to the atmosphere. Plants produce elemental oxygen by photosynthesis and discharge it to the atmosphere. The carbon dioxide produced by combustion in engines, heating appliances, forest fires, and other processes is discharged to the atmosphere and is a significant part of the atmosphere's crucial nitrogen balance. Pollen and combustion particles from natural sources are discharged to the atmosphere and later deposited or washed out with rain. Ever since the beginning of the industrial revolution, the anthrosphere has been a major source of input of often harmful materials to the atmosphere. In addition to carbon dioxide, these have included gaseous nitrogen and sulfur oxides, particles in the form of smoke and fumes, and hydrocarbon products of incomplete combustion, such as auto exhaust by-products that lead to photochemical smog formation.

Not the least of the atmosphere's natural capital is its contribution to esthetics. Clear, clean air unsullied by visibility-obscuring particles has real value to humans. So does air free of acidic gases, aldehydes, and ozone that hinder breathing and irritate eyes. Unlike water, which may come from a mud-laden or polluted sources and brought up to drinking water standards by purification processes, the atmosphere is generally utilized as it comes without the intermediate of artificial purification processes. An atmosphere burdened by fog or persistent drizzling rain resulting from air pollution is unpleasant. Less directly, but still important, is an atmosphere that has not been heated or dehydrated to uncomfortable levels by greenhouse gas emissions.

SUPPLEMENTARY REFERENCES

Barry, Roger G. and Richard J. Chorley, *Atmosphere, Weather, and Climate*, 8th ed., Routledge, New York, 2004.

Calhoun, Yael, *Climate Change*, Chelsea House Publishers, Philadelphia, 2005.

Gallant, Roy A., *Atmosphere: Sea of Air*, Benchmark Books, New York, 2003.

Jennings, Terry, *Atmosphere and Weather*, Smart Apple Media, North Mankato, MN, 2005.

Lutgens, Frederick K. and Edward J. Tarbuck, *The Atmosphere: An Introduction to Meteorology*, 9th ed., Prentice Hall, Upper Saddle River, NJ, 2004.

Michel, David, Ed., *Climate Policy for the 21st Century: Meeting the Long-Term Challenge of Global Warming*, Center for Transatlantic Relations, Washington, D.C., 2004.

Toman, Michael A. and Brent Sohngen, Eds., *Climate Change*, Aldershot Publishing, Burlington, VT, 2004.

QUESTIONS AND PROBLEMS

1. What would you consider to be the two major protective functions of the atmosphere? What would be the status of life on Earth without these two functions?

2. Clouds near ground level are composed of very small droplets of liquid water. What would you expect to be the nature of clouds very high in the troposphere? Why?

3. The reaction, $O + O_2 + M \rightarrow O_3 + M$, is often written for the formation of stratospheric ozone. This reaction releases a lot of energy and M is an energy absorbing third body. Suggest how M enables ozone to form. What species is M most likely to be? What would likely be the second most abundant M species?

4. Dry air is 78.08% N_2 and 20.95% O_2 by volume. What are these percentages for air that is 2.5% H_2O by volume?

5. Calculate the new volumes (V_2) of gas for each of the cases below in which temperature, pressure, or both may be changed:

V_1	T_1	T_2	P_1	P_1	V_2
15.0 l	10°C	115°C	1.00 atm	1.00 atm	_____
15.0 l	210°C	–35°C	1.00 atm	1.00 atm	_____
20.0 l	10°C	10°C	1.00 atm	0.80 atm	_____
20.0 l	10°C	10°C	0.25 atm	1.10 atm	_____
18.0 l	15°C	125°C	1.10 atm	0.85 atm	_____
18.0 l	10°C	115°C	1.00 atm	1.00 atm	_____

6. Assuming that only 30% of solar energy at the full zenith of the sun actually reaches Earth's surface and that it can be converted to electricity by a photovoltaic cell with 15% efficiency, calculate the area of a photovoltaic cell necessary to produce the electricity required to power a 100-W light bulb under those circumstances. List the reasons why photovoltaic electrical power production is limited.

7. Why are average temperatures not exactly what would be expected from the intensity of sunlight that they receive at some distance from the equator? Are they, on average, warmer or colder? Why?

8. Suggest ways in which humans may change the albedo of urban areas to moderate temperatures in these areas.

9. Using the Internet or library resources, look up evidence of past mass extinctions resulting from asteroid impacts. How often do such events occur? What would be the likely result of the impact of an asteroid several kilometers in diameter?

10. What is the likely impact of topography on air pollution phenomena? On rainfall? (In order to answer this question completely, it may be necessary to refer to previous chapters.)

11. How do temperature inversions affect air pollution phenomena?

12. Suggest how the Gulf Stream may have been responsible for significant advances in civilization. An argument can be made that the deployment of air conditioning is responsible for much of the economic progress of the southern United States. Suggest a rationale for this idea.

13. One might think that air heated at the equator would blow straight north in the north of the equator and straight south in the south. Is that what happens? If not, why not?

14. What is the main driving force behind weather?

15. How is water vapor involved in transferring solar energy absorbed by oceans from oceanic regions to land?

16. What is the distinction between wind and air currents? What is the driving force behind these phenomena?

17. What role does wind play in the biosphere? What potentially great role does it have in the anthrosphere?

18. What is the relationship between weather and climate?

19. What is microclimate? How do plants affect microclimate? What is a heat island in microclimate?

20. Describe the ways in which particles influence atmospheric chemistry.

21. List the major natural capital contributions made by the atmosphere.

22. What are the major materials extracted from the atmosphere for use in the anthrosphere?

8. ENVIRONMENTAL CHEMISTRY OF THE ATMOSPHERE

8.1. INTRODUCTION TO ATMOSPHERIC CHEMISTRY

The quality of the air that living organisms breathe, the nature and level of air pollutants, visibility and atmospheric esthetics, and even climate are dependent upon chemical phenomena that occur in the atmosphere. These, in turn are strongly tied to absorption of solar energy, interactions between the gas phase of the atmosphere and small solid particles suspended in it, and interchange of chemical species with the geosphere (see Figure 8.1). Chemical reactions in the atmosphere and the chemical nature of atmospheric chemical species are the topic of atmospheric chemistry, which is introduced in this chapter.

Atmospheric chemistry involves the unpolluted atmosphere, highly polluted atmospheres, and a wide range of gradations in between. The same general phenomena govern all and produce one huge atmospheric cycle, in which there are numerous subcycles. Gaseous atmospheric chemical species fall into the following somewhat arbitrary and overlapping classifications: inorganic oxides (CO, CO_2, NO_2, SO_2); oxidants (O_2); reductants (CO, SO_2, H_2S); organics (in the unpolluted atmosphere, CH_4 is the predominant organic species, whereas alkanes, alkenes, and aromatic compounds are common around sources of organic pollution); photochemically active species (NO_2, formaldehyde); acids (H_2SO_4); bases (NH_3); salts (NH_4HSO_4); and unstable reactive species (electronically excited NO_2, HO• radical). In addition, both solid and liquid particles play a strong role in atmospheric chemistry as sources and sinks for gas-phase species, as sites for surface reactions (solid particles), and as bodies for aqueous-phase reactions (liquid droplets). Two constituents of utmost importance in atmospheric chemistry are radiant energy from the sun, predominantly in the ultraviolet region of the spectrum, and the hydroxyl radical, HO•. The former provides a way to pump a high level of energy into individual gas molecules to start a series of atmospheric chemical reactions, and the latter is the most important reactive intermediate and "currency" of daytime atmospheric chemical phenomena.

An important aspect of atmospheric chemistry is that many of the processes occur in the gas phase where molecules are relatively far apart. Therefore, some

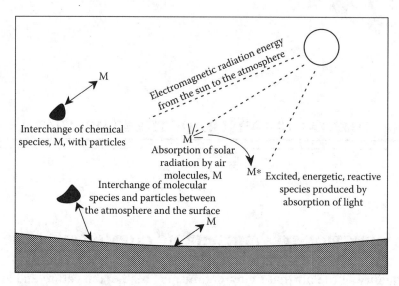

Figure 8.1.   Illustration of atmospheric chemical processes.

reactive species can exist for significantly longer times before reacting than they would in water or in solids. This is especially true in the highly rarefied regions of the stratosphere and above.

8.2. PHOTOCHEMICAL PROCESSES

The absorption of electromagnetic solar radiation ("light," usually in the ultraviolet region of the electromagnetic spectrum) by chemical species may cause **photochemical reactions** to occur. Photochemical reactions give atmospheric chemistry a unique quality and largely determine the nature and ultimate fate of atmospheric chemical species. The ability of electromagnetic radiation to cause photochemical reactions to occur is a function of its energy, E, which increases with increasing frequency (ν) and decreasing wavelength (λ) according to the relationship,

$$E = h\nu \tag{8.2.1}$$

where h is Planck's constant, 6.63×10^{-34} joule-seconds (J/sec). In order for a photochemical reaction to occur, a single unit of photochemical energy called a **quantum** and having an energy of $h\nu$, must be absorbed by the reacting species. If the absorbed light is in the visible region of the sun's spectrum, the absorbing species is colored. Colored NO_2 is a common example of such a species in the atmosphere.

Nitrogen dioxide, NO_2, is one of the most photochemically active species found in a polluted atmosphere. When a molecule such as NO_2 absorbs radiation of energy $h\nu$,

$$NO_2 + h\nu \rightarrow NO_2^* \tag{8.2.2}$$

an **electronically excited molecule** designated in the reaction above by an asterisk (*) may be produced. The photochemistry of nitrogen dioxide is discussed in greater detail later in this chapter.

Electronically excited molecules and atoms are reactive and unstable species in the atmosphere that participate in a wide range of atmospheric chemical processes. Two other generally reactive and unstable species in the atmosphere are **free radicals** composed of atoms or molecular fragments with unshared electrons, and **ions** consisting of charged atoms or molecular fragments. The participation of these three kinds of species in atmospheric chemical processes is discussed in later sections of this chapter.

Electronically excited molecules produced when unexcited ground-state molecules absorb energetic electromagnetic radiation in the ultraviolet or visible regions of the spectrum may possess several possible excited (energized) states. Generally, however, ultraviolet or visible radiation is energetic enough to excite molecules only to several of the lowest energy levels. Because they are energized compared to the ground state, excited chemical species are reactive. Their participation in atmospheric chemical reactions, such as those involved in smog formation, is discussed later in detail.

Electromagnetic radiation absorbed in the infrared region is not sufficiently energetic to break chemical bonds, but does cause the receptor molecules to vibrate and rotate, so it is said that they gain vibrational and rotational energy. The energy absorbed as infrared radiation ultimately is dissipated as heat and raises the temperature of the whole atmosphere. As discussed in Chapter 7, the absorption of infrared radiation is very important in the retention of energy radiated from the Earth's surface.

The reactions that occur following absorption of a photon of light sufficiently energetic to produce an electronically excited species are largely determined by the way in which the excited species loses its excess energy. This may occur by one of several processes divided into two general classes. Of these, **photophysical processes** are those that do not involve chemical bond breakage or loss of electrons and include loss of energy from the excited molecule by electromagnetic radiation as it returns to the ground state (fluorescence or phosphorescence), transfer of energy to other molecules, or transfer of energy within the absorbing molecule; the last two processes result in dissipation of the excess energy as heat. **Photochemical reactions** occur as a result of de-excitation processes that involve chemical bond breakage or ion formation, particularly the following:

- **Photodissociation** of the excited molecule (the process responsible for the predominance of atomic oxygen in the upper atmosphere)

$$O_2{}^* \rightarrow O + O \qquad (8.2.3)$$

- **Direct reaction** with another species

$$O_2^* + O_3 \rightarrow 2O_2 + O \qquad (8.2.4)$$

- **Photoionization** through loss of an electron

$$N_2^* \rightarrow N_2^+ + e^- \qquad (8.2.5)$$

Insofar as gas-phase reactions in the troposphere are concerned, the most important of the processes listed above is photodissociation. This is because photodissociation converts relatively stable and unreactive molecular species to reactive atoms and free radicals that participate in additional reactions, including chain reactions (see Section 8.3).

8.3. CHAIN REACTIONS IN THE ATMOSPHERE: HYDROXYL AND HYDROPEROXYL RADICALS IN THE ATMOSPHERE

As noted in the preceding section, energetic electromagnetic radiation in the atmosphere may produce atoms or groups of atoms with unpaired electrons called free radicals:

$$\overset{\displaystyle O}{\underset{}{H_3C-\overset{\|}{C}-H}} + h\nu \rightarrow H_3C\bullet + H\overset{\bullet}{C}O \qquad (8.3.1)$$

In the formula above, the single dot (•) represents the unpaired electron that makes free radicals so chemically reactive. Free radicals are involved with most significant atmospheric chemical phenomena and are of the utmost importance in the atmosphere. Because of their unpaired electrons and the strong pairing tendencies of electrons under most circumstances, free radicals are highly reactive; therefore, they generally have short lifetimes. It is important to distinguish between high reactivity and instability. A totally isolated free radical or a single atom, such as an O atom, would be quite stable; it wants to react, but there is nothing around for it to react with. Therefore, free radicals and single atoms from diatomic gases (such as O from O_2) tend to persist under the rarefied conditions of very high altitudes because they can travel long distances before colliding with another reactive species. However, unlike free radicals, electronically excited species have finite, generally very short lifetimes because they can lose energy through emission of photons of electromagnetic radiation without having to react with another species.

A key aspect of chemical processes in the atmosphere is that of **chain reactions**. These occur when a series of reactions involving particular reactive intermediates, usually free radicals, goes through a number of cycles. Most commonly, an atmospheric chain reaction begins with the photochemical dissociation of a molecular species to form free radicals, proceeds through a series of reactions that each generates additional free radicals, and ends with a **chain-terminating reaction** when free radicals react to form stable species as shown by the following reaction of two methyl radicals:

$$H_3C\bullet + H_3C\bullet \rightarrow C_2H_6 \tag{8.3.2}$$

An example of an important chain reaction sequence occurs with chlorofluorocarbons (Freons) in the stratosphere. Extremely stable chlorofluorocarbons consisting of carbon atoms to which are bonded fluorine and chlorine atoms were once widely used as refrigerant fluids in air conditioners, as aerosol propellants for products such as hair spray, and for foam blowing to make very porous plastic or rubber foams. Dichlorodifluoromethane, CCl_2F_2, was used in automobile air conditioners. Released to the atmosphere, this compound remained as a stable atmospheric gas until it got to very high altitudes in the stratosphere. In this region, ultraviolet radiation of sufficient energy ($h\nu$) is available to break the very strong C–Cl bonds,

$$CCl_2F_2 + h\nu \rightarrow \bullet CCl_2F_2 + Cl\bullet \tag{8.3.3}$$

releasing extremely reactive Cl• atoms in which the dot represents a single unpaired electron remaining with the Cl atom when the bond in the molecule breaks. As discussed in Section 8.8 and shown by Reaction 8.8.6 and Reaction 8.8.7, there are oxygen atoms and molecules of ozone, O_3, also formed by photochemical processes in the stratosphere. A chlorine atom produced by the photochemical dissociation of CCl_2F_2 as shown in Reaction 8.3.3 can react with a molecule of O_3 to produce O_2 and another reactive free radical species, ClO•. This species can react with free O atoms which are present along with the ozone to regenerate Cl atoms, which in turn can react with more O_3 molecules. These reactions are shown below:

$$Cl\bullet + O_3 \rightarrow O_2 + ClO\bullet \tag{8.3.4}$$

$$ClO\bullet + O \rightarrow O_2 + Cl\bullet \tag{8.3.5}$$

These are chain reactions in which ClO• and Cl• are continually reacting and being regenerated, the net result of which is the conversion of O_3 and O in the atmosphere to O_2. One Cl atom can bring about the destruction of as many as 10,000 ozone molecules! Ozone serves a vital protective function in the atmosphere as a filter for damaging ultraviolet radiation, so its destruction is a very serious problem that has resulted in the banning of chlorofluorocarbon manufacture.

The hydroxyl radical, HO•, which is the single most important reactive intermediate species in atmospheric chemical processes, is formed by several mechanisms. At higher altitudes it is produced by photolysis of water:

$$H_2O + h\nu \rightarrow HO\bullet + H \tag{8.3.6}$$

In the presence of organic matter, hydroxyl radical is produced in abundant quantities as an intermediate in the formation of photochemical smog.

Hydroxyl radical is most frequently removed from the troposphere by reaction with methane or carbon monoxide:

$$CH_4 + HO\bullet \rightarrow H_3C\bullet + H_2O \qquad (8.3.7)$$

$$CO + HO\bullet \rightarrow CO_2 + H \qquad (8.3.8)$$

The reactive methyl radical, $H_3C\bullet$, and the hydrogen atom produced in the preceding reactions undergo additional reactions in the atmosphere. Reaction with hydroxyl radical is the most important means by which greenhouse gas methane and toxic pollutant carbon monoxide are removed from the atmosphere. Other important atmospheric trace species that react with hydroxyl radical and are thus removed from the atmosphere include sulfur dioxide, hydrogen sulfide, nitric oxide, and a variety of hydrocarbons.

8.4. OXIDATION PROCESSES IN THE ATMOSPHERE

The 21% (dry basis) by volume content of molecular O_2 makes the atmosphere thermodynamically oxidizing. One prominent manifestation of this condition is the tendency for oxidizable materials to corrode when exposed to the atmosphere. Iron, for example, exposed to moist air tends to rust:

$$4Fe + 3O_2 + xH_2O \rightarrow 2Fe_2O_3 \bullet xH_2O \qquad (8.4.1)$$

From the standpoint of atmospheric chemistry, however, the oxidizing tendency of the atmosphere is shown by the conversion of reduced molecular species to oxidized forms. It is this feature of the atmosphere exposed to sunlight that results in the formation of photochemical smog. Among the simple molecular species that enter the atmosphere in relatively reduced forms and that are oxidized are the following:

$$2CO + O_2 \rightarrow 2CO_2 \qquad (8.4.2)$$

$$CH_4 + 2O_2 \rightarrow CO_2 + 2H_2O \qquad (8.4.3)$$

$$4NO + 3O_2 + 2H_2O \rightarrow 4HNO_3 \qquad (8.4.4)$$

$$2SO_2 + O_2 + 2H_2O \rightarrow 2H_2SO_4 \qquad (8.4.5)$$

$$H_2S + 2O_2 \rightarrow H_2SO_4 \qquad (8.4.6)$$

Although shown here in a very simple form, these reactions actually represent processes that may involve many steps, photochemistry, and reactive intermediates, particularly hydroxyl radical. Oxidation reactions may also occur on particle surfaces and in solution in aqueous aerosol droplets, which are strongly exposed to atmospheric oxygen. Another aspect of these reactions is that the products are often acidic — mildly acidic CO_2 from carbon-containing species, and strongly acidic nitric and sulfuric acid from nitrogen oxides and gaseous sulfur species, respectively.

Reducing agents, such as those shown above, may be quite stable in dry air that is not exposed to sunlight. However, the absorption of photons from solar radiation starts processes that result in oxidation. As a simple example, the photochemical dissociation of nitrogen dioxide,

$$NO_2 + h\nu \rightarrow NO + O \tag{8.4.7}$$

can produce reactive O atoms that can react with oxidizable molecules,

$$CH_4 + O \rightarrow H_3C\bullet + HO\bullet \tag{8.4.8}$$

to begin the series of reactions that forms the final oxidized products (in this case CO_2 and H_2O). Intermediate hydroxyl radical, HO•, can abstract H atoms from hydrocarbons,

$$CH_4 + HO\bullet \rightarrow H_3C\bullet + H_2O \tag{8.4.9}$$

or add to molecules such as NO_2,

$$HO\bullet + NO_2 \rightarrow HNO_3 \tag{8.4.10}$$

to bring about oxidations. Chain reactions can be involved, such as the following sequence that regenerates NO_2 from NO:

$$H_3C\bullet \text{ (from Reaction 8.4.8)} + O_2 \rightarrow H_3COO\bullet \tag{8.4.11}$$

$$H_3COO\bullet + NO \rightarrow NO_2 \text{ (back to Reaction 8.4.7)} + H_3CO\bullet \tag{8.4.12}$$

The NO_2 product may undergo photochemical dissociation to produce O atoms and again initiate processes that result in oxidation.

A feature of the photochemical atmosphere, particularly when it is polluted by nitrogen oxides and hydrocarbons, is the generation of strong oxidant molecules. The most common example of a strong organic oxidant species is peroxyacetyl nitrate, PAN, formed from the reaction of $H_3CC(O)OO\bullet$ radical with NO_2:

$$H_3CC(O)OO\bullet + NO_2 + M(\text{energy-absorbing third molecule}) \rightarrow$$

$$CH_3C(O)OONO_2 + M \tag{8.4.13}$$

The most prominent inorganic oxidant is ozone, generated by reactions such as,

$$O + O_2 + M(\text{energy-absorbing third molecule}) \rightarrow O_3 + M \tag{8.4.14}$$

One of the ways in which ozone acts as an oxidant is to add to unsaturated compounds to form reactive ozonides:

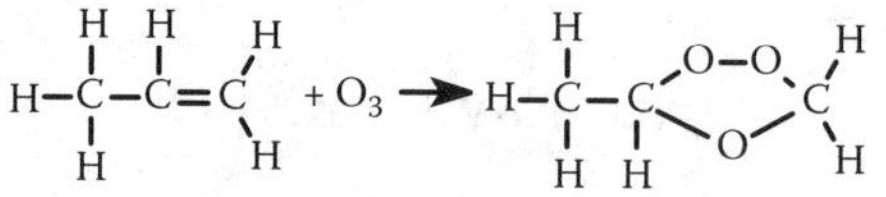

$$\tag{8.4.15}$$

8.5. ACID–BASE REACTIONS IN THE ATMOSPHERE

Acid–base reactions occur between acidic and basic species in the atmosphere. The atmosphere is normally at least slightly acidic because of the presence of a low level of carbon dioxide, which dissolves in atmospheric water droplets and dissociates slightly:

$$CO_2(g) \xrightarrow{\text{water}} CO_2(aq) \tag{8.5.1}$$

$$CO_2(aq) + H_2O \rightarrow H^+ + HCO_3^- \tag{8.5.2}$$

Atmospheric sulfur dioxide forms a somewhat stronger acid when it dissolves in water:

$$SO_2(g) + H_2O \rightarrow H^+ + HSO_3^- \tag{8.5.3}$$

In terms of pollution, however, strongly acidic HNO_3 and H_2SO_4 formed by the atmospheric oxidation of N oxides, O_2, and H_2S (see Reaction 8.4.4 to Reaction 8.4.6) are much more important because they lead to the formation of damaging acid rain.

As reflected by the generally acidic pH of rainwater, basic species are relatively less common in the atmosphere. Particulate calcium oxide, hydroxide, and carbonate can get into the atmosphere from ash and ground rock and can react with acids such as in the following reaction:

$$Ca(OH)_2(s) + H_2SO_4(aq) \rightarrow CaSO_4(s) + H_2O \tag{8.5.4}$$

The most important basic species in the atmosphere is gas-phase ammonia (NH_3). The greatest source of atmospheric ammonia is from biodegradation of nitrogen-containing biological matter and from bacterial reduction of nitrate:

$$NO_3^-(aq) + 2\{CH_2O\}(biomass) + H^+ \rightarrow NH_3(g) + 2CO_2 + H_2O \qquad (8.5.5)$$

Ammonia is particularly important as a base in the atmosphere because it is the only water-soluble base present at significant levels in the atmosphere. Dissolved in atmospheric water droplets, it plays a strong role in neutralizing atmospheric acids:

$$NH_3(aq) + HNO_3(aq) \rightarrow NH_4NO_3(aq) \qquad (8.5.6)$$

$$NH_3(aq) + H_2SO_4(aq) \rightarrow NH_4HSO_4(aq) \qquad (8.5.7)$$

These reactions have three effects: (1) They result in the presence of NH_4^+ ion in the atmosphere as dissolved or solid salts, (2) they serve in part to neutralize acidic constituents of the atmosphere, and (3) they produce relatively corrosive ammonium salts such as NH_4NO_3.

8.6. AIR POLLUTION

A major concern with respect to the atmosphere consists of the effects of inorganic and organic air pollutants of various kinds. Air pollution is addressed in general here and specific aspects in latter parts of this chapter and in Chapter 9.

One of the biggest concerns regarding air pollution is its effect upon plants. In many areas of the world, exposure to air pollutants has significantly reduced yields of crops grown for food and other purposes. Some forests have been seriously damaged by air pollution. An example is the killing of trees in Germany's Black Forest by exposure to acidic precipitation. Therefore, air pollution is one of the more significant threats to sustainability.

Atmospheric particles, often formed from reactions of gaseous pollutants that enter the atmosphere as the result of human activities, constitute one of the main classes of air pollutants. Gaseous air pollutants include CO, SO_2, NO, and NO_2, as well as less abundant NH_3, N_2O, N_2O_5, H_2S, Cl_2, HCl, and HF. Their quantities are relatively small compared to the amount of CO_2 in the atmosphere, which may turn out to be the ultimate air pollutant because of its effects on greenhouse warming.

There is a strong connection between inorganic and organic substances in the atmosphere. For example, inorganic NO_2 photodissociates to start the processes by which organic vapors form aldehydes, oxidants, and other substances in photochemical smog. As another example, smog-generated oxidants convert inorganic SO_2 to much more acidic sulfuric acid, the major contributor to acid precipitation.

Organic pollutants may strongly affect atmospheric quality. Such pollutants may come from both natural and artificial sources. In some cases contaminants from both kinds of sources interact to produce a pollution effect, such as occurs when terpene hydrocarbons from trees interact with NO_x from autos to make photochemical smog.

The effects of atmospheric pollutants may be divided between **direct effects**, from **primary pollutants**, such as cancer caused by exposure to vinyl chloride, and effects resulting from **secondary pollutants** produced by atmospheric reactions of primary pollutants, such as photochemical smog or acid rain. Generally, secondary pollutants are more important. In some localized situations, particularly the workplace, direct effects of organic air pollutants may be equally important.

8.7. ENVIRONMENTAL FATE AND TRANSPORT IN THE ATMOSPHERE

The atmosphere is very much involved in fate and transport processes of environmental chemicals. To understand these processes, it is necessary to consider sources, transport, dispersal, and fluxes of airborne contaminants. Interactions at the atmosphere/surface boundary are important, including flow and dispersal of material in the atmosphere over complex terrain, and around surface obstacles, such as trees and buildings. Interactions and interchange with surface media including rock and soil, water, and vegetation must be considered. Transport and dispersal by advection because of movement of masses of air and diffusive transport are important considerations as are dispersion, and degradation half-life.

Pollutants in the atmosphere may be viewed on local, long-range, and global scales. Local-scale chemical fate and transport may be illustrated by a smokestack point source. As Figure 8.2 illustrates, gases and particles are emitted from a stack and carried and dispersed by wind and air currents while undergoing mixing and dilution. Because stack gases are carried upward by a rising current warmer than the surrounding atmosphere, the effective height of a stack is always greater than its actual height. The farther from the stack source before pollutants reach ground

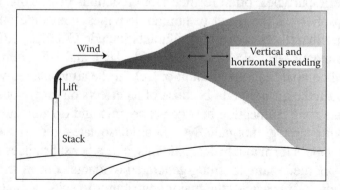

Figure 8.2.   Illustration of localized chemical fate and transport with air pollutants from a point source (smokestack).

level, the more dilute they are. The dispersion of pollutants is strongly influenced by atmospheric conditions such as wind, air turbulence, and the occurrence of temperature inversions (Figure 8.2). High stacks reduce the immediate impact of air pollutants and illustrate the once-prevailing philosophy that "the solution to pollution is dilution."

Long-range transport of species in the atmosphere is an important aspect of air pollution. One illustration of long-range transport of air pollutants was the contamination of much of Europe including northern reaches of Scandinavia by radionuclides emitted in the Chernobyl nuclear reactor meltdown and fire. New England and southeastern Canada are affected by acid rainfall originating from sulfur dioxide emitted by power plants in the U.S. Ohio River valley hundreds of kilometers distant.

Modeling long-range environmental fate and transport is an important exercise in determining sources of pollutants and mitigating their effects. Such models using sophisticated computer programs and high computing capacity must consider all transport and mixing phenomena. Very large areas must be considered and averages taken over long time intervals. Weather conditions are important factors.

Some important atmospheric pollutants must be considered on a global scale. Such pollutants have very long lifetimes so that they persist long enough to mix with and spread throughout the global atmosphere and they are produced from a variety of widely dispersed sources. An example of such a substance is greenhouse gas carbon dioxide emitted to the atmosphere by billions of heating and cooking stoves, millions of automobiles, and thousands of power plants throughout the globe.

Earth's atmosphere cannot be considered as a single large mixing bowl for contaminants on a global scale. Prevailing winds cause relatively rapid mixing within the northern and southern hemisphere, whereas transport of atmospheric constituents across the equator is relatively slow. This phenomenon is illustrated by atmospheric greenhouse gas carbon dioxide. In the northern hemisphere, which has an abundance of photosynthesizing plants, there is a pronounced annual fluctuation of several parts per million in the levels of atmospheric carbon dioxide produced by the annual seasonal cycle of photosynthesis, whereas the fluctuation is much less pronounced in the southern hemisphere, which has much less photosynthetic activity. Mixing between the two hemispheres over a year's period of time is insufficient to dampen the fluctuation, whereas the average concentration of carbon dioxide is essentially the same in both hemispheres reflecting mixing over several years' periods of time.

An interesting aspect of fate and transport involving the atmosphere is provided by the accumulation of semivolatile persistent organic pollutants in polar regions. This phenomenon occurs by a distillation effect in which such pollutants are evaporated in warmer latitudes, carried by air currents toward the poles, and condensed in colder polar regions. As a result, surprisingly high levels of some semivolatile persistent organic pollutants (PCBs in Arctic polar bear fat, for example) have been found in samples from polar regions.

8.8. REACTIONS OF ATMOSPHERIC OXYGEN

Some of the primary features of the exchange of oxygen among the atmosphere, lithosphere, hydrosphere, and biosphere are summarized in Figure 8.3. The oxygen cycle is critically important in atmospheric chemistry, geochemical transformations, and life processes.

Oxygen in the troposphere plays a strong role in processes that occur on the Earth's surface. Atmospheric oxygen takes part in energy-producing reactions, such as the burning of fossil fuels:

$$CH_4 \text{(in natural gas)} + 2O_2 \rightarrow CO_2 + 2H_2O \qquad (8.8.1)$$

Atmospheric oxygen is utilized by aerobic organisms in the degradation of organic material. Some oxidative weathering processes consume oxygen. An important example is the oxidative weathering of mineral iron(II) to iron(III):

$$4FeO + O_2 \rightarrow 2Fe_2O_3 \qquad (8.8.2)$$

Oxygen is returned to the atmosphere through plant photosynthesis:

$$CO_2 + H_2O + h\nu \rightarrow \{CH_2O\} + O_2 \qquad (8.8.3)$$

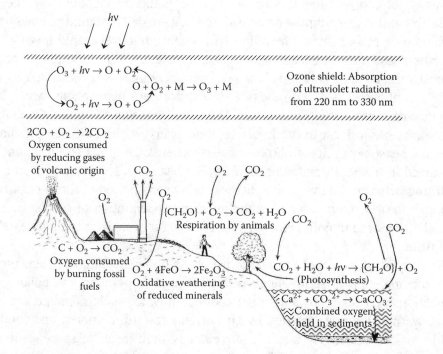

Figure 8.3.   Oxygen exchange among the atmosphere, lithosphere, hydrosphere, and biosphere. Biomass containing carbon is represented as {CH_2O}.

All molecular oxygen in the atmosphere is thought to have originated through the action of photosynthetic organisms, which shows the importance of photosynthesis in the oxygen balance of the atmosphere. It can be shown that most of the carbon fixed by these photosynthetic processes is dispersed in mineral formations as humic material (Section 5.11); only a very small fraction is deposited in fossil fuel beds, large though they are. Therefore, although fossil fuel combustion consumes large amounts of O_2, there is no danger of running out of atmospheric oxygen.

Because of the extremely rarefied atmosphere and the effects of ionizing radiation, elemental oxygen in the upper atmosphere exists to a large extent in forms other than diatomic O_2. In addition to O_2, the upper atmosphere contains oxygen atoms (O); excited oxygen molecules (O_2^*); and ozone (O_3).

Atomic oxygen, O, is stable primarily in the far outer atmosphere, where the air is so rarefied that the three-body collisions necessary for the chemical reaction of atomic oxygen seldom occur (the third body in this kind of three-body reaction absorbs energy to stabilize the products). Atomic oxygen is produced by a photochemical reaction:

$$O_2 + h\nu \rightarrow O + O \tag{8.8.4}$$

The oxygen–oxygen bond is strong and ultraviolet radiation in the wavelength regions 135 to 176 nm and 240 to 260 nm is most effective in causing dissociation of molecular oxygen. Because of photochemical dissociation, O_2 is virtually nonexistent at very high altitudes, and less than 10% of the oxygen in the atmosphere at altitudes exceeding approximately 400 km is present in the molecular form.

Oxygen atoms in the atmosphere can exist in the ground state (O) and in excited states (O*). Excited oxygen atoms are produced by the photolysis of ozone at wavelengths below 308 nm or by highly energetic chemical reactions involving oxygen atoms. Excited atomic oxygen emits visible light at wavelengths of 636, 630, and 558 nm. This emitted light is partially responsible for **airglow**, a very faint electromagnetic radiation continuously emitted by the Earth's atmosphere. Although its visible component is extremely weak, airglow is relatively intense in the infrared region of the spectrum.

Oxygen ion (O⁺), which may be produced by ultraviolet radiation acting upon oxygen atoms,

$$O + h\nu \rightarrow O^+ + e^- \tag{8.8.5}$$

is the predominant positive ion in some regions of the ionosphere.

Ozone and the Ozone Layer

Ozone has an essential protective function because it absorbs harmful ultraviolet radiation in the stratosphere and serves as a radiation shield, protecting living beings on the Earth from the effects of excessive amounts of such radiation. It is produced by a photochemical reaction,

$$O_2 + h\nu(\text{energetic ultraviolet radiation}) \rightarrow O + O \qquad (8.8.6)$$

followed by a three-body reaction,

$$O + O_2 + M \rightarrow O_3 + M(\text{increased energy}) \qquad (8.8.7)$$

in which M is another molecule, usually N_2 or O_2, which absorbs the excess energy given off by the reaction and enables the ozone molecule to stay together. The region of maximum ozone concentration occurs in the stratosphere at an altitude of 25 to 30 km where it may reach levels of 10 ppm.

Ozone absorbs ultraviolet radiation very strongly in the region 220 to 330 nm. If this radiation were not absorbed by ozone, severe damage would result to exposed forms of life on the Earth. Absorption of electromagnetic radiation by ozone converts the radiation's energy to heat and is responsible for the temperature maximum encountered at the boundary between the stratosphere and the mesosphere at an altitude of approximately 50 km. The reason that the temperature maximum occurs at a higher altitude than that of the maximum ozone concentration arises from the fact that ozone is such an effective absorber of ultraviolet light. Therefore, most of this radiation is absorbed in the upper stratosphere, where it generates heat, and only a small fraction reaches the lower altitudes, which remain relatively cool.

Thermodynamically, the overall reaction,

$$2O_3 \rightarrow 3O_2 \qquad (8.8.8)$$

is favored so that ozone is inherently unstable. Its decomposition in the stratosphere is catalyzed by a number of natural and pollutant trace constituents, including NO, NO_2, H, HO•, HOO•, ClO, Cl, Br, and BrO. Ozone decomposition also occurs on solid surfaces such as metal oxides and salts produced by rocket exhausts.

Despite its protective function in the stratosphere, ozone is an undesirable pollutant in the troposphere. It is toxic and a mild overdose causes labored breathing, a feeling of chest pressure, cough, and irritated eyes. In addition to its toxicological effects, which are discussed in Chapter 14, ozone damages materials, such as rubber.

8.9. CARBON OXIDES

Carbon Monoxide

Carbon monoxide (CO) is a toxic gas produced by the partial combustion of carbonaceous fuels. It is generated by internal combustion engines and has been one of the more troublesome air pollutants produced by automobiles. Because of CO emissions from internal combustion engines, the highest levels of this toxic gas tend to occur in congested urban areas at times when the greatest number of people are exposed, such as during rush hours.

In the earlier days of air pollution studies, the fate of atmospheric carbon dioxide was largely a mystery. At one time it was believed that the major sink for CO was metabolism by soil microorganisms. Although some CO is lost from the atmosphere by this route, it is now established that most of it is consumed by reaction with hydroxyl radical as discussed in Section 8.3.

Atmospheric Carbon Dioxide

Although only about 0.038% (380 ppm) of air consists of carbon dioxide, it is the atmospheric "nonpollutant" species of most concern. As mentioned in Section 7.4, carbon dioxide, along with water vapor, is primarily responsible for the absorption of infrared energy reemitted by the Earth such that some of this energy is reradiated back to the Earth's surface. Current evidence suggests that changes in the atmospheric carbon dioxide level will substantially alter the Earth's climate through the greenhouse effect, as discussed in greater detail in Chapter 9.

Valid measurements of overall atmospheric CO_2 can only be taken in areas remote from industrial activity. Such areas include Antarctica and the top of Mauna Loa Mountain in Hawaii. Accurate measurements of carbon dioxide levels in these locations over more than half a century have shown an annual increase in CO_2 of about 1 ppm/year as shown in Figure 8.4.

The most obvious factor contributing to increased atmospheric carbon dioxide is consumption of carbon-containing fossil fuels. In addition, release of CO_2 from the biodegradation of biomass and uptake by photosynthesis are important factors determining overall CO_2 levels in the atmosphere. The role of photosynthesis is illustrated by the seasonal cycle in carbon dioxide levels in the Northern Hemisphere. Maximum values occur in April and minimum values in late September or October. These oscillations are because of the "photosynthetic pulse," influenced most strongly by forests in middle latitudes. Forests have a much greater influence than other vegetation because, in general, forest trees carry out more photosynthesis than other kinds of plants such as prairie grasses. Furthermore, forests store enough fixed, but readily oxidizable carbon in the form of wood and humus to have a marked influence on atmospheric CO_2 content. Thus, during the summer months, photosynthetic activity in forests is sufficient to reduce the atmospheric carbon dioxide content markedly. During the winter, metabolism of biota, such as bacterial decay of humus, releases a significant amount of CO_2. Therefore, the current worldwide trend toward destruction of forests and conversion of forest lands to agricultural uses will contribute substantially to a greater overall increase in atmospheric CO_2 levels.

With current trends, it is likely that global CO_2 levels will double by the middle of the next century, which may well raise the Earth's mean surface temperature by

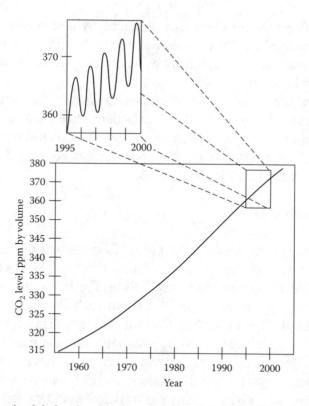

Figure 8.4.   Increase in global carbon dioxide levels showing seasonal fluctuations associated with photosynthesis in the Northern Hemisphere.

1.5 to 4.5°C. Such a change has more potential to cause massive irreversible environmental changes than any other disaster short of global nuclear war or asteroid impact.

Chemically and photochemically, CO_2 is a comparatively insignificant species because of its relatively low concentrations and low photochemical reactivity. The infrared radiation absorbed by carbon dioxide is not energetic enough to cause photochemical reactions to occur.

8.10. REACTIONS OF ATMOSPHERIC NITROGEN AND ITS OXIDES

The 78% by volume of nitrogen contained in the atmosphere constitutes an inexhaustible reservoir of that essential element. The nitrogen cycle and nitrogen fixation by microorganisms and several other microbially mediated reactions important in the nitrogen cycle were discussed in Section 5.6. A small amount of nitrogen is thought to be fixed (chemically bound to other elements) in the atmosphere by lightning, and some is also fixed by combustion processes, as in the internal combustion engine.

Before the use of synthetic fertilizers reached its current high levels, chemists were concerned that denitrification processes in the soil would lead to nitrogen depletion on the Earth. Now, with millions of tons of synthetically fixed nitrogen being added to the soil each year, concern has shifted to possible excess accumulation of nitrogen in soil, fresh water, and the oceans.

Unlike oxygen, which is almost completely dissociated to the monatomic form in higher regions of the thermosphere, molecular nitrogen is not readily dissociated by ultraviolet radiation. However, at altitudes exceeding approximately 100 km, atomic nitrogen is produced by photochemical reactions:

$$N_2 + h\nu \rightarrow N + N \tag{8.10.1}$$

Most stratospheric ozone is probably removed by the action of nitric oxide, which reacts with ozone as follows:

$$O_3 + NO \rightarrow NO_2 + O_2 \tag{8.10.2}$$

$$NO_2 + O \rightarrow NO + O_2 \text{ (regeneration of NO from NO}_2) \tag{8.10.3}$$

Pollutant oxides of nitrogen, particularly NO_2, are key species involved in air pollution and the formation of photochemical smog. For example, NO_2 is readily dissociated photochemically to NO and reactive atomic oxygen by electromagnetic radiation of less than 398 nm wavelength:

$$NO_2 + h\nu \rightarrow NO + O \tag{8.10.4}$$

This reaction is the most important primary photochemical process involved in smog formation.

Air Pollutant Nitrogen Compounds

Nitrogen compounds, especially nitrogen oxides, are among the most significant air pollutants. The three oxides of nitrogen normally encountered in the atmosphere are nitrous oxide (N_2O), nitric oxide (NO), and nitrogen dioxide (NO_2). Microbially generated nitrous oxide is relatively unreactive and probably does not significantly influence important chemical reactions in the lower atmosphere. However, colorless, odorless nitric oxide and pungent red-brown nitrogen dioxide, collectively designated NO_x, are very important in polluted air. Regionally high pollutant NO_2 concentrations can result in severe air quality deterioration. Practically all anthropogenic NO_x enters the atmosphere as NO from the combustion of fossil fuels in both stationary and mobile sources. The contribution of automobiles to nitric oxide production has become significantly lower as newer automobiles with nitrogen oxide pollution controls have become more common.

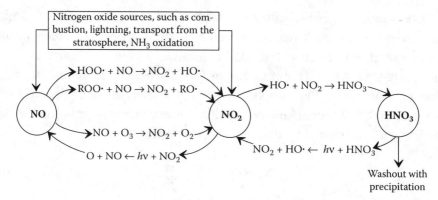

Figure 8.5.   Principal reactions among NO, NO_2, and HNO_3 in the atmosphere. ROO• represents an organic peroxyl radical, such as the methylperoxyl radical, CH_3OO•.

Atmospheric Reactions of NO_x

Nitrogen dioxide is a very reactive and significant species in the atmosphere. It absorbs light throughout the ultraviolet and visible spectrum penetrating the troposphere. The photochemical dissociation of NO_2 (Reaction 8.10.4) and the highly reactive O atom product set off several significant inorganic reactions and many atmospheric reactions involving organic species. Atmospheric chemical reactions convert NO_x to nitric acid, inorganic nitrate salts, organic nitrates, and oxidant peroxyacetyl nitrate as discussed with respect to photochemical smog later in this chapter. These species cycle among each other, as shown in Figure 8.5. Although NO is the primary form in which NO_x is released to the atmosphere, the conversion of NO to NO_2 is relatively rapid in the troposphere.

Harmful Effects of Nitrogen Oxides

Nitric oxide is less toxic than NO_2. (In recent years the role of NO as an important intermediate in biochemical processes has gained recognition.) Acute exposure to NO_2 can be quite harmful to human health and sufficiently high exposures to this gas can be fatal. For exposures ranging from several minutes to one hour, a level of 50 to 100 ppm of NO_2 causes inflammation of lung tissue for a period of 6 to 8 weeks, after which time the subject normally recovers. Exposure of the subject to 150 to 200 ppm of NO_2 causes *bronchiolitis fibrosa obliterans*, a condition fatal within 3 to 5 weeks after exposure. Death generally results within 2 to 10 days after inhalation of air containing 500 ppm or more of NO_2. Although extensive damage to plants is observed in areas receiving heavy exposure to NO_2, most of this damage probably comes from secondary products of nitrogen oxides, such as PAN formed in smog (see Reaction 8.4.13).

Ammonia as an Atmospheric Pollutant

Ammonia is the only nonoxygenated gaseous inorganic nitrogen compound that is likely to be a significant atmospheric pollutant. It was mentioned in Section 8.5 as the only water-soluble base present at significant levels in the atmosphere. It is toxic and can be a significant localized pollutant in specific cases. Its most important effect in the atmosphere is the formation of corrosive pollutant salts including NH_4NO_3, NH_4HSO_4, and $(NH_4)_2SO_4$.

8.11. SULFUR COMPOUNDS IN THE ATMOSPHERE

Figure 8.6 shows the main aspects of the global sulfur cycle. This cycle involves primarily H_2S, SO_2, SO_3, and sulfates. Of these species, SO_2 is the most important because of its high abundance and facile oxidation to highly acidic H_2SO_4.

Sulfur Dioxide Reactions in the Atmosphere

Many factors, including temperature, humidity, light intensity, atmospheric transport, and surface characteristics of particulate matter, may influence the

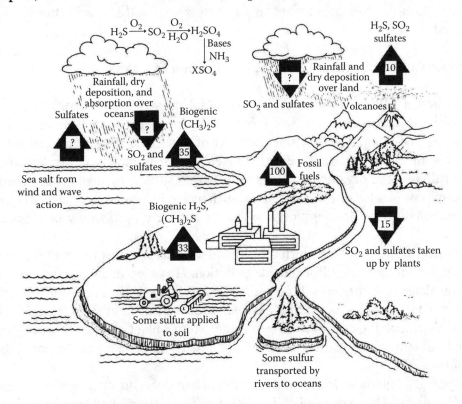

Figure 8.6.   The atmospheric sulfur cycle. Numbers in the arrows are in millions of metric tons per year.

atmospheric chemical reactions of sulfur dioxide. Like many other gaseous pollutants, sulfur dioxide undergoes chemical reactions resulting in the formation of particulate matter. Whatever the processes involved, much of the sulfur dioxide in the atmosphere ultimately is oxidized to sulfuric acid and sulfate salts, particularly ammonium sulfate and ammonium hydrogen sulfate.

Effects of Atmospheric Sulfur Dioxide

Though not terribly toxic to most people, low levels of sulfur dioxide in air do have some health effects. Sulfur dioxide's primary effect is upon the respiratory tract, producing irritation and increasing airway resistance, especially in people with respiratory weaknesses and in sensitized asthmatics. Therefore, exposure to the gas may increase the effort required to breathe. Mucus secretion is also stimulated by exposure to air contaminated by sulfur dioxide.

Atmospheric sulfur dioxide is harmful to plants. Acute exposure to high levels of the gas kills leaf tissue (leaf necrosis). Chronic exposure of plants to sulfur dioxide causes chlorosis, a bleaching or yellowing of the normally green portions of the leaf. Sulfur dioxide in the atmosphere is converted to sulfuric acid, so that in areas with high levels of sulfur dioxide pollution, plants may be damaged by sulfuric acid aerosols. Such damage appears as small spots where sulfuric acid droplets have impinged upon leaves.

Hydrogen Sulfide, Carbonyl Sulfide, and Carbon Disulfide

Hydrogen sulfide is produced by microbial decay of sulfur compounds and microbial reduction of sulfate (see Section 5.6), from geothermal steam, as a by-product of wood pulping, and from a number of miscellaneous natural and anthropogenic sources. Most atmospheric hydrogen sulfide is rapidly converted to SO_2 and to sulfates. The organic homologs of hydrogen sulfide, the mercaptans (hydrocarbon groups bonded to the –SH group, such as methyl mercaptan, H_3CSH), enter the atmosphere from decaying organic matter and have particularly objectionable odors.

Hydrogen sulfide pollution from artificial sources is not as much of an overall air pollution problem as sulfur dioxide pollution. However, there have been several acute incidents of hydrogen sulfide emissions resulting in damage to human health and even fatalities. The most notorious such event occurred in Poza Rica, Mexico, in 1950. Accidental release of hydrogen sulfide from a plant used for the recovery of sulfur from natural gas caused the deaths of at least 22 people and the hospitalization of over 300.

Hydrogen sulfide at levels well above ambient concentrations destroys immature plant tissue. This type of plant injury is readily distinguished from that of other phytotoxins. More sensitive species are killed by continuous exposure to around

3000 ppb H_2S, whereas other species survive with reduced growth, leaf lesions, and defoliation.

Damage to certain kinds of materials is a very expensive effect of hydrogen sulfide pollution. Paints containing basic lead carbonate pigment, $2PbCO_3 \cdot Pb(OH)_2$ (no longer used), were particularly susceptible to darkening by H_2S. A black layer of copper sulfide forms on copper metal exposed to H_2S. Eventually, this layer is replaced by a green coating of basic copper sulfate such as $CuSO_4 \cdot 3Cu(OH)_2$. The green "patina," as it is called, is very resistant to further corrosion. Such layers of corrosion can seriously impair the function of copper contacts on electrical equipment. Hydrogen sulfide also forms a black sulfide coating on silver.

Carbonyl sulfide, COS, is now recognized as a component of the atmosphere at a tropospheric concentration of approximately 500 parts per trillion by volume. It is, therefore, a significant sulfur species in the atmosphere. Both COS and carbon disulfide, CS_2, are oxidized in the atmosphere by reactions initiated by the hydroxyl radical.

8.12. FLUORINE, CHLORINE, AND THEIR GASEOUS INORGANIC COMPOUNDS

Fluorine (F_2), hydrogen fluoride (HF), and other volatile fluorides are produced in the manufacture of aluminum, and hydrogen fluoride is a by-product in the conversion of fluorapatite (rock phosphate) to phosphoric acid, superphosphate fertilizers, and other phosphorus products. Hydrogen fluoride gas is a dangerously toxic substance that is so corrosive that it even reacts with glass. It is irritating to body tissues, and the respiratory tract is very sensitive to it. Brief exposure to HF vapors at the part-per-thousand level may be fatal. The acute toxicity of F_2 is even higher than that of HF. Chronic exposure to high levels of fluorides causes fluorosis, the symptoms of which include mottled teeth and pathological bone conditions.

Plants are particularly susceptible to the effects of gaseous fluorides. Fluorides from the atmosphere appear to enter the leaf tissue through the stomata. Fluoride is a cumulative poison in plants, and exposure of sensitive plants to even very low levels of fluorides for prolonged periods results in damage. Characteristic symptoms of fluoride poisoning are chlorosis (fading of green color because of conditions other than the absence of light), edge burn, and tip burn. Conifers (such as pine trees) afflicted with fluoride poisoning may develop reddish-brown necrotic needle tips at distances of several miles from the pollutant sources.

Chlorine and Hydrogen Chloride

Chlorine gas, Cl_2, can be quite damaging on a local scale. Chlorine was the first poisonous gas deployed in World War I. It is widely used as a manufacturing chemical, in the plastics industry, for example, as well as for water treatment, and as bleach. Therefore, possibilities for its release exist in a number of locations. Highly toxic chlorine is a mucous-membrane irritant, spills of which have caused fatalities among exposed persons. It is very reactive and a powerful oxidizing agent. Chlorine dissolves in atmospheric water droplets, yielding hydrochloric acid and hypochlorous acid, an oxidizing agent:

$$H_2O + Cl_2 \rightarrow H^+ + Cl^- + HOCl \qquad (8.12.1)$$

Hydrogen chloride (HCl) is emitted from a number of sources. Incineration of chlorinated plastics, such as polyvinylchloride, releases HCl as a combustion product.

8.13. WATER IN ATMOSPHERIC CHEMISTRY

Gaseous water in the upper atmosphere is involved in the formation of hydroxyl and hydroperoxyl radicals as mentioned in Section 8.3. Condensed water vapor in the form of very small droplets is important in atmospheric chemistry. The harmful effects of some air pollutants — for instance, the corrosion of metals by acid-forming gases — require the presence of water, which may come from the atmosphere. Atmospheric water vapor has an important influence upon pollution-induced fog formation under some circumstances. Water vapor interacting with pollutant particulate matter in the atmosphere may reduce visibility to undesirable levels through the formation of aerosol particles (see Section 8.13).

Most stratospheric water comes from the photochemical oxidation of methane:

$$CH_4 + 2O_2 + h\nu \xrightarrow{\text{Several Steps}} CO_2 + 2H_2O \qquad (8.13.1)$$

The water thus produced serves as a source of stratospheric hydroxyl radical as shown by the following reaction:

$$H_2O + h\nu \rightarrow HO\bullet + H \qquad (8.13.2)$$

8.14. ATMOSPHERIC PARTICLES AND ATMOSPHERIC CHEMISTRY

Atmospheric particles (Section 7.9), commonly called **particulates**, range in size from about 0.5 mm down to molecular dimensions, and consist of a large variety of solid or liquid materials and discrete objects. Particles are the most visible and obvious form of air pollution. Atmospheric **aerosols** are suspensions in air of solid or liquid particles below 100 μm in diameter. Pollutant particles of 0.001 to 100 μm size are commonly suspended in the air near sources of pollution, such as the urban atmosphere, industrial plants, highways, and power plants. Very small,

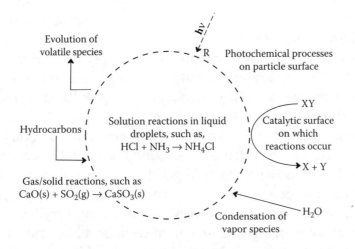

Figure 8.7.   Particles in the atmosphere participate in a number of atmospheric chemical processes.

solid particles include carbon black, silver iodide, combustion nuclei, and sea-salt nuclei formed by the loss of water from droplets of seawater. Larger particles include cement dust, wind-blown soil dust, foundry dust, and pulverized coal.

As summarized in Figure 8.7, particles play an important role in atmospheric chemistry. Neutralization reactions, which occur most readily in solution, may take place in water droplets suspended in the atmosphere. Small particles of metal oxides and carbon have a catalytic effect on oxidation reactions. Particles may also participate in oxidation reactions induced by light.

Formation and Composition of Inorganic Particles

Metal oxides make up a large class of inorganic particles in the atmosphere. They are formed when fuels are burned that contain metals, such as organic vanadium in residual fuel oil or pyrite (FeS_2) in coal:

$$3FeS_2 + 8O_2 \rightarrow Fe_3O_4 + 6SO_2 \tag{8.14.1}$$

A common process for the formation of aerosol mists involves the oxidation of atmospheric sulfur dioxide to H_2SO_4. The sulfuric acid product is a hygroscopic substance that accumulates atmospheric water to form small liquid droplets in which it may react with basic air pollutants to form salts:

$$2SO_2 + O_2 + 2H_2O \rightarrow 2H_2SO_4 \tag{8.14.2}$$

$$H_2SO_4(droplet) + 2NH_3(g) \rightarrow (NH_4)_2SO_4(droplet) \tag{8.14.3}$$

$$H_2SO_4(droplet) + CaO(s) \rightarrow CaSO_4(particle) + H_2O \tag{8.14.4}$$

Under low humidity conditions water is lost from droplets of these salt solutions and a solid aerosol is formed.

The composition of particulate matter reflects both the elemental composition of its source and chemical reactions that may change the composition. Among the constituents of inorganic particulate matter found in polluted atmospheres are salts, oxides, nitrogen compounds, sulfur compounds, various metals, and radionuclides. In coastal areas, sodium and chlorine get into atmospheric particles as sodium chloride from sea spray. The major trace elements that typically occur at levels above 1 µg/m^3 in particulate matter are those largely from terrestrial sources — aluminum, calcium, carbon, iron, potassium, sodium, and silicon. Lesser quantities of copper, lead, titanium, and zinc and even lower levels of antimony, beryllium, bismuth, cadmium, cobalt, chromium, cesium, lithium, manganese, nickel, rubidium, selenium, strontium, and vanadium are commonly observed. Lead is the toxic heavy metal of most concern in particles, although atmospheric lead levels have decreased markedly because leaded gasoline has been phased out of use. Elemental carbon from the partial combustion of fuels is one of the most common constituents of atmospheric particles.

Much of the mineral particulate matter in a polluted atmosphere is in the form of **fly ash** consisting of inorganic material and elemental carbon produced during the combustion of high-ash fossil fuel. Fly ash enters furnace flues and is efficiently collected in a properly equipped stack system. However, some escapes through the stack and enters the atmosphere. Unfortunately, the fly ash thus released tends to consist of smaller particles that do the most damage to human health, plants, and visibility. The constituents of fly ash are oxides of aluminum, calcium, iron, and silicon; elemental carbon (soot, carbon black); and usually minor constituents, including magnesium, sulfur, titanium, phosphorus, potassium, and sodium.

Radioactive Particles

Radionuclides can be significant air pollutants. A major natural source of radionuclides in the atmosphere is **radon**, a noble gas product of radium decay. Radon may enter the atmosphere as either of two isotopes, ^{222}Rn (half-life 3.8 d) and ^{220}Rn (half-life 54.5 sec). Both emit alpha particles (energetic, positively charged helium nuclei) from their atom nuclei in decay chains that terminate with stable isotopes of lead. The initial decay products, ^{218}Po and ^{216}Po, are nongaseous and adhere readily to atmospheric particulate matter, so some of the radioactivity in these particles is of natural origin. Furthermore, cosmic rays act on nuclei in the atmosphere to produce other radionuclides, including ^{7}Be, ^{10}Be, ^{14}C, ^{39}Cl, ^{3}H, ^{22}Na, ^{32}P, and ^{33}P.

The combustion of fossil fuels introduces radioactivity into the atmosphere in the form of radionuclides contained in fly ash. A large coal-fired power plant lacking ash-control equipment may introduce up to several hundred mci (millicuries, a measure of radioactivity) of radionuclides into the atmosphere each year, far more than either an equivalent nuclear or oil-fired power plant.

Before the practice was discontinued, the above-ground detonation of nuclear weapons added large amounts of radioactive particulate matter to the atmosphere. Because of food contamination and biouptake, the most serious fission contaminant products from this source were ^{90}Sr, ^{131}I, and ^{137}Cs. These fission products were widely dispersed in Europe and Scandinavia as the result of the 1986 nuclear reactor explosion and fire at Chernobyl in the Ukraine.

Organic Particles in the Atmosphere

A significant portion of organic particulate matter is produced by complicated processes in internal combustion engines. These products may include nitrogen-containing compounds and oxidized hydrocarbon polymers. Lubricating oil and its additives may also contribute to organic particulate matter. The organic particles of greatest concern are **polycyclic aromatic hydrocarbons** (PAH), which consist of condensed-ring aryl molecules produced by pyrolysis or partial combustion of organic compounds. The most often cited example of a PAH compound is benzo[*a*]pyrene, a compound that the body can metabolize to a carcinogenic form:

Benzo(a)pyrene

Effects of Particles

Atmospheric particles have numerous effects. The most obvious of these are reduction and distortion of visibility. Particles of 0.1 to 1 µm size cause interference phenomena because they are about the same dimensions as the wavelengths of visible light, so their light-scattering properties are especially significant. Particles provide active surfaces on which heterogeneous atmospheric chemical reactions can occur and nucleation bodies for the condensation of atmospheric water vapor, thereby causing significant weather and pollution effects.

Atmospheric particles inhaled through the respiratory tract may damage health. The most dangerous are very small particles, which are likely to reach the lungs and be retained by them. A strong correlation has been found between increases in the daily mortality rate and acute episodes of air pollution, including those in which particulate pollutants are present in high concentrations. The respiratory system may be damaged directly by particulate matter. Substances in inhaled particles may enter the blood system or lymph system in the alveoli of the lungs and be carried throughout the body to act as systemic poisons on other organs.

8.15. ORGANIC COMPOUNDS IN THE ATMOSPHERE

Organic compounds are important atmospheric contaminants, especially because of their role in photochemical smog formation (Section 8.16). Most organics in the atmosphere come from natural sources, with only about 1/7 of the total atmospheric hydrocarbons originating from human activities. This ratio is primarily the result of the huge quantities of methane produced by anaerobic bacteria in the decomposition of organic matter in water, sediments, and soil:

$$2\{CH_2O\}(\text{bacterial action}) \rightarrow CO_2(g) + CH_4(g) \qquad (8.15.1)$$

Flatulent emissions from domesticated animals, arising from bacterial decomposition of food in their digestive tracts, add about 85 million metric tons of methane to the atmosphere each year. Methane is a natural constituent of the atmosphere and is present at a level of about 1.4 ppm in the troposphere. Although its atmospheric chemical effects are minimal, methane is a major and growing contributor to greenhouse gases, and each molecule of methane added to the atmosphere contributes much more to greenhouse warming than does a molecule of carbon dioxide.

Vegetation is the most important natural source of atmospheric hydrocarbons other than methane. Ethene (ethylene), C_2H_4, is released to the atmosphere by a variety of plants, which use ethene as a molecular messenger. Most of the hydrocarbons emitted by plants (predominantly trees, such as citrus and pine trees) are **terpenes**. As exemplified by the structures of α-pinene, isoprene, and limonene,

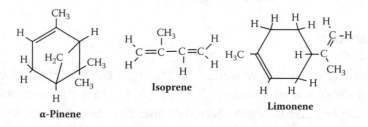

α-Pinene   Isoprene   Limonene

terpenes contain alkenyl (double) bonds, usually two or more per molecule. Therefore, terpenes are among the most reactive compounds in the atmosphere. Terpenes react very rapidly with hydroxyl radical (HO•) and with other oxidizing agents in the atmosphere, particularly ozone (O_3). Such reactions form aerosols, which cause the blue haze in the atmosphere above some heavy growths of vegetation, and terpenes are involved in the formation of photochemical smog.

Perhaps the greatest variety of compounds emitted by plants consist of **esters**, an example of which is coniferyl benzoate, structural formula below:

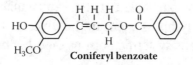

Coniferyl benzoate

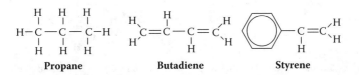

Propane   Butadiene   Styrene

Figure 8.8. Examples of the three common classes of atmospheric pollutant hydrocarbons. Alkanes, such as propane, are saturated and not very reactive, whereas unsaturated alkenes are very reactive in the atmosphere.

However, esters are released in such small quantities that they have little influence upon atmospheric chemistry. Esters are primarily responsible for the fragrances associated with much vegetation.

Pollutant Hydrocarbons

Figure 8.8 shows common atmospheric pollutant hydrocarbons. These include **alkanes** such as propane, **alkenes**, such as butadiene, and **aromatic compounds**, such as styrene (also an alkene). Because of their widespread use in fuels, hydrocarbons predominate among organic atmospheric pollutants. Petroleum products, primarily gasoline, are the source of most of the anthropogenic pollutant hydrocarbons found in the atmosphere. Hydrocarbons from fuel may enter the atmosphere either directly or as by-products of the partial combustion of other hydrocarbons. The latter are particularly important because they tend to be unsaturated and relatively reactive.

Atmospheric alkenes come from a variety of processes, including emissions from internal combustion engines and turbines, foundry operations, and petroleum refining. A number of hydrocarbons, including the ones shown in Figure 8.8, are among the top 50 chemicals produced each year, with worldwide production of several billion kg per year, which adds to the likelihood of their being released to the atmosphere. In addition to the direct release of alkenes in manufacturing processes, now generally well controlled, these hydrocarbons are commonly produced by the partial combustion and "cracking" at high temperatures of alkanes, particularly in internal combustion engines.

Organooxygen Compounds

Organooxygen compounds such as those shown in Figure 8.9 are common species in air. Partial oxidation of hydrocarbons in combustion processes can generate such compounds as can photochemical oxidation processes in the atmosphere. They include a number of important industrial chemicals.

Aldehydes are particularly significant atmospheric organooxygen compounds because they are formed by photochemical processes on hydrocarbons and because they are among the few organic compounds that can undergo direct photodissociation and form reactive free radicals by absorbing photons of electromagnetic radiation in

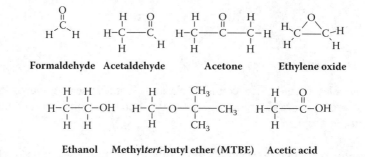

Figure 8.9.   Some of the more common organooxygen compounds found in the atmosphere. Partial oxidation of hydrocarbons produces some organooxygen species.

the atmosphere. Of the aldehydes, the simplest, formaldehyde, is the most significant atmospheric species. Ethanol is used as a supplement to gasoline fuel, acting as an octane booster. As a result, some ethanol gets into the atmosphere, but it is completely soluble in water and is washed out as a consequence. Methyl*tert*-butyl ether, MTBE, has been widely used as an octane booster in gasoline. It is not a serious air pollutant because its vapor pressure is low, but it causes problems of bad taste in water, and is being phased out of gasoline. Organic carboxylic acids, of which acetic acid is an example, are usually the final oxidation products of atmospheric hydrocarbons and readily deposit from or are washed out of the atmosphere.

Organonitrogen Compounds

Figure 8.9 shows some common organic compounds of nitrogen that may be encountered as air pollutants. Lower molecular-mass amines are volatile. These amines are prominent among the compounds giving rotten fish their characteristic odor — an obvious reason why air contamination by amines is undesirable. The simplest and most important aryl amine is aniline, used in the manufacture of dyes, amides, photographic chemicals, and drugs. A number of amines are widely used industrial chemicals and solvents, so that industrial sources have the potential to contaminate the atmosphere with these chemicals. Decaying organic matter, especially protein wastes, produces amines, so that rendering plants, packing houses, and sewage treatment plants are important sources of these substances.

Aromatic amines are of particular concern as atmospheric pollutants, particularly in the workplace, because some, such as 1-naphtylamine, are known to cause urethral tract cancer (particularly of the bladder) in exposed individuals. Some of these compounds, including 1-naphthylamine, are among the few compounds that are known to be human carcinogens based on observations of cancer in humans. This occurred as the result of exposure to workers to the compounds from coal tar used to make dyes in Germany around 1900. Aromatic amines are widely used as chemical intermediates, antioxidants, and curing agents in the manufacture of polymers (rubber and plastics), drugs, pesticides, dyes, pigments, and inks.

Nitriles, which are characterized by the $-C\equiv N$ group, have been reported as air contaminants, particularly from industrial sources. Both acrylonitrile and acetonitrile, CH_3CN, have been reported in the atmosphere as a result of synthetic rubber manufacture. Some nitro compounds, which contain the $-NO_2$ group, are found in polluted atmospheres. Nitromethane is used as a power booster in some high-performance vehicles.

Organohalides

Organohalides consist of halogen-substituted hydrocarbon molecules, each of which contains at least one atom of F, Cl, Br, or I. They may be saturated (**alkyl halides**), unsaturated (**alkenyl halides**), or aryl (**aryl halides**). Organohalides exhibit a wide range of physical and chemical properties. Structural formulas of several organohalides that have the potential to pollute the atmosphere are shown in Figure 8.10.

Organohalides, which have a wide variety of effects, are perhaps the single most important class of organic environmental pollutants with a wide variety of effects. **Dichloromethane** is a volatile liquid with excellent solvent properties for nonpolar organic solutes. It has been used as a solvent for the decaffeination of coffee, in paint strippers, as a blowing agent in urethane polymer manufacture, and to depress vapor

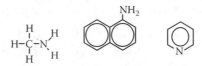

Methylamine   1-Naphthylamine   Pyridine

Acrylonitrile   Nitromethane

Figure 8.10.   Some common organonitrogen compounds that may be atmospheric pollutants.

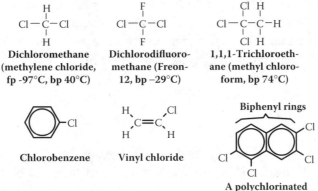

Dichloromethane (methylene chloride, fp -97°C, bp 40°C)

Dichlorodifluoromethane (Freon-12, bp -29°C)

1,1,1-Trichloroethane (methyl chloroform, bp 74°C)

Chlorobenzene

Vinyl chloride

Biphenyl rings

A polychlorinated biphenyl (PCB)

Figure 8.11.   Some examples of organohalides that may be involved in air pollution.

pressure in aerosol formulations. **Dichlorodifluoromethane** is one of the chlorofluorocarbon compounds formerly widely used as a refrigerant and involved in stratospheric ozone depletion discussed in Chapter 9. One of the more common industrial chlorinated solvents is **1,1,1-trichloroethane**. It has been implicated in stratospheric ozone destruction. **Vinyl chloride** is consumed in large quantities as a raw material to manufacture pipe, hose, wrapping, and other products fabricated from polyvinylchloride plastic. This highly flammable, volatile, sweet-smelling gas is one of the few known human carcinogens; it is known to cause angiosarcoma, a rare form of liver cancer. Aromatic halide compounds have many uses that have resulted in substantial human exposure and environmental contamination. The most environmentally significant aromatic halides are polychlorinated biphenyls, PCBs, a group of compounds formed by the chlorination of biphenyl, which have extremely high physical and chemical stabilities and other qualities that once led to their being used in many applications, including heat transfer fluids, hydraulic fluids, and dielectrics. Although not very volatile, PCBs can get into the atmosphere from high temperature sources, such as incinerators, and be transported with atmospheric particles. Although the manufacture of PCBs is now banned, residues of these very persistent compounds are still encountered in the environment.

The environmental implications of organohalide compounds are many and profound and they are arguably the class of organic compounds of most concern with respect to sustainability. The best practice of green technology attempts to eliminate organohalides wherever possible. A major characteristic of organohalides is their generally high **environmental persistence**. As an example, the chlorofluorocarbons are so resistant to chemical and biochemical attack that they persist in the atmosphere for decades, reaching stratospheric levels where they are involved in protective ozone layer depletion. Another example of the persistence of organohalides is provided by the PCBs which, though no longer manufactured, have persisted as environmental pollutants and probably will continue to for decades to come.

Organohalide compounds are **lipophilic**, meaning that they tend to accumulate in animal lipid (fat) tissue. In fact, the lipophilic character of PCBs combined with their extreme persistence means that virtually every human body now contains detectable quantities of at least several PCB compounds. The extreme stability of organohalides and their lipophilic character has resulted in an interesting "environmental distillation" phenomenon as mentioned in Section 8.7 and illustrated in Figure 8.12.

Organosulfur Compounds

Substitution of alkyl or aryl hydrocarbon groups such as phenyl and methyl for H on hydrogen sulfide, H_2S, leads to a number of different **organosulfur compounds**. Substitution for one H yields thiols, or mercaptans, R–SH; substitution for two Hs yields sulfides, also called thioethers, R–S–R. Structural formulas of examples of these compounds are shown in Figure 8.13.

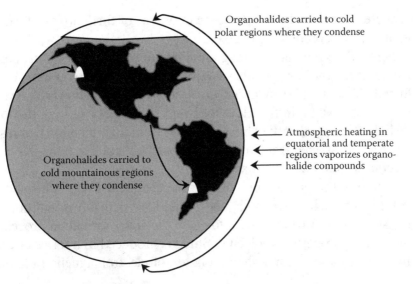

Figure 8.12.   Persistent organohalide compounds undergo a distillation process from warmer tropical and temperate regions to colder polar and mountainous regions.

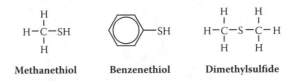

Methanethiol   Benzenethiol   Dimethylsulfide

Figure 8.13.   Examples of organosulfur compounds found in the atmosphere.

The most significant organosulfur compound in the atmosphere is dimethylsulfide, produced in large quantities by marine organisms and responsible for a large fraction of atmospheric sulfur. It is eventually oxidized by atmospheric chemical processes to microscopic droplets of sulfuric acid in the atmosphere that serve as condensation nuclei for the formation of cloud droplets. Other organosulfur compounds, particularly the mercaptans, produce localized air pollution problems because of their disgusting odors. As with all H-containing organic species in the atmosphere, reaction of organosulfur compounds with hydroxyl radical is a first step in their atmospheric photochemical reactions. The sulfur from both mercaptans and sulfides ends up as SO_2 and ultimately as sulfuric acid or sulfate salts.

8.16. PHOTOCHEMICAL SMOG AND ITS HARMFUL EFFECTS

One of the most common urban air pollution problems is the production of **photochemical smog**. This condition occurs in dry, stagnant air masses, usually stabilized by a temperature inversion (see Figure 7.5), that are subjected to intense sunlight. A smoggy atmosphere contains ozone, organic oxidants, nitrogen oxides, aldehydes, and other noxious species. In latter stages of smog formation, visibility in the atmosphere is lowered by the presence of a haze of fine particles formed by the oxidation of organic compounds in smog.

The chemical ingredients of smog are nitrogen oxides and organic compounds, both released from the automobile, as well as from other sources. The driving energy force behind smog formation is electromagnetic radiation with a wavelength at around 400 nm or less, in the ultraviolet region, just shorter than the lower limit for visible light. Energy absorbed by a molecule from this radiation can result in the formation of active species, thus initiating photochemical reactions that lead to the noxious products characteristic of smog.

The outline of the process by which photochemical smog is formed is shown in Figure 8.14. Although methane is one of the least active hydrocarbons in terms of forming smog, it will be used here to show the smog formation process because it is the simplest hydrocarbon molecule. Smog is produced in a series of chain reactions. The first of these occurs when a photon of electromagnetic radiation with a

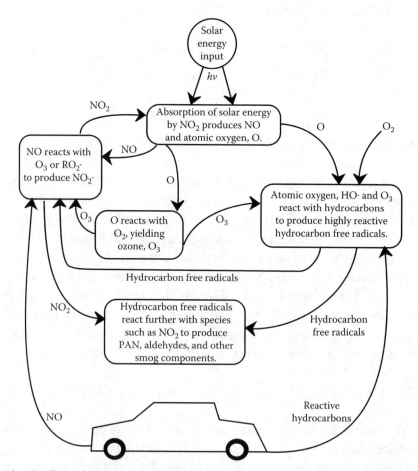

Figure 8.14. Outline of the process by which photochemical smog is formed. "R" represents hydrocarbon groups. The automobile emits nitrogen oxides and hydrocarbons, the two main ingredients required to produce photochemical smog.

wavelength less than 398 nm is absorbed by a molecule of nitrogen dioxide as shown previously in Reaction 8.10.4,

$$NO_2 + h\nu \rightarrow NO + O \qquad (8.10.4)$$

to produce an oxygen atom, O. The oxygen atom is a very reactive species that can abstract a hydrogen atom from methane,

$$CH_4 + O \rightarrow H_3C\bullet + HO\bullet \qquad (8.16.1)$$

to produce a methyl radical ($H_3C\bullet$) and a hydroxyl radical ($HO\bullet$). In these formulas, the dot shows a single unpaired electron. The hydroxyl radical is especially important in the formation of smog and in a wide variety of other kinds of photochemical reactions. The methyl radical can react with an oxygen molecule,

$$H_3C\bullet + O_2 \rightarrow H_3COO\bullet \qquad (8.16.2)$$

to produce a methylperoxyl radical ($H_3COO\bullet$). This is a strongly oxidizing, reactive species. One of the very important reactions of peroxyl radicals is their reaction with NO, produced in the photochemical dissociation of NO_2 (see Reaction 8.10.4 in Section 8.10),

$$NO + H_3COO\bullet \rightarrow NO_2 + H_3CO\bullet \qquad (8.16.3)$$

to regenerate NO_2, which can undergo photodissociation, re-initiating the series of chain reactions by which smog is formed. Literally hundreds of other reactions can occur, leading eventually to oxidized organic matter that constitutes the small particulate matter characteristic of smog.

As the process of smog formation occurs, numerous noxious intermediates are generated. One of the main ones of these is ozone and it is the single species most characteristic of smog. Whereas ozone is an essential species in the stratosphere, where it filters out undesirable ultraviolet radiation, it is a toxic species in the troposphere that is bad for both animals and plants. Another class of materials formed with smog consists of oxygen-rich organic compounds containing nitrogen of which peroxyacetyl nitrate, PAN,

$$
\begin{array}{c}
\ \ \ \ \ \ \text{H}\ \ \ \text{O} \\
\ \ \ \ \ \ | \ \ \ \ || \\
\text{H} - \text{C} - \text{C} - \text{O} - \text{O} - \text{NO}_2 \\
\ \ \ \ \ \ | \\
\ \ \ \ \ \ \text{H}
\end{array}
$$

Peroxyacetyl nitrate (PAN)

is the most common example. This compound and ones similar to it are potent oxidizers and highly irritating to eyes and mucous membranes of the respiratory tract. Also associated with smog are aldehydes, which are irritants to eyes and the respiratory tract. The simplest aldehyde, and one commonly found in smoggy atmospheres, is formaldehyde:

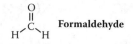

Smog adversely affects human health and comfort, plants, materials, and atmospheric quality. Each of these aspects is addressed briefly here. Ozone is the smog constituent that is generally regarded as being most harmful to humans, plants, and materials, although other oxidants and some of the noxious organic materials, such as aldehydes, are harmful as well.

People exposed to 0.15 ppm of ozone in air experience irritation to the respiratory mucous tissues accompanied by coughing, wheezing, and bronchial constriction. These effects may be especially pronounced for individuals undergoing vigorous exercise because of the large amounts of air that they inhale. On smoggy days, air pollution alerts may advise against exercise and outdoor activities. Because of these effects, the U.S. Environmental Protection Agency has recommended an 8-h standard limit for ozone of 0.08 ppm. In a smoggy atmosphere, the adverse effects of ozone are aggravated by exposure to other oxidants and aldehydes.

Plants are harmed by exposure to nitrogen oxides, ozone, and peroxyacetyl nitrate (PAN, see above), all oxidants present in a smoggy atmosphere. PAN is the most harmful of these constituents, damaging younger plant leaves, especially. Ozone exposure causes formation of yellow spots on leaves, a condition called chlorotic stippling. Some plant species, including sword-leaf lettuce, black nightshade, quickweed, and double-fortune tomato, are extremely susceptible to damage by oxidant species in smog and are used as bioindicators of the presence of smog. Costs of crop and orchard damage by smog run into millions of dollars per year in areas prone to this kind of air pollution, such as southern California.

Materials that are adversely affected by smog are generally those that are attacked by oxidants. The best example of such a material is rubber, especially natural rubber, which is attacked by ozone. Indeed, the hardening and cracking of natural rubber has been used as a test for atmospheric ozone.

SUPPLEMENTAL REFERENCES

Allaby, Michael, *Fog, Smog, and Poisoned Rain*, Facts On File, New York, 2003.

Brasseur, Guy P., Ronald G. Prinn, and Alexander A.P. Pszenny, Eds., *Atmospheric Chemistry in a Changing World*, Springer-Verlag, New York, 2003.

Calhoun, Yael, Ed., *Air Quality*, Chelsea House Publishers, Philadelphia, 2005.

Finlayson-Pitts, Barbara J. and James N. Pitts, *Chemistry of the Upper and Lower Atmosphere: Theory, Experiments, and Applications*, Academic Press, San Diego, CA, 1999.

Fogg, Peter and James Sangster, *Chemicals in the Atmosphere: Solubility, Sources, and Reactivity*, John Wiley & Sons, Hoboken, NJ, 2003.

Godish, Thad, *Air Quality*, 4th ed., CRC Press, Boca Raton, FL, 2004.

Hobbs, Peter V., *Basic Physical Chemistry for the Atmospheric Sciences*, Cambridge University Press, New York, 2000.

Hobbs, Peter V., *Introduction to Atmospheric Chemistry*, Cambridge University Press, New York, 2000.

Jacob, Daniel, *Introduction to Atmospheric Chemistry*, Princeton University Press, Princeton, NJ, 1999.

Jacobson, Mark Z., *Atmospheric Pollution*, Cambridge University Press, New York, 2002.

Kidd, J.S. and Renee A. Kidd, *Air Pollution: Problems and Solutions*, Facts on File, New York, 2005.

Seinfeld, John H. and Spyros N. Pandis, *Atmospheric Chemistry and Physics: From Air Pollution to Climate Change*, 2nd ed., John Wiley & Sons, New York, 2006.

Sherman, Joe, *Gasp!: The Swift and Terrible Beauty of Air*, Shoemaker & Hoard, Washington, D.C., 2004.

Visconti, Guido, *Fundamentals of Physics and Chemistry of the Atmosphere*, Springer-Verlag, New York, 2001.

QUESTIONS AND PROBLEMS

1. What does an asterisk, *, denote after the formula of a species, such as NO_2*, in the atmosphere?

2. In what sense are $h\nu$ and HO• "of utmost importance" in the atmosphere?

3. What does the dot denote after a species such as HO•? Why are such species usually highly reactive?

4. What are the three classes of "relatively reactive and unstable species in the atmosphere"? Describe their characteristics and give examples of each.

5. What are the three main reactions or processes by which photochemically excited species lose their excess energy?

6. What is a chain reaction? What is a common photochemical reaction that initiates chain reactions in the atmosphere?

7. With which two common molecular species does atmospheric hydroxyl radical react leading to its removal from the atmosphere?

8. What kinds of pollutant species are likely to form in an atmosphere contaminated with nitrogen oxides and hydrocarbons and subjected to sunlight?

9. What are three effects of the reaction of NH_3 with acids in the atmosphere?

10. What is the ionosphere? How is it formed? Why was it known to exist long before rockets were developed to reach its altitude?

11. Cite the evidence showing the importance of life on the nature of Earth's atmosphere.

12. How does elemental oxygen react in the stratosphere? What is a very significant product of these reactions? What protective function does it serve?

13. What are three oxygen species other than unexcited O_2 that exist in the upper atmosphere?

14. What is an atmospheric substance that is essential to our well-being in the stratosphere, but toxic in the troposphere? Explain.

15. Explain why there is an annual oscillation in carbon dioxide levels in the atmosphere. These oscillations are more pronounced in the Northern than in the Southern Hemisphere. Offer a possible explanation for that observation.

16. What phenomenon is responsible for the temperature maximum at the boundary of the stratosphere and the mesosphere?

17. Why might it be expected that the reaction of a free radical with NO_2 is a chain-terminating reaction (consider the total number of electrons in NO_2)?

18. Suppose that 22.4 l of dry air at 0°C and 1 atm pressure is used to burn 1.50 g of carbon to form CO_2, and that the gaseous product is adjusted to 0°C and 1 atm pressure. What are the volume and the average molecular mass of the resulting mixture?

19. Of the species O, HO*•, NO_2*, H_3C•, and N^+, which could most readily revert to a nonreactive, "normal" species in total isolation?

20. A 12.0-l sample of air at 25°C and 1.00 atm pressure was collected and dried. After drying, the volume of the sample was exactly 11.50 l. What was the percentage *by mass* of water in the original air sample?

21. At an altitude of 50 km, the average atmospheric temperature is essentially 0°C and pressure is 0.0040 atm. What is the average number of air molecules per cubic centimeter of air at this altitude?

22. Give possible examples of neutralization reactions and oxidation reactions enabled by particles in the atmosphere. What kinds of particles would be required for each of these types of reactions?

23. Define the photochemical phenomena represented by each of the following:

$$(1) \ O_2{}^* \rightarrow O + O$$

$$(2) \ O_2{}^* + O_3 \rightarrow 2O_2 + O$$

$$(3) \ N_2{}^* \rightarrow N_2{}^+ + e^-$$

24. Why are free radicals so highly reactive? Despite their high reactivity, why do free radicals tend to persist for significant lengths of time at high altitudes?

25. What is the kind of reaction below called? Explain.

$$H_3C\bullet + H_3C\bullet \rightarrow C_2H_6$$

9. SUSTAINING AN ATMOSPHERE FOR LIFE ON EARTH

9.1. BLUE SKIES FOR A GREEN EARTH

There is a very strong connection between life forms on Earth and the nature of Earth's atmosphere and climate, which determine its suitability for life. As proposed by James Lovelock, a renowned British environmental chemist, this forms the basis of the **Gaia hypothesis**, which contends that the atmospheric O_2/CO_2 balance established and maintained by organisms determines and stabilizes Earth's climate and other environmental conditions. Living organisms have had a profound influence on the atmosphere in the past. The most massive of the changes caused were those arising from photosynthesis, which utilizes solar energy, $h\nu$, to produce biomass, $\{CH_2O\}$, and molecular oxygen, O_2:

$$CO_2 + H_2O + h\nu \rightarrow \{CH_2O\} + O_2(g) \qquad (9.1.1)$$

This process converted Earth's atmosphere from a chemically reducing to a chemically oxidizing state and precipitated enormous deposits of insoluble oxidized iron:

$$4Fe^{2+} + O_2 + 4H_2O \rightarrow 2Fe_2O_3 + 8H^+ \qquad (9.1.2)$$

In addition to providing O_2 that most nonphotosynthetic organisms use for respiration, the photosynthetically released oxygen formed stratospheric ozone (O_3), which absorbs damaging ultraviolet radiation from the sun, enabling living organisms to move from water onto land where they are directly exposed to sunlight.

Other instances of climatic change and regulation induced by organisms can be cited. An example is the maintenance of atmospheric carbon dioxide at low levels through the action of photosynthetic organisms (note from Reaction 9.1.1 that photosynthesis removes CO_2 from the atmosphere). But, at an ever-accelerating pace during the last 200 years, another organism, humankind, has engaged in a number of activities that are altering the atmosphere profoundly. These are summarized as follows:

- Industrial activities, which emit a variety of atmospheric pollutants including SO_2, particulate matter, photochemically reactive hydrocarbons, chlorofluorocarbons, and inorganic substances (such as toxic heavy metals)

- Burning of large quantities of fossil fuel, which can introduce CO_2, CO, SO_2, NO_x, hydrocarbons (including CH_4), particulate soot, polycyclic aromatic hydrocarbons, and fly ash into the atmosphere

- Transportation practices, which emit CO_2, CO, NO_x, photochemically reactive (smog-forming) hydrocarbons, and polycyclic aromatic hydrocarbons

- Alteration of land surfaces, including deforestation

- Burning of biomass and vegetation, including tropical and subtropical forests and savanna grasses, which produces atmospheric CO_2, CO, NO_x, and particulate soot and polycyclic aromatic hydrocarbons

- Agricultural practices, which produce methane (from the digestive tracts of domestic animals and from the cultivation of rice in waterlogged anaerobic soils) and N_2O from bacterial denitrification of nitrate-fertilized soils

These kinds of human activities have significantly altered the atmosphere, particularly in regard to its composition of minor constituents and trace gases. Major effects have been the following:

- Increased acidity in the atmosphere

- Production of pollutant oxidants in localized areas of the lower troposphere afflicted with photochemical smog

- Elevated levels of infrared-absorbing gases (greenhouse gases)

- Threats to the ultraviolet-filtering ozone layer in the stratosphere

- Increased corrosion of materials induced by atmospheric pollutants

In 1957, photochemical smog was only beginning to be recognized as a serious problem, acid rain and the greenhouse effect were scientific curiosities, and the ozone-destroying potential of chlorofluorocarbons had not even been imagined. In that year, Revelle and Suess[1] prophetically referred to human perturbations of the Earth and its climate as a massive "geophysical experiment."

There is no greater challenge for green science and technology than the maintenance of an atmosphere suitable for life on Earth. Humans may do rather grievous harm to the geosphere (having done so in the past and continuing to do so in the present) but still compensate for the damage by measures such as by utilizing innovative means of raising crops that can grow in impaired soil. Polluted water can be purified, and pure water can be extracted from the limitless resources of Earth's oceans. But, once Earth's atmosphere and nurturing climate are ruined, a tipping point can be reached from which there is little hope of return. Earth's atmosphere is like the roof of a house. If the roof deteriorates, the condition of the rest of the house makes little difference, because, when the roof is destroyed, the rest of the structure will follow. Therefore, the maintenance and enhancement of the atmospheric environment and climate should be at the top of the list of priorities of scientists, engineers, politicians, business interests, and all citizens of Planet Earth.

9.2. GREENHOUSE GASES AND GLOBAL WARMING

It is now generally agreed that the greatest challenge to humankind's relatively comfortable existence on earth today is that of **global warming** associated with emissions of greenhouse gases, especially carbon dioxide, from the anthrosphere to the atmosphere. Ample evidence exists of massive changes in Earth's climate associated with changes in global temperature in times past. Indeed, humankind exists now in an approximately 10,000-year interglacial era called the **holocene**. Evidence from the past suggests that major changes in climate may occur very rapidly, within a few years' period of time.

This section deals with infrared-absorbing trace gases (other than water vapor) in the atmosphere that contribute to global warming and with the influence of particles on temperature. These gases produce a "greenhouse effect" by allowing incoming solar radiant energy to penetrate to the earth's surface while reabsorbing infrared radiation emanating from it. Levels of these "greenhouse gases" have increased at a rapid rate during recent decades and are continuing to do so.

Figure 9.1 shows global temperature trends since 1880. Although an increase in global temperature of only about 1°C over a period of about 1 century as shown

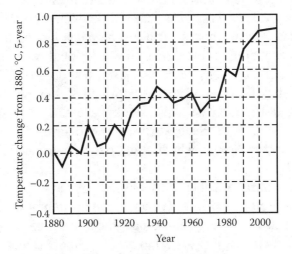

Figure 9.1.   Global temperature trends. Earlier values are less certain because of the lack of sophisticated means of measuring temperature. More recent values are very accurate because of the use of satellite-based technologies for measuring temperature.

in Figure 9.1 may not seem like much, global mean temperature variations of only a few degrees have marked the differences between past ice ages and relatively temperate or tropical climatic conditions. Concern over this phenomenon has intensified since about 1980. This is because ever since accurate temperature records have been kept, the 1980s were the warmest 10-year period recorded and included several record warm years. In general, the warming trend has continued since then.

Although there are uncertainties associated with global warming, several aspects pertaining to the phenomenon are well established. It is known that, along with water vapor, CO_2, and other greenhouse gases, such as CH_4, are primarily responsible for the absorption of infrared energy reemitted by the Earth such that some of this energy is reradiated back to the Earth's surface warming it. The levels of the greenhouse gases have increased markedly since about 1850 as nations have become industrialized and as forest lands and grasslands have been converted to agriculture. As noted in Section 8.9 and plotted in Figure 8.4, global atmospheric levels of carbon dioxide are increasing by about 1 ppm by volume per year and are now approaching 380 ppm.

Current evidence suggests that changes in the atmospheric carbon dioxide level will substantially alter Earth's climate through the greenhouse effect. With current trends, it is likely that global CO_2 levels will double from preindustrial levels later this century, which may well raise Earth's mean surface temperature by 1.5 to 4.5°C or even more. Such a change would cause more irreversible environmental changes than any other disaster short of global nuclear war or asteroid impact.

As shown in Figure 9.2, per capita carbon dioxide emissions are highest for industrialized countries. Furthermore, economic development of countries with high populations, such as China and India, can be expected to add large quantities of carbon dioxide to the atmosphere in the future. Chlorofluorocarbons, which also are greenhouse gases, were not even introduced into the atmosphere until the 1930s. Although trends in levels of these gases are well known, their effects on global temperature and climate are much less certain. The phenomenon has been the subject of much computer modeling. Most models predict global warming of at least 3.0°C and up to 5.5°C occurring over a period of just a few decades. These estimates are sobering because they correspond to the approximate temperature increase since the last Ice Age 18,000 years past, which took place at a much slower pace of only about 1 or 2°C per 1000 years. Such warming would have profound effects on rainfall, plant growth, and sea levels, which might rise as much as 0.5 to 1.5 m, flooding low-lying areas.

Carbon dioxide is responsible for about half of the atmospheric heat retained by trace gases. It is produced primarily by the combustion of fossil fuels, and deforestation accompanied by burning and biodegradation of biomass. On a molecule-for-molecule basis, methane, CH_4, is 20 to 30 times more effective in trapping heat than is CO_2. Other trace gases that contribute are chlorofluorocarbons and N_2O. The potential of such a gas to cause greenhouse warming may be expressed by a **global**

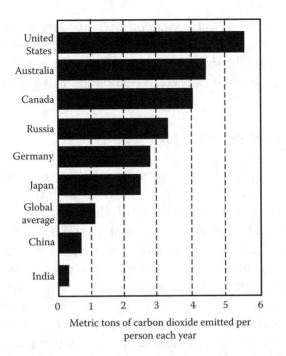

Figure 9.2.   Per capita emissions of carbon dioxide for major countries of the world.

warming potential (GWP), originally defined by the United Nations' Intergovernmental Panel on Climate Change, which is a function of both the infrared absorption characteristics and the lifetime of the gas.

Analyses of gases trapped in polar ice samples indicate that preindustrial levels of CO_2 and CH_4 in the atmosphere were approximately 260 ppm and 0.70 ppm, respectively. Over the last 300 years these levels have increased to current values of around 380 ppm for CO_2 and 1.8 ppm for CH_4, respectively; most of the increase by far has taken place at an accelerating pace over the last 100 years. (A note of interest is the observation based upon analyses of gases trapped in ice cores that the atmospheric level of CO_2 at the peak of the last Ice Age about 18,000 years past was 25% below preindustrial levels.) About half of the increase in carbon dioxide in the last 300 years can be attributed to deforestation, which still accounts for approximately 20% of the annual increase in this gas. Carbon dioxide is increasing by about 1 ppm/year.

Methane

Atmospheric methane, CH_4, is going up at a rate of almost 0.02 ppm/year. The comparatively very rapid increase in methane levels is attributed to a number of factors resulting from human activities. Among these are direct leakage of natural gas, by-product emissions from coal mining and petroleum recovery, and release from the burning of savannas and tropical forests. Biogenic sources resulting from human activities produce large amounts of atmospheric methane. These include methane from bacteria degrading organic matter such as municipal refuse in landfills; methane evolved from anaerobic biodegradation of organic matter in rice paddies; and methane emitted as the result of bacterial action in the digestive tracts of ruminant animals.

Radiative Forcing and Feedback Mechanisms

Radiative forcing is a term used to describe the reduction in infrared radiation penetrating outward through the atmosphere per unit increase in the level of gas in the atmosphere. Radiative forcing of CH_4 is about 25 times that of CO_2. Increases in the concentration of methane and several other greenhouse gases have such a disproportionate effect on retention of infrared radiation because their infrared absorption spectra fill gaps in the overall spectrum of outbound radiation not covered by much more abundant carbon dioxide. Therefore, whereas an increase in carbon dioxide concentration has a comparatively small incremental effect because the gas is already absorbing such a high fraction of infrared radiation in regions of the spectrum where it absorbs, an increase in the concentration of methane, chlorofluorocarbon, or other greenhouse gases has a comparatively much larger effect.

Both positive and negative feedback mechanisms may be involved in determining the rates at which carbon dioxide and methane build up in the atmosphere. Laboratory studies indicate that increased CO_2 levels in the atmosphere cause accelerated uptake of this gas by plants undergoing photosynthesis, which tends to slow the buildup of atmospheric CO_2. Given adequate rainfall, plants living in a warmer climate that would result from the greenhouse effect would grow faster and take up more CO_2. This could be an especially significant effect of forests, which have a high CO_2-fixing ability. However, the projected rate of increase in carbon dioxide levels is so rapid that forests would lag behind in their ability to fix additional CO_2. Similarly, higher atmospheric CO_2 concentrations will result in accelerated sorption of the gas by oceans. The amount of dissolved CO_2 in the oceans is about 60 times the amount of CO_2 gas in the atmosphere. However, the times for transfer of carbon dioxide from the atmosphere to the ocean are of the order of years. Because of low mixing rates, the times for transfer of carbon dioxide from the upper approximately 100-m layer of the oceans to ocean depths is much longer, of the order of decades. Therefore, like the uptake of CO_2 by forests, increased absorption by oceans will lag behind the emissions of CO_2. A concern with increased levels of CO_2 in the oceans is the lowering of ocean water pH that will result. Even though such an effect will be slight, of the order of one tenth to several tenths of a pH unit, it has the potential to strongly impact organisms that live in ocean water. Severe drought conditions resulting from climatic warming could cut down substantially on CO_2 uptake by plants. Warmer conditions would accelerate release of both CO_2 and CH_4 by microbial degradation of organic matter. (It is important to realize that about twice as much carbon is held in soil in dead organic matter — necrocarbon — potentially degradable to CO_2 and CH_4 as is present in the atmosphere.) Global warming might speed up the rates at which biodegradation adds these gases to the atmosphere.

Particles and Global Warming

Whereas the effects of carbon dioxide and other gases on temperature are relatively easy to calculate, the effects of particles are much more complicated. Atmospheric particles have both direct effects exerted by scattering and absorbing radiation and indirect effects in changing the microphysical structure, lifetimes, and quantities of clouds. Particles interact with and scatter most strongly radiation that is of a wavelength similar to the size of the particles. Most of the incoming solar energy is at wavelengths less than 4 μm and most particles are smaller than 4 μm, so the major effect of atmospheric particles is to scatter radiation from incoming solar energy, which has a cooling effect on the atmosphere. Some kinds of particles, such as those composed of black carbon and soot, absorb incoming solar radiation, warming the atmosphere.

Liquid water droplets composing clouds can both scatter incoming radiation and absorb outbound infrared radiation. Clouds at lower altitudes act mainly to lower atmospheric temperature by scattering lower wavelength radiation, whereas clouds at higher altitudes tend to absorb outbound infrared causing temperature increases. Aerosol particles such as sulfate salts that act as cloud condensation nuclei on which atmospheric water vapor condenses tend to increase the number of particles. In general, this has a cooling effect.

Overall, the effects of particles on global temperature are variable and not particularly well understood. Both cooling and warming effects may occur. Modeling these effects is much more challenging than is modeling the effects of gaseous atmospheric constituents such as carbon dioxide.

The Outlook for Global Warming

It is certain that atmospheric CO_2 levels will continue to increase significantly. The degree to which this occurs depends upon future levels of CO_2 production and the fraction of that production that remains in the atmosphere. Given plausible projections of CO_2 production and a reasonable estimate that half of that amount will remain in the atmosphere, projections can be made that indicate that sometime within the next century, the concentration of this gas will reach 600 ppm in the atmosphere, well over twice the levels estimated for preindustrial times. Much less certain are the effects that this change will have on climate. It is virtually impossible for the elaborate computer models used to estimate these effects to accurately take account of all variables, such as the degree and nature of cloud cover. The magnitudes of the effects of clouds reflecting incoming radiation (a cooling effect) and absorbing outgoing radiation (a warming effect) depend upon the degree of cloud cover, brightness, altitude, and thickness. In the case of clouds, too, feedback phenomena occur; for example, warming induces formation of more clouds, which reflect more incoming energy.

Drought (see Section 9.4) is one of the most serious problems that could arise from major climatic change resulting from greenhouse warming. Typically, a 3°C warming would be accompanied by a 10% decrease in precipitation. Water shortages would be aggravated, not just from decreased rainfall, but from increased evaporation as well. Increased evaporation results in decreased runoff, thereby reducing water available for agricultural, municipal, and industrial use. Water shortages, in turn, lead to increased demand for irrigation and to the production of lower quality, higher salinity runoff water and wastewater. In the U.S., such a problem would be especially intense in the Colorado River basin, which supplies much of the water used in the rapidly growing U.S. Southwest.

A variety of other problems, some of them unforeseen as of now, could result from global warming. An example is the effect of warming on plant and animal pests — insects, weeds, diseases, and rodents. Many of these would certainly thrive much better under warmer conditions.

Interestingly, another air pollutant, acid-rain-forming sulfur dioxide (see Section 9.6), may have a counteracting effect on greenhouse gases. This is because sulfur dioxide is oxidized in the atmosphere to sulfuric acid, forming a light-reflecting haze. Furthermore, the sulfuric acid and resulting sulfates act as condensation nuclei on which atmospheric water vapor condenses, thereby increasing the extent, density, and brightness of light-reflecting cloud cover. Sulfate aerosols are particularly effective in counteracting greenhouse warming in central Europe and the eastern U.S. during the summer.

Some evidence of the effects of global warming may have been manifested by powerful El Niño phenomena in recent years. El Niño is the name given to the warming of surface water in the eastern Pacific Ocean that commonly takes place around Christmas time. The 1997/1998 El Niño was particularly powerful and caused many marked weather phenomena. It also increased confidence in global climate models because of the generally accurate forecasts of its effect on climate. Forecasts of a warmer and wetter winter than normal in the continental U.S. with particularly heavy rains in California and the Gulf Coast regions were fulfilled. In fact, Los Angeles experienced more than 33 cm of rainfall during February 1998, setting a new record for the month (since broken in 2005), and the southeastern U.S. had the most rainfall in more than a century of record keeping. Eastern equatorial Africa experienced torrential rains, as did parts of Peru. Indonesia experienced a severe drought that resulted in extensive destruction of forests by fires. The central and northern continental U.S. benefited from a warmer winter than usual with record warmth in some eastern and midwestern regions.

As it affected North America, the stronger El Niño enhanced the intensity of the eastward-flowing high-altitude jet stream, the result of a greater temperature differential between warmer southern tropical waters and the colder northern regions.

The intense jet stream carried storms rapidly across the southern U.S., causing intense rainstorms in these regions and keeping cold Arctic air to the north. Some authorities contend that the effects of the 1997/1998 El Niño were amplified by global warming.

Serious Concern over Changes in Climate

Insurance companies have become quite concerned about the possibility of significant changes in global climate, especially because of potential effects on the frequency and severity of damaging storms. In 1996–1997 there were at least six weather disasters in the U.S. that cost over a billion dollars each. These included (1) a catastrophic drought in the Southern Plains that began in the fall of 1995 and lasted through the summer of 1996; (2) a blizzard followed by flooding that occurred in the northeastern U.S., the mid-Atlantic states, and the Appalachian mountain areas in January of 1996; (3) flooding in the Pacific Northwest in February 1996; (4) Hurricane Fran, which caused 36 deaths and over $5 billion in damage during September 1996; (5) severe flooding in the northern west coast region of the U.S. in December 1996, and January, 1997; and (6) an unprecedented 500-year flood complicated by freezing weather and ice jams that hit the Dakotas and Minnesota in April 1997, virtually wiping out the city of Grand Forks, North Dakota. On May 3, 1999, an F5 tornado, the largest class of this kind of treacherous storm, took a number of lives and caused about $1 billion damage in Central Oklahoma. In the 2004 hurricane season, Florida was hit by four major hurricanes causing record losses. In 2005, a record year for hurricanes, Hurricane Katrina essentially destroyed New Orleans and coastal areas of Louisiana and Mississippi leading to estimates of more than $200 billion to repair the damage. The severity of the 2005 hurricane season was due in large part to very warm temperatures in the Gulf of Mexico, which were at least consistent with global warming.

Although drought is the most frequently mentioned possible effect of greenhouse warming, the frequency and severity of storms, often accompanied by high levels of precipitation, are of particular concern to insurance companies. During the 1990s, "100-year" weather events that are expected to occur statistically only once each century became so common in the U.S. that the term has begun to lose its meaning. As shown in Figure 9.3, the last century has seen a significant increase in precipitation in the lower 48 continental U.S. These observations are consistent with currently accepted models of the weather effects of greenhouse warming, which predict that more precipitation will come in the form of brief, heavy precipitation events such as thunderstorms (heavy convective storms) rather than through gentle, more beneficial, rainfall that comes over a longer time period. The debate continues over whether the apparent weather anomalies observed during recent years denote a marked change in climate or are simply normal fluctuations in weather, but insurance companies are definitely concerned.

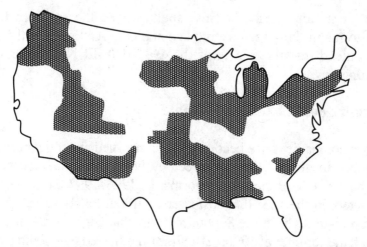

Figure 9.3.   Sections of the lower 48 U.S. states in which precipitation levels have increased by 10–20% since about 1900 (shown as shaded regions). Some areas, particularly North Dakota, eastern Montana, Wyoming, and California have experienced decreases in precipitation of a similar magnitude. This map is based on data gathered by the National Oceanic and Atmospheric Administration's National Climatic Data Center.

9.3. GREEN SCIENCE AND TECHNOLOGY TO ALLEVIATE GLOBAL WARMING

Although there are still "global-warming deniers" who attempt to discredit those who have concern over global warming, the overwhelming consensus of reputable scientists is that global warming is taking place and that greenhouse gases, particularly carbon dioxide, are the main cause. The question for responsible people then becomes what to do about the problem. The possibilities can be divided into three categories: (1) minimization, (2) counteracting measures, and (3) adaptation.

Minimization

Minimization refers to measures taken to reduce emissions of greenhouse gases. Minimization is tied inextricably to energy production and utilization, because most greenhouse gas emissions are because of the burning of fossil fuels. Fortunately, fossil fuel may be consumed much more efficiently utilizing existing and developing energy conservation measures. Efficient utilization of fossil energy can be accomplished without major economic disruption. A prime example would be the conversion of the U.S. private vehicle fleet to hybrid internal combustion/electric propulsion. Hybrid vehicles can be manufactured with the capability to charge their batteries from house electrical current to provide sufficient charge for approximately 30 km of travel before the internal combustion engine even has to be turned on, which would take care of as much as half of routine commuting and travel. As a green fringe benefit without additional charge, employers could provide charging stations in parking facilities so that vehicles could be recharged during the work day. Such a system could readily reduce by 50% the amount of fuel and hence the emissions of greenhouse gases of current private automotive transport systems.

Conversion of the current U.S. freight transportation system from truck to electrified rail to the maximum extent possible would make a further major reduction in greenhouse gas emissions. Since about 1990, there has been a significant shift in the amount of freight transported by rail in the form of shipping containers and truck trailers to the point that some rail freight lines are nearing their maximum capacity. Unlike Europe, most railways in the U.S. are not electrified; although that certainly could be done and the electricity generated by means that do not require burning of fossil fuels. Interstate highway rights of way can serve as routes for new rail lines in some cases.

Utilization of fossil fuels for heating and air conditioning can be done much more efficiently. The best fuel for home heating is natural gas (methane, CH_4). This fuel emits the least amount of greenhouse gas (carbon dioxide) per unit heat delivered because of its relatively high content of hydrogen. Rather than burning methane in a furnace, it can be used as a fuel in a small internal combustion engine connected to a heat pump to pump heat from the outside. The exhaust gases from the engine can be cooled and the water vapor in them condensed to capture additional heat.

One way to reduce the release of carbon dioxide is by using biomass as fuel or raw material for the manufacture of various products. Burning a biomass fuel does release carbon dioxide to the atmosphere, but an exactly equal amount of carbon dioxide was removed from the atmosphere in the photosynthetic process by which the biomass was made, so there is no net addition of CO_2. Unless or until biomass-derived materials used in feedstocks are burned, their use represents a net loss of carbon dioxide from the atmosphere.

Another potential use of green chemistry to prevent addition of carbon dioxide to the atmosphere is through **carbon sequestration** in which carbon dioxide is produced, but is bound in a form such that it is not released to the atmosphere. This approach has the greatest potential in applications where the carbon dioxide is produced in a concentrated form. Carbon from coal can be reacted with oxygen and water to produce elemental hydrogen and carbon dioxide in a coal gasification process. The net reaction for this production is the following:

$$2C + O_2 + 2H_2O \rightarrow 2CO_2 + 2H_2 \tag{9.3.1}$$

The hydrogen generated can be used as a pollution-free fuel in fuel cells or combustion engines. The carbon dioxide can be pumped into deep ocean waters, although this has the potential to lower ocean pH slightly, which would harm marine organisms. Another option is to pump the carbon dioxide deep underground. A side benefit of the latter approach is that in some areas carbon dioxide pumped underground can be used to recover additional crude oil from depleted oil-bearing formations.

The best way to reduce greenhouse gas emissions from home heating is to avoid the use of fossil fuels entirely. Substantial progress has been made in solar heating systems that do not use fossil fuels. Another good measure is the use of electricity for heating and air conditioning, generating the electricity from energy sources other than fossil fuels. Additional options for generating electricity without adding to the burden of greenhouse gases are discussed in Chapter 17.

A green technological approach to the reduction of carbon dioxide emissions is to develop alternative methods of energy production. One thing that would be very beneficial is the development of more efficient photovoltaic cells. These devices have become marginally competitive for the generation of electricity, and even relatively small improvements in efficiency would enable their much wider use, replacing fossil fuel sources of electricity generation. Another device that would be extremely useful is a system for the direct photochemical dissociation of water to produce elemental hydrogen and oxygen, which could be used in fuel cells. An application of green biochemistry that would reduce carbon dioxide emissions is the development of plants with much higher efficiencies for photosynthesis. Plants now are only about 0.5% efficient in converting light energy to chemical energy. Raising this value to only 1% would make a vast difference in the economics of producing biomass as a substitute for fossil carbon.

One of the more contentious issues related to greenhouse gas emissions has been the Kyoto treaty and the refusal of the U.S. to ratify that agreement. This treaty evolved from a 1997 meeting of 160 nations in Kyoto, Japan. The agreement called for stabilization of greenhouse gas emissions to 1990 levels during the period 2008 to 2012, which would have led to 23% less of the emissions below levels projected during the 1990s without remedial action. The U.S. has refused to ratify the agreement, a position reiterated at an international meeting on global warming held in Montreal in December, 2005. The reason given for the U.S. position is that the treaty exempted large developing countries, especially India and China from participating for economic reasons. Examination of Figure 9.2 shows the rationale for this objection in that relatively modest per capita emissions of greenhouse gases in India and China multiplied times their large populations will cause vast increases in the gases emitted. Refusal to participate in the Kyoto agreement of the U.S., the country responsible for more greenhouse gas emissions than any other, has sent a very unpopular message to the rest of the world.

Counteracting Measures

Counteracting measures for reducing global warming consist of schemes such as the injection of light-reflecting particles into the upper atmosphere. The scale required of such measures is so great that they are probably impossible to implement in a meaningful way. One possibility is to increase (or at least not reduce) the amount of sulfur gases emitted to the atmosphere that oxidize to sulfuric acid which serves to generate condensation nuclei that produce light-reflecting clouds.

Another far-out possibility would be to use high-flying tanker aircraft to separately inject hydrogen chloride and ammonia vapors into the atmosphere. Any chemistry student who has had separate beakers of hydrochloric acid and ammonia solution close together knows that HCl and NH_3 react,

$$HCl(g) + NH_3(g) \rightarrow NH_4Cl(s) \qquad (9.3.2)$$

to produce a dense fog of ammonium chloride particles that might serve as cloud condensation nuclei. In addition to the obvious questions regarding air pollution that this remedy raises, it might be difficult to find aircrew willing to fly Airbus A380 jumbo jet aircraft into the lower stratosphere laden with pressurized tanks of corrosive hydrogen chloride and ammonia.

A possibly significant counteracting measure is modification of Earth's surface in a manner to reflect light. This can potentially be done with appropriate kinds of vegetation in the form of forests and grasslands. Again, the scale required to make any significant difference would be huge. Another small effect might be had by designing surfaces (roofs and parking lot surfaces) of anthrospheric structures to maximize reflection of solar radiation. Such a measure would favor reflective aluminum roofs over dark roofing and light-colored concrete over black asphalt for parking lot surfaces.

Adaptation

Because global warming will in fact occur and neither minimization nor counteracting measures will be sufficient to stop it, **adaptation** to climate warming will be required. Other than increases in global temperature, there are many effects of global warming suggesting a variety of adaptations. It may be anticipated that adaptation to global warming will be one of the most significant activities of green science and technology in the future.

Water shortages and drought (see Section 9.4) will constitute perhaps the most troublesome aspects of climate warming. Water, already in short supply in many parts of the world, will become scarcer. It will be necessary to implement more efficient irrigation practices and to grow crops that require less irrigation. One approach with significant promise is to grow crops on arid coastal lands irrigated with sea water.[2] Plants that can grow in seawater are called halophytes and can produce 1 to 2 kg of dry biomass per square meter of field area, which is about the same production as conventional crops such as alfalfa. Some of the most productive plants that grow in seawater have relatively unattractive names including glasswort, sea blite, saltbush, and salt grass. Though not producing grain, some of these plants produce abundant forage that can be eaten by animals. One small problem is that because of the high salt content of the forage, animals that consume it must drink significantly greater amounts of fresh water. One saltwater plant that does produce abundant seeds is *Salicornia bigelovii* that rapidly colonizes new mud flats. With a salt content of less than 3%, the seeds are 35% protein and 30% highly polyunsaturated oil similar to safflower oil in composition. The seeds contain bitter saponins, which limit somewhat the amount of seed or meal left after extracting oil that can be fed to animals. Especially for oil production, growing oil-producing algae in saltwater ponds is a promising approach.

One of the greatest adverse effects of global warming results from the effects of high temperatures on people, particularly the elderly who are vulnerable. This was illustrated tragically in Europe in August, 2003. The highest temperatures ever recorded in the U.K. occurred on August 10, 2003, with temperatures of 38°C (100.4°F) at London's Heathrow airport and 38.1°C at Gravesend, Kent. Over 1000 people died from the heat wave in the U.K. However, the greatest toll was in France where about 15,000 people, mostly elderly, died from the heat in August, 2003. The problem was exacerbated by the fact that France is not used to extreme heat and the custom is for many people, including government ministers and physicians who would have been involved in remedial measures, to take August off for vacation. Funeral homes were overwhelmed and refrigerated warehouses had to be used to store bodies until they could be identified and buried. A major concern was that the nuclear reactors that provide much of France's electrical power could not be adequately cooled and in some cases had to be cooled by spraying with hoses. Other countries were adversely affected. Portugal lost 10% of its forests in forest fires. On the positive side, the hot dry summer yielded grapes with very high sugar contents resulting in one of the best years ever for French wines.

As global warming occurs, a major adaptation will need to be the installation of air conditioning and other cooling measures in regions of the world where air conditioning in homes and commercial buildings has been uncommon. This is particularly true in Europe where periods of hot weather will become more common, though of shorter duration than in much of the U.S., for example. In addition to the installation of air conditioning, provision will need to be made to provide sustainable power for it. For example, there may be more of a need for fuel turbine peaking facilities for electrical power generation and for greater reserves of cooling water for nuclear power reactors.

9.4. DROUGHT AND DESERTIFICATION

Drought is a condition of water shortage resulting from a long-term deficiency in rainfall. In many parts of the world, cycles of drought are natural and expected, and their adverse effects are minimized with proper planning. In the continental U.S., for example, periods of drought lasting several years occurred in the 1870s, 1900s, 1930s (the most severe on record), 1950s, and 1970s. Record droughts afflicted the southwestern U.S. during the early 2000s. The "dust bowl" years of the 1930s were characterized by devastating dust storms that ripped topsoil from the land and virtually ruined huge areas of once productive prairie land. This disaster gave rise to one of the earliest environmental movements and resulted in huge federal programs designed to protect irreplaceable farmland from drought and wind erosion. Shelter belts of trees were planted to slow the sweep of wind across the land. Farming practices were altered such that strips of land were left uncultivated in alternate years with crop stubble exposed to catch and hold sparse rain and winter snow. Use of the plow, which severely disturbs soil to considerable depth and buries crop residues under a layer of dirt, gave way to tillage methods that leave a surface residue of crop biomass, which captures precipitation and anchors soil against wind erosion.

Whereas periodic droughts are normal and largely manageable events, a much more serious condition can occur through human mismanagement. This phenomenon is **desertification** in which vegetation is removed from once fertile land, streams and groundwater sources dry up, and the atmospheric, terrestrial, and living environments assume characteristics of desert conditions. There are several causes of desertification. The most troubling potential cause is greenhouse warming discussed in the preceding section. "Slash-and-burn" agriculture by which trees and other vegetation are stripped from rain forests for short-term production of pasture and agricultural land is currently the largest contributor to desertification. Excessive grazing is ruinous to soil and pushes land along the path to desert formation. Desertification has a strong tendency toward positive feedback, meaning that it feeds on itself. Decreased plant cover leads to erosion and rapid loss of water from soil, which in turn further decreases the capacity of land to support plant life. Obviously, desertification is a major environmental problem that must be dealt with firmly and vigorously, if Earth is to sustain its present populations.

9.5. GREEN SCIENCE AND TECHNOLOGY TO COMBAT DESERTIFICATION

One of the first points to recognize with respect to desertification is that deserts are the natural state of much of Earth's surface area. Some substantial problems have occurred because of humankind's insistence on trying to convert desert lands to cropland, lawns, golf courses and other applications unsuitable for regions in which deserts are the natural state of land.

In areas where land has been converted to desert from grassland or cropland, for example, it is important to stop doing the things that cause desertification. This means, particularly, not cultivating land that should not be cultivated and to avoid overgrazing land with ruminant animals.

In the U.S., alone, several million acres of once productive rangeland has deteriorated to desert conditions by overgrazing so that **rangeland restoration** is the most important countermeasure to desertification. A key aspect of rangeland restoration is to remove cattle and other grazing animals from rangeland for periods of up to several years to give grasses and other native plants the opportunity to be reestablished. Another helpful measure is to plant riparian (stream bank) areas with hardy trees, such as aspens, cottonwoods, willows, and fast-growing hybrid willows in order to stabilize streams and reduce erosion.

Once rangeland restoration is under way, it is critical to manage the grazing for maximum sustainability. This requires consideration of the feeding and social habits of livestock. For example, allowing pasturage from May to mid-July enables utilization of bunchgrass when it is most palatable. After mid-July, cattle tend to congregate in the cool shade under trees and to feed on new willow and aspen sprouts and seedlings, which has the undesirable effect of preventing propagation of these trees. By removing cattle in mid-July, new growth of trees is allowed and bunchgrass growth is restored by fall.

Perennial grasses are preferable to annual grasses for rangeland restoration. In general, perennial grasses are more productive, allow for longer grazing and haying seasons, and are better for weed control than annual grasses. Prolonged growth of perennial grasses generally improves soil quality. Whereas the root zones of annual grasses normally extend to depths of only about 20 cm, the roots of perennial grasses can extend to 10 times that depth. This enables perennial grasses to recapture nutrients and water that have leached to greater depth.

Broad-leaved flowering plants called **forbs** are also important in rangeland restoration and are important indicators of rangeland health. Although forbs are often considered to be weeds, many are legumes that harbor nitrogen-fixing bacteria and many species serve as forage plants for range animals. Of the many common forbs, an interesting one is goats rue (*Tephrosia virginiana*), a legume that is a source of the natural pesticide rotenone used by Native Americans to kill fish.

In the U.S., a program that can significantly reduce the deterioration of marginal agricultural areas tending toward desertification is the USDA's Conservation Reserve Program. This program now includes 34 million acres — an area equal to that of New York State (!) — of land in Kansas, North Dakota, eastern Washington, and other agricultural states in which land owners are paid to simply leave usually highly erodable land alone to revert to natural vegetation. The program costs about $2 billion per year, about 8% of total agricultural subsidies in the U.S., but less than the subsidies would be on the same land for crop production. The program has been highly beneficial to wildlife, and it is estimated that it produces 2 million additional ducks per year by providing habitat and safe nesting regions. Farmers in the program have made additional income by selling hunting rights on their reserve lands. One criticism has been that prohibitions on grazing have led to predominance of just a few kinds of very tall grass, whereas grazing would allow much more diverse populations of shorter plants. Another problem is with invasive trees, and it has been proposed that such lands should be burned every three years to prevent growth of undesirable trees.

In addition to more natural approaches to combat desertification, such as rangeland restoration described above, there are potentially more proactive measures that humans can take to restore desert regions. Modifications of terrain to construct terraces and water impoundments can be performed to conserve water. Mulching with biomass can be used to improve soil quality. One interesting approach has been to spread biosolids from municipal wastewater treatment onto desertified land to add biomass and promote plant growth, although the generally large distances from biosolids production sources to receptor lands is a deterrent to this measure. Nitrogen, phosphorus, and potassium fertilizers can be applied to promote the growth of ground cover. It may be possible to genetically engineer plant species that are particularly well adapted to colonizing impaired desert soil in order to establish conditions conducive to the later growth of other plant species.

9.6. ACID PRECIPITATION

Precipitation made acidic by the presence of acids stronger than $CO_2(aq)$ is commonly called **acid rain** or **acid precipitation**, a term that applies to all kinds of acidic aqueous precipitation, including fog, dew, snow, and sleet. In a more general sense, **acid deposition** refers to the deposition on the Earth's surface of aqueous acids, acid gases (such as SO_2), and acidic salts (such as NH_4HSO_4). Therefore, deposition in solution form is acid precipitation, and deposition of dry gases and compounds is dry deposition. Sulfur dioxide, SO_2, contributes more to the acidity of precipitation than does CO_2 present at higher levels in the atmosphere for two reasons. The first of these is that sulfur dioxide is significantly more soluble in water, and SO_2 is a stronger acid with a greater tendency to react with water to produce H^+ ion than does CO_2:

$$SO_2(aq) + H_2O \rightarrow H^+ + HSO_3^-$$ (9.6.1)

Although acid rain can originate from the direct emission of strong acids, such as HCl gas or sulfuric acid mist, most of it is a secondary air pollutant produced by the atmospheric oxidation of acid-forming gases such as the following:

$$SO_2 + \tfrac{1}{2}O_2 + H_2O \xrightarrow[\text{step reaction}]{\text{Overall multi-}} \{2H^+ + SO_4^{2-}\}(aq)$$ (9.6.2)

$$2NO_2 + \tfrac{1}{2}O_2 + H_2O \xrightarrow[\text{step reaction}]{\text{Overall multi-}} 2\{H^+ + NO_3^-\}(aq)$$ (9.6.3)

Chemical reactions such as these play a dominant role in determining the nature, transport, and fate of acid precipitation. As the result of such reactions, the chemical properties (acidity, and ability to react with other substances) and physical properties (volatility and solubility) of acidic atmospheric pollutants are altered drastically. For example, even the small fraction of NO that does dissolve in water does not react significantly. However, its ultimate oxidation product, HNO_3, though volatile, is highly water-soluble, strongly acidic, and very reactive with other materials. Therefore, it tends to be removed readily from the atmosphere and to do a great deal of harm to plants, corrodible materials, and other things that it contacts.

Although emissions from industrial operations and fossil fuel combustion are the major sources of acid-forming gases, acid rain has also been encountered in areas far from such sources. This is due in part to the fact that acid-forming gases are oxidized to acidic constituents and deposited over several days, during which time the air mass containing the gas may have moved a thousand kilometers or more. It is likely that the burning of biomass, such as is employed in "slash-and-burn" agriculture, evolves the gases that lead to acid formation in more remote areas. In arid regions, dry acid gases or acids associated with particles may be deposited, with effects similar to those of acid rain deposition.

Acid rain spreads out over areas of several hundred to several thousand kilometers. This classifies it as a regional air pollution problem compared to a largely local air pollution problem for smog and a global one for ozone-destroying chlorofluorocarbons and greenhouse gases. Regional air pollution problems include those caused by soot, smoke, and fly ash from combustion sources and fires (forest fires). Nuclear fallout from weapons testing or from reactor fires (of which, fortunately, there has been only one major incident to date — the one at Chernobyl in the Soviet Union) may also be regarded as a regional phenomenon.

Acid precipitation shows a strong geographic dependence, as illustrated in Figure 9.4, representing the pH of precipitation in the continental U.S. The preponderance of acidic rainfall in the northeastern U.S. is obvious. Analyses of the movements of air masses have shown a correlation between acid precipitation and prior movement of an air mass over major sources of anthropogenic sulfur and nitrogen oxides emissions. This is particularly obvious in southern Scandinavia, which receives a heavy burden of air pollution from densely populated, heavily industrialized areas in Europe.

Acid rain is not a new phenomenon; it has been observed for well over a century, with many of the older observations from Great Britain. The first manifestations of this phenomenon were elevated levels of SO_4^{2-} in precipitation collected in industrialized areas. More modern evidence was obtained from analyses of precipitation in Sweden in the 1950s and of U.S. precipitation a decade or so later.

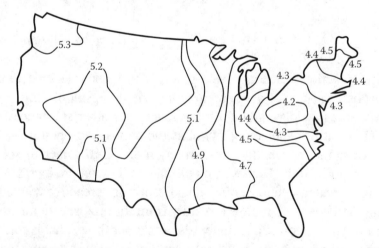

Figure 9.4.   Isopleths of pH illustrating a hypothetical precipitation–pH pattern in the lower 48 continental U.S. Actual values found vary with the time of year and climatic conditions.

In cases of moderate to severe acid precipitation, the pH of the rainwater is typically around 4 (neutral is pH 7.00 and lower values are more acidic). The predominant cation in acid rain is H^+, and NH_4^+ is also frequently encountered. Sulfate, SO_4^{2-}, is usually the predominant anion reflecting the role of sulfuric acid in acid precipitation, and there are usually lesser amounts of NO_3^- and Cl^- because of the presence of nitric and hydrochloric acids, respectively.

Ample evidence exists of the damaging effects of acid precipitation. The major such effects are the following:

- Direct phytotoxicity to plants from excessive acid concentrations (evidence of direct or indirect phytotoxicity of acid rain is provided by the declining health of eastern U.S. and Scandinavian forests and especially by damage to Germany's Black Forest)

- Phytotoxicity from acid-forming gases, particularly SO_2 and NO_2, that accompany acid rain

- Indirect phytotoxicity, such as from Al^{3+} liberated from soil by acidic rainwater

- Destruction of sensitive forests

- Respiratory effects on humans and other animals

- Acidification of lake water with toxic effects to lake flora and fauna, especially fish fingerlings

- Corrosion to exposed structures, electrical relays, equipment, and ornamental materials. Because of the effect of hydrogen ion, limestone, $CaCO_3$, is especially susceptible to damage from acid rain:

$$2H^+ + CaCO_3(s) \rightarrow Ca^{2+} + CO_2(g) + H_2O$$

- Associated effects, such as reduction of visibility (increased haziness) by acidic sulfate aerosols and the influence of sulfate aerosols on physical and optical properties of clouds (as mentioned in Section 9.2, intensification of cloud cover and changes in the optical properties of cloud droplets — specifically, increased reflectance of light — resulting from acid sulfate in the atmosphere may even have a mitigating effect on greenhouse warming of the atmosphere).

Soil sensitivity to acid precipitation can be estimated from cation exchange capacity, the ability of soil to bind with cations, specifically H^+. Soil with a high cation exchange capacity can exchange ions such as Ca^{2+} ion on soil for H^+ ions in acidic water, thus reducing the acidity of the water. Soil is generally insensitive if free carbonates are present or if it is flooded frequently.

Forms of precipitation other than rainfall, such as snow, may be acidic. Acidic fog can be especially damaging because it is very penetrating. Water collected from acid fog has exhibited pH values as low as 1.7.

9.7. GREEN REMEDIES FOR ACID PRECIPITATION

Excessively acidic precipitation is generally relatively easy to eliminate by removing combustion sources of acid. The most common of these is sulfur dioxide, which forms sulfuric acid and nitrogen oxides, which produce nitric acid. The most common nitrogen oxide entering the atmosphere is nitric oxide, NO. Sulfur dioxide comes from sulfur in fuels, especially coal and, to a lesser extent, sulfur-containing fuel oils. Nitric oxide can come from nitrogen in fuels or by the reaction of molecular N_2 and O_2 under the high pressure and temperature conditions that occur in an internal combustion engine.

The greenest remedy for acid precipitation from sulfur and nitrogen is to avoid using fossil fuels that contain these elements. Because it will likely be some time before that remedy is widely practiced, measures must be taken to eliminate these pollutants from the fuels, from the gaseous combustion products, or, in the case of nitrogen oxides, during the combustion process in the internal combustion engine.

Significant progress has been made to remove sulfur from coal and fuel oil prior to combustion. About half of the sulfur in coal is in the form of mineral pyrite, FeS_2, and the remainder of the sulfur is bound to the organic coal molecule. Various washing techniques are effective in removing pyrite from coal. Little can be done to reduce the amount of organically bound sulfur in coal except in cases where the coal is processed into a completely different fuel. Coal can be treated with hydrogen, steam, and oxygen to produce a combustible synthesis gas mixture of H_2 and CO. Coal gasification converts the sulfur in coal to hydrogen sulfide, H_2S, which is readily removed from the gas product. The technology for removing hydrogen sulfide from gas has long been practiced with "sour" natural gas containing a large fraction of H_2S. This is a very green process in that part of the hydrogen sulfide is burned,

$$2H_2S + 3O_2 \rightarrow 2SO_2 + 2H_2O \tag{9.7.1}$$

and the sulfur dioxide product is reacted with H_2S to generate elemental sulfur (the Claus reaction):

$$2H_2S + SO_2 \rightarrow 3S + 2H_2O \tag{9.7.2}$$

This process is the main source of commercial sulfur, one of the most widely used industrial raw materials, particularly for the production of sulfuric acid.

The technology for removing sulfur dioxide from flue gas produced by combustion of sulfur-containing coal is widely practiced. The most common process for doing this is scrubbing of stack gas with basic materials. Slurries in water of lime, $Ca(OH)_2$, work well for this purpose:

$$Ca(OH)_2 + SO_2 \rightarrow CaSO_3 + H_2O \qquad (9.7.3)$$

The calcium sulfite product can be oxidized to calcium sulfate,

$$CaSO_3 + \tfrac{1}{2}O_2 + 2H_2O \rightarrow CaSO_4 \cdot 2H_2O \qquad (9.7.4)$$

which in the hydrated form is called Plaster of Paris. In Kalundborg, Denmark, often cited as a good example of the practice of industrial ecology (see Section 17.8) this by-product of flue gas desulfurization is used to produce wallboard.

As sulfur emissions leading to acid precipitation in the form of sulfuric acid are increasingly controlled, it has become more important to control emissions of nitrogen oxides that can cause nitric acid pollution of the atmosphere. NO production in combustion is favored by high temperatures and by high excess oxygen concentrations. Therefore, measures taken to reduce these conditions are used to lower NO production. Reduction of flame temperature to prevent NO formation is accomplished by adding recirculated exhaust gas, cool air, or inert gases. Low excess-air firing used to reduce NO_x emissions during the combustion of fossil fuels employs the minimum amount of excess air required for oxidation of the fuel, so that less oxygen is available for the reaction

$$N_2 + O_2 \rightarrow 2NO \qquad (9.7.5)$$

in the high-temperature region of the flame. To minimize production of NO, a two-stage combustion process may be used. The first stage is fired at a relatively high temperature with a substoichiometric amount of air (insufficient to completely burn the fuel), and NO formation is limited by the absence of excess oxygen. In the second stage, burnout of hydrocarbons, soot, and CO is completed at a relatively low temperature in excess air, the low temperature preventing formation of NO.

9.8. STRATOSPHERIC OZONE DESTRUCTION

As mentioned in Section 9.1, stratospheric ozone serves as a shield to absorb harmful ultraviolet radiation in the stratosphere, protecting living beings on the Earth from the effects of excessive amounts of such radiation. The two reactions by which stratospheric ozone are produced are

$$O_2 + h\nu \rightarrow O + O \qquad (\lambda < 242.4 \text{ nm}) \qquad (9.8.1)$$

$$O + O_2 + M \rightarrow O_3 + M \text{ (energy-absorbing } N_2 \text{ or } O_2) \qquad (9.8.2)$$

and is destroyed by photodissociation

$$O_3 + h\nu \rightarrow O_2 + O \qquad (\lambda < 325 \text{ nm}) \qquad (9.8.3)$$

and a series of reactions from which the net result is the following:

$$O + O_3 \rightarrow 2O_2 \qquad (9.8.4)$$

Ozone in the stratosphere is present at a steady-state concentration resulting from the balance of ozone production and destruction by the above processes. The quantities of ozone involved are interesting. A total of about 350,000 t of ozone are formed and destroyed daily. Ozone never makes up more than a small fraction of the gases in the ozone layer. In fact, if the atmosphere's entire ozone were in a single pure layer of ozone at surface temperature and pressure conditions of approximately 273 K and 1 atm, it would be only 3 mm thick!

Ozone absorbs ultraviolet radiation very strongly in the region 220 to 330 nm. Therefore, it is effective in filtering out dangerous UV-B radiation, 290 nm $< \lambda <$ 320 nm. (UV-A radiation, 320 to 400 nm, is relatively less harmful and UV-C radiation, < 290 nm, does not penetrate to the troposphere.) If UV-B were not absorbed by ozone, severe damage would result to exposed forms of life on the Earth. Absorption of electromagnetic radiation by ozone converts the radiation's energy to heat and is responsible for the temperature maximum encountered at the boundary between the stratosphere and the mesosphere at an altitude of approximately 50 km. The reason that the temperature maximum occurs at a higher altitude than that of the highest ozone concentration is that ozone is so effective in absorbing ultraviolet radiation that most of this radiation is absorbed in the upper stratosphere, where it generates heat, and only a small fraction reaches the lower altitudes, which remain relatively cool.

Increased intensities of ground-level ultraviolet radiation caused by stratospheric ozone destruction would have some significant adverse consequences. One major effect would be on plants, including crops used for food. The destruction of microscopic plants that are the basis of the ocean's food chain (phytoplankton) could severely reduce the productivity of the world's seas. Human exposure would result in an increased incidence of cataracts. The effect of most concern to humans is the elevated occurrence of skin cancer in individuals exposed to ultraviolet radiation. This is because UV-B radiation is absorbed by cellular DNA (see Section 3.16) resulting in photochemical reactions that alter the function of DNA so that the genetic code is improperly translated during cell division. This can result in uncontrolled cell division leading to skin cancer. People with light complexions lack protective melanin, which absorbs UV-B radiation, and are especially susceptible to its effects. The most common type of skin cancer resulting from ultraviolet exposure is squamous cell carcinoma, which forms lesions that are readily removed and has little tendency to spread (metastasize). Readily metastasized malignant melanoma caused by absorption of UV-B radiation is often fatal. Fortunately, this form of skin cancer is not very common, although becoming more so.

The major culprit in ozone depletion consists of chlorofluorocarbon (CFC) compounds, commonly known as "freons." These volatile compounds have been used and released to a very large extent in recent decades. The major use associated with CFCs is as refrigerant fluids. Other applications have included solvents, aerosol propellants, and blowing agents in the fabrication of foam plastics. The same extraordinarily high chemical stability that makes CFCs nontoxic enables them to persist for years in the atmosphere and to enter the stratosphere. In the stratosphere the photochemical dissociation of CFCs by intense ultraviolet radiation,

$$CF_2Cl_2 + h\nu \rightarrow Cl\bullet + CClF_2\bullet \qquad (9.8.5)$$

yields chlorine atoms, each of which can go through chain reactions involving first the reaction of atomic chlorine with ozone:

$$Cl\bullet + O_3 \rightarrow ClO\bullet + O_2 \qquad (9.8.6)$$

In the most common sequence of reactions involved with stratospheric ozone destruction, the $ClO\bullet$ radicals react to form a dimer, which then reacts to regenerate Cl atoms (where M in the reactions below is an energy-absorbing third body, such as an N_2 molecule), which in turn react with ozone to regenerate $ClO\bullet$ in the following reaction sequence:

$$ClO\bullet + ClO\bullet \rightarrow ClOOCl \qquad (9.8.7)$$

$$ClOOCl + h\nu \rightarrow ClOO\bullet + Cl\bullet \qquad (9.8.8)$$

$$ClOO\bullet + M \rightarrow Cl\bullet + O_2 + M \qquad (9.8.9)$$

$$2Cl\bullet + 2O_3 \rightarrow 2ClO\bullet + 2O_2 \qquad (9.8.10)$$

$$\rule{6cm}{0.4pt}$$

$$2O_3 \rightarrow 3O_2 \text{ (net reaction)} \qquad (9.8.11)$$

The net effect of these reactions is catalysis of the destruction of several thousand molecules of O_3 for each Cl atom produced. Because of their widespread use and persistency, the two CFCs of most concern in ozone destruction are CFC-11 and CFC-12, $CFCl_3$, and CF_2Cl_2, respectively. Even in the intense ultraviolet radiation of the stratosphere, the most persistent chlorofluorcarbons have lifetimes of the order of 100 years.

The Antarctic Ozone Hole

The most prominent instance of ozone layer destruction is the so-called "Antarctic ozone hole" that was first firmly established in 1985 by the British Antarctic Survey and observed with great alarm in subsequent years. This phenomenon is manifested by the appearance during the Antarctic's late winter and early spring months of September and October of severely depleted stratospheric ozone (up to 50%) over the polar region. The reasons why this occurs are related to the normal effect of NO_2 in limiting Cl-atom-catalyzed destruction of ozone by combining with ClO,

$$ClO + NO_2 \rightarrow ClONO_2 \tag{9.8.12}$$

During the winter in the polar regions, particularly Antarctica, at temperatures below $-70°C$, NO_x gases are removed along with water by freezing to produce ice crystals or aerosols composed of liquid supercooled ternary mixtures of HNO_3, H_2SO_4, and H_2O in which chlorine originally from chlorofluorocarbons is held in the form of $ClONO_2$ and HCl in polar stratospheric clouds. The reaction of HCl (which comes primarily from the reaction of stratospheric methane, CH_4, with Cl• atoms produced from chlorofluorocarbons) with $ClONO_2$

$$ClONO_2 + HCl \rightarrow Cl_2 + HNO_3 \tag{9.8.13}$$

releases Cl_2. Under the conditions of low temperature and sunlight that prevail in the lower stratosphere above Antarctia in spring, the Cl_2 released and the HOCl produced by the reaction of Cl_2 with H_2O undergo photodissociation,

$$Cl_2 + h\nu \rightarrow 2Cl• \tag{9.8.14}$$

$$HOCl + h\nu \rightarrow HO• + Cl• \tag{9.8.15}$$

to produce Cl atoms that can undergo the sequence of chain reactions (Reaction 9.8.7 to Reaction 9.8.10) leading to ozone destruction. The preceding reactions are aided by the tendency of the HNO_3 product to become hydrogen-bonded with water in the cloud particles. The result of these processes is that over the winter months photoreactive Cl_2 and HOCl accumulate in the Antarctic stratospheric region in the absence of sunlight, and then undergo a burst of photochemical activity when spring arrives leading to stratospheric ozone destruction and the formation of the Antarctic ozone hole.

The story of the discovery of the Antarctic ozone hole is an interesting one. Depletion of lower atmosphere ozone in Antarctica was first observed in the 1970s and the first accurate measurements were taken in 1985. The drop in ozone levels during the Antarctic spring of 1985 was so dramatic that the scientists measuring it assumed that their instruments were faulty and had new instruments assembled and flown in, which confirmed the low ozone levels. The TOMS satellite designed to provide stratospheric ozone measurements did not pick up the ozone hole because the software analyzing the satellite data was designed to throw out very low readings as faulty! Subsequent analysis of the data showed that the Antarctic ozone hole did in fact occur, and it has been mapped accurately every year since 1985.

The Antarctic ozone hole that developed in 2002 was the smallest since 1988, covering about 15.5 million square kilometers. Furthermore, it was split into two parts compared to the single ozone hole that is normally observed. These observations have been attributed to unusual stratospheric weather patterns in 2002 that resulted in warmer than normal temperatures in the polar vortex that forms over Antarctica. However, the 2003 Antarctic ozone hole was very large, covering an area of 28.2 million square kilometers on September 11, 2003, second only to the all-time record of 29.8 million square kilometers reached on September 10, 2000, and reaching the southern parts of South America including the city of Ushuaia, Argentina. In 2004, the ozone hole reached a maximum area of about 19 million square kilometers, significantly below the average for the previous decade. Several times in September, 2004, the edge of the ozone hole passed over southern South America and the Falkland Islands. The 2005 Antarctic ozone hole reached a maximum area of 27 million square kilometers on September 19, 2005.

The Nobel Prize in Environmental Chemistry

A richly deserved Nobel Prize in Chemistry, the first ever for environmental chemistry, was awarded to three scientists, Paul J. Crutzen, Mario J. Molina, and F. Sherwood Rowland, for their work on the role of chlorofluorocarbons in ozone depletion in 1995. In 1970, Dr. Crutzen, Director of the Department of Atmospheric Chemistry at Max Planck Institute for Chemistry, Mainz, Germany, showed that nitrogen oxides are involved with the balance of levels of upper atmospheric ozone, suggesting that catalytic substances from anthropogenic sources, such as NO emitted by high-flying supersonic aircraft, could accelerate the natural destruction of stratospheric ozone, lowering levels of this essential substance. Drs. Molina and Rowland working at the University of California Irvine established that photodissociation of stratospheric chlorofluorocarbon contaminants could put catalytic amounts of atomic Cl into the stratosphere, which would be extraordinarily effective in destroying ozone. Their work provided the basis on which the United Nations Environment Program (UNEP) arrived at the Montreal Protocol of 1987 through which production and use of chlorofluorocarbons were to be phased out.

9.9. GREEN SOLUTIONS TO STRATOSPHERIC OZONE DESTRUCTION

In a sense, chlorofluorocarbons were an example of green chemistry, developed in the 1930s long before the concept of green technology was even imagined. The fluids that they replaced, sulfur dioxide and ammonia, are quite toxic and caused fatalities when leaked from refrigerators in homes. The chlorofluorocarbon replacements performed ideally and were remarkably nontoxic. Several related compounds, such as halothane, 2-bromo-2-chloro-1,1,1-trifluoroethane, have been used as anesthetics. It was not until the 1970s and later that the analytical capability became available to show that chlorofluorocarbons had become spread throughout the global atmosphere and are far from green when considering the global environment as a whole.

The solutions to the problem of stratospheric ozone depletion posed by chlorofluorocarbons provide a good example of green chemistry and green technology, taking advantage of fundamental knowledge regarding the properties and behavior of chemicals. The reason that chlorofluorocarbons are so stable and do not break down at all until they have entered the stratosphere — and then only slowly — is the extreme stability of the C–Cl and C–F bonds. Essentially all anthropogenic chemical species that are broken down in the troposphere are attacked by hydroxyl radical, HO• (section 8.3), which is abundant in the troposphere. This reactive species attacks and breaks C–H bonds, but is not reactive enough to break C–Cl and C–F bonds. So the solution to the problems posed by chlorofluorocarbons has been to develop **hydrohaloalkanes** that contain at least one C–H bond per molecule that is susceptible to attack by HO• radical in the troposphere, thereby eliminating the compound with its potential to produce ozone-depleting Cl atoms before it reaches the stratosphere. The substitutes are either hydrochlorofluorocarbons (HCFCs) or hydrofluorocarbons (HFCs). The latter are especially desirable because they do not contain Cl atoms that can be involved in stratospheric ozone destruction; F atoms do not have that effect. The most commonly used CFC substitute is HFC-134a (CH_2FCF_3). Other compounds used or proposed for use include HCFC-22 ($CHClF_2$), HCFC-123 ($CHCl_2CF_3$), HCFC-141b (CH_3CCl_2F), HCFC-124 ($CHClFCF_3$), HCFC-225ca ($CHCl_2CF_2CF_3$), HCFC-225cb ($CHFClCF_2CF_2Cl$), HCFC-142b (CH_3CClF_2), and HFC-152a (CH_3CHF_2).

Ozone depletion potentials of HCFCs and HFCs are compiled to express potential likelihood for the destruction of stratospheric ozone relative to a value of 1.0 for CFC-11, a non-hydrogen-containing chlorofluorocarbon with a formula of $CFCl_3$. Low ozone depletion potential correlates with short tropospheric lifetime, which means that the compound is destroyed in the troposphere before migrating to the stratosphere. The ozone-depletion potentials of some of the substitutes mentioned above are HCFC-22, 0.030, HCFC-123, 0.013, HCFC-141b, 0.10, HCFC-124, 0.035, and HCFC-142b, 0.038. Low ozone depletion potential correlates with short tropospheric lifetime, which means that the compound is destroyed in the troposphere before migrating to the stratosphere. The ozone depletion potential of HFC-134a is zero since it does not contain any Cl. The major concern with its use is that it acts as a greenhouse gas.

9.10. PHOTOCHEMICAL SMOG

Photochemical smog is a major local or regional air pollution phenomenon characterized by oxidants, irritating vapors, and visibility-obscuring particles that occurs in urban areas where the combination of pollution-forming emissions and appropriate atmospheric conditions are right for its formation. Though not a threat to the global atmosphere as such, in some urban areas photochemical smog is highly detrimental to health and to the quality of life. *Smog* originally was used to describe the unpleasant combination of smoke and fog laced with sulfur dioxide, a chemically reducing atmosphere, which was formerly prevalent in London when high-sulfur coal was the primary fuel used in that city. However, the photochemical smog discussed in this section is chemically quite different because of its oxidizing qualities, and the formation of oxidants in the air, particularly ozone, is indicative of smog formation.

Photochemical smog has a long history. Exploring what is now southern California, in 1542 Juan Rodriguez Cabrillo named San Pedro Bay "The Bay of Smokes" because of the heavy haze that covered the area. Complaints of eye irritation from anthropogenically polluted air in Los Angeles were recorded as far back as 1868. Characterized by reduced visibility, eye irritation, cracking of rubber, and deterioration of materials, smog became a serious nuisance in the Los Angeles area during the 1940s. It is now recognized as a major air pollution problem in many areas of the world.

The species in the atmosphere that give smog its noxious character do not enter the atmosphere directly, but are produced by photochemical processes acting on precursor atmospheric pollutants. Because of this, the constituents of smog are **secondary pollutants**, in contrast to a material such as sulfur dioxide, which is a primary air pollutant. The three ingredients required to generate photochemical smog are ultraviolet light, reactive hydrocarbons, and nitrogen oxides, the latter two of which are produced as emissions from internal combustion engines. Although the automobile is the major source of these pollutants, hydrocarbons may come from biogenic sources, of which α-pinene and isoprene (structural formulas shown in Section 8.15) from trees are the most abundant. In order for high levels of smog to form, relatively stagnant air must be subjected to sunlight under low humidity conditions in the presence of pollutant nitrogen oxides and hydrocarbons.

The atmospheric chemistry of smog formation was discussed in Section 8.16. The driving force behind smog formation is the tendency for hydrocarbons to be eliminated from the atmosphere by a number of chemical and photochemical reactions. Starting from relatively innocuous hydrocarbon precursors, these reactions are responsible for the formation of many noxious secondary pollutant products and intermediates that make up photochemical smog. The processes by which this occurs are driven by the natural tendency for the oxygen-rich atmosphere to be oxidizing, particularly through photochemical processes. The oxidation process terminates with formation of CO_2, solid organic particulate matter that settles from the atmosphere, or water-soluble products (for example, acids, aldehydes), which are removed by rain. Inorganic species such as ozone or nitric acid are by-products of these reactions.

The urban atmosphere when held in place for relatively long times by meteorologic conditions and subjected to sunlight functions as a massive solar-powered chemical reactor. In this reactor, hydrocarbons, oxides of nitrogen and sulfur, and oxygen naturally present undergo vast numbers of photochemical reactions to synthesize aldehydes, ozone, organic oxidants, acids, particulate matter, and other noxious air pollutants. Although not as great a threat to the global atmosphere as some of the other air pollutants discussed in this chapter, smog does pose significant hazards to living things and materials in local urban areas in which millions of people are exposed.

The two most significant classes of inorganic products from smog are sulfates and nitrates. Inorganic sulfates and nitrates, along with sulfur and nitrogen oxides can contribute to acidic precipitation, corrosion, reduced visibility, and adverse health effects. Nitric acid formed by chemical processes in a smoggy atmosphere reacts with ammonia in the atmosphere to form ammonium nitrate:

$$NH_3 + HNO_3 \rightarrow NH_4NO_3 \qquad (9.10.1)$$

Other nitrate salts may also be formed.

Nitric acid and nitrates are among the more damaging end products of smog. In addition to possible adverse effects on plants and animals, they cause severe corrosion problems. Electrical relay contacts and small springs associated with electrical switches are especially susceptible to damage from nitrate-induced corrosion.

Effects of Smog

The harmful effects of smog occur mainly in the areas of (1) human health and comfort, (2) damage to materials, (3) effects on the atmosphere, and (4) toxicity to plants. The exact degree to which exposure to smog affects human health is not known, although substantial adverse effects are suspected. Pungent-smelling, smog-produced ozone is known to be toxic. Ozone at 0.15 ppm causes coughing, wheezing, bronchial constriction, and irritation to the respiratory mucous system in healthy, exercising individuals as well as more adverse effects on those in poor health, particularly those with weakened respiratory systems. Peroxyacyl nitrates and aldehydes found in smog are eye irritants. Materials are adversely affected by some smog components. Rubber has a high affinity for ozone and is cracked and aged by it. Indeed, the cracking of rubber used to be employed as a test for the presence of ozone.

Even lightly populated nonindustrial areas are subject to the effects of smog brought about by human activities. Particularly, the practice of burning savanna grasses for agricultural purposes causes smog. This burning produces NO_x, and reactive hydrocarbons that are required for smog formation. Furthermore, these grasses grow in tropical regions, which have the intense sunlight required for smog formation. The net result is rapid development of smoggy conditions as manifested by ozone levels several times normal background values.

The Urban Aerosol and Acid Fog

The most apparent manifestation of smog is visibility-obscuring **urban aerosol**. Many of the particles composing this aerosol are made from gases by chemical processes and are therefore quite small, usually less than 2 μm. Particles of such a size are especially harmful because they scatter light most efficiently and are the most respirable. Aerosol particles formed from smog often contain toxic constituents, such as respiratory tract irritants and mutagens. The urban aerosol also contains particle constituents that originate from processes other than smog formation, among which are sulfuric acid droplets, salts, metals, and polycyclic aromatic hydrocarbons. Highly corrosive ammonium salts, such as NH_4HSO_4, are common constituents of urban aerosol particles. Water is always present, even in low humidity atmospheres, and is usually a constituent of urban aerosol particles. Carbon and polycyclic aromatic hydrocarbons from partial combustion and diesel engine emissions are generally abundant constituents, and particulate elemental carbon is usually most responsible for absorbing light in the urban aerosol.

A kind of urban aerosol particulate matter formed under smoggy conditions that is of particular concern is **acid fog**, which may have pH values below 2 because of the presence of H_2SO_4 or HNO_3. This material is part of the acid rain phenomenon discussed in Section 9.6. Acid fog formation occurs because the gas-phase oxidation of SO_2 and NO_x under the strongly oxidizing conditions in smog produces strong acids, which form very small aerosol particles. These, in turn, act as condensation nuclei for water vapor. Acid–base phenomena occur in the droplets, and they act as scavengers to remove ionic species from air. Because fog aerosol particles form in areas of intense acid gas pollution near the surface, the concentrations of acids and ionic species in fog aerosol droplets tend to be much higher than in cloud aerosol droplets at higher altitudes.

Effects of Smog on Plants and Crops

In view of worldwide shortages of food, the known harmful effects of smog on plants are of particular concern. These effects are largely because of oxidants in the smoggy atmosphere. The three major oxidants involved are ozone, PAN (see Reaction 8.4.13), and nitrogen oxides. Of these, PAN has the highest toxicity to plants,

Figure 9.5.   Representation of ozone damage to a lemon leaf. In color, the spots appear as yellow chlorotic stippling on the green upper surface caused by ozone exposure.

attacking younger leaves and causing "bronzing" and "glazing" of their surfaces. Exposure for several hours to an atmosphere containing PAN at a level of only 0.02 to 0.05 ppm will damage vegetation. The sulfhydryl group of proteins in organisms is susceptible to damage by PAN, which reacts with such groups as both an oxidizing agent and an acetylating agent. Fortunately, PAN is usually present at only low levels. Nitrogen oxides occur at relatively high concentrations during smoggy conditions, but their toxicity to plants is relatively low. The low toxicity of nitrogen oxides and the usually low levels of PAN leave ozone as the greatest smog-produced threat to plant life.

In addition to health effects and damage to materials, one of the greater problems caused by smog is destruction of crops and reduction of crop yields. The annual cost of these effects in California alone is about $15 billion. Typical of the phytotoxicity of O_3, ozone damage to a lemon leaf is typified by chlorotic stippling (characteristic yellow spots on a green leaf), as represented in Figure 9.5. Reduction in plant growth may occur without visible lesions on the plant. Brief exposure to approximately 0.06 ppm of ozone may temporarily cut photosynthesis rates in some plants in half. Crop damage from ozone and other photochemical air pollutants in California alone is estimated to cost millions of dollars each year. The geographic distribution of damage to plants in California is illustrated in Figure 9.6.

Figure 9.6.   Geographic distribution of plant damage from smog in California.

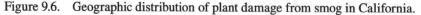

Green Solutions to the Smog Problem

Green technology has been widely used and very successful in combating photochemical smog. Internal combustion engines used in automobiles and trucks produce reactive hydrocarbons and nitrogen oxides, two of the three key ingredients required for smog to form. Therefore, control of automotive emissions is a key to reducing photochemical smog.

Figure 9.7 illustrates the basic operation of the internal combustion engine, which is the predominant power source for vehicles of various kinds. The four steps in the total cycle are (1) **intake** of air or air–fuel mixture as the piston moves down with the intake valve open, (2) **compression** of the charge as the piston moves up, (3) ignition of the fuel–air mixture creating pressure that forces the piston down in the **power stroke**, and (4) an **exhaust** stroke in which the exhaust gases are forced out as the piston moves upward with the exhaust valve open. In the case of a direct fuel injection engine, fuel is injected into the cylinder during (or in the case of a diesel engine, at the top of) the compression stroke.

The conditions inside the cylinders of a gasoline engine as it operates are conducive to production of nitrogen oxides and unburned hydrocarbons. The nitrogen and oxygen that produce nitrogen oxides come from air sucked into the cylinder during intake. At the high-pressure and high temperature conditions that prevail immediately following ignition, nitrogen and oxygen react,

$$N_2 + O_2 \rightarrow 2NO \qquad (9.10.2)$$

generating nitric oxide, NO. The rapid expansion and cooling of the combustion gas mixture during the power stroke prevents some of the NO from reverting back to N_2

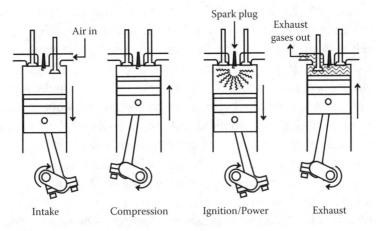

Figure 9.7.   Operation of the four-cycle piston internal combustion engine, which is responsible for most of the unburned hydrocarbons and nitrogen oxides that are the raw ingredients for photochemical smog formation.

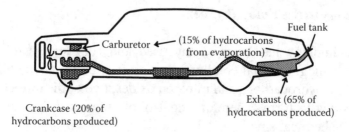

Figure 9.8.   Potential sources of pollutant hydrocarbons from an automobile without pollution control devices typical of automobiles prior to the 1960s.

and O_2 with the result that some NO is ejected in the exhaust gas. Lowering the peak combustion temperatures in an internal combustion engine reduces NO emissions. This is commonly done by recirculating exhaust gas into the engine with an exhaust gas recirculation valve.

At the high temperature and high pressure conditions in an internal combustion engine, products of incompletely burned gasoline undergo chemical reactions, which produce several hundred different hydrocarbons. Many of these are highly reactive in forming photochemical smog. As shown in Figure 9.8, the automobile has several potential sources of hydrocarbon emissions other than the exhaust. The first of these to be controlled was the mist of hydrocarbons composed of lubricating oil and "blowby" emanating from the engine crankcase. The latter consists of exhaust gas and unoxidized fuel–air mixture that enters the crankcase from the combustion chambers around the pistons. This mist is destroyed by recirculating it through the engine intake manifold by way of the positive crankcase ventilation (PCV) valve.

A second major source of automotive hydrocarbon emissions is the fuel system, from which hydrocarbons are emitted through fuel tank and carburetor vents. When the engine is shut off and the engine heat warms up the fuel system, gasoline may be evaporated and emitted to the atmosphere. In addition, heating during the daytime and cooling at night causes the fuel tank to breathe and emit gasoline fumes. Such emissions are reduced by fuel formulated to reduce volatility. Automobiles are equipped with canisters of carbon, which collect evaporated fuel from the fuel tank and fuel system, to be purged and burned when the engine is operating.

Modern automobiles use sophisticated computer control to reduce emissions of nitrogen oxides, hydrocarbons, and carbon monoxide. This is done in part by exacting control of the proportions of air and fuel into the engine combustion chambers, requiring a balance between a fuel-rich mixture favoring carbon monoxide and hydrocarbon emissions and an air-rich mixture favoring nitrogen oxide emissions, both of which function in the catalytic converter (below) to reduce emissions. Ignition timing is also a key to reduced emissions. Some of the most sophisticated automobile engines even employ valve timing to reduce exhaust emissions.

Currently, automobiles employ **catalytic converters** to destroy pollutants in exhaust gases. The most commonly used automotive catalytic converter is the three-way conversion catalyst, so called because a single catalytic unit destroys all three of the main class of automobile exhaust pollutants — hydrocarbons, carbon monoxide, and nitrogen oxides. This catalyst depends upon accurate sensing of oxygen levels in the exhaust combined with computerized engine control which cycles the air–fuel mixture several times per second back and forth between slightly lean and slightly rich relative to the stoichiometric ratio. Under these conditions carbon monoxide, hydrogen, and hydrocarbons (C_cH_h) are oxidized.

$$CO + \tfrac{1}{2}O_2 \rightarrow CO_2 \qquad\qquad (9.10.3)$$

$$H_2 + \tfrac{1}{2}O_2 \rightarrow H_2O \qquad\qquad (9.10.4)$$

$$C_cH_h + (c + {}^h\!/_4)O_2 \rightarrow cCO_2 + {}^h\!/_2H_2O \qquad\qquad (9.10.5)$$

Nitrogen oxides are reduced on the catalyst to N_2 by carbon monoxide, hydrocarbons, or hydrogen as shown by the following reduction with CO:

$$CO + NO \rightarrow \tfrac{1}{2}N_2 + CO_2 \qquad\qquad (9.10.6)$$

Automotive exhaust catalysts are dispersed on a high-surface-area substrate, most commonly consisting of cordierite, a ceramic composed of alumina (Al_2O_3), silica, and magnesium oxide. The substrate is formed as a honeycomb type structure providing maximum surface area to contact exhaust gases. The support needs to be mechanically strong to withstand vibrational stresses from the automobile, and it must resist severe thermal stresses in which the temperature may rise from ambient temperatures to approximately 900°C over an approximately two-minute period during "light-off" when the engine is started. The catalytic material, which composes only about 0.10 to 0.15% of the catalyst body, consists of a mixture of precious metals. Platinum and palladium catalyze the oxidation of hydrocarbons and carbon monoxide, and rhodium acts as a catalyst for the reduction of nitrogen oxides; presently, palladium is the most common precious metal in exhaust catalysts.

Because lead can poison automotive exhaust catalysts, automobiles equipped with catalytic exhaust-control devices require lead-free gasoline, which has become the standard motor fuel. Sulfur in gasoline is also detrimental to catalyst performance, and sulfur levels in gasoline and, more recently, diesel fuel have been lowered markedly.

Hybrids

The internal combustion automobile engine has been developed to an extremely high degree in terms of its emissions. Developments continue with this remarkable piece of machinery. Some of the most important ongoing advances are in the area of fuel economy forced by high fuel prices, but also lowering pollutant emissions. In general, less fuel used means lower emissions. As mentioned earlier, newly developed hybrid automobiles combining an internal combustion engine with an electric motor/generator and enabling the internal combustion engine to run evenly under optimum operating conditions promise to lower emissions even further. The ultimate such system employs a diesel engine coupled to an electrical system. Because of its higher peak combustion temperatures, the diesel engine is inherently more efficient that the conventional gasoline engine. Whereas the gasoline engine used in a hybrid turns off automatically when the vehicle is stopped and the battery is not charging, a diesel engine can idle with remarkably little fuel consumption and can be left running. This has the advantage of keeping the exhaust gas catalyst system hot and functioning at maximum efficiency. Problems with particulate and nitrogen oxide emissions from diesel engines are being solved, and new requirements for low-sulfur diesel fuels in the U.S. will prevent sulfur-poisoning of exhaust gas catalyst systems.

Green Transportation Alternatives

No aspect of modern civilization has contributed more to environmental degradation of the atmosphere and other parts of the environment than have transportation systems based on private automobiles, buses and trucks. Therefore, it is important to think of alternative systems of transport, of which there are excellent examples. One of these is that of the Washington, D.C., Metrorail system, which has saved the area from catastrophic environmental devastation while acting as a tremendous engine for relatively more sustainable economic growth. Such systems should be a top priority for urban areas throughout the world.

9.11. CATASTROPHIC ATMOSPHERIC EVENTS

Greenhouse warming, stratospheric ozone depletion, and photochemical smog are long-term problems that have developed over many years. The possibility also exists of sudden, catastrophic damage to the atmosphere that could cause irreversible environmental damage. Two possibilities are discussed here.

Nuclear Winter

Nuclear winter is a term used to describe a catastrophic atmospheric effect that might occur after a massive exchange of nuclear firepower between major powers. The heat from the nuclear blasts and from resulting fires would result in powerful updrafts carrying combustion products to stratospheric regions. The reflection and scattering of sunlight by particles carried into the stratosphere would result in several years of much lower temperatures and freezing temperatures even during summertime. Such conditions occurred in 1816, "the year without a summer," following the astoundingly massive Tambora, Indonesia, volcanic explosion of 1815. Brutally cold years around 210 BC that followed a similar volcanic incident in Iceland were recorded in ancient China. In addition to the direct suffering caused, massive starvation would result from crop failures accompanying years of nuclear winter. The incidents cited above clearly illustrate the climatic effects of huge quantities of particulate matter ejected high into the atmosphere.

Evidence exists to suggest that military explosives can result in the introduction of large quantities of particulate matter into the atmosphere. For example, carpet bombings of cities, such as the tragic, militarily pointless fire-bombing of Dresden, Germany, near the end of World War II, produced huge firestorms that created their own wind causing a particle-laden updraft into the atmosphere. Of course, the effect of a full-scale nuclear exchange would be manyfold higher.

An idea of the potential climatic effect resulting from a full-scale nuclear exchange may be obtained by considering the magnitude of the blasts that might be involved. Only two nuclear bombs have been used in warfare, both dropped on cities in Japan in 1945. The Hiroshima fission bomb had the explosive force of 12 kilotons of TNT explosive. Its blast, fireball, and instantaneous emissions of neutrons and gamma radiation, followed by fires and exposure to radioactive fission products, killed about 100,000 people and destroyed the city on which it was dropped. By comparison with this 12-kiloton bomb, modern fusion bombs are typically rated at 500 kilotons, and 10-megaton weapons are common. A full-scale nuclear exchange might involve a total of the order of 5000 megatons of nuclear explosives. As a result, unimaginable quantities of soot from the partial combustion of wood, plastics, paving asphalt, petroleum, forests, and other combustibles would be carried to the stratosphere. At such high altitudes, tropospheric removal mechanisms are not effective because there is not enough water in the stratosphere to produce rainfall to wash particles from the air, and convection processes are very limited. Much of the particulate matter would be in the micrometer size range in which light is reflected, scattered, and absorbed most effectively and settling is very slow. Therefore, vast areas of Earth would be overlain by a stable cloud of particles, and the fraction of sunlight reaching Earth's surface would be drastically reduced, resulting in a dramatic cooling effect. There would be other effects as well. The extreme heat and pressure in the fireball would result in the fixation of nitrogen as ozone-destroying nitrogen oxides:

$$O_2 + N_2 \rightarrow 2NO \qquad (9.11.1)$$

The timing and location of nuclear blasts are very important in determining their climatic effects. Atmospheric testing of nuclear weapons, including a 58-megaton monster detonated by the Soviet Union, has had little atmospheric effect. Such tests were carried out at widely spaced intervals on deserts, small tropical islands, and other places with little combustible matter. In contrast, military use of nuclear weapons would involve a high concentration of firepower, both in time and in space, on industrial and military targets consisting largely of combustibles. Furthermore, destruction of hardened military sites requires blasts that disrupt large quantities of soil, rock, and concrete, which are pulverized, vaporized, and blown into the atmosphere.

On a hopeful note, the East–West conflict that dominated world politics and threatened nuclear war from the mid-1900s until about 1990 has now abated, and the probability of nuclear warfare seems to have diminished. However, war is still common and more nations continue to develop nuclear arsenals. An especially frightening trend of the times is the willingness of fanatics to commit suicide in terrorist attacks. The carnage, including the deaths of over 50 people and the wounding of hundreds, resulting from four suicide attacks on the London public transportation system in July, 2005, was horrifying. Each of the attackers carried just a few kg of explosives and one can imagine what would occur with a similar attack involving a nuclear bomb carried in a suitcase.

In 1998, both India and Pakistan tested nuclear weapons and by 2002 both countries had arsenals consisting of a number of nuclear bombs. A dispute over the province of Kashmir in 2002 resulted in great concern over the possibility that the two nations would engage in a nuclear exchange that could kill perhaps millions of people. Fortunately, the dispute was resolved without resort to nuclear warfare, but it cannot be assumed that such will always be the case. The willingness of leaders of countries with much larger nuclear arsenals, who should know better, to go to war with less than convincing justification, heightens concern over possible devastation from nuclear warfare.

Visitors from Outer Space

Of all the possible atmospheric catastrophes that can occur, arguably the most threatening would be one caused by collision of a large asteroid or comet with Earth. Convincing evidence now exists that mass extinctions of species in the past have resulted from Earth being hit by asteroids several kilometers in diameter. Such an event would cause much the same effects as those from "nuclear winter" described in the preceding section, though with a large asteroid the effects would be much more pronounced.

The objects most likely to collide with Earth are comets and asteroids that have been guided to the vicinity of Earth's orbit by the gravitational attraction of nearby planets. **Comets** consist primarily of water ice with embedded dust particles formed in cold outer planetary systems. **Asteroids** are rocky bodies formed closer to the sun between the orbits of Mars and Jupiter and are essentially rocky debris left over from the formation of the solar system about 4.6 billion years ago.

An illustration of the potential of comets and asteroids to cause damage occurred on June 30, 1908, in a remote region of Siberia with a massive explosion that leveled trees over a distance of many kilometers. The few witnesses to the explosion described a bright flash and very loud noise. The consensus of opinion regarding this event is now that it was caused by an asteroid with a diameter of 50 to 60 m hitting the atmosphere at a velocity of 12 to 20 km/sec causing the asteroid to fragment into dust and gravel. The energy released was equivalent to about 60 Hiroshima-sized nuclear fission bombs or a very large hydrogen fusion bomb. Considering that such damage was done by an object about the size of a football field and that there are asteroids in near-Earth orbits that are a kilometer or more in size illustrates the potential danger posed by these objects.

9.12. ARE THERE GREEN REMEDIES FOR ATMOSPHERIC CATASTROPHES?

What, if anything, can be done about potential atmospheric catastrophes that could wipe out or significantly diminish life on Earth? In some cases, perhaps distressingly little can be done. This is especially true of natural disasters such as asteroid impacts or massive volcanic eruptions, which would have much the same effects. Fortunately, the probability of a truly enormous natural catastrophe happening during the lifetime of any reader of this book is very low. Programs to track comets and asteroids have shown that catastrophic impact is unlikely in the foreseeable future. If such impact appears to be likely sometime several decades in the future, it is plausible that an object could be nudged from its orbit by landing a rocket-like device on it and operating the device for long enough to have the desired effect. Certainly, an object of the size of the one that hit Siberia in 1908 could be blasted into relatively harmless dust with a massive thermonuclear fusion device.

The probability of severe environmental harm from an exchange of nuclear weapons is certainly higher than that of asteroid impact. In this case there are potential technical, political, and social remedies. It may be possible to develop "Star Wars" technologies to destroy nuclear weapons before they reach their targets. Continued efforts to prevent proliferation of nuclear weapons are imperative. The social and economic conditions that tend to lead to conflict must be addressed. Education is of the utmost importance. People the world over must be made aware of the possibilities for ruining Planet Earth and of the measures and policies required to prevent that from happening.

9.13. SENSIBLE MEASURES

As discussed in this chapter, there are numerous threats to the global atmosphere that could adversely affect life on Earth. One of the greatest challenges facing humankind in the modern age is to avoid conditions that could do grievous harm to the global atmosphere. Preventive measures are often opposed by entrenched economic interests. For example, the refusal of the U.S. to ratify the Kyoto treaty to limit greenhouse gas emissions has been rationalized on the basis of its potential harm to the free market economy.

A sensible approach, sometimes called a "**tie-in strategy**" advocates taking measures consisting of "high-leverage actions," which are designed to prevent problems from occurring and which have substantial merit even if the major problems that they are designed to avoid do not materialize. An example is implementation of environmentally sound substitutes for fossil fuels to lower atmospheric CO_2 output and prevent greenhouse warming. Even if it turns out that the greenhouse effect is exaggerated, such substitutes would save the Earth from other kinds of environmental damage, such as disruption of land by strip mining coal or preventing oil spills from petroleum transport. Definite economic and political benefits would also accrue from lessened dependence on uncertain, volatile petroleum supplies. Increased energy efficiency would diminish both greenhouse gas and acid rain production, while lowering costs of production and reducing the need for expensive and environmentally disruptive new power plants. The implementation of these kinds of tie-in strategies requires some degree of incentive beyond normal market forces, and, therefore, is opposed by some on ideological grounds. A good example is opposition to mandatory fuel mileage standards for automobiles, which many view as unjustified interference with their personal freedom to have as large and wasteful a vehicle as their finances can stand. However, the question may be raised whether any economic system that does not take account of potential environmental harm and that is not sustainable is truly a free market system.

LITERATURE CITED

1. Revelle, Roger and Hans E. Suess, Carbon dioxide exchange between atmosphere and ocean and the question of an increase of atmospheric CO_2 during the past decades, *Tellus*, **9**, 18, 1957.

2. Edward P., J. Jed Brown, and James W. O'Leary, Irrigating Crops with Seawater, *Scientific American*, August, 1998, pp. 76–81.

SUPPLEMENTARY REFERENCES

Adger, W. Neil and Katrina Brown, *Land Use and the Causes of Global Warming*, John Wiley & Sons, New York, 1995.

Allaby, Michael, *Fog, Smog, and Poisoned Rain*, Facts On File, New York, 2003.

Andersen, Stephen O. and K. Madhava Sarma, *Protecting the Ozone Layer: The United Nations History*, Sterling, VA: Earthscan, London, 2004.

Bell, Randall and Donald T. Phillips, *Disasters: Wasted Lives, Valuable Lessons*, Tapestry Press, Irving, TX, 2005.

Burroughs, William J., *Weather Cycles*, 2nd ed., Cambridge University Press, New York, 2003.

Burroughs, William J., *Climate change: A Multidisciplinary Approach*, Cambridge University Press, Cambridge, U.K., 2001.

Burroughs, William J., *Climate: Into the 21st Century*, Cambridge University Press, Cambridge, U.K., 2003.

Challen, Paul C., *Drought and Heat Wave Alert!*, Crabtree, New York, 2005.

Chambers, Frank and Michael Ogle, Eds., *Climate Change: Critical Concepts in the Environment and Physical Geography*, Routledge, New York, 2002.

Chehoski, Robert, *Critical Perspectives on Cimate Disruption*, Rosen Publishing Group, New York, 2006.

Council for Agricultural Science and Technology, *Climate Change and Greenhouse Gas Mitigation: Challenges and Opportunities for Agriculture*, Iowa State University, Ames, IA, 2004.

Dewet, Andrew, *Whole Earth: Earth System Science and Global Change*, W. H. Freeman & Company, New York, 2004.

DuTemple, *Acid Rain*, Lucent Books, San Diego, CA, 2001.

Duursma, Egbert. K., *Ozone Hole(s) 2000–2100*, Heineken Foundation for the Environment, Amsterdam, 2000.

Farhana Yamin, Ed., *Climate Change and Carbon Markets: A Handbook of Emission Reduction Mechanisms*, Earthscan, London, 2005.

Frumkin, Howard, Ed., *Environmental Health: from Global to Local*, Jossey-Bass, San Francisco, CA, 2005.

Graedel, T.E. and Paul J. Crutzen, *Atmospheric Change: An Earth System Perspective*, W. H. Freeman and Co., New York, 1993.

Hardy, John T., *Climate Change: Causes, Effects and Solutions*, John Wiley & Sons, New York, 2003.

Hoffmann, Matthew J., *Ozone Depletion and Climate Change: Constructing a Global Response*, State University of New York Press, Albany, New York, 2005.

Houghton, John T., *Global Warming: The Complete Briefing*, 3rd ed., Cambridge University Press, Cambridge, U.K., 2004.

Hunter, Robert, *Thermageddon, Countdown to 2030*, Arcade Publ., New York, 2003.

Ingram, W. Scott, *The Chernobyl Nuclear Disaster*, Facts On File, New York, 2005.

Jacobson, Mark Z., *Atmospheric Pollution: History, Science, and Regulation*, Cambridge University Press, New York, 2002.

Kirill, Ya. Kondratyev, Alexei A. Grigoryev, and Costas A. Varotsos, *Environmental Disasters: Anthropogenic and Natural*, Springer-Verlag, New York, 2002.

Kovats, Sari, Ed., *Climate Change and Stratospheric Ozone Depletion: Early Effects on Our Health in Europe*, World Health Organization, Regional Office for Europe, Copenhagen, 2000.

Kump, Lee R., James F. Kasting, James F., and Robert G. Crane, *The Earth System*, Prentice Hall, Upper Saddle River, NJ, 2003.

Lane, Carter N., Ed., *Acid Rain: Overview and Abstracts*, Nova Science Publishers, New York, 2003.

Leygraf, Christofer, and Thomas Graedel, *Atmospheric Corrosion*, Wiley-Interscience, New York, 2000.

Lewis, John S., *Rain of Iron and Ice: The Very Real Threat of Comet and Asteroid Bombardment*, Perseus Press, Addison-Wesley, Reading, MA, 1996.

Maslin, Mark, *Global Warming Very Short Introduction*, Oxford University Press, New York, 2004.

McElroy, Michael B., *The Atmospheric Environment: Effects of Human Activity*, Princeton University Press, Princeton, NJ, 2002.

Middlebrook, Ann M. and Margaret A. Tolbert, *Stratospheric Ozone Depletion*, University Science Books, Sausalito, CA, 2000.

Milne, Antony, *Doomsday: The Science of Catastrophic Events*, Praeger, Westport, CT, 2000.

O'Hare, Greg, John Sweeney, and Rob Wilby. *Weather, Climate, and Climate Change: Human Perspectives*, Prentice Hall, Upper Saddle River, NJ, 2005.

Olson, Nathan, *Droughts*, Capstone Press, Mankato, MN, 2006.

Oppenländer, Thomas, *Photochemical Purification of Water and Air*, Wiley-VCH, Weinheim, Germany, 2003.

Parker, Larry and Wayne A. Morrissey, *Stratospheric Ozone Depletion*, Novinka Books, New York, 2003.

Parks, Peggy J., *Global Warming*, Lucent Books, San Diego, CA, 2004.

Parson, Edward. A., *Protecting the Ozone Layer: Science and Strategy*, Oxford University Press, New York, 2003.

Pittock, A. Barrie, *Climate Change: Turning Up the Heat*, Sterling, VA: Earthscan, London, 2005.

Seinfeld, John H. and Spyros N. Pandis, *Atmospheric Chemistry and Physics*, John Wiley & Sons, New York, 1998.

Smith, Jim and Nicholas A. Beresford, *Chernobyl — Catastrophe and Consequences*, Springer-Verlag, New York, 2005.

Smith, Trevor, *Earth's Changing Climate*, Weigl Publishers, Calgary, Canada, 2004.

Spangenburg, Ray and Kit Moser, *If an Asteroid Hit Earth*, Franklin Watts, New York, 2000.

Turco, Richard P., *Earth Under Siege: From Air Pollution to Global Change*, Oxford University Press, New York, 1996.

Watts, Claire, *Heat Hazard Droughts*, Raintree, Chicago, IL, 2005.

Weart, Spencer R., *The Discovery of Global Warming*, Harvard University Press, Cambridge, MA, 2003.

Willis, Henry, *Earth's Future Climate*, Llumina Press, Coral Springs, FL, 2003.

Wise, William, *Killer Smog: The World's Worst Air Pollution Disaster*, iUniverse, Lincoln, NE, 2001.

QUESTIONS AND PROBLEMS

1. How do modern transportation problems contribute to the kinds of atmospheric problems discussed in this chapter?

2. What is the rationale for classifying most acid rain as a secondary pollutant?

3. Distinguish among UV-A, UV-B, and UV-C radiation. Why does UV-B pose the greatest danger in the troposphere?

4. How does the extreme cold of stratospheric clouds in Antarctic regions contribute to the Antarctic ozone hole?

5. How does the oxidizing nature of ozone from smog contribute to the damage that it does to cell membranes?

6. What may be said about the time and place of the occurrence of maximum ozone levels from smog with respect to the origin of the primary pollutants that result in smog formation?

7. What is the basis for "nuclear winter"?

8. What is meant by a "tie-in strategy"?

9. List two ways in which modern agricultural practices contribute to the production of atmospheric methane.

10. Describe how humans have been conducting a "massive geophysical experiment" with Earth.

11. Describe how cloud formation may exercise a degree of self-correction on the greenhouse effect.

12. Explain how acid rain is a regional air pollution problem compared to a local or global problem.

13. What is phytotoxicity? Give an example of indirect phytotoxicity from acid rain.

14. What reactive species is produced from chlorofluorocarbons that react with stratospheric ozone? What is the reaction? How does it lead to ozone layer destruction?

15. What are the conditions that lead to the formation of photochemical smog? How is photochemical smog manifested?

16. Other than ozone, what are two major inorganic products from smog? What are their effects?

17. In what respect, especially in respect to a particular significant constituent, is the composition of gases in the troposphere not uniform?

18. Cite an atmospheric chemical condition or phenomenon that shows that the O_2 molecule is easier to break apart than the N_2 molecule.

19. In what respect is atmospheric carbon dioxide essential to life on Earth? Why may it end up being the "ultimate air pollutant?"

10. THE GEOSPHERE

10.1. THE GEOSPHERE

The preservation of the **geosphere** — that part of the solid Earth on which humans live and from which they extract most of their food, minerals, and fuels — is one of the greater challenges affecting humankind today. Humans greatly alter the geosphere. Billions of tons of Earth material are mined or otherwise disturbed each year in the extraction of minerals and coal. Excess atmospheric carbon dioxide and acid rain (see Chapter 9) may cause major changes in the geosphere. Too much carbon dioxide in the atmosphere may cause global heating (greenhouse effect), which could significantly alter rainfall patterns and turn currently productive areas of the Earth into desert regions. Acidic rainfall can change the solubilities and oxidation-reduction rates of minerals. Erosion caused by intensive cultivation of land is washing away vast quantities of topsoil from fertile farmlands each year. The geosphere has been a dumping ground for large quantities of toxic, persistent chemicals.

Environmental geology deals with the relationship of the geosphere to the other environmental spheres that it influences and is influenced by, including humankind and its technology. Included in environmental geology are the following:

- Ways in which human activities and technology impact the geosphere, and the manner in which such impacts may be minimized or be made beneficial

- Utilization of resources from the geosphere, such as minerals, fossil fuels, groundwater, and rock

- Evaluation, prediction, and minimization of natural hazards, including earthquakes, landslides, floods, and volcanoes

A particularly important, pertinent, and controversial aspect of environmental geology is that of **land use** consisting of the ways in which land is employed for the purposes of humankind. Land abuse includes construction of shopping centers and residential developments on prime agricultural land, poorly restored strip mines, and loss of topsoil from improperly cultivated farmland. Biblical accounts (from ancient times) of lands that abounded with crops and vineyards, described areas in present-day Syria, Lebanon, and Palestine that have lost their productive capacity because of misuse of the land, erosion, overgrazing, and poor agricultural practices that have led to desertification as discussed in Chapter 9 and Chapter 11. Agricultural and other green technologies are now being used to restore some of these areas to productivity. A major challenge of land use planning is to utilize the principles of environmental geology to minimize such abuses in the future.

Earth Science

Earth science considers the intimate connections and influences with each other — of the geosphere with the hydrosphere and atmosphere (Figure 10.1). The study of water as it interacts with the solid Earth may be divided among **hydrology**, pertaining to nonoceanic liquid water above and below ground; **glaciology**, dealing with ice and snowpack on Earth's surface; and **oceanography**. **Meteorology** addesses phenomena in the atmosphere, and the science of climate is **climatology**. **Geology** deals with the solid Earth as a whole. As covered later in this chapter, geology is itself divided into several major categories.

As the medium on which plants grow, and virtually all terrestrial organisms depend on for their existence, the most important part of the geosphere for life on Earth is soil, the subject of Chapter 11. The geosphere is crucial in protecting water resources by providing watersheds on which water is collected and by its ability to assimilate and store groundwater in aquifers. It is essential to avoid geospheric pollutants that may contaminate water. The geosphere has a good capability to absorb and neutralize a variety of pollutants, although it is important to avoid discarding to the geosphere refractory substances such as heavy metals and poorly biodegradable organics.

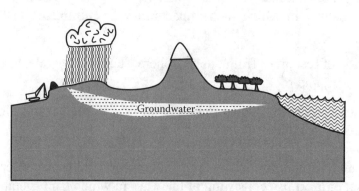

Figure 10.1.   The geosphere has a very close relationship with the hydrosphere, the atmosphere, and the biosphere and is strongly affected by human activities in the atmosphere.

Sustainability of the Geosphere and Geodiversity

The sustenance of the geosphere as a medium hospitable to humans and other organisms is one of the primary goals of sustainability. The importance of biodiversity has long been accepted and now geodiversity is increasingly being recognized as an essential support for biodiversity. **Geodiversity** is the maintenance of the geospheric environment, including landforms, rocks, sediments, soils, fossils, aquifers, and all other aspects of the geosphere. Geodiversity is maintained through **geoconservation**, through which the geosphere is preserved for its intrinsic ecological and other values. Geoconservation preserves geospheric processes, such as cycles involving surface water and groundwater, as well as geospheric sites and features.

Geodiversity is very much affected by human activities. The greatest effects have been through the conversion of large areas of forest and prairie lands to agricultural production. One unfortunate result of this conversion has been loss of enormous quantities of soil from erosion. Large areas of the geosphere have been rearranged to construct roads, airports, manufacturing facilities, and shopping areas. These activities have disrupted and covered topsoil, disturbed springs, and adversely affected surface water infiltration to underground aquifers. Therefore, it is crucial to consider geodiversity in carrying out anthrospheric activities. In addition to avoiding harm from damaging or destroying geospheric artifacts, humans can even restore or improve parts of the geosphere by, for example, constructing wetlands.

10.2. THE NATURE OF SOLIDS IN THE GEOSPHERE

The Earth is divided into layers, including the solid iron-rich inner core, molten outer core, mantle, and crust. Environmental science is most concerned with the **lithosphere**, which consists of the outer mantle and the **crust**. The latter is the Earth's outer skin that is accessible to humans. It is extremely thin compared to the diameter of the Earth, ranging from 5 to 40 km thick.

Most of the solid earth crust consists of rocks. Rocks are composed of minerals, where a **mineral** is a naturally occurring inorganic solid with a definite internal crystal structure and chemical composition. A **rock** is a solid, cohesive mass of pure mineral or an aggregate of two or more minerals.

Igneous, Sedimentary, and Metamorphic Rock

At elevated temperatures deep beneath Earth's surface, rocks and mineral matter melt to produce a molten substance called **magma**. Cooling and solidification of magma produces **igneous rock**. Common igneous rocks are granite, basalt, quartz (SiO_2), feldspar ($(Ca,Na,K)AlSi_3O_8$), magnetite (Fe_3O_4), and micas, a large group of minerals containing high proportions of silicon and oxygen along with fluoride and a wide variety of metals, such as sodium, calcium, and titanium. **Lava** is igneous

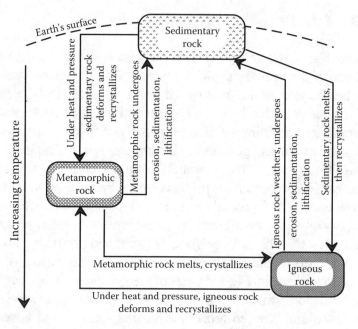

Figure 10.2.   The rock cycle.

rock which has flowed onto Earth's surface in a molten state and solidified rapidly so that it consists of very fine crystals or a form of natural glass. Exposure of igneous rocks to water from the hydrosphere and air from the atmosphere causes the rocks to disintegrate by a process called **weathering**. Erosion from wind, water, or glaciers picks up materials from weathering rocks and deposits it as **sediments** or **soil**. A process called **lithification** describes the conversion of sediments to relatively porous, soft, chemically reactive **sedimentary rocks**, which may contain particles eroded from igneous rocks. **Organic sedimentary rocks** contain residues of plant and animal remains. Carbonate minerals of calcium and magnesium — **limestone** or **dolomite** — are especially abundant in sedimentary rocks. **Metamorphic rock** is formed by the action of heat and pressure on sedimentary, igneous, or other kinds of metamorphic rock that are not in a molten state. These changes are illustrated by the **rock cycle** in Figure 10.2.

Structure and Properties of Minerals

A **mineral** is a geospheric solid with a defined chemical formula, and a specific crystal structure. Although over 2000 minerals are known, only about 25 **rock-forming minerals** make up most of Earth's crust. The chemical composition of minerals reflects the chemical composition of Earth's crust, which, in descending order of elemental composition is oxygen (49.5%), silicon (25.7%), aluminum (7.4%), iron (4.7%), calcium (3.6%), sodium (2.8%), potassium (2.6%), magnesium (2.1%), and other (1.6%). The most abundant minerals are **silicates** such as quartz, SiO_2, or potassium feldspar, $KAlSi_3O_8$.

Secondary minerals are formed by alteration of parent mineral matter. The most important class of secondary minerals is composed of **clays**, a group of microcrystalline minerals consisting of hydrous aluminum silicates that have sheetlike structures and often are present as very small colloidal particles. Clays are particularly important sedimentary minerals in soil and in the sediments of bodies of water. Clays predominate in the inorganic components of most soils and are very important in holding water and in plant nutrient cation exchange. Clays may hold pollutant compounds within their layered structures. A typical clay mineral is kaolinite ($Al_2Si_2O_5(OH)_4$) formed by the chemical weathering of potassium feldspar rock ($KAlSi_3O_8$):

$$2KAlSi_3O_8(s) + 2H^+ + 9H_2O \rightarrow$$

$$Al_2Si_2O_5(OH)_4(s) + 2K^+(aq) + 4H_4SiO_4(aq)$$

$$(10.2.1)$$

10.3. THE RESTLESS EARTH

Far from being a static, unchanging mass of rocks and soil, the geosphere has a highly varied, constantly changing physical form. Most of the Earth's landmass is contained in several massive continents separated by vast oceans. Towering mountain ranges spread across the continents and, in some places, the ocean bottom is at extreme depths. Earthquakes, which often cause great destruction and loss of life, and volcanic eruptions, which sometimes throw enough material into the atmosphere to cause temporary changes in climate, serve as reminders that the Earth is a dynamic, living body that continues to change. There is convincing evidence, such as the close fit between the western coast of Africa and the eastern coast of South America, that widely separated continents were once joined and have moved relative to each other. This ongoing phenomenon is known as **continental drift**. It is now believed that 200 million years ago, much of Earth's landmass was all part of a supercontinent, now called Gowandaland. This continent split apart to form the present-day continents of Antarctica, Australia, Africa, and South America, as well as Madagascar, the Seychelle Islands, and India.

The observations described in the preceding paragraph are explained by the theory of **plate tectonics** (Figure 10.3). This theory views Earth's solid surface as consisting of several rigid plates that move relative to each other at an average rate of several centimeters per year atop a relatively weak, partially molten layer that is part of Earth's upper mantle called the **asthenosphere**. The science of plate tectonics explains the large-scale phenomena that affect the geosphere, including the creation and enlargement of oceans as the ocean floors open up and spread, the collision and breaking apart of continents, the formation of mountain chains, volcanic activities, the creation of islands of volcanic origin, and earthquakes.

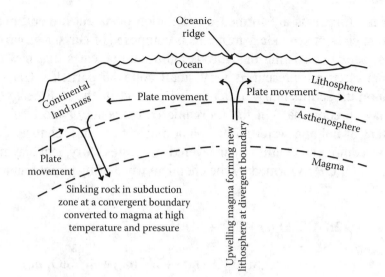

Figure 10.3.   Illustration of the tectonic cycle in which upwelling magma along a boundary where two plates diverge creates new lithosphere on the ocean floor, and sinking rock in a subduction zone is melted to form magma.

The boundaries between these plates are where most geological activity, such as earthquakes and volcanic activity occur. These boundaries are of the three following types:

- **Divergent boundaries** where the plates are moving away from each other. Occurring on ocean floors, these are regions in which hot magma flows upward and cools to produce new solid lithosphere. This new solid material creates **ocean ridges**.

- **Convergent boundaries** where plates move toward each other. One plate may be pushed beneath the other in a **subduction zone** in which matter is buried in the asthenosphere and eventually remelted to form new magma. When this does not occur, the lithosphere is pushed up to form mountain ranges along a collision boundary.

- **Transform fault boundaries** in which two plates slide past each other. These boundaries create faults that result in earthquakes.

Structural Geology

Earth's surface is constantly being reshaped by geological processes. **Structural geology** addresses the geometric forms of geologic structures over a wide range of sizes, the nature of structures formed by geological processes, and the formation of folds, faults, and other geological structures. **Primary structures** are those that have resulted from the formation of a rock mass from its parent materials. Primary structures are modified and deformed to produce **secondary structures**.

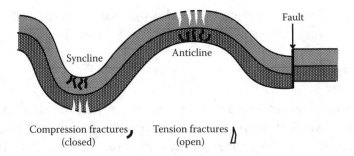

Figure 10.4.   Folds (syncline and anticline) are formed by the bending of rock formations. Faults are produced by rock formations moving vertically or laterally with respect to each other.

A basic premise of structural geology is that most layered rock formations were deposited in a horizontal configuration, then later moved by tectonic forces. Cracking of such a formation without displacement of the separate parts of the formation relative to each other produces a **joint**, whereas displacement produces a **fault** (see Figure 10.4).

Related to the physical configuration of the geosphere are several major kinds of processes that can change this configuration and that have the potential to cause damage and even catastrophic effects. These processes, mentioned here and addressed in more detail in Chapter 12, can be divided into the two main categories of **internal processes** (that arise from phenomena located significantly below the Earth's surface) and **surface processes** (that occur on the surface).

Internal Processes

The two main types of internal processes are earthquakes and volcanoes. These natural disasters are covered in more detail in Chapter 12. **Earthquakes** occur as motion of ground resulting from the release of energy that accompanies an abrupt slippage of rock formations subjected to stress along a fault. In addition to shaking of ground, which can be quite violent, earthquakes can cause the ground to rupture, subside, or rise. A devastating phenomenon that sometimes follows an earthquake consists of **tsunamis** — large ocean waves resulting from earthquake-induced vertical movement of ocean floor. Tsunamis sweeping onshore have destroyed many homes and taken many lives, often large distances from the epicenter of the earthquake itself. On December 26, 2004, a powerful earthquake just off the Indonesia coast produced a deadly tsunami that travelled across the Indian ocean at several hundred kilometers per hour devastating coastal areas of Indonesia, Sri Lanka, Somalia, Sumatra, and other countries. At least 150,000 people were killed, millions lost their homes and possessions, and the total cost ran into multiple billions of dollars.

In addition to earthquakes, the other major subsurface process that has the potential to massively affect the environment consists of emissions of molten rock (lava), gases, steam, ash, and particles because of the presence of magma near the

Earth's surface. This phenomenon is called a **volcano**. As discussed in Chapter 12, volcanoes can be very destructive and damaging to the environment.

Surface Processes

Surface geological features are formed by upward movement of materials from the Earth's crust. With exposure to water, oxygen, freeze–thaw cycles, organisms, and other influences on the surface, surface features are subject to two processes that largely determine the landscape — weathering and erosion. As noted earlier in this chapter, weathering consists of the physical and chemical breakdown of rock, whereas erosion is the removal and movement of weathered products by the action of wind, liquid water, and ice. Weathering and erosion work together in that one augments the other in breaking down rock and moving the products. Weathered products removed by erosion are eventually deposited as sediments and may undergo diagenesis and lithification to form sedimentary rocks.

Some surface processes can be very damaging and even hazardous to humans. Of these, the most harmful consists of **landslides** that occur when soil or other unconsolidated materials slide down a slope. Related phenomena include rockfalls, mudflows, and snow avalanches. These phenomena are discussed in more detail in Chapter 12.

Subsidence occurs when the surface level of earth sinks over a significant area. The most spectacular evidence of subsidence is manifested as large sinkholes that may form rather suddenly, sometimes swallowing trees, automobiles, and even whole buildings in the process. Overall, much more damage is caused by gradual and less extreme subsidence, which may damage structures as it occurs or result in inundation of areas near water level. Such subsidence is frequently caused by the removal of fluids, such as petroleum, from below ground.

10.4. SEDIMENTS

Vast areas of land, as well as lake and stream sediments, are formed from sedimentary rocks. The properties of these masses of material depend strongly upon their origins and transport. Water is the main vehicle of sediment transport, although wind can also be significant. Hundreds of millions of tons of sediment are carried by major rivers each year. The study of sediments and the environments in which they are formed is the science of **sedimentology**.

The action of flowing water in streams cuts away stream banks and carries sedimentary materials for great distances. Sedimentary materials may be carried by flowing water in streams as dissolved load, suspended load, or bed load. **Dissolved load** consists of solutions of minerals, such as calcium bicarbonate, which can precipitate to form a sediment of solid calcium carbonate:

$$Ca^{2+} + 2HCO_3^- \rightarrow CaCO_3(s) + CO_2(g) + H_2O \qquad (10.4.1)$$

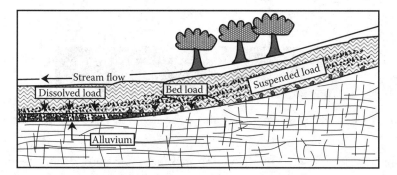

Figure 10.5.   Streams carry sediment-forming materials as bed load, suspended load, and dissolve matter and deposit them in sediments (alluvium).

Most flowing water containing dissolved load originates underground, where it dissolves minerals from the rock strata that it flows through. **Suspended load** consists primarily of finely divided silt, clay, or sand originating from sources such as soil eroded from land or finely divided rock released by melting glaciers. **Bed load** is made up of larger particles dragged along the bottom of the stream channel (Figure 10.5).

Typically, about ⅔ of the sediment carried by a stream is transported in suspension, about ¼ in solution, and the remaining relatively small fraction as bed load. The ability of a stream to carry sediment increases with both the overall rate of flow of the water (mass per unit time) and the velocity of the water. Both of these are higher under flood conditions, so floods are particularly important in the transport of sediments. Streams mobilize sedimentary materials through **erosion**, **transport** materials along with stream flow, and release them in a solid form during **deposition**. Deposits of stream-borne sediments are called **alluvium**.

10.5. INTERACTION WITH THE ATMOSPHERE AND HYDROSPHERE

The geosphere interacts strongly with the other spheres of the environment. It is strongly influenced by them and, in turn, has a strong influence on each. These interactions are briefly addressed here.

The hydrosphere has more influence than any other on the geosphere. Ocean water covers more of the geosphere than does land. Freshwater from precipitation falls on land and produces streams, vast rivers, and lakes. Liquid water as groundwater occurs in huge quantities beneath the surface of the land. Water in the solid state as glacial ice was a major force in determining much of the geomorphology of Earth's surface during the Ice Ages and continues to do so in colder climates today. The nature of solids in the geosphere largely determines the distribution and fate of surface water and groundwater. Porous, permeable rock formations may be conducive to the inflow and storage of groundwater. The nature and elevation of surrounding geological strata determine whether a stream is an **influent stream** in which water is lost to the ground to recharge groundwater supplies in underground

Figure 10.6. Continued erosion of mountains by water wears them down and rounds their features.

aquifers, or an **effluent stream** in which groundwater enters the stream from surrounding aquifers to maintain stream flow.

One of the largest influences of water on the geosphere arises from erosion of land surfaces by water. Flowing water dislodges weathered rock, carries it some distance from its source, and deposits it as sediments. Water then plays a major role in the chemical processes that convert these sediments to sedimentary rock. Relatively young mountain formations tend to have sharp features. Through the erosive action of liquid water and ice, these features become rounded, and eventually the mountain formations are worn down (Figure 10.6).

The most important role of the atmosphere in shaping the geosphere arises from water carried from oceans onto land in the hydrologic cycle. Climate, resulting largely from atmospheric conditions, has a strong influence on the geosphere, particularly in determining the amount of precipitation falling on land. One of the most direct atmospheric forces that influences the geosphere is from wind. Wind can be a major factor in erosion, particularly of disturbed soil under dry conditions. In addition, wind can transport and deposit solids that make up parts of the geosphere. One such solid material consists of **sand dunes**, composed of sand carried just above the surface of the ground by wind. Active sand dunes that are still moving can cause severe problems with roads and other structures. Stabilized sand dunes should be treated with care to prevent their becoming active and mobile. The other major kind of solid material carried by wind is **loess** consisting of silt, defined as finely divided sediment in a size range of $^1/_{256}$ to $^1/_{16}$ mm in diameter, carried and deposited by wind. Loess deposits in the U.S. are near major rivers. These deposits were formed from rock ground by glaciers during the Pleistocene Ice Ages and deposited in large areas along the river floodplains. As the river flows subsided when the glaciers retreated, large areas of sediment were left dry without much vegetative cover. Winds carried this material away and deposited it as loess.

The interactions of the geosphere and the anthrosphere are many. Mountainous terrain may make it impossible to grow enough food to support a significant human population. Weak geological strata can make the construction and maintenance of buildings very difficult. Earth is the source of the minerals — metal ores, fossil fuels, stone — required to sustain an economy. Human activities, in turn, have a tremendous influence on the geosphere. This is seen when one observes Earth's surface from above while travelling by airplane. Highways cut across terrain, dams interrupt the flow of rivers, and vast expanses of land have been converted from forests and prairies to cultivated land. Increasingly, human endeavors are being used to enhance the geosphere that has been ruthlessly exploited, especially during the last two centuries. Some of the greatest success has been in terracing farmland to minimize water erosion and topsoil loss. In some cases, streams that were channelized into straight ditches are being restored with constructed bends and meanders.

10.6. LIFE SUPPORT BY THE GEOSPHERE

The connection between the biosphere and the geosphere is obvious. Most plants exist on soil, as do other forms of life, such as earthworms, fungi, and bacteria. Some kinds of rock have a biological origin. Deposits of limestone and silicaceous deposits from the shells of aquatic organisms originated through life processes. The production of oxygen by photosynthesis resulted in the oxidation of soluble iron to insoluble iron oxides, thus producing iron ore deposits. Coal, kerogen (the organic matter in oil shale), and petroleum are of biological origin.

Much of what is known about geological history and about the evolution of life on Earth is based on **paleontology**, the study of fossils in rock. Similar life-forms have existed at different times, so it is possible to use observations of fossils to determine the relative ages of rock strata. Fossil records extend back approximately 3 billion years, so that very long time spans can be addressed.

Trace-level elements in the geosphere play an important role in the health of living organisms. Fluorine as fluoride ion prevents tooth decay and strengthens bone when ingested at relatively low levels, whereas at somewhat higher levels it has detrimental effects on tooth and bone. Goiter, a condition manifested by enlargement of the thyroid gland, is caused by a deficiency of iodine. Selenium at very low levels is required in the diet of animals, but at levels above only a few parts per million selenium is toxic. Sickness in animals has been shown to result from either too little or too much selenium in the soil on which animal feed is grown and on the selenium content of the water that they drink. Zinc is an essential nutrient for both plants and animals. It is required by animals, for example, for wounds to heal properly. Too much zinc in soil, such as soil treated with excessive amounts of zinc-laden sewage sludge, can be phytotoxic (toxic to plants).

10.7. GEOCHEMISTRY

Geochemistry deals with chemical species, reactions, and processes in the lithosphere and their interactions with the atmosphere and hydrosphere. The branch of geochemistry that explores the complex interactions among the rock/water/air/life (and human) systems that determine the chemical characteristics of the surface environment is **environmental geochemistry**. Obviously, geochemistry and its environmental subdiscipline are very important in environmental science. Geochemistry addresses a large number of chemical and related physical phenomena. Some of the major areas of geochemistry are the following:

- The chemical composition of major components of the geosphere, including magma and various kinds of solid rocks

- Processes by which elements are mobilized, moved, and deposited in the geosphere through a cycle known as the **geochemical cycle**

- Chemical processes that occur during the formation of igneous rocks from magma

- Chemical processes that occur during the formation of sedimentary rocks

- Chemistry of rock weathering

- Chemistry of volcanic phenomena

- Role of water and solutions in geological phenomena, such as deposition of minerals from hot brine solutions

- The behavior of dissolved substances in concentrated brines

An important consideration in geochemistry is that of the interaction of life-forms with geochemical processes addressed as **biogeochemistry** or **organic geochemistry**. The deposition of biomass and the subsequent changes that it undergoes have led to the formation of huge deposits of petroleum, coal, and oil shale. Chemical changes induced by photosynthesis have resulted in massive deposits of calcium carbonate (limestone). Deposition of the biochemically synthesized shells of microscopic animals have led to the formation of large masses of calcium carbonate and silica. Biogeochemistry is closely involved with elemental cycles, such as those of carbon.

Physical and Chemical Aspects of Weathering

Defined in Section 10.2, weathering is discussed here as a geochemical phenomenon. Rocks tend to weather more rapidly when there are pronounced differences in physical conditions — alternate freezing and thawing and wet periods alternating with severe dryness. Other mechanical aspects are swelling and shrinking of minerals with hydration and dehydration as well as growth of roots through cracks in rocks. An important aspect of weathering is **exfoliation** through which outer layers of rock peel away as the result of influences such as freezing and thawing. The rates of chemical reactions involved in weathering increase with increasing temperature.

As a chemical phenomenon, **chemical weathering** can be viewed as the result of the tendency of the rock/water/mineral system to attain equilibrium. This occurs through the usual chemical mechanisms of dissolution/precipitation, acid–base reactions, complexation, hydrolysis, and oxidation-reduction. Biological processes,

biogeochemical weathering, such as production of weathering agents by lichen growing on rock surfaces, can be involved as well.

Weathering is very slow in dry air. Water increases the rate of weathering by many orders of magnitude for several reasons. Water, itself, is a chemically active substance in the weathering process. Furthermore, water holds weathering agents in solution such that they are transported to chemically active sites on rock minerals and contact the mineral surfaces at the molecular and ionic level. Prominent among such weathering agents are CO_2, O_2, organic acids, sulfur acids ($SO_2(aq)$, H_2SO_4), and nitrogen acids (HNO_3, HNO_2). Water provides the source of H^+ ion needed for acid-forming gases to act as acids as shown by the following:

$$CO_2 + H_2O \rightarrow H^+ + HCO_3^- \tag{10.7.1}$$

$$SO_2 + H_2O \rightarrow H^+ + HSO_3^- \tag{10.7.2}$$

Rainwater is essentially free of mineral solutes. It is usually slightly acidic because of the presence of dissolved carbon dioxide or more highly acidic because of acid-rain-forming constituents. As a result of its slight acidity and lack of alkalinity and dissolved calcium salts, rainwater is chemically aggressive toward some kinds of mineral matter, which it breaks down by a process of chemical weathering. Because of this process, river water has a higher concentration of dissolved inorganic solids than does rainwater.

A typical chemical reaction involved in weathering is the dissolution of calcium carbonate (limestone) by water containing dissolved carbon dioxide:

$$CaCO_3(s) + H_2O + CO_2(aq) \rightarrow Ca^{2+}(aq) + 2HCO_3^-(aq) \tag{10.7.3}$$

Weathering may also involve oxidation reactions, such as occurs when pyrite (FeS_2) dissolves:

$$4FeS_2(s) + 15O_2(g) + (8 + 2x)H_2O \rightarrow$$

$$2Fe_2O_3 \cdot xH_2O + 8SO_4^{2-}(aq) + 16H^+(aq) \tag{10.7.4}$$

Isotopic Geochemistry

The measurement of subtle differences in the ratios of naturally occurring elemental isotopes found in rocks and minerals can yield significant information about geochemical phenomena. The activity of radioactive carbon-14, which occurs in atmospheric carbon dioxide and is incorporated into biomass by photosynthesis, has been widely used to determine the age of biological materials. The half-life of carbon-14 is 5570 years, so this method can be used to determine the ages of organic materials less than about 30,000 years old. Small differences in the ratios of oxygen-16 to oxygen-18 in calcium carbonate deposited by the metabolic activities of marine organisms can be used to estimate the temperatures at which these deposits formed. Such information can be used to estimate past climatic conditions and in predicting future climates. Radiometric age dating is based upon the decay of radioisotopes and the formation of daughter products. The most abundant isotope of uranium, uranium-238, decays eventually to lead-206, so that determination of the uranium-238/lead-206 ratio in a rock can be used to estimate the time from which the original uranium was deposited.

10.8. WATER ON AND IN THE GEOSPHERE AND WATER WELLS

Groundwater (Figure 10.7) is a vital resource in its own right that plays a crucial role in geochemical processes, such as the formation of secondary minerals. The nature, quality, and mobility of groundwater are all strongly dependent upon the rock formations in which the water is held. Physically, an important characteristic of such formations is their **porosity**, which determines the percentage of rock volume available to contain water. A second important physical characteristic is **permeability**, which describes the ease of flow of the water through the rock. High permeability is usually associated with high porosity. However, clays tend to have low permeability even when a large percentage of the volume is filled with water.

Most groundwater originates as **meteoric** water from precipitation in the form of rain or snow. If water from this source is not lost by evaporation, transpiration, or to stream runoff, it may infiltrate into the ground. Initial amounts of water from precipitation onto dry soil are held very tightly, as a film on the surfaces and in the

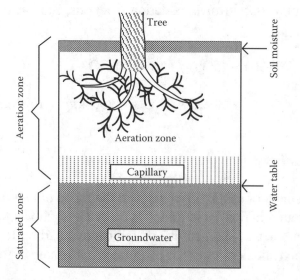

Figure 10.7.   Some major features of the distribution of water underground.

micropores of soil particles in a **belt of soil moisture**. At intermediate levels, the soil particles are covered with films of water, but air is still present in larger voids in the soil. The region in which such water is held is called the **unsaturated zone** or **zone of aeration** and the water present in it is **vadose water**. At lower depths in the presence of adequate amounts of water, all voids are filled to produce a **zone of saturation**, the upper boundary of which is the **water table**. Water present in a zone of saturation is called **groundwater**. Because of its surface tension, water is drawn somewhat above the water table by capillary-sized passages in soil in a region called the **capillary fringe**.

The water table is crucial in explaining and predicting the flow of wells and springs and the levels of streams and lakes. It is also an important factor in determining the extent to which pollutant and hazardous chemicals underground are likely to be transported by water. The water table can be mapped by observing the equilibrium level of water in wells, which is essentially the same as the top of the saturated zone. As shown in Figure 10.8, the water table is usually not level, but tends to follow the general contours of the surface topography. It also varies with differences in permeability and water infiltration. The water table is at surface level in the vicinity of swamps and frequently above the surface where lakes and streams are encountered. The water level in such bodies may be maintained by the water table. **Influent** streams or reservoirs are located above the water table; they lose water to the underlying aquifer and cause an upward bulge in the water table beneath the surface water.

Groundwater **flow** is an important consideration in determining the accessibility of the water for use and transport of pollutants from underground waste sites. Various parts of a body of groundwater are in hydraulic contact so that a change in pressure at one point will tend to affect the pressure and level at another point. For example, infiltration from a heavy, localized rainfall may affect the water table at a point remote from the infiltration. Groundwater flow occurs as the result of the natural tendency of the water table to assume even levels by the action of gravity.

Groundwater flow is strongly influenced by rock permeability. Porous or extensively fractured rock is relatively highly **pervious**, meaning that water can migrate through the holes, fissures, and pores in such rock. Because water can be extracted from such a formation, it is called an **aquifer**. By contrast, an **aquiclude** is a rock formation that is too impermeable or unfractured to yield groundwater. Impervious

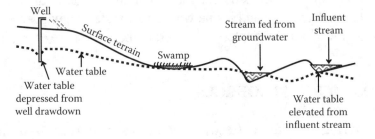

Figure 10.8.   The water table and influences of surface features and terrain on it.

rock in the unsaturated zone may retain water infiltrating from the surface to produce a **perched water table** that is above the main water table and from which water may be extracted. However, the amounts of water that can be extracted from such a formation are limited, and the water is vulnerable to contamination.

Most groundwater is tapped for use by water wells drilled into the saturated zone. The use and misuse of water from this source has a number of environmental implications. In the U.S. about ⅔ of the groundwater pumped is consumed for irrigation; lesser amounts of groundwater are used for industrial and municipal applications.

As water is withdrawn, the water table in the vicinity of the well is lowered. This **drawdown** of water creates a **zone of depression**. In extreme cases, groundwater is severely depleted, and surface land levels can even subside (which is one reason that Venice, Italy, is now very vulnerable to flooding). Heavy drawdown can result in infiltration of pollutants from sources such as septic tanks, municipal refuse sites, and hazardous waste dumps. Mineral deposits, such as oxides of iron and manganese, can form from well water exposed to air as it drains from the aquifer into the well. Such deposits may seriously limit the flow of water into the well.

10.9. ECONOMIC GEOLOGY

Technological applications of geology have several important economic benefits. Geology is used to find and develop essential resources of raw materials extracted from Earth's crust. These include minerals, uranium used to fuel nuclear reactors, natural gas, petroleum, and coal. Knowledge of subsurface strata is essential to the siting and construction of buildings and other structures. Geological principles are used to prevent and warn of hazards, such as earth slides.

Geological science, now equipped with a wide array of high-technology tools, is used to locate and develop deposits of metal ores; nonmetal minerals, such as aggregate, gypsum, clay and salt; and fossil fuels. Most economic geologists work in the petroleum industry in which one of their major functions is to maximize the probability of striking economically viable deposits of crude oil or natural gas when exploratory wells are drilled. Other important aspects of petroleum geology include structural geology, stratigraphy (the science of stratified rock formations), and sedimentary petrology.

The provision of mineral resources for human use is one of the most important aspects of economic geology and is one with a multitude of environmental implications. The remainder of this chapter is devoted to the geosphere as a green resource for minerals.

10.10. GEOSPHERIC RESOURCES

Manufacturing requires a steady flow of raw materials — minerals, fuel, wood, and fiber. These can be provided from either **extractive** (nonrenewable) and

renewable sources. The remainder of this chapter addresses the extractive industries in which irreplaceable resources are taken from the Earth's crust. Therefore, it deals with a particularly important part of Earth's support system on which humankind depends — geospheric resources. As such, the chapter emphasizes mineral resources, nonrenewable materials that are extracted from Earth's crust for use in manufacturing, agriculture, and other applications. Addressed in other chapters are related matters pertaining to soil resources (Chapter 11), the use of resources in manufacturing and energy utilization (Chapter 17), and utilization of renewable materials and energy (Chapter 17 and Chapter 18).

The utilization of mineral resources is strongly tied with technology, energy, and the environment. Perturbations in one usually cause perturbations in the others. For example, reductions in automotive exhaust pollutant levels with the use of catalytic devices, discussed in Chapter 9, have resulted in increased demand for platinum and other precious metals used in catalysts. The availability of many metals depends upon the quantity of energy used and the amount of environmental damage tolerated in the extraction of low-grade ores. Many other such examples could be cited. Because of these intimate interrelationships, technology, resources, and energy must all be considered when green science and technology are discussed.

The quantities of available minerals, particularly metals in limited supply, are subject to political influences and financial manipulation. Cartels, such as the once powerful tin cartel, have been organized to attempt to raise prices. Governments may limit exports to influence foreign policy, or subsidize prices to discourage competition and increase foreign exchange. Prices fluctuate with costs of labor, materials, property, and pollution controls. During the early 1980s, speculators cornered the market for silver, causing panic buying by interests fearing shortages and raising prices severalfold. As often happens, this excursion was followed by a crash in prices. Increased industrial demand, such as accompanies economic development in developing countries, can cause metal prices to increase significantly. As of 2005, this had occurred for several important minerals with the rapidly developing Chinese economy.

In discussing minerals and fossil fuels in the remainder of this chapter, two terms related to available quantities are used and should be defined. The first of these is **resources**, defined as quantities that are estimated to be ultimately available. The second term is **reserves**, which refers to well-identified resources that can be profitably utilized with existing technology.

10.11. GEOSPHERIC SOURCES OF USEFUL MINERALS

There are numerous kinds of mineral deposits that are used in various ways. The most available deposits have already been exploited. A challenge now is to find sustainable ways to utilize available resources in a cost-effective manner consistent with maximum conservation of the resource, environmental protection, and material recycling.

Geological and geochemical factors are crucial in locating deposits of crucial metals. Deposits of metals often occur in masses of igneous rock that have been extruded in a solid or molten state into the surrounding rock strata; such masses are called **batholiths**. Other geological factors to consider include age of rock, fault zones, and rock fractures. The crucial step of finding ore deposits falls in the category of mining geology.

Deposits from Igneous Rocks and Magmatic Activity

Magmatic activity of molten rock giving rise to the formation of igneous rock formations is responsible for forming numerous kinds of useful mineral deposits. In addition to deposits formed directly from solidifying magma, associated deposits are produced by water in association with magma. Both of these kinds of deposits tend to occur in subduction zones where the lithosphere beneath the ocean is being forced under another continental or oceanic plate.

As masses of magma cool and solidify, crystals of minerals often form and, depending upon their densities relative to the magma, either rise or sink in the molten magma, forming layers enriched in specific minerals. This material called **pegmatite** may contain desired minerals as coarse-grained material in the form of relatively large crystals that can be isolated by physical processes to extract the desired material. Such a mineral is magnetite, Fe_3O_4, which is more dense than the comparatively light silicates prevalent in magma; because of its lower density, it settles as a layer when the magma solidifies. Another such dense, settling mineral is chromite, an oxide mineral containing Cr, Mg, Fe, and Al and mined as a source of chromium. Of course, the more valuable a mineral, the more worthwhile it is to separate it from dispersed sources, even if it has not been isolated well within a pegmatite deposit. This is the case with gold, platinum, palladium, and diamond.

Deposits from Hydrothermal Activity

Hot aqueous solutions associated with magma can form rich deposits of minerals. These solutions, either associated with the magma as it comes to the surface or produced by water coming into contact with hot magma, can carry dissolved minerals out of the magma or can pick up minerals from surrounding strata as the magma cools. As the solutions cool, dissolved minerals crystallize from solution, forming **hydrothermal** mineral deposits. Several important metals, including lead, zinc, and copper, are often associated with hydrothermal deposits. The hydrothermal waters are chemically reducing, so that sulfur associated with them is in the reduced sulfide form, S^{2-}. Most metal sulfides are insoluble and can produce sulfide deposits as shown by the following reaction:

$$Pb + S^{2-} \rightarrow PbS \ (\textit{solid}, \text{galena}) \qquad (10.11.1)$$

In addition to galena, two other representatives of the numerous metal sulfide deposits are CuS, HgS (cinnabar), and ZnS (sphalerite).

The ocean floor in some locations is an especially rich source of hydrothermal deposits consisting of metal sulfides. One such region is the Juan de Fuca ridge in the Pacific Ocean off the northwest coast of the U.S., where significant hydrothermal deposits of silver and zinc sulfides are located on the ocean floor. Similar deposits, including sulfides of copper and lead, are found at the bottom of the Red Sea.

Deposits Formed by Sedimentary or Metamorphic Processes

Some useful mineral deposits are formed as **sedimentary deposits** in association with the formation of sedimentary rocks (see Chapter 13). **Evaporites** are produced when seawater is evaporated. In addition to the obvious example of NaCl deposits of halite, evaporite deposits include sodium carbonates, potassium chloride, gypsum ($CaSO_4 \cdot 2H_2O$), and magnesium salts. Many significant iron deposits consisting of hematite (Fe_2O_3) and magnetite (Fe_3O_4) were formed as sedimentary bands when Earth's atmosphere was changed from reducing to oxidizing as photosynthetic organisms produced oxygen. Oxidation reactions, such as

$$4Fe^{2+} + O_2 + 4H_2O \rightarrow 2Fe_2O_3(s) + 8H^+ \qquad (10.11.2)$$

precipitated soluble iron(II) ion in ancient oceans to form insoluble deposits of oxidized iron oxides.

Deposition of suspended rock solids by flowing water can cause segregation of the rocks according to differences in rock size and density. This can result in the formation of useful deposits that are enriched in desired minerals. Originally the minerals were removed from rock deposits upstream by weathering processes and transported by flowing water to the location at which they were deposited as **placer** deposits. During the weathering, transport, and deposition processes that lead to useful placer deposits, the desired minerals are enriched and sorted, whereas impurity constituents are dissolved or washed out by the flowing water. Gravel, sand, and some other minerals, such as gold, often occur in placer deposits.

Some significant placer deposits are now located beneath coastal ocean waters. These were formed during glacial times by streams carrying glacial melt and were deposited on what was then land because of the much shallower ocean depths (much more of Earth's water was tied up as ice on land at the time). Later, as the glacial ice melted and ocean levels raised, these deposits were immersed by seawater, where they remain today as potential mineral sources.

Metamorphic deposits of some minerals have been observed. These occur when minerals, usually of sedimentary origin, become buried and subjected to high pressures and extreme heat. Mineable deposits of graphite, a useful, "slick" form of carbon used as a lubricant, have been formed by metamorphic action on fossil carbon, such as coal.

Some mineral deposits are formed by the enrichment of desired constituents when other fractions are weathered or leached away. The most common example of such a deposit is bauxite, Al_2O_3, remaining after silicates and other more soluble constituents have been dissolved by the weathering action of water under the severe conditions of hot tropical climates with very high levels of rainfall. This kind of material is called a **laterite.**

10.12. EVALUATION OF MINERAL RESOURCES

To make its extraction worthwhile, a mineral must be enriched at a particular location in Earth's crust relative to the average crustal abundance. Normally applied to metals, such an enriched deposit is called an **ore**. The value of an ore is expressed in terms of a **concentration factor**:

$$\text{Concentration factor} = \frac{\text{Concentration of material in ore}}{\text{Average crustal concentration}} \qquad (10.12.1)$$

Obviously, higher concentration factors are always desirable. Required concentration factors decrease with average crustal concentrations and with the value of the commodity extracted. A concentration factor of 4 might be adequate for iron, which makes up a relatively high percentage of Earth's crust. Concentration factors must be several hundred or even several thousand for relatively low-value metals that are not present at very high percentages in Earth's crust. However, for an extremely valuable metal, such as platinum, a relatively low concentration factor is acceptable because of the high financial return obtained from extracting the metal.

Acceptable concentration factors are a sensitive function of the price of a metal. Shifts in price can cause significant changes in which deposits are mined. If the price of a metal increases by, for example, 50%, and the increase appears to be long term, it becomes profitable to mine deposits that had not been mined previously. The opposite can happen, as is often the case when substitute materials are found, or when newly discovered, richer sources go into production.

In addition to large variations in the concentration factors of various ores, there are extremes in the geographic distribution of mineral resources. The U.S. is perhaps about average for all nations in terms of its mineral resources, possessing significant resources of copper, lead, iron, gold, and molybdenum, but virtually without resources of some important strategic metals, including chromium, tin, and platinum-group metals. For its size and population, South Africa is particularly blessed with some important metal mineral resources.

10.13. EXTRACTION AND MINING

Minerals are usually extracted from Earth's crust by various kinds of mining procedures, but other techniques may be employed as well. The raw materials so obtained include inorganic compounds, such as phosphate rock; sources of metal, such as lead sulfide ore; clay used for firebrick; and structural materials, such as sand and gravel.

Surface mining, which can consist of digging large holes in the ground to remove copper ore, or strip mining, is used to extract minerals that occur near the surface. A common example of surface mining is quarrying of rock. Vast areas have been dug up to extract coal. Because of past mining practices surface mining got a well-deserved bad name. When strip mining was employed, the common practice was to dump waste overburden in rather randomly constructed **spoil banks** consisting of poorly compacted, steeply sloped piles of finely divided material. The rock and soil on spoil banks was highly susceptible to erosion and physical and chemical weathering. No effort was made to replace topsoil on the surface of the spoil banks, and that which was there quickly eroded away, so that vegetation on these unsightly piles was sparse. Now, however, with modern reclamation practices, topsoil is first removed and stored to place on top of overburden that is replaced such that it has gentle slopes and proper drainage. Topsoil spread over the top of the replaced spoil, often carefully terraced to prevent erosion, is seeded with indigenous grass and other plants, fertilized, and watered, if necessary, to provide vegetation. The end result of carefully done **mine reclamation** projects is a well-vegetated area suitable for wildlife habitat, recreation, forestry, and other beneficial purposes.

A particularly controversial practice that has developed in recent years is **mountaintop removal strip mining** that is being practiced in West Virginia and to a lesser extent in Kentucky and Virginia. This procedure involves blasting the tops off mountains and pushing the overburden into valleys to get to coal seams that are then dug up with huge draglines and shipped for use in power plants. Proponents contend that the practice does minimal harm and even provides flat land in areas notably short of level areas. Opponents cite destruction of hardwood forests and damage to water sources as major problems.

Extraction of minerals from placer deposits formed by deposition from water has obvious environmental implications. Mining of placer deposits can be accomplished by dredging from a boom-equipped barge. Another means that can be used is hydraulic mining with large streams of water. One interesting approach for more coherent deposits is to cut the ore with intense water jets, then suck up the resulting small particles with a pumping system.

For many minerals, underground mining is the only practical means of extraction. An underground mine can be very complex and sophisticated. The structure of the mine depends upon the nature of the deposit. It is, of course, necessary to have a shaft that reaches to the ore deposit. Horizontal tunnels extend out into the deposit, and provision must be made for sumps to remove water and for ventilation. Factors that must be considered in designing an underground mine include the depth, shape, and orientation of the ore body, as well as the nature and strength of the rock in and around it; thickness of overburden; and depth below the surface.

Usually, significant amounts of processing are required before a mined product is used or even moved from the mine site. Such processing, and by-products, can have significant environmental effects. Even rock to be used for aggregate and for road construction must be crushed and sized, a process that has the potential to emit air-polluting dust particles to the atmosphere. Crushing is also a necessary first step for further processing of ores. Some minerals occur to an extent of a few percent or even less in the rock taken from the mine and must be concentrated on site so that the residue does not have to be hauled far. For metals mining, these processes, as well as roasting, extraction, and similar operations are covered under the category of **extractive metallurgy**.

One of the more environmentally troublesome by-products of mineral refining consists of waste **tailings**. By the nature of the mineral processing operations employed, tailings are usually finely divided, and therefore subject to chemical weathering processes. Heavy metals associated with metal ores can be leached from tailings, producing water runoff contaminated with cadmium, lead, and other pollutants. Adding to the problem are some of the processes used to refine ore. Large quantities of cyanide solution are used to extract low levels of gold from ore, posing obvious toxicological hazards.

10.14. METALS

With an adequate supply of all of the important elements and energy, almost any material can be manufactured. Most of the elements, including practically all of those likely to be in short supply, are metals. Some metals are considered especially crucial because of their importance to industrialized societies, uncertain sources of supply, and price volatility in world markets. One of these is antimony, used in automotive batteries, fire-resistant fabrics, and rubber. Chromium, another crucial metal, is used to manufacture stainless steel (especially for parts exposed to high temperatures and corrosive gases), jet aircraft, automobiles, hospital equipment, and mining equipment. The platinum-group metals (platinum, palladium, iridium, and rhodium) are used as catalysts in the chemical industry, in petroleum refining, and in automobile exhaust antipollution devices. Around 90% of these metals used in the U.S. is imported, with the remainder from recycling.

Mining and processing of metal ores involve major environmental concerns, including disturbance of land, air pollution from dust and smelter emissions, and water pollution from disrupted aquifers. This problem is aggravated by the fact that the general trend in mining involves utilization of less rich ores. This is illustrated in Figure 10.9, showing the average percentage of copper in copper ore mined since

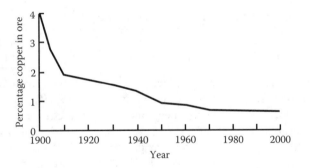

Figure 10.9.   Average percentage of copper in ore that has been mined.

1900. The average percentage of copper in ore mined in 1900 was about 4%, but by 1982 it was about 0.6% in domestic ores and 1.4% in richer foreign ores. Ores as low as 0.1% copper may eventually be processed. Increased demand for a particular metal, coupled with the necessity to utilize lower grade ores, has a vicious multiplying effect upon the amount of ore that must be mined and processed, and accompanying environmental consequences.

Metals exhibit a wide variety of properties and uses. They come from a number of different compounds; in some cases two or more compounds are significant mineral sources of the same metal. Usually these compounds are oxides or sulfides. However, other kinds of compounds and, in the cases of gold and platinum-group metals, the elemental (native) metals themselves serve as metal ores. Table 10.1 lists the important metals, their properties, major uses, and sources.

10.15. NONMETAL MINERAL RESOURCES

A number of minerals other than those used to produce metals are important resources. There are so many of these that it is impossible to discuss them all in this chapter; however, mention will be made of the major ones. As with metals, the environmental aspects of mining many of these minerals are quite important. Typically, even the extraction of ordinary rock and gravel can have important environmental effects.

Clays mentioned in this chapter and in Chapter 11 are secondary minerals that occur as suspended and sedimentary matter in water and as secondary minerals in soil. Other than stone tools, clays were the first minerals used by humans, especially in making pottery containers. Various clays are used for clarifying oils, as catalysts in petroleum processing, as fillers and coatings for paper, and in the manufacture of firebrick, pottery, sewer pipe, and floor tile. The main types of clays that have industrial uses are shown in Table 10.2. U.S. production of clay is about 60 million metric tons per year, and global and domestic U.S. resources are abundant.

Fluorine compounds are widely used in industry. Large quantities of fluorspar (CaF_2) are required as a flux in steel manufacture. Synthetic and natural cryolite (Na_3AlF_6) is used as a solvent for aluminum oxide in the electrolytic preparation of

Table 10.1.   Worldwide and Domestic Metal Resources

Metals	Properties[a]	Major Uses	Ores, Aspects of Resources[b]
Aluminum	mp 660°C, bp 2467°C, sg 2.70, malleable, ductile	Metal products, including autos, aircraft, electrical equipment; conducts electricity better than copper per unit mass and is used in electrical transmission lines	From bauxite ore containing 35-55% Al_2O_3; about 60 million metric tons of bauxite is produced worldwide per year; U.S. resources of bauxite are 40 million metric tons, world resources about 15 billion metric tons
Chromium	mp 1903°C, bp 2642°C, sg 7.14, hard, silvery color	Metal plating, stainless steel, wear-resistant and cutting tool alloys, chromium chemicals, including chromates	From chromite, an oxide mineral containing Cr, Mg, Fe, Al; resources of 1 billion metric tons in South Africa and Zimbabwe, large deposits in Russia, virtually none in U.S.
Cobalt	mp 1495°C, bp 2880°C, sg 8.71, bright, silvery	Manufacture of hard, heat-resistant alloys, permanent magnet alloys, driers, pigments, glazes, animal feed additive	From a variety of minerals, such as linnaeite, Co_3S_4, and as a by-product of other metals; abundant global and U.S. resources
Copper	mp 1083°C, bp 2582°C, sg 8.96, ductile, maleable	Electrical conductors, alloys, chemicals; many uses	Occurs in low percentages as sulfides, oxides and carbonates. U.S. consumption 1.5 million metric tons per year; world resources of 344 million metric tons, including 78 million in U.S.
Gold	mp 1063°C, bp 2660°C, sg 19.3	Jewelry, basis of currency, electronics, increasing industrial uses	In various minerals at only around 10 ppm for ores currently processed in the U.S.; by-product of copper refining; world resources of 1 billion oz, 80 million in U.S.
Iron	mp 1535°C, bp 2885°C, sg 7.86, silvery metal, in (rare) pure form	Most widely produced metal, usually as steel, a high-tensile-strength material containing 0.3–1.7% C; made into many specialized alloys	Occurs as hematite (Fe_2O_3), goethite, ($Fe_2O_3 \cdot H_2O$), and magnetite (Fe_3O_4), abundant global and U.S. resources

Table 10.1. (continued)   Worldwide and Domestic Metal Resources

Metals	Properties[a]	Major Uses	Ores, Aspects of Resources[b]
Lead	mp 327°C, bp 1750°C, sg 11.35, silvery color	Fifth most widely used metal, storage batteries, chemicals; uses in gasoline, pigments, and ammunition largely eliminated for environmental reasons	Major source is galena, PbS; worldwide consumption about 3.5 million metric tons, 1/3 in U.S.; global reserves about 140 million metric tons, 39 million metric tons U.S.
Manganese	mp 1244°C, bp 2040°C, sg 7.3, hard, brittle, gray-white	Sulfur and oxygen scavenger in steel, manufacture of alloys, dry cells, gasoline additive, chemicals	Found in several oxide minerals; about 20 million metric tons per year produced globally, 2 million consumed in U.S., no U.S. production, world reserves 6.5 billion metric tons
Mercury	mp −38°C, bp 357°C, sg 13.6, shiny, liquid metal	Instruments, electronic apparatus, electrodes, chemicals; decreasing use because of toxicity	From cinnabar, HgS; annual world production 11,500 metric tons, 1/3 used in U.S.; world resources 275,000 metric tons, 6600 U.S.
Molybdenum	mp 2620°C, bp 4825°C, sg 9.01, ductile, silvery-gray	Alloys, pigments, catalysts, chemicals, lubricants	Molybdenite (MoS_2) and wulfenite ($PbMoO_4$) are major ores; about 2/3 global Mo production in U.S., large global resources
Nickel	mp 1455°C, bp 2835°C, sg 8.90, silvery color	Alloys, coins, storage batteries, catalysts (such as for hydrogenation of vegetable oil)	Found in ores associated with iron; U.S. consumes 150,000 metric tons per year, 10% from domestic production, large domestic reserves of low-grade ore
Silver	mp 961°C, bp 2193°C, sg 10.5, shiny metal	Photographic film, electronics, sterling ware, jewelry, bearings dentistry	Found with sulfide minerals, by-product of Cu, Pb, Zn; annual U.S. consumption of 150 million troy ounces
Tin	mp 232°C, bp 2687°C, sg 7.31	Coatings, solders, bearing alloys, bronze, chemicals, organometallic biocides	Many forms associated with granitic rocks and chrysolites; global consumption 190,000 metric tons/year, U.S. 60,000 metric tons/year, world resources 10 million metric tons

Table 10.1. (continued)   Worldwide and Domestic Metal Resources

Metals	Properties[a]	Major Uses	Ores, Aspects of Resources[b]
Titanium	mp 1677°C, bp 3277°C, sg 4.5, silvery color	Strong, corrosion-resistant, used in aircraft, valves, pumps, paint pigments	Commonly as TiO_2, ninth in elemental abundance, no shortages
Tungsten	mp 3380°C, bp 5530°C, sg 19.3, gray	Very strong, high boiling point, used in alloys, drill bits, turbines, nuclear reactors, to make tungsten carbide	Found as tungstates, such as scheelite ($CaWO_4$); U.S. has 7% world reserves, China 60%
Vanadium	mp 1917°C, bp 3375°C, sg 5.87, gray	Used to make strong steel alloys	In igneous rocks, primarily a by-product other metals, U.S. consumption of 5000 metric tons/year equals production
Zinc	mp 420°C, bp 907°C, sg 7.14, bluish-white	Widely used in alloys (brass), galvanized steel, paint pigments, chemicals. Fourth in world metal production	Found in many ore minerals; world production is 5 million metric tons per year (10% from U.S.), U.S. consumption is 1.5 million metric tons; world resources 235 million metric tons, 20% in U.S.

[a] Abbreviations: mp, melting point; bp, boiling point; sg, specific gravity.

[b] All figures are approximate; quantities of minerals considered available depend upon price, technology, recent discoveries, and other factors, so that quantities quoted are subject to fluctuation.

aluminum metal. Sodium fluoride is added to water to help prevent tooth decay, a measure commonly called water fluoridation. World reserves of high-grade fluorspar are around 190 million metric tons, about 13% of which is in the U.S. This is sufficient for several decades at projected rates of use. A great deal of by-product fluorine is recovered from the processing of fluorapatite ($Ca_5(PO_4)_3F$), used as a source of phosphorus.

Micas are complex aluminum silicate minerals, which are transparent, tough, flexible, and elastic. Muscovite ($K_2O \cdot 3Al_2O_3 \cdot 6\,SiO_2 \cdot 2\,H_2O$) is a major type of mica. Better grades of mica are cut into sheets and used in electronic apparatus, capacitors, generators, transformers, and motors. Finely divided mica is widely used in roofing, paint, welding rods, and many other applications. Sheet mica is imported into the U.S., and finely divided "scrap" mica is recycled domestically. Shortages of this mineral are unlikely.

In addition to consumption in fertilizer manufacture (Section 11.6), phosphorus is used for supplementation of animal feeds, synthesis of detergent builders, and preparation of chemicals such as pesticides and medicines. The most common phosphate minerals are hydroxyapatite ($Ca_5(PO_4)_3(OH)$) and fluorapatite ($Ca_5(PO_4)_3F$). Ions of Na, Sr, Th, and U are found substituted for calcium in apatite minerals. Small

Table 10.2.   Major Types of Clays and Their Uses in the U.S.

Type of Clay	Percentage Use	Composition	Applications
Miscellaneous	72	Variable	Filler, brick, tile, portland cement, others
Fireclay	12	Variable; can be fired at high temperatures without warping	Refractories, pottery, sewer pipe, tile, brick
Kaolin	8	$Al_2(OH)_4Si_2O_5$; white and can be fired without losing shape or color	Paper filler, refractories, pottery, dinnerware, petroleum-cracking catalyst
Bentonite and fuller's earth	7	Variable	Drilling muds, petroleum catalyst, carriers for pesticides, sealers, clarifying oils
Ball clay	1	Variable, very plastic	Refractories, tile, whiteware

amounts of PO_4^{3-} may be replaced by AsO_4^{3-}, and the arsenic must be removed for food applications. Approximately 17% of world phosphate production is from igneous minerals, primarily fluorapatites. About three-fourths of world phosphate production is from sedimentary deposits, generally of marine origin. Vast deposits of phosphate, accounting for approximately 5% of world phosphate production, are derived from guano (droppings of seabirds and bats). Current U.S. production of phosphate rock is around 40 million metric tons per year, most of it from Florida. Tennessee and several of the western states are also major producers of phosphate. Reserves of phosphate minerals in the U.S. amount to 10.5 billion metric tons containing approximately 1.4 billion metric tons of phosphorus. Identified world reserves of phosphate rock are approximately 6 billion metric tons.

Pigments and fillers of various kinds are used in large quantities. The only naturally occurring pigments still in wide use are those containing iron. These minerals are colored by limonite, an amorphous brown-yellow compound with the formula $2Fe_2O_3 \cdot 3H_2O$, and hematite, composed of gray-black Fe_2O_3. Along with varying quantities of clay and manganese oxides, these compounds are found in ocher, sienna, and umber. Manufactured pigments include carbon black, titanium dioxide, and zinc pigments. About 1.5 million metric tons of carbon black, manufactured by the partial combustion of natural gas, are used in the U.S. each year, primarily as a reinforcing agent in tire rubber.

Over 7 million metric tons of minerals are used in the U.S. each year as fillers for paper, rubber, roofing, battery boxes, and many other products. Among the minerals used as fillers are, carbon black, diatomite, barite, fuller's earth, kaolin, mica, limestone, pyrophyllite, and wollastonite ($CaSiO_3$).

Although sand and gravel are the cheapest of mineral commodities per ton, the average annual dollar value of these materials is greater than all but a few mineral products because of the huge quantities involved. In tonnage, sand and gravel production is by far the greatest of nonfuel minerals. Almost 1 billion tons of sand and gravel are employed in construction in the U.S. each year, largely to make concrete structures, road paving, and dams. Slightly more than that amount is used to manufacture portland cement and as construction fill. Although ordinary sand is predominantly silica, SiO_2, about 30 million tons of a more pure grade of silica are consumed in the U.S. each year to make glass, high-purity silica, silicon semiconductors, and abrasives.

At present, old river channels and glacial deposits are used as sources of sand and gravel. Many valuable deposits of sand and gravel are covered by construction and lost to development. Transportation and distance from source to use are especially crucial for this resource. Environmental problems involved with defacing land can be severe, although bodies of water prized for fishing and other recreational activities frequently are formed by removal of sand and gravel.

The biggest single use for sulfur is in the manufacture of sulfuric acid. However, the element is employed in a wide variety of other industrial and agricultural products. Current consumption of sulfur amounts to approximately 10 million metric tons per year in the U.S. The most important sources of sulfur are (in decreasing order) sulfur recovered from sour natural gas (as H_2S) and petroleum, deposits of elemental sulfur mostly mined with hot water (Frasch process), and sulfur recovered from pyrite (FeS_2) and sulfide metal ores (such as galena, PbS). Supply of sulfur is no problem either in the U.S. or worldwide. The U.S. has abundant deposits of elemental sulfur, and sulfur recovery from fossil fuels as a pollution control measure could even result in surpluses of this element.

Sodium chloride, gypsum, and potassium salts are all important minerals that are recovered as evaporites (remaining from the evaporation of seawater). Sodium chloride, in the form of mineral halite is used as a raw material for the production of industrially important sodium, chlorine, and their compounds. It is used directly to melt ice on roads, in foods, and in other applications. Potassium salts are, of course, essential ingredients of fertilizers, and have some industrial applications as well. Gypsum, hydrated calcium sulfate, is used to make plaster and wallboard and is an ingredient for the manufacture of portland cement.

10.16. HOW LONG WILL ESSENTIAL MINERALS LAST?

During about a 30-year period following World War II, demand for most important mineral commodities increased at a very rapid rate. Coinciding roughly with the "energy crisis" of the early to mid-1970s, demand slowed. Now, however, with the emergence of newly developing economies, particularly those in the highly populous countries of China and India, it may be assumed that demand for minerals will increase sharply.

To a degree, the economic demand for and price of a resource determine its availability. Higher prices lead to greater exploration, exploitation of less available resources, and often spectacular increases in supply. This phenomenon has led to misinterpretation of the resource supply and demand equation by authorities whose understanding of economics is limited to only conventional monetary supply and demand models, without due appreciation of sustainability aspects. The fact is that the total available amounts of most resources are limited, painfully so in terms of the time span over which they will be needed by humankind. Although higher prices and improved technologies can increase supplies of critical resources significantly, the ultimate result will be the same — the resource will run out. Furthermore, exploitation of lower grades of resources results in ever-increasing environmental disruption, adding significantly to the cost, considering environmental economics.

Mineral resources may be divided into several categories based upon current production, consumption, and known reserves. In the first category are those that are in relatively comfortable supply, with supply of at least 100 to several hundred years. Minerals in this category include bauxite, the source of aluminum, iron ore, platinum-group metals, and potassium salts. In an intermediate category are minerals with a current projected lifetime supply of 25 to 100 years. These include chromium, cobalt, copper, manganese, nickel, gypsum, phosphate minerals, and sulfur. The most critical group consists of minerals for which the supply based upon current rates of consumption and known reserves is 25 years or less. Among these minerals are sources of lead, tin, zinc, gold, and silver. Only a few years ago mercury was on this list, but concerns over its toxicity have led to greatly reduced use such that at times it has been in surplus recently.

The U.S. is essentially without economic reserves of a number of essential minerals. These include aluminum, antimony, chromium, cobalt, manganese, tantalum, niobium, platinum, nickel, and tin. Reserves of asbestos, fluorine, and vanadium are quite limited. Larger, but still limited domestic supplies are available of gold, potash, silver, mercury, tungsten, sulfur, and zinc. As far as the U.S. is concerned, metals of most concern are chromium, manganese, and cobalt. These substances are essential for a modern industrialized economy. Although global supplies are adequate for the immediate future, they are threatened by the potential instability of the countries from which they come — Zaire, Zambia, South Africa, Russia, and other countries in the former Soviet Union.

The world economy will never totally run out of any of the minerals listed above. However, severely constrained supplies of any one or several of them will have some marked effects. For example, world food production now depends on fertilizers, which require phosphorus, of which resources are limited. Within the next century, a food crisis related to phosphate shortages may be anticipated.

10.17. GREEN SOURCES OF MINERALS

One of the most crucial aspects of green technology is the sustainable utilization of minerals. In a sense, the concept of sustainable mineral utilization is an oxymoron, because minerals removed from the geosphere are not replaced. However, the idea of sustainability can greatly extend supplies of minerals. This section addresses the approach to sustainability in obtaining minerals. The broader questions of green utilization of materials, substitution of materials, and recycling are discussed in more detail in Chapter 17 and Chapter 18.

As discussed later in this book, modern technology and human ingenuity are very effective in alleviating shortages of important minerals. Applications of materials science (see Chapter 17) continue to produce substances made from readily available materials that provide good substitutes for more scarce resources. For example, concrete covered by strong layers of composite materials can readily substitute for iron in construction. Ceramics with special heat- and abrasion-resistant qualities are being used where high-temperature metal alloys were formerly required.

As minerals become less available, one of the measures to be taken is one that has been taken historically — find more. Modern technology provides a number of useful tools for finding new mineral deposits. Arguably, the most useful approach to finding new mineral deposits is through the applications of geology. Recent advances in plate tectonics, for example, have contributed understanding of likely locations of significant mineral deposits.

Slight differences in magnetic field, gravity, and electrical conductivity can be detected very sensitively, and these reflect differences in density, magnetic properties, and electrical properties that indicate the presence of remote mineral deposits. These techniques are in the realm of **geophysical prospecting**. Another useful technique for finding mineral deposits is **geochemical prospecting**. As its name implies, this method depends upon detecting the presence of chemical species, usually specific metals. In addition to finding such substances directly in rock, geochemical prospecting can reveal evidence of minerals in water some distance from the mineral sources. Even gas analysis can be indicative of some minerals, such as volatile mercury or sulfur compounds from sulfur deposits. Plants, particularly those that concentrate some elements, such as copper, can be analyzed to indicate the presence of minerals, a process called **biogeochemical prospecting**.

Photography and the measurement of light and infrared radiation from aircraft and from satellites has greatly increased human understanding of Earth and its resources. These **remote sensing** techniques make it unnecessary to go into remote, poorly accessible, dangerous regions. With satellite measurements, it is possible to cover huge areas of Earth's surface in a reasonable period of time. The most ambitious program of remote sensing for mineral exploration is the Landsat satellite system, first launched in 1972 and subsequently followed by other launches from the U.S. and other countries. The sensors on these satellites measure visible and infrared radiation. The Landsat images reveal numerous features of Earth's surface including abundance and types of vegetation, soil and rock type, and moisture. Such features in turn may reflect the presence of various kinds of mineral deposits with potential for exploitation. An interesting aspect of satellite-based minerals exploration is geobotany in which, for example, trees poisoned by heavy metals indicative of metal deposits may be observed. More subtle changes, such as the timing by which deciduous trees start to produce leaves in the spring and change color and lose leaves in the fall may provide clues to mineral deposits.

Exploitation of Lower Grade Ores

Modern technology enables exploitation of lower grade ores, thus significantly increasing supplies. A striking example of this phenomenon has been provided by copper. About 100 years ago, the average copper content of ore mined in the U.S. was around 5%; now it is only about 1/10 that figure for copper ore mined globally. Despite the decline in copper ore quality, during the last 50 years, known copper reserves have increased about 5-fold and, adjusted for inflation, the price of copper is now less than it was a century ago.

The ability to exploit much less rich sources of ores has resulted from improved technologies. Of particular importance have been advances in the means of moving huge quantities of rock, essential for the exploitation of lower grade ores. Earth-moving equipment has greatly increased in size and versatility during the last several decades. There has been an environmental cost, of course, for these advances. As an approximation, for each 10-fold decrease in mineral content, it is necessary to move 10 times as much material to obtain the same amount of metal. In addition to disruption of land, disturbed material is more prone to erosion, landslides, and water pollution. Much more energy is required, as is more water for those mining operations that use large quantities of water. Not the least of the factors required for exploiting lower grade resources is the need for additional capital and operating investment, which may be in short supply.

Remote Sources of Minerals

All the rich mineral ores in readily accessible areas have already been found and exploited. Therefore, any rich deposits will be found in remote locations and hostile environments, such as deep under the ocean. One major possibility is Antarctica, a remote continent noted for its ice, wind, and generally hostile conditions. It is very likely that rich mineral deposits are buried beneath the thick Antarctic ice sheet. However, the probability of severe environmental damage from extracting minerals there is very high, even if the extreme climate conditions can be overcome. In recognition of that concern 26 nations involved in Antarctic exploration signed a treaty in 1991 banning mineral extraction for 50 years. However, if shortages of crucial minerals become severe, it is likely that efforts will be made to find and extract them in Antarctica.

Exciting possibilities exist for the extraction of minerals from very hostile places, especially at great depths. One possibility is ultradeep mining under conditions too severe to enable human participation. It may one day be possible to use robots to mine deposits several kilometers deep, where extreme pressures and heat would make it impossible for humans to work.

Another potentially abundant remote source of minerals is on ocean floors, which remain today largely unexplored. Here, relatively new technologies, such as remote-controlled submarines capable of withstanding crushing pressure have opened new possibilities for exploration and resource utilization. Large areas of the ocean are covered with manganese-rich lumps called **manganese nodules**. In addition to manganese, these lumps also contain other metals, including valuable platinum, copper, and nickel. Extraction of these metals as by-products adds to the economic attractiveness of mining manganese nodules.

Waste Mining

Waste mining is the term given to the extraction of useful materials from waste streams. Sulfur is one of the best examples of waste mining. Technology has been developed for the removal of sulfur from flue gas in nonferrous metal smelters. The recovery of sulfur from the smelting of lead, copper, zinc, and other metals also provides an example of policy-driven waste mining, because it is mandated by law in most developed countries. Waste mining is employed to recover some metals as part of the production of other metals. By-product metals can be recovered from the gangue remaining from beneficiation of ore, slag from smelting, or dust collected from flues in metal-refining operations. Arsenic and cadmium are recovered from the production of copper and cadmium, respectively. Coal ash is a huge untapped resource for waste mining of aluminum and ferrosilicon. Factoring in the costs of waste disposal and potential environmental degradation can make the economics of waste mining relatively more attractive. Some caution is suggested in that policy-driven waste mining of some substances creates a need to market them, sometimes to the detriment of the environment. Cadmium and arsenic are both examples of substances recovered from waste mining that should not be used any more than necessary because of their toxicities.

Related to waste mining is the utilization of scrap materials, especially scrap metal. There are two major categories of scrap. **New scrap** consists of materials that are reclaimed during the manufacture of an item, such as metal shavings from machining. **Old scrap** consists of material that has been in products used in the consumer market and reclaimed as scrap material. The quality of new scrap can be carefully controlled and it can be reclaimed very quickly and efficiently. However, old scrap has a recycling time that depends upon the life of the product in which it is contained and it is difficult to control its quality. Furthermore, the percentage return of material is lower from old scrap because of the products that are discarded and not recycled.

The anthrosphere provides a large reservoir of materials that eventually can be recycled. A prime example of such a material is that of copper contained in copper wiring and plumbing in buildings, electrical lines, and other anthrospheric structures. The relatively high price of copper metal makes it attractive to reclaim from these sources. Another example is the large amount of iron contained in vehicles, machinery, rail lines, bridges, and other artifacts.

Landfills constitute another anthrospheric construct that contains large amounts of materials, including metals. Unfortunately, the usable materials in landfills are usually too dispersed and mixed with other materials to be reclaimable. Any metals that are put in landfills should be put in segregated areas from which they may later be reclaimed.

Recycling

Recycling should be practiced for all major mineral commodities. Both economic and environmental concerns have resulted in vastly increased efforts to recycle materials in recent years. The largest quantity of metal that is recycled consists of ferrous metals (iron). During the last 30 years, electric-arc furnaces for iron have become commonplace. Fortunately, these devices require scrap iron as feedstock, and have resulted in a continuing market for recyclable iron scrap. Aluminum ranks next to iron in quantity of metal recycled; at least ⅓ of aluminum is recycled in the U.S. and globally. Particular success has been achieved in the recycling of aluminum beverage cans. The refining of aluminum metal from bauxite ore is particularly energy consumptive, so a big advantage of aluminum recycling is reduced energy consumption. In addition, recycling produces only about 5% the amount of wastes and potential pollutants as are generated by refining aluminum from ore. Cost savings are huge as well. Other metals that are largely recycled are copper and copper alloys, cadmium, lead, tin, mercury, zinc, silver, and, of course, gold and platinum.

A crucial consideration in recycling is the nature of source material. Copper is relatively easy to recycle because it is often found in a relatively pure form in wire, pipe, and electrical apparatus. Lead in lead storage batteries can simply be melted down and recast into battery electrodes. Large amounts of aluminum are available from waste cans and structural materials. Although iron is largely recycled, it often occurs as specialized alloys containing varying contents of other metals, such as titanium or tungsten. The contents of these elements complicate the utilization of scrap iron.

SUPPLEMENTARY REFERENCES

American Geological Institute Staff, *Environmental Geology*, McGraw-Hill Higher Education, Burr Ridge, IL, 2001.

Anderson, Ewan W. and Liam D. Anderson, *Strategic Minerals: Resource Geopolitics and Global Geo-Economics*, John Wiley & Sons, New York, 1998.

Auty, Richard M. and Raymond F. Mikesell, *Sustainable Development in Mineral Economies*, Clarendon, Oxford, U.K., 1998.

Azcue, Jose M., Ed., *Environmental Impacts of Mining Activities: Emphasis on Mitigation and Remedial Measures*, Springer-Verlag, Berlin, 1999.

Brownlow, Arthur H., *Geochemistry*, 2nd ed., Prentice Hall, Upper Saddle River, NJ, 1996.

Carlson, Diane H., Charles C. Plummer, and David McGeary, *Physical Geology: Earth Revealed*, 6th ed., McGraw-Hill, Boston, MA, 2006.

Chamley, Herve, *Geosciences, Environment and Man*, Elsevier, New York, 2003.

Ehlers, Eckart and Thomas Krafft, *Earth System Science in the Anthropocene: Emerging Issues and Problems*, Springer-Verlag, New York, 2005.

Erickson, Jon, *Environmental Geology: Facing the Challenges of Our Changing Earth*, Facts on File Inc., New York, 2002.

Farndon, John, *Rocks and Mineral*, DK Publishing, New York, 2005.

Faure, Gunter, *Principles and Applications of Geochemistry: A Comprehensive Textbook for Geology Students*, Prentice Hall, Upper Saddle River, NJ, 1998.

Fortey, Richard A., *The Earth: An Intimate History*, Harper Collins, London, 2004.

Gallant, Roy A., *Minerals*, Benchmarks Books, Tarrytown, New York, 2001.

Gallios, G.P. and K.A. Matis, Eds., *Mineral Processing and the Environment*, Kluwer Academic Publishers, Boston, MA, 1998.

Gray, Murray, *Geodiversity: Valuing and Conserving Abiotic Nature*, John Wiley & Sons, Hoboken, NJ, 2004.

Hamblin, W. Kenneth and Christiansen, Eric H., *Earth's Dynamic Systems*, Prentice Hall, Upper Saddle River, NJ, 2003.

Johnsen, Ole, *Minerals of the World*, Princeton University Press, Princeton, NJ, 2002.

Keller, Edward A., *Introduction to Environmental Geology*, 3rd ed., Pearson Prentice Hall, Upper Saddle River, NJ, 2005.

Leeder, Mike R. and Marta Perèz-Arlucea, *Physical Processes in Earth and Environmental Sciences*, Blackwell Publishing, Malden, MA, 2006.

Lutgens, Frederick K. and Edward J. Tarbuck, *Essentials of Geology*, 9th ed., Pearson Prentice Hall, Upper Saddle River, NJ, 2006.

Lutgens, Frederick K. and Edward J. Tarbuck, *Earth Science*, 11th ed., Pearson Prentice Hall, Upper Saddle River, NJ, 2006.

Marshak, Stephen, *Earth: Portrait of a Planet*, 2nd ed. W.W. Norton, New York, 2005.

Marshall, Clare P. and Rhodes Whitmore Fairbridge, Eds., *Encyclopedia of Geochemistry (Encyclopedia of Earth Sciences)*, Kluwer Academic Publishing Co., Hingham, MA, 1998.

McGraw-Hill Concise Encyclopedia of Earth Science, McGraw-Hill, New York, 2005.

McSween, Harry Y., Steven M. Richardson, and Maria E. Uhle, *Geochemistry: Pathways and Processes*, 2nd ed., Columbia University Press, New York, 2003.

Monroe, James S. and Reed Wicander, *Physical Geology: Exploring the Earth*, 5th ed., Thomson Brooks/Cole, Belmont, CA, 2005.

Montgomery, Carla W., *Environmental Geology*, 7th ed., McGraw-Hill, Boston, MA, 2006.

Moon, Charles J., Michael K.G. Whateley, and Anthony M. Evans, Eds., *Introduction to Mineral Exploration*, 2nd ed., Blackwell Publishing, Malden, MA, 2006.

Oreskes, Naomi, *Plate Tectonics: An Insider's History of the Modern Theory of the Earth*, Westview Press, New York, 2003.

Ottonello. Giulio, *Principles of Geochemistry*, Columbia University Press, New York, 1997.

Pipkin, Bernard W., D. D. Trent, and Richard Hazlett, *Geology and the Environment*, 4th ed., Thomson Brooks/Cole, Belmont, CA, 2005.

Plummer, Charles C., David McGeary, and Diane H. Carlson, *Physical Geology*, McGraw-Hill, Boston, MA, 2005.

Rowe, R. Kerry, Ed., *Geotechnical and Geoenvironmental Engineering Handbook*, Kluwer Academic, Boston, MA, 2001.

Selley, Richard C., Ed., *Encyclopedia of Geology*, Elsevier, San Diego, CA, 2004.

Skinner, H. Catherine W., and Anthony R. Berger, Eds., *Geology and Health: Closing the Gap*, Oxford University Press, New York, 2003.

Skinner, Brian J., Stephen C. Porter, and Jeffrey Park, *Dynamic Earth: An Introduction to Physical Geology*, John Wiley & Sons, Hoboken, NJ, 2004.

Tarbuck, Edward J. and Frederick J. Lutgens, *Earth: Introduction to Physical Geology*, Pearson/Prentice Hall, Upper Saddle River, NJ, 2005.

Wicander, Reed and James S. Monroe, *Essentials of Geology*, 4th ed., Thomson-Brooks/Cole, Belmont, CA, 2006.

Wicander, Reed and James S. Monroe, *Historical Geology: Evolution of Earth and Life through Time*, 4th ed. Thomson-Brooks/Cole, Belmont, CA, 2004.

QUESTIONS AND PROBLEMS

1. Of the following, the one that is not a manifestation of desertification is (1) declining groundwater tables, (2) salinization of topsoil and water, (3) production of deposits of MnO_2 and $Fe_2O_3 \cdot H_2O$ from anaerobic processes, (4) reduction of surface waters, and (5) unnaturally high soil erosion.

2. Why do silicates and oxides predominate among Earth's minerals?

3. Explain how the following are related: weathering, igneous rock, sedimentary rock, and soil.

4. Match the following:

   a. Metamorphic rock
   b. Chemical sedimentary rocks
   c. Detrital rock
   d. Organic sedimentary rocks

   i. Produced by the precipitation or coagulation of dissolved or colloidal weathering products
   ii. Contain residues of plant and animal remains
   iii. Formed from action of heat and pressure on sedimentary rock
   iv. Formed from solid particles eroded from igneous rocks by weathering

5. Where does most flowing water that contains dissolved load originate? Why does it tend to come from this source?

6. What is engineering geology? How can it be related to environmental improvement?

7. As related to minerals, what is the distinction between resources and reserves?

8. In what sense do humans have more direct knowledge of outer space than they do of Earth's lithosphere? Why is this so?

9. What are the three major classes of rocks? How are they involved in the rock cycle?

10. How is the presence of sulfide related to some hydrothermal deposits of metals?

11. What is the distinction between a mineral evaporite and a sublimate?

12. What is meant by plate tectonics, and how does this concept relate to continental drift?

13. What is the distinction between a joint and a fault in rock formations?

14. What does lava manifest? What is the origin of lava?

15. What is the role of flowing water in the formation of placer deposits?

16. How are weathering and leaching related to the formation of bauxite deposits? Which metal is extracted from bauxite?

17. What is meant by concentration factor in minerals and how does it relate to their economic recovery?

18. What are mine tailings? What are some of the special environmental problems associated with tailings? Why are they particularly susceptible to weathering?

19. Discuss how technology that enables utilization of less rich ores relate to environmental and energy considerations.

20. Match the following pertaining to metals:

    a. Chromium   i. Used in organometallic biocides
    b. Lead   ii. Fourth in world metal production
    c. Tin   iii. South Africa and Russia are major sources
    d. Zinc   iv. From galena

21. What are some of the main uses of clays?

22. How are fluorine compounds used in metal manufacture?

23. What are some of the major applications of phosphate minerals?

24. What is the largest use for sulfur?

25. What is the overall effect of the development of the economies of highly populated countries on the demand for mineral resources?

26. How can modern materials science help alleviate mineral shortages?

27. Explain how modern high technology may assist in finding new mineral sources.

28. How does larger and more sophisticated earth-moving equipment increase the supply of minerals? What are the environmental implications, both good and bad?

29. What are two largely unexplored parts of the Earth where significant mineral resources may yet be found?

30. Explain why the nature of source material is particularly important in recycling.

11. SOIL, AGRICULTURE, AND FOOD PRODUCTION

11.1. AGRICULTURE

For most of its lifetime on Earth, the human species was preoccupied with hunting and gathering in a never-ending quest for food. This enterprise required the full-time efforts of all members of a society, leaving little time for other pursuits. Food was where they found it, often requiring that they travel large distances in pursuit of game and edible plants. Storing food was very difficult, and people often perished during bitter winters or severe droughts. Some large animals that were abundant sources of meat objected strenuously to being used for that purpose and often the hunter lost the battle to secure a tasty meal of fresh meat and ended up as fresh meat for some carnivore. Life was hard, the trappings of civilization were minimal, and the human populations that could be supported remained small. Humankind's imprint on the environment remained small, although some animal species were hunted to extinction and, in some areas, woodland was deliberately burned to provide grazing grasslands for game animals.

Approximately 10,000 years ago, the harsh circumstances described above changed when humans in the Fertile Crescent (the Middle East) learned to cultivate certain grasses that produced grain for food. Furthermore, this grain could be stored for long periods of time in dry granaries providing a stable source of food. About that time as well, humans domesticated some animals including sheep that provided wool and meat, goats that provided milk and meat, and donkeys that could be used for transport and to provide power to cultivate land. Humans had discovered agriculture, the production of food and fiber by raising plants and animals. The changes brought about by agriculture were many and profound. It meant that human populations needed to stay in particular locations conducive to the growing of crops. It freed significant numbers of people from the task of getting food so that human ingenuity could be devoted to other pursuits, such as the development of wheeled vehicles, the construction of sailing boats, and the discovery of writing.

Crop and Livestock Farming

The two basic categories of agriculture are (1) crop farming to produce edible substances from photosynthetically generated biomass, and (2) livestock farming for the production of meat, milk, wool, hide, and other animal products. Both crops and livestock were developed from wild ancestors by early farmers. Particularly in the case of crops, output has been increased markedly during the last century by developing hydrids from crossing two or more true-breeding strains. Now recombinant DNA technology and genetic engineering are being used to revolutionize agriculture through the production of higher yielding crops and animals, hormones to increase milk production, engineering of crops resistant to herbicides applied to kill competing weeds, and similar developments. The modern agricultural enterprise is remarkably productive, not only in the production of the top five plants that provide food consumed by humans — wheat, corn, rice, potatoes, and soybeans — but other delectable foods as well.

Influence of Agriculture on the Environment

Agriculture has a tremendous influence on the environment and has a significant potential for environmental harm. In addition to direct effects from the cultivation of land, there are indirect effects from irrigation and other measures used to increase agricultural yield. The rearing of domestic animals may have environmental effects. For example, The Netherlands' pork industry has been so productive that accumulations of hog manure and its by-products have caused serious problems. Goats and sheep have destroyed pastureland in the Near East, Northern Africa, Portugal, and Spain. Of particular concern are the environmental effects of raising cattle. Significant amounts of forestland have been converted to marginal pastureland to raise beef. Production of 1 lb of beef requires about four times as much water and more than twice as much feed as does production of 1 lb of chicken and much more than to produce an equivalent amount of vegetable protein. An interesting aspect of the problem is emission of greenhouse gas methane by anaerobic bacteria in the digestive systems of cattle and other ruminant animals; cattle rank right behind wetlands and rice paddies as producers of atmospheric methane. However, cattle and other ruminant livestock do have a positive environmental/resource impact because they can use cellulose from plants as a food source to produce meat and milk, the result of the action of specialized bacteria in the stomachs of ruminant animals.

Agriculture and Sustainability

The agricultural sector is obviously of enormous importance to sustainability. This is because, along with water, food is the most basic requirement for human existence. As human populations grow, the amount of land to grow food for each human decreases, a problem exacerbated by the collapse of world fisheries from poor management and overexploitation. Agriculture interacts strongly with all spheres of the environment. It is of the utmost importance for humans to properly manage the agricultural enterprise, not only with the goal of increasing the quality and quantity of production, but also to integrate agriculture with the other sectors of the environment so that they work to mutual advantage.

This chapter deals with agriculture and the production of food. The first sections of the chapter discuss the most basic requirement for agriculture, soil. Later sections of the chapter discuss other aspects of agriculture and food production.

11.2. SOIL: ESSENTIAL FOR LIFE, KEY TO SUSTAINABILITY

Apart from dwindling resources of food from the ocean, humans and most other living things are dependent on soil, humble dirt, for their existence. As human populations grow, the area of soil that provides the food for each person continues to diminish. Reserves of food are low; one growing season with minimal food production, such as might be the result of climate disruption from an asteroid impact, for example, could result in massive starvation around the world. On a longer time scale, changes in climate from global warming could have a similar effect. On a positive note, agriculturists have long recognized the importance of soil and have been leaders in conserving this essential resource and in sustainability. In Europe, farmlands have been productive for centuries. In the U.S., concern over soil loss starting in the late 1800s gave rise to soil conservation programs with generous government support throughout the following century that have largely reversed soil loss and degeneration. Now a greater threat is that posed by urbanization and, especially, suburbanization that are covering large areas of productive farmland with houses, parking lots and roads.

What Is Soil?

Soil is one of the most variable materials on the face of Earth. Just a partial list of soil types includes alfisols, andisols, aridisols, entisols, inceptisols, histisols, mollisols, oxisols, spodosols, ultisols, and vertisols. A general definition is that soil is a relatively incohesive material on Earth's surfaces consisting of particles that make up a variable mixture of minerals, organic matter, and water, capable of supporting plant life. It is the final product of the weathering action of physical, chemical, and biological processes on rocks, which largely produces clay minerals. The solid fraction of typical productive soil is approximately 5% organic matter and 95% inorganic matter. Some soils, such as peat soils, may contain as much as 95% organic material. Other soils, particularly sandy soils, contain as little as 1% organic matter. The organic portion of soil consists of plant biomass in various stages of decay. High populations of bacteria, fungi, and animals such as earthworms may be found in

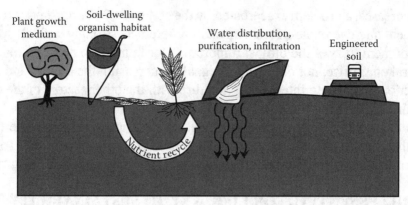

Figure 11.1.   Five ecological roles of soil are (1) as a medium for plant growth, (2) as a habitat for soil-dwelling organisms, (3) as a medium for decay of biomass leading to recycle of nutrients, (4) as a key component of the hydrologic cycle in water transfer and purification, and (5) as a key component of the anthrosphere in engineered soil.

soil. Soil contains air spaces and generally has a loose texture. Engineers who work with earthen materials view soil as divided earthen materials that can be moved without blasting.

Although the most obvious use of soil is for plant growth leading to food production, it serves many functions in the maintenance of sustainability (Figure 11.1). It holds water, regulates water supplies, and serves as a medium to filter and conduct water from precipitation into groundwater aquifers. It serves to recycle raw materials and nutrients. It is a habitat for a large variety of organisms, especially fungi and bacteria. Soil interfaces with the anthrosphere as an engineering medium that is dug up, moved, and smoothed over to make roads, dams, and other engineering constructs.

The study of soil is called pedology or, more simply, soil science. To humans and most terrestrial organisms, soil is the most important part of the geosphere. Though only a tissue-thin layer compared to the Earth's total diameter, soil is the medium that produces most of the food required by most living things. Good soil — and a climate conducive to its productivity — is the most valuable asset a society can have.

Soils exhibit a large variety of characteristics that are used to classify them for various purposes, including crop production, road construction, and waste disposal. The parent rocks from which soils are formed obviously play a strong role in determining the composition of soils. Other soil characteristics include strength, workability, soil particle size, permeability, and degree of maturity.

11.3. SOIL FORMATION AND HORIZONS

In most areas, at depths ranging from the surface to many meters below the surface, lie underlying strata of hard, unweathered, consolidated rock. Above this rock there is usually a layer of variable thickness from zero (exposed bedrock) to several tens of meters of unconsolidated rock debris formed by physical and chemical processes operating on the bedrock and called the **regolith**. The regolith is commonly derived from the underlying rock formations, but may also have been deposited from elsewhere by phenomena such as glacial action. Wind transport of eolian materials including dune sand, loess, and volcanic ash can be a major source of parent material in the regolith. (It is believed that wind-borne nutrients, such as calcium, blowing across the Atlantic Ocean from the Sahara Desert are a significant source of fertility for the severely leached soils of the Amazon rain forest.) Soil is the part of the regolith formed by weathering of the unconsolidated rock and deposition of organic matter from decaying plants. Soil forms as a generally porous material composed of mineral and organic solids, with air spaces and varying amounts of water. Soil forms at the interface of the atmosphere and the regolith. The process of soil formation is influenced by interacting factors of climate, topography, biological activity, chemical factors, and time. The most common indicator of soil formation from parent rocks is the appearance of distinct layers called horizons, as discussed in the next paragraph.

Soil Horizons

As a result of the manner in which it forms and is transformed and the complex interactions that occur among weathering processes, soil generally is divided into layers called **soil horizons** (Figure 11.2). Different soils have different horizons, often several in the same soil. Rainwater percolating through soil carries dissolved

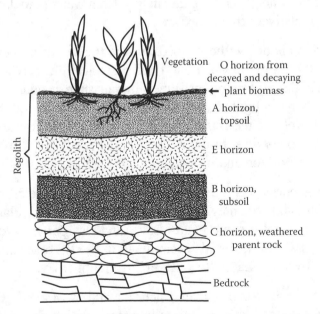

Figure 11.2.   Soil is divided into layers called horizons, the most common of which are shown here. Not all of the horizons shown may be present. Different kinds of horizons may be present and there may be subhorizons or transitional horizons as well.

and colloidal solids to lower horizons where they are deposited. Biological processes, such as bacterial decay of residual plant biomass, produce slightly acidic CO_2, organic acids, and metal-binding complexing compounds that are carried by rainwater to lower horizons where they interact with clays and other minerals, altering the properties of the minerals. There are five master horizons that are recognized. These are listed below in order from the highest horizon:

- In undisturbed soils, there may be a relatively thin **O horizon** consisting mostly of plant debris in various stages of decay. The surface of the O horizon is composed of leaves, pine needles, twigs, and stems in the initial stages of decay, whereas at slightly greater depths, the readily metabolized cellulose in these materials has degraded leaving a residue of decay-resistant biomass. Commonly missing from grassland soils, the O horizon is formed in forests and is sometimes called the *forest floor.* One of the major objectives of the practice of soil sustainability is the maintenance of a substantial O horizon by returning crop by-product biomass to the soil surface by practices such as conservation tillage.

- The next horizon is the A horizon, commonly called topsoil, which is usually rich in organic matter and humus and the site of much biological activity including that from bacteria, fungi, plant roots, microorganisms associated with roots, and larger organisms, such as earthworms. In grasslands, the A horizon may be relatively thick — up to around 1 m — because of thick mats of plant roots that are in it. The A horizon is subject to leaching of minerals and organic matter from water percolating through it and has a relatively coarse texture.

- The E horizon is below the A horizon. It is the site of maximum eluviation (from the Latin "to wash out") and is depleted of clay and oxides of aluminum and iron, leaving resistant minerals such as quartz. The E horizon normally has a light color. It may be severely weathered, leached, and bleached, largely by the action of organic acids formed by fungi metabolizing acidic forest litter in the A horizon that carry brownish iron oxide through the E horizon and into the lower B horizon.

- Below the A horizon is the B horizon (sometimes imprecisely called the subsoil). The B horizon may be regarded as a zone of accumulation because it is a repository of organic matter, salts, and clay particles leached from higher layers. These materials accumulate in the B horizon, largely from illuviation (from the Latin "to wash in") from upper layers.

- The C horizon is composed of fractured and weathered parent rocks, unconsolidated and loose enough to be moved with a shovel, from which the soil originated. It is below the area of maximum biological activity.

There are many subcategories of soil horizons that are recognized in various locations. For example, the O horizon may contain subcategories of Oi denoting slightly decomposed plant matter, Oe for moderately decomposed organic matter, and Oa for highly decomposed plant matter. As another example, the designation EB, indicates a soil horizon that is in transition between E and B horizons, but more like horizon E than horizon B.

11.4. SOIL MACROSTRUCTURE AND MICROSTRUCTURE

Soil Macrostructure

Soil horizons discussed in the preceding section are part of **soil macrostructure**. Another important aspect of soil macrostructure is **soil topography**, which refers to how the soil lies and how much slope it has, as shown in Figure 11.3. On steeper slopes, the soil is more prone to erosion, the effective rainfall is less, and rainwater runs off more quickly. The result of these factors is that soil on slopes is relatively thin and unproductive. In flatter areas, the weathered regolith layer is thicker, and the soil formation process has progressed farther with less erosion leading to thicker soil. In areas that are too flat without drainage, wetland soil conditions develop that are not conducive to high soil productivity.

Humans can alter topography to a degree as shown on the right of Figure 11.3. In this case, the contour of sloping land has been modified to construct relatively level areas alternating with very steep slopes held in place with rock and perennial plants. Such modifications are readily done with modern earth-moving equipment; the ancient South American Incas did the same thing with human labor. Climatic and soil conditions would have to be very favorable to crop productivity to justify the cost of such topographical alteration.

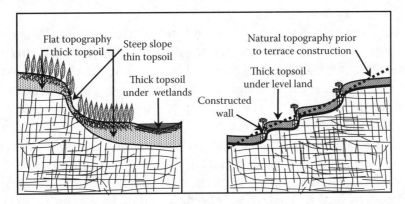

Figure 11.3.   Topsoil is thicker on level areas and beneath wetlands and swamps (left). Construction of terraces with walls held in place by rocks and perennial plants can enable retention of thicker topsoil under level areas.

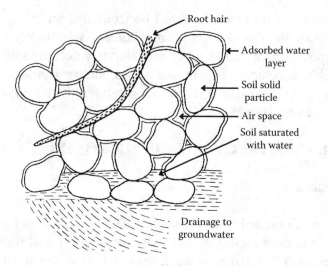

Figure 11.4.   Microstructure of typical health soil showing solid particles, air spaces, water, and biological material (a root hair).

Soil Microstructure

Figure 11.4 shows major aspects of **soil microstructure**. A normal healthy soil consists of small particles of solids, some of which may be largely organic matter, air spaces, root hairs and other biological features, and water held to varying degrees.

Water in Soil

Water is part of the three-phase, solid–liquid–gas system making up soil. It is the basic transport medium for carrying essential plant nutrients from solid soil particles into plant roots and to the farthest reaches of the plant's leaf structure (Figure 11.5). The water enters the atmosphere from the plant's leaves, a process called *transpiration*.

Normally, because of the small size of soil particles and the presence of small capillaries and pores in the soil, the water phase is not totally independent of soil solid matter. Water present in larger spaces in soil is relatively more available to plants and readily drains away. Water held in smaller pores, or between the unit layers of clay particles, is held much more strongly. Water in soil interacts strongly with organic matter and with clay minerals.

As soil becomes waterlogged (water-saturated), it undergoes drastic, mostly detrimental, changes in physical, chemical, and biological properties. Oxygen in such soil is rapidly used up by the respiration of microorganisms that degrade soil organic matter. In such soils, the bonds holding soil colloidal particles together are broken, which causes disruption of soil structure. Thus, the excess water in such soils is detrimental to plant growth, and the soil does not contain the air required by most plant roots. Most useful crops, with the notable exception of rice, cannot

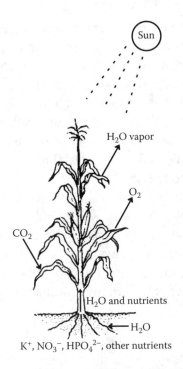

Figure 11.5.   Transpiration in a corn plant. Water from soil is carried by capillary action to the leaf surfaces of the plant from where it evaporates into the atmosphere. The water carries nutrients with it. On a hot summer's day, a field of corn transfers vast quantities of water from soil to the atmosphere.

grow on waterlogged soils. The exclusion of air from waterlogged soil results in the establishment of chemically reducing conditions that cause reduction of insoluble manganese and iron oxides (MnO_2 and Fe_2O_3) to soluble Mn^{2+} and Fe^{2+} species, which are phytotoxic and can lead to death of plants.

The **soil solution** is the aqueous portion of soil that contains dissolved matter from soil, chemical and biochemical processes in soil, and from exchange with the hydrosphere and biosphere. This medium transports chemical species to and from soil particles and provides intimate contact between the solutes and the soil particles. In addition to providing water for plant growth, it is an essential pathway for the exchange of plant nutrients between roots and solid soil.

11.5. INORGANIC AND ORGANIC MATTER IN SOIL

The Inorganic Components of Soil

The most abundant elements in the Earth's crust are oxygen, silicon, aluminum, iron, calcium, sodium, potassium, and magnesium, so minerals composed of these elements — particularly silicon and oxygen — constitute most of the mineral fraction of the soil. Common soil mineral constituents are finely divided quartz (SiO_2), epidote ($4CaO \cdot 3(AlFe)_2O_3 \cdot 6SiO_2 \cdot H_2O$), albite ($NaAlSi_3O_8$), orthoclase ($KAlSi_3O_8$), geothite (FeO(OH)), magnetite (Fe_3O_4), calcium and magnesium carbonates ($CaCO_3$, $CaCO_3 \cdot MgCO_3$), and oxides of manganese and titanium.

The weathering of parent rocks and minerals to form the inorganic soil components results ultimately in the formation of inorganic colloids. These colloids are repositories of water and plant nutrients, which may be made available to plants as needed. Inorganic soil colloids often absorb toxic substances in soil, thus playing a role in detoxification of substances that otherwise would harm plants. The abundance and nature of inorganic colloidal material in soil are obviously important factors in determining soil productivity.

The uptake of plant nutrients by roots may involve complex biological, physical, and chemical processes involving soil, water, and inorganic phases. For example, a nutrient held by inorganic colloidal material has to traverse the mineral–water, and then the water–root interfaces. This process is often strongly influenced by the ionic structure of soil inorganic matter. Excessively dry or wet soil conditions, soil compaction, and temperature extremes may inhibit nutrient uptake and transport in plants.

Organic Matter in Soil

Though typically comprising less than 5% of a productive soil, organic matter largely determines soil productivity. It serves as a source of food for microorganisms; undergoes chemical reactions such as ion exchange; and influences the physical properties of soil. Some organic compounds even contribute to the weathering of mineral matter, the process by which soil is formed. For example, oxalate ion ($C_2O_4{}^{2-}$) produced as a soil fungi metabolite and present in the soil solution, dissolves minerals, thus speeding the weathering process and increasing the availability of nutrient ion species. Some soil fungi and bacteria produce citric acid and other chelating organic acids, which react with silicate minerals and release potassium and other nutrient metal ions held by these minerals.

The accumulation of organic matter in soil is strongly influenced by temperature and by the availability of oxygen. Because the rate of biodegradation decreases with decreasing temperature, organic matter does not degrade rapidly in colder climates and tends to build up in soil. In water and in waterlogged soils, decaying vegetation does not have easy access to oxygen, and organic matter accumulates. The organic content may reach 90% or more in areas where plants grow and decay in soil saturated with water. This results in the formation of peat soils. Such soils are classified according to the material from which they are formed. Moss peat is produced from mosses, especially sphagnum. Cattails, reeds, sedges, and other herbaceous plants produce herbacerous peat. The residues of woody shrubs and trees yield woody peat. Residues of aquatic plants and remains and feces of aquatic animals produce sedimentary peat.

Soil Humus

Soil humus is by far the most significant organic constituent of soil. Humus, composed of a base-soluble fraction of **humic** and **fulvic acids** and an insoluble fraction called **humin**, is the residue left from plant biodegradation. Humus is largely the partial biodegradation product of lignin which, along with readily degraded cellulose, makes up the bulk of plant biomass. The process by which humus is formed is called **humification**. Part of each molecule of humic substance is nonpolar and hydrophobic, and part is polar and hydrophilic.

Humic substances influence soil properties to a degree out of proportion to their small percentage in soil. They strongly bind metals, and serve to hold micronutrient metal ions in soil. Because of their acid-base character, humic substances serve as buffers in soil. The water-holding capacity of soil is significantly increased by humic substances. These materials also stabilize aggregates of soil particles and increase the sorption of organic compounds by soil.

11.6. NUTRIENTS AND FERTILIZERS IN SOIL

One of the most important functions of soil in supporting plant growth is to provide essential plant nutrients — macronutrients and micronutrients. **Macronutrients** are those elements that occur in substantial levels in plant materials or in fluids in the plant. **Micronutrients** are elements that are essential only at very low levels and generally are required for the functioning of essential enzymes.

The elements generally recognized as essential macronutrients for plants are carbon, hydrogen, oxygen, nitrogen, phosphorus, potassium, calcium, magnesium, and sulfur. Carbon, hydrogen, and oxygen are obtained from the atmosphere. The other essential macronutrients must be obtained from soil. Of these, nitrogen, phosphorus, and potassium are the most likely to be lacking and are commonly added to soil as fertilizers.

Calcium-deficient soils are relatively uncommon. Treatment of soil with calcium carbonate lime to neutralize excess soil acidity usually remedies any calcium deficiency. Magnesium in soil is generally available to plants and is held by ion-exchanging organic matter or clays. Soils deficient in sulfur do not support plant growth well, and sulfur may need to be added with fertilizers in some cases.

Nitrogen, Phosphorus, and Potassium in Soil

Nitrogen, phosphorus, and potassium are plant nutrients that are obtained from soil. They are so important for crop productivity that they are commonly added to soil as fertilizers.

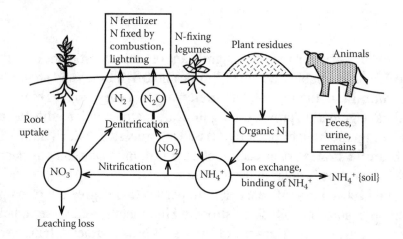

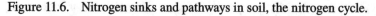

Figure 11.6.   Nitrogen sinks and pathways in soil, the nitrogen cycle.

Nitrogen is an essential component of proteins and other constituents of living matter. Plants and cereals grown on nitrogen-rich soils not only provide higher yields, but are often substantially richer in protein and, therefore, more nutritious.

Figure 11.6 summarizes the primary sinks and pathways of nitrogen in soil. In most soils, over 90% of the nitrogen content is organic. This organic nitrogen is primarily the product of the biodegradation of dead plants and animals. It is eventually hydrolyzed to NH_4^+, which can be oxidized to NO_3^- by the action of bacteria in the soil.

Nitrogen-fixing organisms ordinarily cannot supply sufficient nitrogen to meet peak demand. Inorganic nitrogen from fertilizers and rainwater is often largely lost by leaching. Soil humus, however, serves as a reservoir of nitrogen required by plants. It has the additional advantage that its rate of decay, hence its rate of nitrogen release to plants, roughly parallels plant growth — rapid during the warm growing season, slow during the winter months.

Nitrogen is most generally available to plants as nitrate ion, NO_3^-. Some plants such as rice may utilize ammonium nitrogen; however, other plants are poisoned by this form of nitrogen. When nitrogen is applied to soils in the ammonium form, nitrifying bacteria perform an essential function in converting it to available nitrate ion.

Nitrogen fixation is the process by which atmospheric N_2 is converted to nitrogen compounds available to plants. Prior to the widespread introduction of nitrogen fertilizers, soil nitrogen was provided primarily by **legumes**. These are plants such as soybeans, alfalfa, and clover, which contain on their root structures bacteria capable of fixing atmospheric nitrogen. Leguminous plants have a symbiotic (mutually advantageous) relationship with the bacteria that provide their nitrogen. The nitrogen-fixing bacteria in legumes exist in special structures on the roots called *root nodules* (see Figure 11.7). The rod-shaped bacteria that fix nitrogen are members of a special genus called *Rhizobium*. These bacteria fix nitrogen in symbiotic combination with plants.

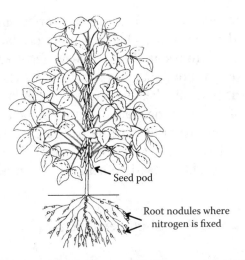

Seed pod

Root nodules where
nitrogen is fixed

Figure 11.7.   A soybean plant, showing root nodules where nitrogen is fixed.

Nitrate pollution of some surface waters and groundwater is a significant problem in some agricultural areas. Although fertilizers have been implicated in such pollution, there is evidence that feedlots are a major source of nitrate pollution. The growth of livestock populations and the concentration of livestock in feedlots have aggravated the problem. Such concentrations of cattle, coupled with the fact that a steer produces approximately 18 times as much waste material as a human, have resulted in high levels of water pollution in rural areas with small human populations. Streams and reservoirs in such areas, frequently, are just as polluted as those in densely populated and highly industrialized areas.

Nitrate in farm wells is a common and especially damaging manifestation of nitrogen pollution from feedlots because of the susceptibility of ruminant animals to nitrate poisoning. The stomach contents of ruminant animals such as cattle and sheep constitute a reducing medium (low pE) and contain bacteria capable of reducing nitrate ion to toxic nitrite ion (NO_2^-). This species oxidizes the iron(II) to iron(III) in the animals' blood hemoglobin, producing methemoglobin, which does not carry oxygen. The origin of most nitrate produced from feedlot wastes is organically bound nitrogen present in nitrogen-containing waste products, which gets converted to ammonium ion (NH_4^+), then nitrate ion (NO_3^-), by soil bacteria.

Although the percentage of phosphorus in plant material is relatively low, it is an essential component of plants. Phosphorus, like nitrogen, must be present in a simple inorganic form before it can be taken up by plants. In the case of phosphorus, the utilizable species is some form of orthophosphate ion. In the pH range that is present in most soils, $H_2PO_4^-$ and HPO_4^{2-} are the predominant orthophosphate species. Because of the formation of poorly soluble species, especially hydroxyapatite ($Ca_5(PO_4)_3OH$), a little phosphorus applied to soil as fertilizer or from the decay of organic matter is leached from soil.

Relatively high levels of potassium are utilized by growing plants. Potassium activates some enzymes and plays a role in the water balance in plants. It is also essential for some carbohydrate transformations. Crop yields are generally greatly reduced in potassium-deficient soils. The higher the productivity of the crop, the more potassium is removed from soil. When nitrogen fertilizers are added to soils to increase productivity, removal of potassium is enhanced. Therefore, potassium may become a limiting nutrient in soils heavily fertilized with other nutrients.

Potassium is one of the most abundant elements in the Earth's crust, of which it makes up 2.6%; however, much of this potassium is not easily available to plants. For example, some silicate minerals such as leucite ($K_2O \cdot Al_2O_3 \cdot 4SiO_2$), contain strongly bound potassium. Exchangeable potassium held by clay minerals is relatively more available to plants.

Nitrogen, Phosphorus, and Potassium Fertilizers

Crop fertilizers contain nitrogen, phosphorus, and potassium as major components. Magnesium, sulfate, and micronutrients may also be added. Fertilizers are designated by numbers, such as 6-12-8, showing the respective percentages of nitrogen expressed as N (in this case 6%), phosphorus as P_2O_5 (12%), and potassium as K_2O (8%). Farm manure corresponds to an approximately 0.5-0.24-0.5 fertilizer, so it is not a very effective fertilizer. Such organic fertilizers must biodegrade to release the simple inorganic species (NO_3^-, $H_xPO_4^{-3}$, K^+) assimilable by plants.

Most modern nitrogen fertilizers are made by the Haber process, in which N_2 and H_2 are combined over a catalyst at temperatures of approximately 500°C and pressures up to 1000 atm:

$$N_2 + 3H_2 \rightarrow 2NH_3 \qquad (11.6.1)$$

The **anhydrous ammonia** product has a very high nitrogen content of 82%. It may be added directly to the soil, for which it has a strong affinity because of its water solubility and formation of ammonium ion, NH_4^+, in contact with soil water. Special equipment is required to apply anhydrous NH_3, because ammonia gas is toxic. **Aqua ammonia**, a 30% solution of NH_3 in water, may be used with much greater safety. It is sometimes added directly to irrigation water.

Ammonium nitrate, NH_4NO_3, is a common solid nitrogen fertilizer. It is made by oxidizing ammonia over a platinum catalyst, converting the nitric oxide product to nitric acid, and reacting the nitric acid with ammonia. Although convenient to apply to soil, ammonium nitrate requires considerable care during manufacture and storage because it is explosive. Ammonium nitrate also poses some hazards. It is mixed with fuel oil to form an explosive that serves as a substitute for dynamite in quarry blasting and construction. This mixture was used as the explosive agent in the tragic 1995 bombing of the Federal Building in Oklahoma City.

By-products of coal processing can be important sources of ammonium fertilizer. Coking of coal (heating in the absence of air to produce a carbon residue) required for making metallic iron from iron ore yields significant amounts of ammonium sulfate and serves as a source of sulfur, which may be deficient in commercial fertilizers. The gasification of coal to produce synthesis gas for fuel and chemical manufacture can yield large quantities of aqueous ammonia.

Phosphate minerals used to make **phosphate fertilizers** are found in several states, including Idaho, Montana, Utah, Wyoming, North Carolina, South Carolina, Tennessee, and Florida. The principal mineral is insoluble fluorapatite ($Ca_5(PO_4)_3F$) and it must be treated with phosphoric or sulfuric acids to make superphosphates consisting of relatively more soluble compounds, such as $Ca(H_2PO_4)_2 \cdot H_2O$. Toxic hydrogen fluoride (HF) is produced as a by-product of superphosphate production and must be contained to prevent air pollution problems.

Potassium fertilizer components consist of potassium salts, generally KCl. Such salts are found as deposits in the ground or may be obtained from some brines. Very large deposits are found in Saskatchewan, Canada. These salts are all quite soluble in water. One problem encountered with potassium fertilizers is the luxury uptake of potassium by some crops, which absorb more potassium than is really needed for their maximum growth. In a crop where only the grain is harvested, leaving the rest of the plant in the field, luxury uptake does not create much of a problem because most of the potassium is returned to the soil with the dead plant. However, when hay or forage is harvested, potassium contained in the plant as a consequence of luxury uptake is lost from the soil.

Micronutrients in Soil

Boron, chlorine, cobalt, copper, iron, manganese, molybdenum (for N-fixation), nickel, and zinc are considered essential **plant micronutrients**. These elements are needed by plants only at very low levels and frequently are toxic at higher levels. Most of these elements function as components of essential enzymes. Manganese, iron, chlorine, and zinc may be involved in photosynthesis. Though not established for all plants, it is possible that sodium, silicon, and cobalt may also be essential plant nutrients.

Iron and manganese occur in a number of soil minerals. Sodium and chlorine (as chloride) occur naturally in soil and are transported as atmospheric particulate matter from marine sprays. Some of the other micronutrients and trace elements are found in primary (unweathered) minerals that occur in soil.

Soil trace elements may be coprecipitated with secondary minerals that are involved in soil formation. Such secondary minerals include oxides of aluminum, iron, and manganese (precipitation of hydrated oxides of iron and manganese very efficiently removes many trace metal ions from solution), calcium and magnesium carbonates, smectites, vermiculites, and illites.

Some plants accumulate extremely high levels of specific trace metals. Those accumulating more than 1.00 mg/g of dry weight are called **hyperaccumulators**. There are reportedly around 450 hyperaccumulators ranging from low-growing ground cover to trees in size. These plants have evolved in areas enriched in or polluted by particular metals. *Aeolanthus biformifolius DeWild* growing in copper-rich regions of Shaba Province, Zaire, contains up to 1.3% copper (dry weight) and is known as "copper flower." Hyperaccumulators are disliked by farmers because their metal-laden biomass harms animals that eat the plants. There is considerable interest in using hyperaccumulators to remediate waste sites contaminated with toxic metals.

Adjustment of Soil Acidity

Cation exchange in soil is the mechanism by which potassium, calcium, magnesium, and essential trace-level metals are made available to plants. When nutrient metal ions are taken up by plant roots, hydrogen ion is exchanged for the metal ions. This process, plus the leaching of calcium, magnesium, and other metal ions from the soil by water containing carbonic acid, tends to make the soil acidic:

$$\text{Soil}\}Ca^{2+} + 2CO_2 + 2H_2O \rightarrow \text{Soil}\}(H^+)_2 + Ca^{2+}(root) + 2HCO_3^- \quad (11.6.2)$$

Soil acts as a buffer, that is, it resists changes in pH. The buffering capacity depends upon the type of soil.

Most common plants grow best in soil with a pH near neutrality (pH 7). The pH of humid region mineral soils is slightly acidic in a range of 5 to 7, whereas arid soils tend to have a higher pH in a range of 7 to 9. Acid peat and acid-sulfate soils may have a very low pH around 3, whereas alkali mineral soils may have pH values up to 10 to 11. If the soil becomes too acidic for optimum plant growth by the process shown in Reaction 11.6.2 or by input of acid from an external source, it may be restored to productivity by **liming**, ordinarily through the addition of calcium carbonate:

$$\text{Soil}\}(H^+)_2 + CaCO_3 \rightarrow \text{Soil}\}Ca^{2+} + CO_2 + H_2O \quad (11.6.3)$$

11.7. SOIL AND THE BIOSPHERE

Soil is strongly related to the biosphere. The most obvious such relationship is with plants rooted in soil and growing on the soil surface. Plants are most intimately bound to soil in the **rhizosphere**, the region in which plant roots are anchored and extract water and nutrients from soil. The rhizosphere has a much greater activity of microorganisms (fungi and bacteria) than do other areas of soil. Root hairs provide a hospitable biological surface for colonization of microorganisms. Epidermal cells are sloughed from roots as they grow and carbohydrates, amino acids, and root-growth-lubricant mucigel secreted from roots provide nutrients for microorganisms. Because of the high microbial activity, the rhizosphere is a region in which soil pollutants are readily biodegraded.

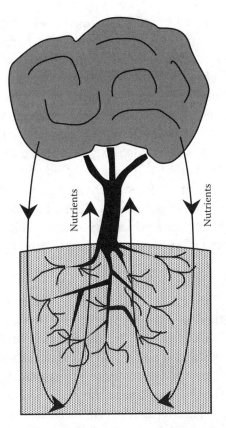

Figure 11.8.   Nutrients taken into roots of trees from lower levels of soil are carried into the plant. When leaves and other biomass fall to the soil surface and decay, nutrients are released into the topsoil fertilizing the tree. Maple, for example, recycles nutrient calcium in this manner.

As shown in Figure 11.8, trees are very much involved in nutrient cycling. Roots reaching relatively deep into soil can draw nutrients, such as calcium, into the leaves of the tree. When leaves decay, nutrients are restored to the upper layers of soil. The same thing happens when the whole tree or other vegetation dies and decays. This is one of the ways that a plant/soil ecosystem is self-sustainable.

Animals can have strong effects on soil. Earthworms and termites burrow through soil mixing it. Ants and termites build mounds that bring organic matter and nutrients from lower to higher regions. The holes that they make aerate the soil and allow water infiltration. Earthworms aerate soil and the organic matter and nutrients passing through their bodies improve soil quality and provide plant nutrients.

11.8. WASTES AND POLLUTANTS IN SOIL

Soil receives large quantities of waste products. Much of the sulfur dioxide emitted in the burning of sulfur-containing fuels ends up in soil as sulfates. Atmospheric nitrogen oxides are converted to nitrates in the atmosphere, and the nitrates eventually are deposited on soil. Soil sorbs NO and NO_2 readily, and these gases are oxidized to nitrate in the soil. Carbon monoxide is converted to CO_2 and possibly to biomass by soil bacteria and fungi. Particulate lead from automobile exhausts is found at elevated levels in soil along heavily traveled highways. Elevated levels of lead from lead mines and smelters are found on soil near such facilities.

Soil is the receptor of many hazardous wastes from landfill leachate, lagoons, and other sources. In some cases, land farming of degradable hazardous organic wastes is practiced as a means of disposal and degradation. The degradable material is worked into the soil, and soil microbial processes bring about its degradation. Sewage and fertilizer-rich sewage sludge may be applied to soil.

Volatile organic compounds (VOC), such as benzene, toluene, xylenes, dichloromethane, trichloroethane, and trichloroethylene may contaminate soil in industrialized and commercialized areas. One of the more common sources of these contaminants is leaking underground storage tanks. Landfills built before current stringent regulations were enforced and improperly discarded solvents are also significant sources of soil VOCs.

Soil receives enormous quantities of pesticides as an inevitable result of their application to crops. Approximately $20 billion are spent each year on 2.5 million tons of agricultural pesticides, whereas in the U.S. the corresponding figures are around $4 billion and 500,000 tons. The degradation and eventual fate of these enormous quantities of pesticides on soil largely determine their ultimate environmental effects. Detailed knowledge of these effects are now required for licensing of a new pesticide (in the U.S. under the Federal Insecticide, Fungicide, and Rodenticide Act, FIFRA). Among the factors to be considered are the sorption of the pesticide by soil, leaching of the pesticide into water, as related to its potential for water pollution, effects of the pesticide on microorganisms and animal life in the soil, and possible production of relatively more toxic degradation products.

Adsorption by soil is a key step in the degradation of a pesticide. The degree of adsorption and the speed and extent of ultimate degradation are influenced by a number of factors, including solubility, volatility, charge, polarity, and molecular structure and size. Adsorption of a pesticide by soil components may have several effects. Under some circumstances, it retards degradation by separating the pesticide from the microbial enzymes that degrade it, whereas under other circumstances the reverse is true. Purely chemical degradation reactions may be catalyzed by adsorption. Loss of the pesticide by volatilization or leaching is diminished. The toxicity of a herbicide to plants may be strongly affected by soil sorption.

Degradation of Pesticides in Soil

The three primary ways in which pesticides are degraded in or on soil are chemical degradation, photochemical reactions, and, most important, biodegradation. Various combinations of these processes may operate in the degradation of a pesticide.

Chemical degradation of pesticides has been observed experimentally in soils and clays sterilized to remove all microbial activity. Of the chemical degradation reactions, probably the most common are hydrolytic reactions of pesticides in which the molecules split apart with the addition of molecules of H_2O.

Many pesticides have been shown to undergo photochemical reactions, that is, chemical reactions brought about by the absorption of light. Many of the studies reported apply to pesticides in water or on thin films, and the photochemical reactions of pesticides on soil and plant surfaces remain largely a matter of speculation.

Biodegradation and the Rhizosphere

Although insects, earthworms, and plants may be involved to a minor extent in the biodegradation of pesticides and other pollutant organic chemicals, microorganisms have the most important role. The rhizosphere (see Section 11.7), the layer of soil in which plant roots are most active, is a particularly important part of soil in respect to biodegradation of wastes. It is a zone of increased biomass and is strongly influenced by the plant root system and the microorganisms associated with plant roots. The rhizosphere may have more than 10 times the microbial biomass per unit volume compared to nonrhizospheric zones of soil. This population varies with soil characteristics, plant and root characteristics, moisture content, and exposure to oxygen. If this zone is exposed to pollutant compounds, microorganisms adapted to their biodegradation may also be present.

The biodegradation of a number of synthetic organic compounds has been demonstrated in the rhizosphere. Understandably, studies in this area have focused on herbicides and insecticides that are widely used on crops, and many of these substances have exhibited enhanced biodegradation in the rhizosphere. It is interesting to note that enhanced biodegradation of partial combustion product polycyclic aromatic hydrocarbons (PAH) has been observed in the rhizospheric zones of prairie grasses. This observation is consistent with the fact that in nature such grasses burn regularly and significant quantities of PAH compounds are deposited on soil as a result.

11.9. SOIL LOSS AND DETERIORATION

There are two ways by which more food can be grown on soil. The first of these is to bring more soil into production from fragile lands by measures such as clearing forest lands, as has occurred with the cultivation of Amazon rain forests, cultivating grasslands with marginal rainfall, and cultivating areas on relatively steep slopes. The second approach is to increase the cropping intensity of existing lands. These approaches have serious implications for the maintenance of soil quality and sustainability. Tropical rain forests are fragile ecosystems in which essential nutrients are maintained largely within the plant biomass and the upper layers of relatively thin soil; therefore, clearing of the forests results in rapid and largely irreversible loss of productivity. The cultivation of arid grasslands leads to wind erosion and conversion of the land to deserts. Farming of steeply sloping land causes severe water erosion and soil loss.

Significant success has been achieved with more intensive utilization of existing soils. The green revolution dating back to the 1950s used newly developed high-yielding varieties of wheat, rice, and corn along with intensive irrigation, use of pesticides, and heavy applications of fertilizer to dramatically increase crop yields. These advances not only prevented the widespread starvation forecast around 1950, but enabled improved nutrition for large numbers of people. In some cases, two or even three crops per year of rapidly-maturing, high-yielding monoculture varieties became possible. Such intensive cultivation, though not necessarily devastating to soil quality, has often resulted in loss of the natural ability of soil to sustain crops, requiring application of increasing amounts of pesticides and fertilizer in some cases in order to maintain productivity. Another problem has been accumulation of salt in some irrigated soils. There is no inherent reason for intensively cultivated land to lose soil quality, but proper measures must be taken to make sure that sustainable practices are used.

Soil is a fragile resource that can be lost by erosion or become so degraded that it is no longer useful to support crops. There are several physical, chemical, and biological indicators of soil health and quality. These are discussed briefly here.

With respect to physical indicators of soil quality, one of the most important is texture and bulk density, which determines such important properties as soil's ability to anchor roots, its resistance to erosion, and its ability to retain and transport water and chemicals. Soil depth is important in determining its productivity, ability to anchor roots, to provide water during dry periods, and resist erosion. Water-holding capacity is important in retaining and transporting water and determining tendency toward erosion.

Among chemical indicators of soil quality are factors related to both organic and inorganic matter. Soil organic matter relates to fertility, structural stability, and ability to serve as a food source for soil organisms including microbes and earthworms. The pH of soil is indicative of excessive acidity and alkalinity, both of which are detrimental to soil productivity. Electrical conductivity is indicative of the availability of nutrient salts. Extractable nitrogen, phosphorus, and potassium reflect the availability of these essential plant nutrients.

Healthy soils have favorable biological properties and relatively high levels of biological activity, often reflected in microbial biomass C and N. Mineralizable N is indicative of the availability of essential nitrogen for biological activity. The activity of microorganisms is measured by a parameter called **specific respiration** related to oxygen consumption by microbes per unit volume of soil. Macroorganism numbers are indicators of nonmicrobial soil organisms such as earthworms.

Factors in Soil Sustainability

Two of the major factors in soil sustainability are soil resistance and soil resilience. Soil resistance refers to soil's capacity to resist detrimental effects. For example, some crops, such as hay, tend to remove nutrient potassium from soil. If the soil contains mineral sources of potassium, it is readily replenished, whereas if such sources are not available, soil productivity, its ability to grow crops, and its ability to restore organic matter content are seriously compromised. **Soil resilience** refers to the ability of soil to recover from insults. The removal of forests or plowing of grasslands on soil can be detrimental to soil quality. If the soil is readily converted back to forest or grassland, it has a high resilience, whereas, if it is not, the resilience is low.

Desertification refers to the process associated with drought and loss of fertility by which soil becomes unable to grow significant amounts of plant life. Desertification caused by human activities is a common problem globally, occurring in diverse locations, such as Argentina, the Sahara, Uzbekistan, the U.S. Southwest, Syria, and Mali. It is a very old problem dating back many centuries to the introduction of domesticated grazing animals to areas where rainfall and groundcover were marginal. The most notable example is desertification aggravated by domesticated goats in the Sahara region. Desertification involves a number of interrelated factors, including erosion, climate variations, water availability, loss of fertility, loss of soil humus, and deterioration of soil chemical properties. An important contributor to desertification is **salinization**, the accumulation of salt in irrigated soils. Salinization is actually a very old problem; soils in parts of ancient Mesopotamia in the Middle East were afflicted by it centuries ago.

A related problem is **deforestation** consisting of loss of forests. The problem is particularly acute in tropical regions, where the forests contain most of the existing plant and animal species. In addition to extinction of these species, deforestation can cause devastating deterioration of soil through erosion and loss of nutrients.

Erosion

Soil erosion can occur by the action of both water and wind, although water is the primary source of erosion. Cultivation of hilly land in Greece and Rome more than 2000 years ago caused severe erosion problems from which the soil has not yet recovered. Millions of tons of topsoil are carried by the Mississippi River and swept from its mouth each year. About one-third of U.S. topsoil has been lost since cultivation began on the continent. At present, approximately one-third of U.S. cultivated land is eroding at a rate sufficient to reduce soil productivity. It is estimated that 48 million acres of land, somewhat more than 10% of that under cultivation, is eroding at unacceptable levels, taken to mean a loss of more than 14 tons of topsoil per acre each year. Specific areas in which the greatest erosion is occurring include northern Missouri, southern Iowa, west Texas, western Tennessee, and the Mississippi Basin. Figure 11.9 shows the pattern of soil erosion in the lower 48 continental U.S. states.

Water erosion is responsible for most of the erosion that occurs. As shown in Figure 11.9, water erosion in the continental U.S. tends to be concentrated in agriculturally productive areas located in watersheds of major rivers.

Wind erosion, such as occurs on the generally dry, high plains soils of eastern Colorado, poses another threat. After the Dust Bowl days of the 1930s, much of this

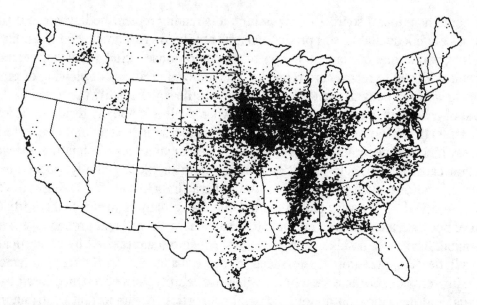

Figure 11.9.   Pattern of soil erosion in the continental U.S. Most of the erosion is from water and tends to be concentrated in the productive farmlands of the Mississippi, Ohio, Missouri, and Platte river valleys.

land was allowed to revert to grassland, and the topsoil was held in place by the strong root systems of the grass cover. However, in an effort to grow more wheat and improve the sale value of the land, much of it has been cultivated in later years. Although freshly cultivated grassland may yield well for 1 or 2 years, the nutrients and soil moisture are rapidly exhausted, and the land becomes very susceptible to wind erosion.

11.10. SOIL CONSERVATION AND RESTORATION

The preservation of soil from erosion is commonly termed **soil conservation**. There are a number of solutions to the soil erosion problem. Some are old, well-known agricultural practices, such as terracing, contour plowing (Figure 11.10), and

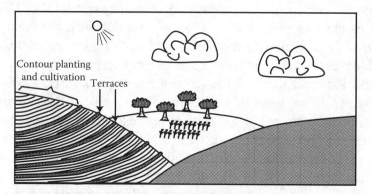

Figure 11.10.   Construction of terraces on the contour of land and planting crops on the contour are practices that have been very effective in reducing soil erosion.

periodically planting fields with cover crops, such as clover. For some crops, **conservation tillage** (no-till agriculture) greatly reduces erosion. This practice consists of planting a crop among the residue of the previous year's crop, without plowing. Weeds are killed in the newly planted crop row by application of a herbicide prior to planting. The surface residue of plant material left on top of the soil prevents erosion.

Another, more experimental, solution to the soil erosion problem is the cultivation of perennial plants, which develop a large root system and come up each spring after being harvested the previous fall. For example, a perennial corn plant has been developed by crossing corn with a distant, wild relative, teosinte, which grows in Central America. Unfortunately, the resulting plant does not give outstanding grain yields. It should be noted that an annual plant's ability to propagate depends upon producing large quantities of seeds, which is why plants harvested for their grain (seeds) are annual plants. In contrast, a perennial plant must develop a strong root system with bulbous growths called rhizomes, which store food for the coming year. However, it is possible that the application of genetic engineering (see Section 11.16) may result in the development of perennial crops with good seed yields. The cultivation of such a crop would cut down on a great deal of soil erosion.

The best known perennial plants are trees, which are very effective in stopping soil erosion. Wood from trees can be used as biomass fuel, as a source of raw materials, and as food (see next paragraph). There is a tremendous unrealized potential for an increase in the production of biomass from trees. In the past, trees were often allowed to grow naturally with native varieties and without the benefit of any special agricultural practices, such as fertilization. The productivity of biomass from trees can be greatly increased with improved varieties, including those that are genetically engineered, and with improved cultivation and fertilization.

The most important use for wood is, of course, as lumber for construction. This use will remain important as higher energy costs increase the costs of other construction materials, such as steel, aluminum, and cement. Wood is about 50% cellulose, which can be hydrolyzed by rapidly improving enzyme processes to yield glucose sugar. The glucose can be used directly as food, fermented to ethyl alcohol for fuel (gasohol), or employed as a carbon and energy source for protein-producing yeasts. Given these and other potential uses, the future of trees as an environmentally desirable and profitable crop is very bright.

Soil Restoration

Soil can be impaired by loss of fertility, erosion, buildup of salinity, contamination by phytotoxins, such as zinc from sewage sludge, and other insults. Like all natural systems, soil has a degree of resilience and can largely recover whenever the conditions leading to its degradation are removed. However, in many cases, more active measures called **soil restoration** are required to restore soil productivity, through the application of restoration ecology. Measures taken in soil restoration may include physical alteration of the soil to provide terraces and relatively flat areas not subject to erosion. Organic matter can be restored by planting crops the residues of which are cultivated into the soil for partially decayed biomass. Nutrients may be added and contaminants neutralized. Excess acid or base can be neutralized, and salinity can be leached from the soil. As the demand for food increases and damage to soil becomes more evident, soil restoration will become a very important endeavor.

Water Resources and Soil

The conservation of soil and the protection of water resources are strongly interrelated. Most freshwater falls initially on soil, and the condition of the soil largely determines the fate of the water and how much is retained in a usable condition. The land area on which rainwater falls is called a **watershed**. In addition to collecting the water, the watershed determines the direction, rate of flow, and the degree of water infiltration into groundwater aquifers. Excessive rates of water flow prevent infiltration, lead to flash floods, and cause soil erosion. Measures taken to enhance the utility of land as a watershed also fortunately help prevent erosion. Some of these measures involve modification of the contour of the soil, particularly terracing, construction of waterways, and construction of water-retaining ponds. Waterways are planted with grass to prevent erosion, and water-retaining crops and bands of trees can be planted on the contour to achieve much the same goal. Reforestation and control of damaging grazing practices conserve both soil and water.

11.11. SHIFTING CULTIVATION: SLASH AND BURN

Shifting cultivation refers to the practice of clearing natural vegetation from an area, growing crops on it for several years until the soil is depleted, then moving on to a new area. The formerly cultivated plot then becomes repopulated with native plants and after 15 to 20 years may become available for cultivation again. The most common shifting cultivation practice is the **slash-and-burn technique** in which the bark of forest trees is cut to kill them and the dead trees are burned to clear the soil for cultivation.

Slash-and-burn cultivation techniques are now practiced on approximately 30% of the world's arable land, supporting approximately 300 million people, primarily in tropical and subtropical regions. Because of high demand for food, the period during which the land remains fallow before returning to cultivation has been reduced from the more sustainable 15 to 20 years that used to be the norm to around 5 years. As a result, the land has become much less productive and has suffered increased erosion.

In addition to its adverse effects on soil productivity, slash-and-burn agriculture has caused additional harm. It is the dominant factor in deforestation accounting for about 70% of deforestation in Africa. The release of carbon dioxide and, to a lesser extent, methane, from slash-and-burn agriculture is a significant factor in greenhouse gas climate warming.

A potential remedy for the problems posed by slash-and-burn agriculture is to grow food within the forests. This is already practiced to some extent with coffee trees that grow within the shelter of larger forest trees. By clearing rows within the forest, other crops can be grown in proximity to forest trees. After a period of several years, the cultivated land can be replanted to trees. Rather than simply destroying the forest trees, in the land in which cultivation is to be practiced again, they can be harvested for their fuel, which is often in short supply for cooking in tropical and semitropical regions, or the wood can be gasified to produce synthetic fuels or hydrogen for ammonia production.

Grasslands are often subjected to a form of rotation between crops and native grasses. Although 2 or 3 years of good cereal production can often be obtained from freshly plowed grassland, cultivation results in nutrient depletion and erosion. Furthermore, the productivity of grassland returned to its native state rarely approaches the levels that it had prior to cultivation.

11.12. PROCESS INTENSIFICATION IN AGRICULTURE

One of the basic ideas of green technology is **process intensification**, which refers to increased production from smaller facilities. Agriculture and food production have provided one of the best examples of process intensification. From 1950 until the present, more food has been produced from agriculture than was produced throughout the history of agriculture up to 1950, a period of 10,000 years!

Process intensification in agriculture is commonly called the **green revolution**. The green revolution was the result of intensified management of crops, soil, and water. Except for sub-Saharan Africa, food production increased dramatically from 1950 to 1990, with especially dramatic advances in developing countries of Asia. Grain production increased threefold, more food (particularly cereals) became available per capita, and food prices actually fell.

Key to the green revolution were high-yielding hybrid and dwarf varieties of wheat, rice, and corn. These were grown in monoculture systems and, in climates free of freezing weather, with two or three crop cycles per year. Key to the increased yields was enhanced availability of water and fertilizers.

Aside from the obvious benefits of process intensification in making more food available and averting starvation in some areas, there have been other benefits as well. One of these has been increased recycling of crop residues to soil. These materials add organic matter essential for soil quality, provide soil cover, reduce erosion, and serve to bind some essential nutrients. Another benefit has been the reduction of demands on fragile lands. Without process intensification on prime agricultural land, marginal lands located on sloping erosion-prone terrain, often with insufficient rainfall would have been developed, ultimately leading to irreversible soil degradation. Another advantage of the green revolution has been higher efficiency of nutrient utilization by some grain varieties. For example, prior to 1950, wheat varieties yielded about 45 kg more of grain per kilogram of additional nitrogen fertilizer applied. With improved varieties of wheat, by 1990, the increase in yield had risen to about 70 kg of wheat per additional kilogram of nitrogen. These benefits are observed only up to a point, beyond which increased application of nitrogen yields little increase in grain yield.

There have been some adverse effects of process intensification in agriculture. In some cases, increased crop growth from improved varieties, application of fertilizers, and irrigation have resulted in depletion of micronutrients, such as sulfur and essential metals. Application of too much fertilizer can result in excessive accumulation of nutrients, such as nitrogen and phosphorus, causing eutrophication (see Chapter 5 and Chapter 6) of bodies of water receiving soil runoff. Irrigation has resulted in accumulation of salts on soil, a process called *salinization*. Pesticides, particularly herbicides, may accumulate on intensively cultivated soil. The cultivation of just a few varieties of crops can result in reduced biodiversity. For example, particular varieties may become subject to diseases and, with a depleted gene pool of alternate varieties, disease-resistant substitutes may not be available. The intensive cultivation of monoculture crops without crop rotation can increase the occurrence of crop diseases.

One of the major effects of process intensification in agriculture can be adverse effects on diet. The kinds of crops most amenable to process intensification are cereal crops, particularly wheat, rice, and corn. Also needed for proper nutrition is consumption of vegetables, fruits, and protein-rich beans, peas, and lentils (pulses). The green revolution has given comparatively little attention to these kinds of foods, which are often not consumed in sufficient quantities in human diets. Agricultural process intensification has often given insufficient attention to potential deficiencies of micronutrients. As a result, diseases because of lack of nutrient iron, zinc, vitamin A, and other micronutrients have been observed.

11.13. SUSTAINABLE AGRICULTURAL MANAGEMENT

An important challenge to modern agriculture is **agricultural management for sustainability** involving both soil and crop management techniques. The major aspects of this approach are the following:

1. Increase biological productivity and diversity

2. Prevent soil degradation including erosion, salinization, and desertification

3. Reduce pollution of soil and other environmental spheres

4. Decrease quantities of nutrients and water used per unit of production by increasing efficiency of nutrient and water utilization

5. Increase amounts and quality of soil organic matter

6. Increase desirable biological activity in the soil subsurface by earthworms, plant roots, nitrogen-fixing bacteria and other organisms

The biological productivity of soil has been greatly enhanced in recent decades through improved crop varieties (particularly hybrids), increased use of fertilizers, and increased irrigation. Although total productivity has increased, diversity has decreased with intensive cultivation of monoculture. Measures are needed to ensure cultivation of diverse crop varieties that may be relatively less productive but that need to be preserved to ensure diversity of the gene pool. Although very difficult to realize in practice, development of perennial crop systems that grow each year without seeding would be highly desirable. This is currently the case with fruit-bearing trees and berry bushes and may one day be feasible with cereal grains.

Prevention of soil degradation is a key aspect of sustainable agricultural management. Erosion can be greatly reduced by conservation tillage techniques that avoid plowing or otherwise significantly disturbing soil. Irrigation must be practiced in ways that prevent salinization of soil by, for example, applying enough water to ensure runoff of excess salts. Desertification can be prevented by proper cultivation and irrigation techniques. A key to preventing soil degradation is the maintenance of soil cover with perennial plants or with crop residues.

It is important to manage agricultural production in a manner that avoids pollution of soil and the other environmental spheres. The greatest potential for soil pollution is from the application of herbicides to kill weeds. Herbicides are needed that are biodegradable within a few weeks of application. They should be applied in minimum amounts only where and when needed. Pollution of water can occur from runoff containing pesticides and fertilizer nutrients that cause eutrophication of receiving waters. Such runoff should be minimized by measures such as minimum application of pesticides and fertilizers. Pesticide runoff can be reduced by using substances that have a low water solubility and high affinity for soil solids. Ammonium nitrogen fertilizers are preferred over nitrates because soil binds to the NH_4^+ ion, but not to anionic nitrate. Nitrogen can be applied as organically bound nitrogen that is slowly released as the organic matter decays.

Quantities of nutrients and water used per unit of production can be increased by careful control of times and amounts of application of these materials. Also, plants can be bred that require minimum amounts of water and that are particularly efficient in utilization of fertilizer. Adding fertilizer with irrigation water is a particularly good method of maintaining optimum rates of application of fertilizer.

Soil organic matter can be enhanced by returning crop residues to soil in optimum amounts. Agricultural practice used to call for plowing harvested fields to bury organic matter under a layer of topsoil. Modern conservation tillage techniques do not use deep plowing, but instead leave crop residues on top of the soil where the plant biomass partially decays and is gradually incorporated into the soil. A key aspect of modern sustainable soil management is to keep the soil surface covered with a substantial amount of crop residues and organic mulch. This reduces runoff, prevents erosion, reduces evaporation loss of moisture, increases desirable microbial activity, and provides a reservoir of gradually released nutrients — especially nitrogen — required for optimum plant growth. In some cases, crops are grown to produce "green manure" in the form of plant biomass. Sweet clover is particularly productive of biomass and can be grown as a source of biomass. Because of the nitrogen-fixing bacteria on the roots of clover, its cultivation also increases levels of soil nitrogen.

Crop rotation in which different crops are alternated or are planted adjacent to each other in strips has long been recognized as a beneficial agricultural practice. Legumes, which have nitrogen-fixing bacteria on their roots, are especially beneficial in crop rotation. Planting forage crops in rotation and allowing animals to graze on these crops enables decentralized production of livestock and fertilization of the soil from the urine and manure of the animals. In some cases, the same crop can be used for both forage and cereal production. Wheat planted in the fall can be grazed by cattle after growth is established until the point at which grain-bearing stalks are ready to be established.

11.14. AGROFORESTRY

A promising alternative in sustainable agriculture is **agroforestry** in which crops are grown in strips between rows of trees as shown in Figure 11.11. The trees stabilize the soil, particularly on sloping terrain. By choosing trees with the capability to fix nitrogen, the system can be self-sufficient in this essential nutrient.

The mode of crop growth shown in Figure 11.11 is called "**alley cropping across the slope**." Fast-growing, nitrogen-fixing trees hold the sloping soil in place. In between crop seasons, the trees are pruned and the nutrient-rich prunings are spread on the soil where crops are grown, fertilizing the soil, adding organic matter, and holding the soil in place. The trees potentially have economic value in providing

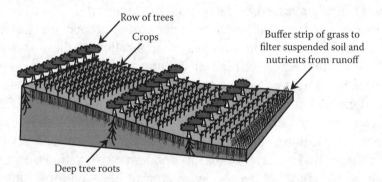

Figure 11.11.   Alley cropping with crops between rows of trees running across sloping land can be an effective means of practicing agroforestry sustainably.

wood for construction, firewood for cooking, fruit, and nuts. Genetically engineered trees may even provide high-value pharmaceuticals and specialty chemicals in the future. At the bottom of the slope, a buffer strip of grass can serve to filter nutrients and suspended soil from runoff from the fields. Potentially, rich topsoil collected by the buffer strip can be returned to higher levels to enrich the soil.

11.15. PROTEIN FROM PLANTS AND ANIMALS

The provision of adequate amounts of protein is the greatest challenge associated with feeding world population today. Protein can be obtained from both plant and animal sources. Figure 11.12 shows the relative amounts of grain required to produce equivalent amounts of food from grain, itself, and three kinds of meat produced by animals fed grain. It is obvious that direct consumption of grain is the most efficient means of getting required nutrition. However, it is important to note that, as discussed below, meat is a much more balanced form of protein than that from grain. Furthermore, ruminant animals, such as cattle, have digestive systems in which low-grade plant biomass, such as grass or ensilage from fermented, chopped cornstalks, is converted to food material by the action of specialized bacteria. Therefore, cattle, sheep, and goats can convert plant biomass worthless for human nutrition to high-protein-quality meat.

In considering various food sources, it is important to know that foods differ significantly in their protein quality. The proteins in the human body are composed of 20 different amino acids. Some of these can be biosynthesized in the body, but eight are essential amino acids that are required in food. All of these proteins are present in animal sources including meat and eggs, but one or more are lacking in individual plant sources of protein. Typically, grains are deficient in lysine amino acid whereas pulses (see following paragraph) lack methionine. Therefore, a vegetarian diet normally requires at least two sources of plant protein to provide a proper

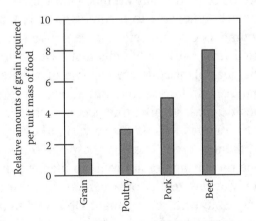

Figure 11.12.   Relative quantities of grain, poultry, pork, and beef required for equivalent amounts of food.

amino acid balance. An exception is soya protein from soya beans, which may be regarded as a complete protein source by itself.

Of particular importance with respect to supplying vegetable protein are the **pulses**. These are seeds from the family *Leguminosae*, a family of about 13,000 species, the second largest in the plant kingdom, distinguished by a characteristic pod that protects the seeds during their formation. Legumes have high economic value, providing, in addition to food, chemicals, pharmaceuticals, oils, dyes, and wood. Many members of the family have the capability of fixing their own fertilizer nitrogen from the atmosphere by virtue of *Rhizobium* bacteria growing on their roots. As a food source, pulses are a particularly rich source of protein.

11.16. AGRICULTURAL APPLICATIONS OF GENETICALLY MODIFIED ORGANISMS

Genes composed of deoxyribonucleic acid, DNA, located in the nuclei of cells direct cell reproduction and synthesis of proteins and generally direct the organism activities. Plant scientists are now able to modify DNA by processes called *recombinant DNA technology*. (Recombinant DNA technology is also being applied to animals, but to a lesser extent than with plants.) Recombinant DNA technology normally involves taking a single characteristic from one organism — the ability to produce a bacterially synthesized insecticide, for example — and splicing it into another organism. By so doing, for example, corn and cotton have been genetically engineered to produce their own insecticide. Plants produced by this method are called **transgenic** plants. During the 1970s, the ability to manipulate DNA through genetic engineering became a reality, and during the 1980s, it became the basis of a major industry. This technology promises some exciting developments in agriculture and, indeed, is expected to lead to a "second green revolution." Direct manipulation of DNA can greatly accelerate the process of plant breeding to give plants that are much more productive, resistant to disease, and tolerant to adverse conditions. In the future, entirely new kinds of plants may even be engineered.

Plants are particularly amenable to recombinant DNA manipulation. In part, this is because huge numbers of plant cells can be grown in appropriate media, and mutants can be selected from billions of cells that have desired properties, such as virus resistance. Individual plant cells are capable of generating whole plants, so cells with desired qualities can be selected and allowed to grow into plants, which may have the qualities desired. Ideally, this accomplishes in weeks what conventional plant breeding techniques would require years to do.

There are many potential green chemistry aspects from genetic engineering of agricultural crops. One promising possibility is to increase the efficiency of photosynthesis, which is only a few tenths of a percent in most plants. Doubling this efficiency should be possible with recombinant DNA techniques, which might significantly increase the production of food and biomass by plants. For example, with some of the more productive plant species, such as fast-growing hybrid poplar trees and sugarcane, biomass is close to becoming economical as a fuel source. A genetically engineered increase in photosynthesis efficiency could enable biomass to economically replace expensive petroleum and natural gas for fuel and raw material. A second possibility with genetic engineering is the development of the ability to support nitrogen-fixing bacteria on plant roots in plants that cannot do so now. If corn, rice, wheat, and cotton could be developed with this capability it could save enormous amounts of energy and natural gas (a source of elemental hydrogen) now consumed to make ammonia synthetically.

Transgenic crops have many detractors; demonstrations have broken out and test plots of crops have been destroyed by people opposed to what they call "Frankenfoods." There is some evidence to suggest that bacterial insecticide produced by transgenic corn kills beautiful Monarch butterflies that have contacted the corn pollen. In year 2000, a lot of concern was generated over the occurrence of transgenic corn in taco shells made for human consumption, and a large recall of the product from supermarket shelves occurred. Opposition to transgenic foods has been especially strong in Europe, and the European Commission, the executive body of the European Union, has disallowed a number of transgenic crops. Despite these concerns, transgenic crops are growing in importance and there is a lot of interest in them in highly populated countries, particularly China, where they are seen as a means of feeding very large populations.

The Major Transgenic Crops and Their Characteristics

The two characteristics most commonly developed in transgenic crops are tolerance for herbicides that kill competing weeds and resistance to pests, especially insects, but including microbial pests (viruses) as well. The most common transgenic crop grown in the U.S. is the soybean, of which about 56 million acres consisted of herbicide-resistant transgenic varieties in 2005. About 30 million acres of transgenic corn, about 38% of the total corn acreage and consisting of varieties that produce Bt insecticide (see following paragraph) effective against the corn borer were planted in the U.S. in 2003. In 2001, it was estimated that 69% of the cotton grown in the U.S. was transgenic. The other major transgenic crop is canola. Only small fractions of the potato, squash, and papaya crops are transgenic. In 2005, an estimated 222 million acres in 17 countries were planted to transgenic crops. The U.S. is the leading producer of transgenic crops with 123 million acres planted in 2005, followed by Argentina, Brazil, Canada, China, Paraguay, India, and South Africa.

The overwhelming majority of characteristics spliced into transgenic crops consist of herbicide tolerance and resistance to insects. Insect resistance has been imparted by addition of a gene from *Bacillus thuringiensis* (Bt) that causes the plant to produce a natural insecticide in the form of a protein that damages the digestive systems of insects, killing them. Of the acreages of transgenic crops planted in 1999, 70.2% were herbicide tolerant, 22.2% were Bt insect resistant, 7.3% were both herbicide tolerant and insect resistant, and 0.3% were virus resistant.

The disruption of natural ecosystems by cultivation of land and planting agricultural crops provides an excellent opportunity for opportunistic plants — weeds — to grow in competition with the desired crops. To combat weeds, farmers use large quantities of a variety of herbicides. The heavy use of herbicides poses a set of challenging problems. In many cases, to be effective without causing undue environmental damage, herbicides must be applied in specified ways and at particular times. Collateral damage to crop plants, environmental harm, and poor biodegradation leading to accumulation of herbicide residues and contamination of water supplies are all problems with herbicides. A number of these problems can be diminished by planting transgenic crops that are resistant to particular herbicides. The most common such plants are those resistant to Monsanto's Roundup herbicide (glyphosate, structural formula is as follows):

$$
\begin{array}{c}
\text{O H H H O} \\
\text{HO--C--C--N--C--P--OH} \\
\text{H   H OH}
\end{array}
\quad \text{Glyphosate, Roundup herbicide}
$$

This widely used compound is a broad-spectrum herbicide, meaning that it kills most plants that it contacts. One of its advantages from an environmental standpoint is that it rapidly breaks down to harmless products in soil, minimizing its environmental impact and problems with residue carryover. By using "Roundup Ready" crops, of which by far the most common are transgenic soybeans, the herbicide can be applied directly to the crop, killing competing weeds. Application when the crop plants are relatively small, but after weeds have had a chance to start growing, kills weeds and enables the crop to get a head start. After the crop has developed significant size, it deters the growth of competing weeds by shade that deprives the weeds of sunlight.

Aside from weeds, the other major class of pests that afflict crops consists of a variety of insects. Two of the most harmful of these are the European corn borer and the cotton bollworm, which cost millions of dollars in damage and control measures each year and can even threaten an entire year's crop production. Even before transgenic crops were available, Bt was used to control insects. This soil-dwelling bacterium produces a protein called delta-endotoxin. Ingested by insects, delta-endotoxin partially digests the intestinal walls of insects causing ion imbalance, paralyzing the system, and eventually killing the insects. Fortunately, the toxin does not affect mammals or birds. Bt has been a popular insecticide, because as a natural product, it degrades readily and has gained the acceptance often accorded to "natural" materials (many of which are deadly).

Genetic engineering techniques have enabled transplanting genes into field crops that produce Bt. This is an ideal circumstance in that the crop being protected is generating its own insecticide, and the insecticide is not spread over a wide area. There are several varieties of insecticidal Bt, each produced by a unique gene. Several insecticidal pests are well controlled by transgenic Bt. In addition to the European corn borer mentioned above, these include the southwestern corn borer and corn earworm. Cotton varieties that produce Bt are resistant to cotton bollworm. Bt-producing tobacco resists the tobacco budworm. Potato varieties have been developed that produce Bt to kill the Colorado potato beetle, although this crop has been limited because of concerns regarding Bt in the potato product consumed directly by humans. Although human digestive systems are not affected adversely by Bt, there is concern over its being an allergen because of its proteinaceous nature.

The greatest success to date with Bt crops has occurred with cotton, which has saved as much as a half million kilograms of synthetic insecticides in the U.S. each year. The benefits of Bt corn are less certain. One of the concerns with Bt corn is the production of the insecticide on pollen, which spreads from the corn plants. Some studies have suggested that this pollen deposited on milkweed that is the natural source of food for Monarch butterflies is a serious threat to this beautiful migratory insect. Another concern with all Bt crops is the potential to develop resistance in insects through the process of natural heredity. To combat resistance, farmers are required to plant a certain percentage of each field to non-Bt crops with the idea that insects growing in these areas without any incentive to develop resistance will crossbreed with resistant strains, preventing them from becoming dominant.

Virus resistance in transgenic crops has concentrated on papaya. This tropical fruit is an excellent source of Vitamin A and Vitamin C and is an important nutritional plant in tropical regions. The papaya ringspot virus is a devastating pest for papaya, and transgenic varieties resistant to this virus are now grown in Hawaii. One concern with virus-resistant transgenic crops is the possibility of transfer of genes responsible for the resistance to wild relatives of the plants that are regarded as weeds, but are now kept in check by the viruses. For example, it is possible that virus-resistant genes in transgenic squash may transfer to competing gourds, which would crowd out the squash grown for food.

Future Crops

The early years of transgenic crops can be rather well summarized by soybeans, corn, and cotton resistant to herbicides and insects. In retrospect, these crops will almost certainly seem rather crude and unsophisticated. In part, this lack of sophistication is because of the fact that the genes producing the desired qualities are largely expressed by all tissues of the plants and throughout their growth cycle, giving rise to problems, such as the Bt-contaminated corn pollen that may threaten Monarch butterflies or Bt-containing potatoes that may not be ideal for human consumption. It is anticipated that increasingly sophisticated techniques will overcome these kinds of problems and will lead to much improved crop varieties in the future.

A wide range of other transgenic crops are under development. One widely publicized crop is "golden rice," which incorporates β-carotene in the grain, which is therefore yellow, rather than the normal white color of rice. The human body processes β-carotene to Vitamin A, the lack of which impairs vision and increases susceptibility to maladies including respiratory diseases, measles, and diarrhea.

Because rice is the main diet staple in many Asian countries, the widespread distribution of golden rice could substantially improve health. As an example of the intricacies of transgenic crops, two of the genes used to breed golden rice were taken from daffodil and one from a bacterium! Some investigators contend that humans cannot consume enough of this rice to provide a significant amount of Vitamin A.

One of the first transgenic crops designed for human consumption was a variety of tomato that ripened slowly and could be left on the vine longer than conventional tomatoes, thus developing a better flavor than other varieties, which are normally picked while still green. Unfortunately the genetically engineered variety, which was given the brand name of FlavrSavr, did not have other desirable characteristics and failed. Work is continuing on delayed-ripening tomatoes and on improving the nutritional value of tomatoes, such as by raising the content of lycopene, which is involved with the production of Vitamin A.

Work continues on improved transgenic oilseed crops especially canola, which produces canola oil. Efforts are under way to modify the distribution of oils in canola to improve the nutritional value of the oil. Another possibility is increased Vitamin E content in transgenic canola. Sunflower, another source of vegetable oils, is the subject of efforts to produce improved transgenic varieties. Herbicide tolerance and resistance to white mold are among the properties that are being developed in transgenic sunflowers.

Decaffeinated coffee and tea have become important beverages. Unfortunately, the processes that remove caffeine from coffee beans and tea leaves also remove flavor, and some such processes use organic solvents that may leave undesirable residues. The genes that produce caffeine in coffee and tea leaves have now been identified, and it is possible that they may be removed or turned off in the plants to produce coffee beans and tea leaves that would give full-flavored products without the caffeine. Additional efforts are under way to genetically engineer coffee trees in which all the beans ripen at once, thereby eliminating the multiple harvests that are now required because of the beans ripening at different times.

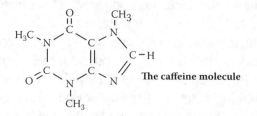

The caffeine molecule

Although turf grass for lawns would not be regarded as an essential crop, enormous resources in the form of water and fertilizers are consumed in maintaining lawns and grass on golf courses and other locations. Healthy grass certainly contributes to the "green" esthetics of a community. Furthermore, herbicides, insecticides, and fungicides applied to turf grass leave residues that can be environmentally harmful. So the development of improved transgenic varieties of grass and other groundcover crops can be quite useful. There are many desirable properties that can benefit grass. Included are tolerances for adverse conditions of water and temperature, especially resistance to heat and drought. Disease and insect resistance are desirable. Reduced growth rates can mean less mowing, saving energy. For grass used on waterways constructed to drain excess rain runoff from terraced areas (see Figure 11.10) a tough, erosion-resistant sod composed of masses of grass roots is very desirable. Research is under way to breed transgenic varieties of grass with some of these properties. Also, grass is being genetically engineered for immunity to the effects of Roundup herbicide, which is environmentally more benign than some of the herbicides, such as 2,4-D currently used on grass.

An interesting possibility for transgenic foods is to produce foods that contain vaccines against disease. This is possible because genes produce proteins that resemble the proteins in infectious agents, causing the body to produce antibodies to such agents. Diseases for which such vaccines may be possible include cholera, hepatitis B, and various kinds of diarrhea. The leading candidate as a carrier for such vaccines is the banana. This is because children generally like this fruit and bananas are readily grown in some of the tropical regions where the need for vaccines is the greatest.

SUPPLEMENTARY REFERENCES

Alef, Kassem and Paolo Nannipieri, *Methods in Applied Soil Microbiology and Biochemistry,* Academic Press, Orlando, FL, 1995.

Arntzen, Charles J. and Ellen M. Ritter, Eds., *Encyclopedia of Agricultural Science,* Academic Press, Orlando, FL, 1994.

Board on Agriculture, *Ecologically Based Pest Management: New Solutions for a New Century,* National Academy Press, Washington, D.C., 1996.

Brown, Lester R., *Tough Choices: Facing the Challenge of Food Scarcity,* Worldwatch Institute, Washington, D.C., 1996.

Brussaard, Lijbert and Ronald Ferrara-Cerrato, *Soil Ecology in Sustainable Agriculture Systems,* CRC Press/Lewis Publishers, Boca Raton, FL, 1997.

Ellis, Boyd G. and Henry D. Foth, *Soil Fertility,* 2nd ed., CRC Press/Lewis Publishers, Boca Raton, FL, 1997.

Filson, Glen C., Ed., *Intensive Agriculture and Sustainability: A Farming Systems Analysis,* UBC Press, Vancouver, 2004.

Franz, John E., Michael K. Mao, and James A. Sikorski, *Glyphosate: A Unique Global Herbicide,* American Chemical Society, Washington, D.C., 1997.

Hedin, Paul A., Ed., *Phytochemicals for Pest Control,* American Chemical Society, Washington, D.C., 1997.

Lal, Rattan, W.H. Blum, and C. Valentin, *Methods for Assessment of Soil Degradation,* CRC Press/Lewis Publishers, Boca Raton, FL, 1997.

McBaride, Murray B., *Environmental Chemistry of Soils,* OUP, New York, 1994.

Marschner, Horst, *Mineral Nutrition of Higher Plants,* 2nd ed., Academic Press, Orlando, FL, 1995.

Montgomery, John H., Ed., *Agrochemicals Desk Reference,* 2nd ed., CRC Press/Lewis Publishers, Boca Raton, FL, 1997.

Paul, Eldor A. and Francis E. Clark, *Soil Microbiology and Chemistry,* 2nd ed., Academic Press Textbooks, San Diego, CA, 1995.

Prakash, Anand and Jagadiswari Rao, *Botanical Pesticides in Agriculture,* CRC Press/Lewis Publishers, Boca Raton, FL, 1997.

Prasad, Rajendra and James F. Power, *Soil Fertility Management for Sustainable Agriculture,* CRC Press/Lewis Publishers, Boca Raton, FL, 1997.

Raman, Saroja, *Agricultural Sustainability: Principles, Processes, and Prospects,* Food Products Press, New York, 2006.

Rechcigl, Jack E. and Nancy A. Rechcigl, *Environmentally Safe Approaches to Crop Disease Control,* CRC Press/Lewis Publishers, Boca Raton, FL, 1997.

Sparks, Donald L., *Environmental Soil Chemistry,* Academic Press, Orlando, FL, 1995.

Tan, Kim H., *Environmental Soil Science,* Marcel Dekker, Inc., New York, 1994.

Wu, Felicia and William Butz, *The Future of Genetically Modified Crops: Lessons from the Green Revolution,* RAND, Santa Monica, CA, 2004.

QUESTIONS AND PROBLEMS

1. Justify the statement that "soil and soil systems are highly complex and variable."

2. Suggest a phenomenon by which heavy crop growth during the summer may have a severe drying effect on soil.

3. List the functions and explain the importance of the soil solution.

4. What is the most significant organic constituent of soil? How is it produced? What does it do in soil?

5. Which macronutrients are most likely to be lacking in soil? How may they be replenished?

6. Some kinds of plants can be "self-fertilizing" with nitrogen. Explain how this works and how a symbiotic relationship with another kind of organism is involved.

7. What is the purpose of treating phosphate minerals with sulfuric or phosphoric acids to make phosphate fertilizers?

8. Under the U.S. FIFRA Act what are some of the factors that must be considered and studied when licensing a new fertilizer?

9. What is conservation tillage? How are herbicides essential for the practice of this environmentally friendly technique?

10. What was the "green revolution"? How might advances in genetic engineering with recombinant DNA lead to a second, even greater "green revolution?"

11. Suggest how soil might act on pollutants to reduce their harmful effects.

12. How is soil divided physically? Which is the top one of these divisions?

13. What is humification, and what does it have to do with soil?

14. What is water in soil called? Give the name of the process by which this water enters the atmosphere by way of plants.

15. In what respects is conservation tillage consistent with the practice of green science and technology?

16. Explain why corn is especially amenable to the production of hybrids.

17. How do human activities affect the nitrogen cycle?

18. Name a gaseous, liquid, and solid form of fixed nitrogen used as fertilizer.

19. How are phosphate minerals treated to make the phosphorus more available to plants?

20. Name a pollutant that was once commonly transferred from the atmosphere through plants to soil. Why is this pollutant no longer such a problem?

21. Explain what is meant by desertification.

22. What is the good news in the U.S. regarding deforestation?

23. What is the potential use of perennial plants in grain production?

24. Give the meaning of transgenic.

25. What are the two main qualities currently developed in transgenic field crops? What are some other possibilities?

26. Explain the importance of *Bacillus thuringiensis* and glyphosate in transgenic crops.

27. Why is the potato not a very good candidate for Bt insecticide?

28. What is a potential environmental problem with Bt corn?

29. Name a concern with transgenic crops, such as squash, that are virus resistant.

30. How might transgenic crops be used to produce vaccines?

12. GEOSPHERIC HAZARDS AND SUSTAINING A GREEN GEOSPHERE

12.1. MANAGING THE GEOSPHERE

Humankind has an often uneasy relationship with the geosphere on which it dwells. Frequently, there is a need to modify the geosphere to construct dwellings and other structures, to build roads and railroads, to impound water, or for a number of other purposes. Humans are often unpleasantly surprised by the results of their efforts as destructive landslides form on sloping ground, dams collapse releasing destructive floods, and other unforeseen consequences result. Some natural geospheric processes are quite destructive of property and even human life. The two most dangerous of these are earthquakes and volcanoes. These are internal phenomena that result from changes deep underground. Surface processes, including landslides and ground subsidence, can be very destructive to property. Though usually not threatening to human life, surface processes can cause fatalities. Perhaps the most dangerous such processes are mudslides following extremely high rainfalls that sometimes bury whole villages entombing their residents. Cases have occurred in which vast amounts of earthen material have slid into reservoirs, causing them to overflow violently drowning many people downstream.

This chapter discusses two major related aspects of the geosphere. It considers destructive geospheric phenomena, especially volcanic eruptions, earthquakes, and landslides. Secondly, the chapter discusses preservation of the geosphere and modifications to it that can preserve and enhance geospheric quality, such as restoration of impaired land and conversion of contaminated areas back to safe and productive uses. In so doing, it considers the strong relationship of the geosphere to the other environmental spheres.

The Angry Earth

Although we usually regard Earth as *terra firma*, a safe surface on which we can rest securely and safely, that is not always the case. It is important to consider that just a few kilometers below the surface, rocks are hot and plastic and subject to movement. At somewhat greater depths, rock is so hot that it is liquid. So, the firm surface of Earth actually floats on a vast ocean of hot liquid rock. As a consequence, the plates of solid rock floating on this sea of molten rock can move, sometimes suddenly and violently, causing devastating destruction on the surface in the form of massive earthquakes. In some locations, the molten subsurface rock, known as lava, finds its way to the surface, spewing out as volcanic eruptions, which are, at best, spectacular displays of Earth's pent up power or, at worst, terribly destructive catastrophic events.

Earth's surface is in a constant process of seeking equilibrium with its surroundings. Igneous rock thrust to the surface from a hot, dry, oxygen-free environment encounters cool, moist, oxygen-rich surroundings. Chemically, the rock becomes modified to reach equilibrium with its new surroundings. Physically, the rock thrust to the surface, often to considerable heights, is broken into smaller particles by processes such as freeze/thaw and hydration/dehydration phenomena and is carried to lower levels. This usually occurs through gradual erosive processes, but can take place in mountainous regions as sudden destructive rockslides.

The processes described above can be extremely destructive. Hardly a year passes without several massive earthquakes on Earth that take thousands or even tens of thousands of lives, and past earthquakes have been recorded that killed close to a million people. Volcanic eruptions routinely displace thousands of people, sometimes killing a few, and causing millions of dollars in damage to property and agricultural areas. Compared to earthquakes, the death toll from volcanic eruptions is relatively low, in part because volcanoes give warning of upcoming eruptions, whereas earthquakes strike suddenly in full force. Landslides can be very destructive of both life and property. Such destructive events are usually associated with heavy rainfalls that saturate soil on sloping land, which may suddenly slide downslope, burying homes or even entire settlements in the process.

12.2. EARTHQUAKES

Earthquakes are internal phenomena that usually arise from plate tectonic processes and originate along plate boundaries when tectonic plates move relative to each other. Earthquakes are manifested on the surface as motion of ground resulting from the release of energy that accompanies an abrupt slippage of rock formations subjected to stress along a fault. Basically, two huge masses of rock tend to move relative to each other, but are locked together along a fault line. This causes deformation of the rock formations, which increases with increasing stress. Eventually, the friction between the two moving bodies is insufficient to keep them locked in place, and movement occurs along an existing fault, or a new fault is formed. Freed from constraints on their movement, the rocks undergo elastic rebound, causing the earth to shake. Serious earthquake damage may ensue.

The location of the initial movement along a fault that causes an earthquake to occur is called the **focus** of the earthquake. The surface location directly above the focus is the **epicenter**. Energy is transmitted from the focus by **seismic waves**. Seismic waves that travel through the interior of the Earth are called **body waves** and those that traverse the surface are **surface waves**. Body waves are further categorized as **P-waves**, compressional vibrations that result from the alternate compression and expansion of geospheric material, and **S-waves**, consisting of shear waves manifested by sideways oscillations of material. The motions of these waves are detected by a **seismograph**, often at great distances from the epicenter. The two types of waves move at different rates, with P-waves moving faster. From the arrival times of the two kinds of waves at different seismographic locations, it is possible to locate the epicenter of an earthquake.

The scale of earthquakes can be estimated by the degree of motion that they cause and by their destructiveness. The former is termed the **magnitude** of an earthquake and is commonly expressed by the **Richter scale**. The Richter scale is open-ended, and each unit increase in the scale reflects a tenfold increase in magnitude. Several hundred thousand earthquakes with magnitudes from two to three occur each year; they are detected by seismographs, but are not felt by humans. Minor earthquakes range from four to five on the Richter scale, and earthquakes cause damage at a magnitude greater than about five. Great earthquakes, which occur about once or twice a year, register over eight on the Richter scale.

The shaking and movement of ground are the most obvious means by which earthquakes cause damage. In addition to shaking it, earthquakes can cause the ground to rupture, subside, or rise. **Liquefaction** is an important phenomenon that occurs during earthquakes with ground that is poorly consolidated and in which the water table may be high. Liquefaction results from separation of soil particles accompanied by water infiltration such that the ground behaves like a fluid.

Distance from the epicenter, the nature of underlying strata, and the types of structures affected may all result in variations in intensity from the same earthquake. In general, structures built on bedrock will survive with much less damage than those constructed on poorly consolidated material. Displacement of ground along a fault can be substantial, for example, up to 6 m along the San Andreas fault during the 1906 San Francisco earthquake. Such shifts can break pipelines and destroy roadways. Highly destructive surface waves can shake vulnerable structures apart.

The loss of life and destruction of property by earthquakes makes them some of nature's more damaging natural phenomena. The destructive effects of an earthquake are because of the release of energy. The released energy moves from the quake's focus as seismic waves, discussed above. Literally millions of lives have been lost in past earthquakes, and damage from an earthquake in a developed urban area can easily run into billions of dollars. As examples, a massive earthquake in Egypt and Syria in 1201 A.D. took over 1 million lives, one in Tangshan, China, in 1976 killed about 650,000, and the 1989 Loma Prieta earthquake in California cost about 7 billion dollars. An earthquake registering 7.6 on the Richter scale devastated parts of northern Pakistan and a section of Kashmir on the morning of October 8,

2005. An estimated 80,000 people were killed and around 3.5 million people were left homeless.

Tsunamis from Earthquakes

Earthquakes may cause catastrophic secondary effects, especially large, destructive ocean waves called **tsunamis**. Sweeping onshore at speeds up to 1000 km/h, tsunamis have destroyed many homes and taken many lives, often long distances from the epicenter of the earthquake itself. This effect occurs when a tsunami approaches land and forms huge breakers, some as high as 10 to 15 m, or even higher. On April 1, 1946, an earthquake off the coast of Alaska generated a tsunami estimated to be more than 30 m high that killed 5 people on a nearby lighthouse. About 5 h later a tsunami generated by the same earthquake reached Hilo, Hawaii, and killed 159 people with a wave exceeding 15 m high. The March 27, 1964, Alaska earthquake generated a tsunami over 10 m high that hit a freighter docked at Valdez, tossing it around like matchwood. Miraculously, nobody on the freighter was killed, but 28 people on the dock died.

The most destructive tsunami in recent times occurred as the result of a great earthquake that struck at 6:58 a.m. local time, on Sunday, December 26, 2004, off the west coast of Northern Sumatra. This magnitude 9.0 quake occurring at a depth of 10 km was the largest since the 1964 Prince William Sound, Alaska earthquake and was the fourth largest earthquake in the world since 1900. The quake caused a massive tsunami that struck coastal areas throughout the Indian Ocean and was especially devastating to low-lying areas of Sri Lanka. In a much diminished form, the tsunami spread across the Pacific Ocean and was noted along the coasts of South and North America. The death toll from it reached approximately 200,000.

Mitigating Earthquake Effects

Unfortunately, earthquakes cannot be predicted or prevented. An earthquake can strike at any time — during the calm of late night hours or in the middle of busy rush hour traffic. Although the exact prediction of earthquakes has so far eluded investigators, the locations where earthquakes are most likely to occur are much more well known. These are located in lines corresponding to boundaries along which tectonic plates collide and move relative to each other, building up stresses that are suddenly released when earthquakes occur. Such interplate boundaries are locations of preexisting faults and breaks. Occasionally, however, an earthquake will occur within a plate, made more massive and destructive because, for it to happen, the thick lithosphere composing the plate has to be ruptured.

Accurate prediction would be a tremendous help in lessening the effects of earthquakes, but so far has been generally unsuccessful. Most challenging of all is the possibility of preventing major earthquakes. One unlikely possibility would be to detonate nuclear explosives deep underground along a fault line to release stress before it builds up to an excessive level. Fluid injection to facilitate slippage along a fault has also been considered.

Adaptation measures may be taken to lessen the effects of earthquakes. Significant progress has been made in designing structures that are earthquake-resistant. As evidence of that, during a 1964 earthquake in Niigata, Japan, some buildings tipped over on their sides because of liquefaction of the underlying soil, but remained structurally intact! The great death toll of the 2005 Pakistani quake was due largely to the stone dwellings that collapsed on the victims. Replacement of these structures with more flexible wooden buildings that would flex, but not collapse, during an earthquake could reduce future death tolls. Other areas of endeavor that can lessen the impact of earthquakes are the identification of areas susceptible to earthquakes, discouraging development in such areas, and educating the public about earthquake hazards.

12.3. VOLCANOES

In addition to earthquakes, the other major subsurface process that has the potential to massively affect the environment consists of emissions of molten rock (lava), gases, steam, ash, and particles due to the presence of molten rock magma near the Earth's surface. This phenomenon is called a **volcano** (Figure 12.1). Volcanoes can be very destructive and damaging to the environment.

On May 18, 1980, Mount St. Helens, a volcano in Washington State erupted, blowing out about 1 km^3 of material. This massive blast spread ash over half the U.S., causing about $1 billion in damages and killing an estimated 62 people, many of whom were never found. Many volcanic disasters have been recorded throughout history. Perhaps the best known of these is the 79 A.D. eruption of Mount Vesuvius, which buried the Roman city of Pompeii with volcanic ash.

Volcanoes take on a variety of forms, which are beyond the scope of this chapter to cover in detail. Basically, they are formed when magma rises to the surface. This frequently occurs in subduction zones created where one plate is pushed beneath another (see Figure 10.3). The downward movement of solid lithospheric material subjects it to high temperatures and pressures that cause the rock in it to melt and

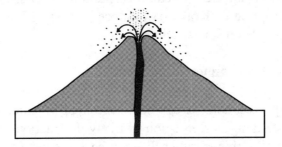

Figure 12.1.   Volcanoes come in many shapes and forms. A classically shaped volcano may be a cinder cone formed by ejection of rock and lava, called pyroclastics, from the volcano to produce a relatively uniform cone.

rise to the surface as magma. Molten magma issuing from a volcano at temperatures usually in excess of 500°C and often as high as 1,400°C, is called **lava**, and is one of the more common manifestations of volcanic activity.

Volcanoes can be very destructive in their immediate vicinities. Flows of lava as hot as 1400°C may destroy everything in their paths, causing buildings and forests to burn and burying them under rock that cools and becomes solid. Often more dangerous than a lava flow are the **pyroclastics** produced by volcanoes and consisting of fragments of rock and lava. Some of these particles are large and potentially very damaging, but they tend to fall quite close to the vent. Ash and dust may be carried for large distances and, in extreme cases, as was the case in ancient Pompei, may bury large areas to some depth with devastating effects.

A special kind of particularly dangerous pyroclastic consists of **nuée ardente**. This term, French for "glowing cloud," refers to a dense mixture of hot toxic gases and fine ash particles reaching temperatures of 1000°C that can flow down the slopes of a volcano at speeds of up to 100 km/h. In 1902, a nuée ardente was produced by the eruption of Mont Pelée on Martinique in the Caribbean. Of as many as 40,000 people in the town of St. Pierre, the only survivor was a terrified prisoner shielded from the intense heat by the dungeon in which he was imprisoned.

One of the more spectacular and potentially damaging volcanic phenomena is a **phreatic eruption** that occurs when infiltrating water is superheated by hot magma and causes a volcano to literally explode. This happened in 1883 when uninhabited Krakatoa in Indonesia blew up with an energy release of the order of 100 megatons of TNT. Dust was blown 80 km into the stratosphere.

Volcanoes may have effects on the environment thousands of kilometers from an eruption, and in severe cases may cause global climate effects lasting several years. Some of the most damaging health and environmental effects of volcanic eruptions are caused by gases and particulate matter. The explosion of the Tambora volcano in Indonesia, in 1815, blew out about 30 km³ of solid material, some of which reached the stratosphere. The ejection of so much solid into the atmosphere had such a devastating effect on global climate that the following year was known as "the year without a summer," causing widespread hardship and hunger because of global crop failures; a perceptible climatic cooling was noted for the next 10 years. As is the case with earthquakes, volcanic eruptions may cause the devastating tsunamis. The 1883 eruption of Krakatoa produced a tsunami 40 m high that killed 30,000 to 40,000 people on surrounding islands.

Mitigating Effects of Volcanoes

As with earthquakes, nothing can be done to stop volcanic eruptions. The best approach is to avoid living in the region of likely destruction around a volcano. In some parts of the world, this is rather difficult because some volcanoes are located near highly populated regions. Furthermore, volcanic eruptions may expose material that evolves into fertile soil that is attractive for agriculture. It is difficult to prohibit people from settling in areas where the last volcanic eruption may have been centuries past and the next one, though inevitable, may be centuries in the future. Dwellings in the vicinity of volcanoes should be fire-resistant, if at all possible, avoiding roofs made of flammable materials. Although challenging, it may be feasible, in some cases, to construct diversion dams to direct future lava flows away from settled areas.

12.4. SURFACE PROCESSES

Surface geological features are formed by upward movement of materials from the Earth's crust. With exposure to water, oxygen, freeze–thaw cycles, organisms, and other influences on the surface, surface features are subject to two processes that largely determine the landscape — weathering and erosion. As noted in Chapter 9, *weathering* consists of the physical and chemical breakdown of rock, and *erosion* is the removal and movement of weathered products by the action of wind, liquid water, and ice. Weathering and erosion work together in that one augments the other in breaking down rock and moving the products. Weathered products removed by erosion are eventually deposited as sediments and may undergo diagenesis and lithification to form sedimentary rocks.

Though natural phenomena, surface processes on the geosphere can be very damaging. Often the ill effects of destructive surface processes are aggravated by human mismanagement. Sediments washed into waterways by cultivation of soil can make bodies of water too shallow and be detrimental to aquatic life. Removal of vegetation from the surface can make it prone to destructive erosion. Improperly designed excavations, such as cuts constructed for roads, may have banks that are too steep and unstable, therefore prone to collapse. On the other hand, proper management of land surfaces can prevent harmful effects. As discussed in Chapter 11, this is particularly true of soil management in which construction of terraces and waterways and the practice of conservation tillage preserve soil from erosion and loss.

Mass Movements and Landslides

Mass movements are the result of gravity acting upon rock and soil on Earth's surface. This produces a shearing stress on earthen materials located on slopes that can exceed the shear strength of the material and produce often destructive phenomena involving the downward movement of geological materials. Such phenomena are affected by several factors, including the kinds and, therefore, strengths of materials, slope steepness, and degree of saturation with water. Usually, a specific event initiates mass movement. This can occur when excavation by humans steepens the slopes, by the action of torrential rains, or by earthquakes.

One of the most common mass movements that can adversely affect humans consists of **landslides** that occur when soil or other unconsolidated materials slide

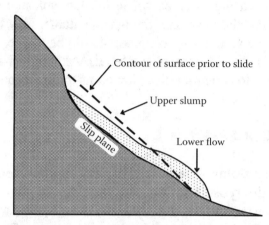

Figure 12.2.   A landslide occurs when earth moves along a slip plane. Typically, a landslide consists of an upper slump and lower flow. The latter serves to stabilize the slide, and when it is disturbed, such as by cutting through it to construct a road, the earth may slide farther.

down a slope. Related phenomena include rockfalls, mudflows, and snow avalanches. As shown in Figure 12.2, a landslide typically consists of an upper slump that is prevented from sliding farther by a mass of material accumulated in a lower flow. Figure 12.2 illustrates what commonly happens in a landslide when a mass of earth moves along a slip plane under the influence of gravity. The stability of earthen material on a slope depends upon a balance between the mass of slope material and the resisting force of the shear strength of the slope material. Landslides occur when material resting on a slope at an **angle of repose** is acted upon by gravity to produce a **shearing stress**. This stress may exceed the forces of friction or **shear strength**. The shear strength is, of course, a function of the geological material along the slip plane and may be affected by other factors as well, such as the presence of various levels of water and the degree and kinds of vegetation growing on the surface. Weathering, fracturing, water, and other factors may induce the formation of **slide planes** or **failure planes** such that a landslide results.

The tendency of landslides to form is influenced by a number of outside factors. Climate is important because it influences the accumulation of water that often precedes a landslide, as well as the presence of plants that can also alter soil stability. Although it would seem that plant roots should stabilize soil, the ability of some plants to add significant mass to the slope by accumulating water and to destabilize soil by aiding water infiltration may have an opposite effect. Disturbance of earth by road or other construction may cause landslides to occur. Earth may be shaken loose by earthquakes, causing landslides to happen.

Landslides can be very dangerous to human life and their costs in property damage can be enormous. In 1970, a devastating avalanche of soil, mud, and rocks initiated by an earthquake slid down Mt. Huascaran in Peru killing an estimated 20,000 people. Landslides (mudslides) following extreme rainfalls from hurricanes and killing dozens of people at a time occur somewhere on Earth almost every year. In addition to destroying structures located on the surface of sliding land or covering structures or people with earth, landslides can have catastrophic indirect effects. For example, those that dump huge quantities of earth into reservoirs can raise water levels almost instantaneously and cause devastating waves and floods. In 1963, a total of 2600 people were killed near the Vaiont Dam in Italy, when a sudden landslide filled the reservoir behind the dam with earthen material. Although the dam held, the displaced water spilled over its abutments as a wave 90 m high, wiping out structures and lives in its path.

Although often not properly considered by developers, the tendency toward landslides is predictable and can be used to determine areas in which homes and other structures should not be built. Slope stability maps based upon the degree of slope, the nature of underlying geological strata, climatic conditions, and other factors can be used to assess the risk of landslides. Evidence of a tendency for land to slide can be observed from effects on existing structures, such as walls that have lost their alignment, cracks in foundations, and poles that tilt. The likelihood of landslides can be minimized by moving material from the upper to the lower part of a slope, avoiding the loading of slopes, and avoiding measures that might change the degree and pathways of water infiltration into slope materials. In cases where the risk is not too severe, retaining walls may be constructed that reduce the effects of landslides.

Several measures can be used to warn of impending landslides. Simple visual observations of changes in the surface can be indicative of an impending landslide. More sophisticated measures include tilt meters and devices that sense vibrations accompanying the movement of earthen materials.

In addition to landslides, there are several other kinds of mass movements that have the potential to be damaging. **Rockfalls** occur when rocks fall down slopes so steep that at least part of the time the falling material is not in contact with the ground. The fallen material accumulates at the bottom of the fall as a pile of **talus**. A much less spectacular event is **creep**, in which movement is slow and gradual. The action of frost — frost heaving — is a common form of creep. Though usually not life-threatening, over a period of time creep may ruin foundations and cause misalignment of roads and railroads with significant property damage often the result.

Subsidence

Subsidence occurs when the surface level of earth sinks over a significant area. The most spectacular evidence of subsidence is manifested as large sinkholes that may form rather suddenly, sometimes swallowing trees, automobiles, and even whole buildings in the process. Sinkholes normally form in areas where large cavities have been produced in limestone formations by the action of dissolved carbon dioxide in water. Eventually the cavity may collapse allowing the overlying land to fall in. Sinkholes may also be formed by loss of underground water during drought or from heavy pumping, thus removing support that previously kept soil and rock from collapsing, heavy underground water flow, and other factors that remove solid material from underground strata. Overall, much more damage is caused by gradual and less extreme subsidence, which may damage structures as it occurs or result in inundation of areas near water level. Such subsidence is frequently caused by the removal of water or petroleum from below ground.

Expansive Soil

Some types of soils, particularly so-called expansive clays, expand and shrink markedly as they become saturated with water and dry out. Although essentially never life-threatening, the movement of structures and the damage caused to them by expansive clays can be very high. Aside from years when catastrophic floods and earthquakes occur, the monetary damage done by the action of expansive soil exceeds that of earthquakes, landslides, floods, and coastal erosion combined!

Permafrost

Special problems are presented by permanently frozen ground in arctic climates such as Alaska or Siberia. In such areas the ground may remain permanently frozen, thawing to only a shallow depth during the summer. This condition is called **permafrost**. Permafrost poses particular problems for construction, especially where the presence of a structure may result in thawing such that the structure rests in a pool of water-saturated muck on top of a slick surface of frozen water and soil. The construction and maintenance of highways, railroads, and pipelines, such as the Trans-Alaska pipeline in Alaska, can become quite difficult in the presence of permafrost. Permafrost is discussed further in Section 12.7.

12.5. THE VULNERABLE COASTS

Coastal areas are among the most vulnerable regions with respect to natural disasters. This vulnerability is highly increased by the tendencies of people to live near coasts and to place structures as close as possible to water. About 75% of the U.S. population lives in coastal states, and the total coastline (including the Great Lakes) is about 150,000 km. Furthermore, tropical storms, which are one of nature's most damaging phenomena, are at their most destructive along coasts and lose their destructive powers rather abruptly as they move inland. The infamous 2005 hurricane Katrina, followed by hurricane Rita virtually wiped out New Orleans and leveled structures along the coasts of Mississippi, Louisiana, and Texas that had stood for more than a century, emphasizing the vulnerability of these areas to natural disasters. Insurance costs to replace structures destroyed by coastal storms have reached prohibitive levels. Intelligent land use regulations are desperately needed for these regions.

Tropical Cyclones

There are several main areas of coastal vulnerability. In some areas, most notably in the Gulf Coast of the U.S., the greatest potential for damage is from **tropical cyclones**, commonly called **hurricanes** in the Atlantic and the Gulf of Mexico, and **typhoons** in the Pacific and Indian Oceans. Commonly developing between 15° south and 8° north of the equator, tropical cyclones gain their enormous energy from warm tropical seawater. Water vapor from evaporated seawater is contained in the atmospheric body of the cyclone. When this vapor reaches higher, cooler altitudes, it condenses to form torrential rain, releasing tremendous amounts of latent heat contained in the water vapor. This hot air rises creating air currents moving vertically and strong winds exceeding 100 km/h, often approaching 150 km/h, and getting as high as 300 km/h. The cyclone develops a region of very low atmospheric pressure, and the storm assumes a characteristic spiral pattern of clouds with a small, calm "eye" in the center. When a tropical storm passes over land, it loses contact with the warm seawater that is its main energy source and rapidly loses its energy and intensity.

The year 2005 set records for the number and severity of hurricanes in the Gulf of Mexico and the Carribean. This is believed to be because of higher-than-normal temperatures in these oceanic waters, and it is entirely possible that global warming was a contributing factor. The 10 to 20 years following the 2005 season should provide further information regarding the role of global warming in the formation of tropical cyclones, though at a high cost in storm damage and human lives!

The most devastating effects of tropical cyclones occur when they hit coastal areas at high tide accompanied by a **storm surge**. Storm surges develop as the consequence of low pressures in cyclones that raise water levels by several meters. Such surges combined with high tide may reach 10 m as they sweep ashore accompanied by high winds. In 1900, a "perfect surge" from a powerful Gulf hurricane swept across the coastal city of Galveston, Texas, killing at least 6,000. Even though ship-to-shore radio communication that might have warned the city of the approaching disaster was not available, weather experts on the island of Cuba, over which the storm had passed on its way to Galveston tried to warn the U.S. Weather Service. But, "What did they know?," and the Weather Service personnel in their hubris even shut down telegraphic communication with the Cuban authorities to not hear their warnings!

In addition to damage from storm surges and high winds, coastal areas hit by tropical cyclones are subject to flooding from the extreme rainfall from the storms. Rainfall amounts approaching a meter over 2 or 3 d have been reported, causing great problems from the resulting flooding.

Tsunamis

Tsunamis, discussed in Section 12.2, consist of ocean surges often many meters high generated by seismic activity. These potentially devastating events strike with much less warning than tropical cyclones, often under nonthreatening weather conditions (many coastal areas devastated by the great December 26, 2004, Indian Ocean tsunami were experiencing beautiful conditions with many of the victims enjoying a nice day on the beach). Although tsunami surges may travel at speeds of hundreds of kilometers per hour, barely noticeable on open ocean water, they can be detected with the appropriate monitoring equipment and warnings sounded in time to save many lives.

Coastal Erosion

The geosphere in coastal regions takes a severe beating because of normal water and wind processes aggravated by unwise human management of the shoreline. As shown in Figure 12.3, a coastline typically consists of a bank of earthen material, a sea cliff, of some height below which is a sandy beach. In many cases, such as a major portion of the Gulf Coast region ravished by Hurricane Katrina in 2005, the coastal land is not much higher than the sea that it borders, which makes it susceptible to flooding by tidal surges.

Shorelines are subject to destructive forces that erode the banks. Houses originally constructed at a comfortable distance from the edge may eventually end up perched precariously above the beach, often supported by poles extending downslope. Such structures frequently collapse following an extreme weather event or normal

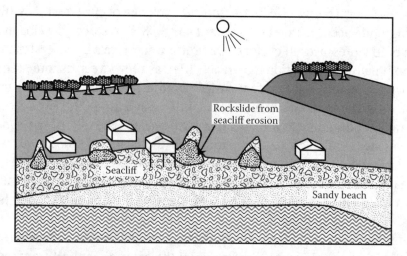

Figure 12.3.   Coastal regions are subject to damage from storm surges, normal erosive processes, and human influences. Much of the damage and loss stems from unwise construction practices along the coastline.

coastline erosion. Usually during the winter, heavy storm waves pound the base of a sea cliff washing sand away from it and exposing it to damage from the waves. In addition to wave action, the sea cliff may be eroded by a number of other factors including rainwash, weathering, the action of some organisms such as boring mollusks, and tree roots that penetrate the sea cliff rock and force it apart. Anthrospheric constructs can exacerbate sea cliff erosion. Paving of surface areas without proper diversion of rainwater runoff can cause erosive flows of water that erode the sea cliff.

The most valued part of a coastline is often the sandy beach. Sand is washed into the ocean as the result of weathering of silicon-rich quartz and feldspar rocks upstream and distributed along the shoreline by the action of ocean currents and waves. Human intervention has reduced the amount of sand available. Dams built on streams trap sand as sediment depriving the ocean of its source of sand. Improper beach management can result in loss of beach sand. The term **littoral cell** applies to a region of shoreline that includes inflow of sand from a stream or from sea cliff erosion, transport of sand generally parallel to the coast by ocean currents and wave action, and eventual loss of sand from the coastal region.

Preserving the Coastline

A number of measures may be taken to preserve the coastline. Foremost among these is to avoid doing those things that result in damage to the shoreline and destruction along it. Dwellings and other structures subject to damage by natural forces along the coast simply should not be placed where they are likely to be harmed. Damaging rain runoff can be diverted to drainage pipes extending to the base of a sea cliff. Water infiltrating from excessive watering of vegetation along the edge of a sea cliff can infiltrate and flow out of the base as seeps and springs that weaken the shoreline rock.

Areas that have lost beach sand are sometimes restored by expensive programs of **beach nourishment** to haul in more sand. This is usually a losing proposition in that the forces that caused the loss of beach sand originally will simply wash away the new sand. Walls composed of large rock or concrete constructed along a shoreline can be used in attempts to enhance beach areas. A common such structure consists of **groins** composed of dams perpendicular to the coastline that intercept the flow of water and suspended sediment in a littoral cell (see preceding subsection). These are generally constructed in groups constituting a groin field in a littoral cell. Sand accumulates in the upstream direction from each groin leading eventually to a segment of beach extending out to the end of the cell. However, erosion of the beach occurs downstream from the groin leading to an irregular beach.

Perhaps the most massive undertakings to prevent coastline damage are high **seawalls** that are designed to intercept ocean surges and prevent damage onshore. Following the deadly 1900 Galveston, Texas, hurricane, the city undertook construction of a concrete seawall 17 ft high, 16 ft thick at the base, and 7 mi long. In addition, sand was pumped into the city to raise the grade substantially. Since construction of the seawall, Galveston has avoided major damage from seawater surges. A disadvantage of seawalls is the loss of beach sand that they often cause.

12.6. ENGINEERING GEOLOGY

Engineering geology addresses the ways in which geological materials and structures are used and dealt with technologically in respect to structures, materials science, and other engineering aspects.[1] It therefore deals with the interface of the geosphere with the anthrosphere. Practically anything that is built on or near the surface of the Earth requires consideration of geological engineering aspects. For example, if a structure is to be built on a slope, it is important to consider the geological properties of the surface on which it is to be built. The degree of the slope and the nature of the material on which it is located can be used to predict the likelihood of earth slides and to design structures to avoid these natural disasters. Engineering geology is used to design rock quarries and to determine the most efficient means of removing rocks from the ground.

Large public works projects are the human endeavors most likely to affect the geosphere because they entail earth moving, digging, boring, and other operations performed on the geosphere. In turn, the nature, costs, and safety of structures constructed as part of huge public works projects are all highly dependent on the characteristics of the geological formations on which they are built. Therefore, public works projects require a high degree of sophisticated geological engineering throughout their planning and construction stages, and their maintenance and operation require consideration of geological engineering as well. Large public works projects include dams, roads, railroads, airports, pipelines, canals, tunnels, and large structures. Much of the engineering that goes into such projects involves evaluation of the geologic strata on which the structures are located and development of measures to prevent problems. For example, dams should be located on and anchored to strong formations of rock, preferably igneous or metamorphic rock. Fractures in the rock formations along which leaks may develop should be detected and filled with a wet mixture of cement and sediment called **grout**. Numerous factors must be considered in highway construction. To a greater extent than with most structures, highways must make use of surrounding geological materials, which must be evaluated for their suitability for fill and roadbed. Topography (surface configuration) and slope are crucial for grading and drainage. Geologic engineering is a crucial consideration in the siting and construction of large structures, such as buildings or nuclear or fossil-fueled power plants. The underlying strata must be carefully evaluated for its load-bearing capacity and to discover unexpected features, such as faults or fractures that might shift or caverns that might cause subsidence.

12.7. THE CRYSOSPHERE AND VANISHING PERMAFROST

Based upon the Greek word *kryos* for cold, the **cryosphere** refers to large areas of Earth consisting of snow, glaciers, sea ice, freshwater ice, frozen ground, and permafrost. Depending as it does on low temperatures, the cryosphere is very sensitive to global warming and shrinkage, and disappearance of elements of the cryosphere are indicative of global warming.

A particularly important part of the crysphere is **permafrost** consisting of frozen soil, rock, or sediment remaining at or below 0°C for a period of at least 2 years. From 12 to 18% of the Northern Hemisphere's exposed land surface is underlain by permafrost including vast areas of Siberia, Alaska, and northern Scandinavia. In many areas, permafrost is the ground on which structures stand. The huge Alaskan crude oil pipeline rests on permafrost. It is essential to consider the nature of permafrost in siting and maintaining buildings on it. The properties of permafrost and changes observed in these properties are key to understanding major shifts in climate, such as global warming.

Permafrost is a relatively unstable material that often exists at a temperature near its melting point and often is just about to melt. Permafrost can be frozen for thousands of years or is newly formed. Much permafrost is covered by an active layer that melts at least once each year. Melting permafrost can damage or destroy anthrospheric structures on it and alter natural topography. Melting of permafrost on sloping ground can cause landslides.

Some of the most convincing evidence of global warming during recent years has been the thawing of permafrost in a number of areas.[2] Typical of the effects of this thawing is the retreating shoreline of the Arctic village of Bykovsky on Russia's northeast coast where the shoreline is retreating by 5 to 6 m/year, threatening homes and heating oil tanks. On a nearby island, the thaw in the permafrost has exposed large numbers of mammoth bones and tusks, the latter of which sell for $25 to $50/lb as ivory, bringing some temporary prosperity to people willing to collect them. Thawing along the coast of Alaska has forced the costly relocation of several Inuit villages so far, and more will have to be moved if the warming trend continues.

A vast portion of Russia's area lies above the Arctic Circle. Because of rich resources in some of these areas, Soviet-era Russia constructed small cities, oil wells, pipelines, and other structures in these areas using techniques by which these structures were stabilized on the permafrost base. The thawing of the permafrost has created engineering nightmares that will be very difficult to solve. Such problems have been especially severe in the Russian Arctic city of Vorkuta, a coal-mining city of 130,000 constructed on permafrost. Apartment buildings in the city have suffered warped, cracked walls and ceilings and cracked window and door frames as the result of shifting because of melting permafrost.

The Inuit (Eskimo) peoples of Greenland and northern Canada and Alaska have lived in the frozen regions of the cryosphere for 5000 years making use of sea ice as a medium on which to travel, as construction material, and as hunting platforms. With warming of Arctic regions, this lifestyle is becoming much more difficult to maintain. Many areas of ice can no longer be depended upon as solid surfaces to support travel. Canadian Inuit hunters have reported that diminished ice cover has led to an inability of polar bears to put on fat on a heavy seal diet leading to emaciation of the bear population. Especially startling to some Inuit boaters have been attempts by walruses to climb aboard their white boats, which the huge animals have mistaken for ice floes in the increasingly ice-deficient Arctic waters!

Adding to the permafrost problems in Arctic regions are environmental threats from increasing petroleum and natural gas production. Movement of these and other commodities by ship in Arctic waters increases the probability of oil spills, which can be particularly damaging in cryospheric surroundings. Oil shipments from the White Sea and Barents Sea areas reached 20 million tons in 2005, a figure originally not projected until 10 years later, and are likely to reach 100 million tons within a decade or two.

Permafrost has a strong connection to the biosphere and it is affected by and, in turn, influences global warming. Much of the Northern Hemisphere's frozen ground supports boreal evergreen forests, which play an important role in removing carbon dioxide from the atmosphere. Carbon from dead trees and other plant biomass that falls onto these forest floors decays extremely slowly such that it is estimated that about one third of Earth's stored soil carbon is contained in Arctic regions. Warming of these regions will likely lead to increased carbon release as carbon dioxide that could exacerbate global warming or, in a more favorable case, increase photosynthetic fixation of atmospheric carbon.

12.8. CONSTRUCTION ON THE GEOSPHERE

The anthrosphere rests on the geosphere. Before placing a structure on or below the surface of the Earth, it is important to consider a number of factors regarding the suitability of the construction, a process called **site evaluation**. A number of factors must be considered in site evaluation. Site topography is important; for some applications, steep slopes may be detrimental. Present and former use of the site must be considered. For example, if the site was formerly employed in an application that might involve leakage and infiltration of wastes, extensive site remediation must be considered. The physical, chemical, and engineering properties of the earthen material must be known. In many cases, structures should rest on bedrock, so depth to bedrock should be measured. Natural hazards have to be evaluated. These include possibilities of earthquake (locating a critical structure directly on a seismic fault can be particularly troublesome) or volcanic activity. The potential for less spectacular, but much more common hazards should be evaluated including tendency to landslides and erosion. Potential interactions with the hydrosphere are crucial. These include surface runoff characteristics, surface water infiltration underground, flooding potential, depth of the water table, and groundwater flow.

A number of different kinds of structures are located on or below the surface of the geosphere, most requiring some degree of excavation. The most common of these are the following:

- Highways: Important considerations are topography considering limits to permissible gradients, excavation required such as in constructing cuts through hills, stability of slopes related to the likelihood of landslides onto the highway, strength of base rock, potential for flooding, and availability of construction materials including rock and sand to use in making paving.

- Railroads: In general, the same considerations given to highway construction apply to railroads. Gradient is much more important with railroads and must be much lower than for highways. This requirement leads to consideration of the need for cuts through hills and tunnels.

- Bridges: Both highways and railroads require bridges to cross rivers, estuaries, and deep valleys. In planning a bridge, it is essential to assess the earthen material on which it rests. In some cases, pilings must be driven to bedrock and in other cases grout, a fluid cement material, must be pumped underground to fill voids and provide stability.

- Airports: Important considerations are topography including the presence or potential for essentially level areas of sufficient size to construct long runways, soil characteristics including good load-bearing capacity, surface drainage and absence of flooding potential, and availability of rock and sand to use in construction.

- Tunnels: An important consideration is whether the tunnel is constructed through cohesionless earth requiring structures to prevent flow of earth into the tunnel or solid rock. It is common for both types of material to be encountered in the same tunnel. Rock structure and rock fractures must be considered. Also important is the potential for water infiltration and for means to remove water.

- Buildings: The larger the building, the more important are the considerations given to siting it. Often, extensive drilling and coring are performed to evaluate the suitability of the material on which the building will rest. In areas where underground voids may exist, it is crucial to make sure that none are in the vicinity of the building. In some areas, the potential to make structures earthquake-resistant is an important consideration. Soil that can undergo liquefaction (see Section 12.2) during earthquakes must be avoided if at all possible. Another consideration is the potential for soil to expand and contract, which may crack or even destroy basement walls.

The likelihood of groundwater infiltration must be considered; in some unfortunate cases, building excavations have intercepted springs or aquifers with a high flow potential.

- Dams: Both concrete and earth dams require careful assessment of the geological strata involved. A dam is almost always placed across a valley, which then fills upstream with water. The susceptibility of the reservoir slopes to landslides must be considered; such slides may be more likely as the bottom parts of the slope become constantly saturated with water from the reservoir. The rate of sediment accumulation in the reservoir should be evaluated. The strata on which the dam rests are very important and its strength, stability, and potential to develop routes for leakage of water must be evaluated. Especially in the case of earthen dams, the suitability of earthen material in the vicinity for construction is important.

- Mines: There is not much choice of the locations of mines because they must be where the desired minerals are. However, evaluation of the geosphere at the location of the mine can be used to determine its construction, a major issue being the choice between an underground mine and an open pit. The type of mine construction depends strongly upon the type and stability of the earthen material that must be excavated. Water infiltration from groundwater sources is a very important consideration and, for some underground mines, the potential for infiltration of explosive methane must be known. Also important is consideration of locations for placement of mine spoils (wastes).

12.9. DIGGING IN THE DIRT

One of the most common human interactions with the Earth is to excavate it to extract materials, construct structures such as roadways, construct basements for buildings, alter surface slope, or simply put holes in the ground for uses such as harbors. The ease with which earthen material is excavated and moved and the measures required to do so depend strongly upon the nature of the material excavated. Earthen material may vary in hardness from dirt through rock of varying degrees of hardness and consolidation to extremely hard rock. The easiest means of excavation is performed by simply scooping up unconsolidated material composed of relatively small particles. When this is not possible, the earthen material can be loosened by a process called *ripping*. A ripper consists of a strong, vertical blade mounted on the back of a heavy tracked tractor that extends a meter or more below ground and is pulled through the earth so that it breaks up the earthen material. Material that cannot be broken down by ripping may require blasting in which holes are drilled at intervals and packed with explosives that are detonated to break the hard rock into small pieces that can be removed with a scoop.

The properties of an excavation may deteriorate with time. One way in which this occurs is by slumping of walls, which is more likely and more severe with more steeply angled walls. The removal of earthen material reduces the mass resting on the floor of an excavation so that it may have a tendency to rise. Some of the greater challenges in maintaining an excavation occur when it extends to below the water table. Tapping a large, highly permeable aquifer can result in an excavation rapidly filling with water, which can be a major problem unless the objective was to have a water reservoir. In some cases, problems with water can be solved by surrounding the excavation with wells that are constantly pumped, which not only prevents water infiltration, but also stabilizes the floor and walls of the excavation.

Numerous means are employed to stabilize slopes to prevent slumping. Rock bolts and rock anchors can be installed in holes drilled into the slope to hold it in place. Anchors embedded deep within a slope can be used to fasten concrete panels to the sloping surface preventing material from sliding down. Often, a slope face loses its integrity as the result of the weathering action of water. This can be mitigated to a large extent by cement applied pneumatically to the surface that cures to provide a relatively impervious material that resists weathering.

Excavations below the Surface

The most challenging excavations are those that are made subsurface for tunnels, mines, subways, and underground caverns. Among the challenges presented by such excavations are the integrity and hardness of the earthen material, presence of faults, and groundwater infiltration. Assessing these conditions prior to construction can be done by indirect methods from the surface, such as seismic testing, or by direct methods including drilling, coring, and pilot tunnels.

Seismic testing using vibrational waves generated on the surface to assess subsurface conditions is the most commonly used indirect method of examining subsurface strata. Seismic testing involves generation of shock waves on the surface by an explosive charge or mechanically by dropping a heavy weight and monitoring the resulting seismic P-waves and S-waves (discussed under earthquakes in Section 12.2). The waves generated are monitored by geophones placed appropriately relative to the **shot point** where the waves are initiated. Times at which the waves reflected by the underlying strata are received are very accurately monitored along with their intensities by the geophones. Computer analysis of the data enables evaluation of the geometry and lithology (hardness, integrity, and strength) of the underlying strata. Faults, joints, synclines, and anticlines can also be mapped by seismic testing.

Figure 12.4 illustrates a subsurface excavation for a water tunnel used for water supply to a large urban area. In addition to a tunnel to conduct water, such an installation has shafts extending from the surface to the tunnel. Shafts are excavated from the surface; when they are excavated from below the surface upward, they are called *raises*. A large water system may also have caverns consisting of large voids hollowed out of the underground strata for water storage.

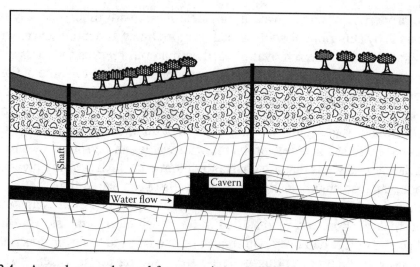

Figure 12.4.   An underground tunnel for conveying water. Underground caverns can be used for water storage. Access to the tunnel and caverns is through vertical shafts.

Green Underground Storage

Underground storage facilities offer a number of advantages and rank high in sustainability. Such facilities must be located in suitable strata free from major water infiltration. They will maintain a particular temperature within a relatively narrow range regardless of outside conditions. With suitable insulation, they can be maintained at almost any reasonable temperature, including low temperatures for cold storage.

The most commonly used underground storage facilities are located in limestone formations. In some cases, natural caves have been used, but underground limestone quarries are usually superior. One of the largest such installations was the Atchison Storage Facility located along the Missouri River about 2 mi southeast of Atchison, KS. This huge facility was mined to obtain limestone from a river bluff. It covers 127 acres and consists of a series of caverns with a limestone ceiling supported by 78 pillars of undisturbed limestone, each around 10 m in diameter. During World War II, the U.S. War Food Administration leased the Atchison Storage Facility and converted part of it to a refrigerated food storage facility maintained at 0°C for the storage of sides of beef, eggs, vegetables, fruits, butter, lard, and salt pork. An initial delivery of 12 railcars of dried eggs in 1944 expanded to 8900 tons of eggs along with 48 tons of skim milk, 1000 tons of raisins and 20,000 tons of prunes by 1949. Considering that the war ended in 1945 and that dried eggs and prunes were very low on the soldiers' dietary requests, it is likely that much of this food was eventually discarded.

During the 1950s, parts of the Atchison Storage Facility were converted to warehouse space for specialized military hardware. An ammonia-to-brine dehumidification system was installed to reduce the humidity to 42% relative humidity and to maintain the temperature of the facility to a range of 18°C to 22°C. Later, the facility was used for record storage.

Salt Dome Storage

Salt domes were formed when deposits of salts (usually sodium chloride) were left from the evaporation of water from vast inland seas. Eventually, the salt deposits became covered with sediments. Being less dense than the sedimentary material that covered them, the salt deposits rose over time forming domes 1 to 10 km in diameter and up to 6.5 km thick. The salt in salt domes is generally dry and impermeable.

Initial interest in salt domes was as sources of salt that could be extracted by forcing water through wells into the dome, pumping out the resulting brine, and evaporating the water solvent to obtain a solid salt residue. It was also found that petroleum deposits are often encountered above the domes, the dominant source of petroleum along the Gulf of Mexico. More recently, salt domes have come into use for petroleum storage, especially for the U.S. Strategic Petroleum Reserve (SPR). Located in salt domes in Texas and Louisiana, the SPR has a capacity of about 727 million barrels (116 million m^3) of petroleum. Figure 12.5 shows a basic salt dome facility with a cavern in the dome for petroleum storage. In such a facility, the crude oil floats on top of a layer of brine (salt solution). When the crude oil is to be removed, brine is pumped in from a surface reservoir, and the oil is forced out. The process is reversed to place petroleum back into the storage cavity.

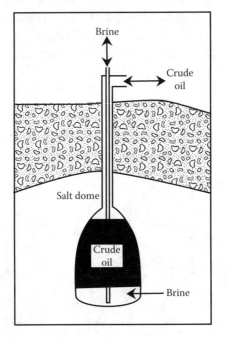

Figure 12.5.   Salt dome storage of petroleum such as those used for the U.S. National Strategic Petroleum Reserve. A cavity is formed in the dome by the action of water. Petroleum floats on top of a layer of brine and is pumped out of the storage cavern by pumping in more brine.

12.10. MODIFYING THE GEOSPHERE TO MANAGE WATER

Some of the most extensive modifications of the geosphere are those under-taken to manage water. The interaction of water with the geosphere is important and is addressed by the science of **hydrogeology**. Water is encountered both on top of the geosphere in streams, lakes, and reservoirs and beneath the geosphere surface as groundwater contained in aquifers. This section addresses water in the geosphere other than saline water in oceans and seas.

For the most part, water falls on the geosphere as precipitation (rain, snow, and sleet) from the atmosphere. Much of this water flows along the surface in streams from which it may be held in soil, evaporate into the atmosphere, or infiltrate into the ground. Water falling on the surface of the geosphere may infiltrate downward; if it reaches a zone of saturation, the process is called **percolation**. Percolation is an important process by which **groundwater recharge** of underground aquifers occurs. Groundwater recharge is essential to maintaining crucial underground water resources. Paving land in urban areas is detrimental to groundwater recharge, and in some cases artificial recharge is employed to pump water into aquifers. In some places, paving materials are now made of permeable solids that allow surface water to penetrate and eventually recharge groundwater reservoirs.

One of the most common human modifications of the geosphere to manage water is the construction of **reservoirs** in which flowing water is retained as a water supply, for recreational areas, to generate hydroelectric power, and for flood control. Figure 12.6 shows the basic components of a reservoir. It is seen that a reservoir is constructed when a dam is placed across a stream. The streambed and parts of the valley occupied by the stream compose the bed of the reservoir. The dam contains a sluiceway through which water from the reservoir can be drained to a very low volume, or the sluiceway may serve as a conduit to hydroelectric turbines attached

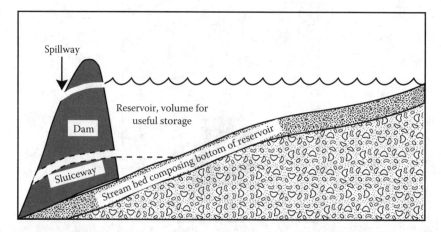

Figure 12.6.   A dam placed across a stream impounds water in the stream valley above the dam producing a reservoir. The maximum level of water in the reservoir is at the level of the spillway, and it can be lowered to the level of the sluiceway.

to generators for hydroelectric power. Somewhat below the top of the dam is a spillway through which excess water automatically overflows when the water level in the reservoir reaches that of the spillway. Water flowing over the top of a dam, particularly an earthen one, can destroy it rather quickly and is to be avoided. In addition to water held behind the dam, water in bank storage in rock formations along the edge of the reservoir may flow into it as the water level is lowered adding to the useful storage volume of the reservoir.

Many factors must be considered in constructing a dam to make a reservoir. These include stream flow, rainfall, geologic conditions, and topographic conditions. Ideally, dams are constructed in locations where the stream is constricted and where the banks are high and steep so that a small dam will impound a large reservoir volume. It is particularly important to have relatively watertight reservoir walls so that there is no excessive loss of water. One of the major sources of dam failure has been dam construction downstream from underground channels through which leakage can take place to below the dam. Eventually such leakage can erode the earth on which the dam rests, leading to its collapse. In some cases, channels and voids can be filled with fluid cement (grout) prior to dam construction.

An often-cited case of reservoir failure occurred with the Baldwin Hills Reservoir near Los Angeles, in 1963. This rectangular basin had been carved on top of a hill in 1951 and rested on poorly consolidated earthen material consisting primarily of silt, sand, and clay. Pumping of oil in the vicinity had led to ongoing settling of the land. Movement of faults beneath the reservoir dam led to an abrupt loss of water, erosion of material from beneath the dam, and a sudden breach of the dam. The sudden rush of water from the reservoir that resulted killed 5 people and caused 15 million dollars of damage.

12.11. DERELICT LANDS AND BROWNFIELDS: RECYCLING LAND

Derelict lands, commonly called **brownfields**, consist of properties that have been damaged by anthrospheric activities and are generally unsuitable for further use without restoration. Often such lands are contaminated with potentially hazardous substances and require cleanup and decontamination. Generally, brownfields are the result of abandoned industrial enterprises and mining. In the latter case, subsidence into voids excavated underground can be a major problem. Chemical contamination can result from industrial activity and mining.

Vast amounts of land are covered by brownfields that were formerly sites of enterprises such as factories, mills, quarries, petroleum refineries, trucking depots, and railyards. The Environmental Protection Agency (EPA) estimates that there are 450,000 brownfield sites in the U.S., with many more in other countries. It has been estimated that England has approximately 66,000 hectares of brownfield land. Some of the most challenging brownfield sites are located in nations of the former Soviet Union where the collapse of former state-owned manufacturing enterprises combined with a disregard of environmental considerations have given rise to many impaired areas with severe contamination problems. Some specific examples of brownfields include the following:

- Collinwood Rail Yard in Cleveland, a railroad maintenance facility and industrial site since 1873, abandoned in 1980. Problems with this site included oil and grease contamination of soil and groundwater, underground storage tanks, asbestos and lead contamination of abandoned buildings, and PCB contamination.

- An old trucking depot and lock manufacturing site in Anaheim, CA

- An oil tank site in Huntington Beach, CA, contaminated by petroleum products

- An abandoned textile mill in Wilmington, DE

- An abandoned grist and munitions mill in Gladwyne, PA

- A former shoe manufacturing plant with contaminated soil near Binghamton, NY

- An area of hundreds of square kilometers around the Russian nickel smelter of Monchegorsk South of Murmansk, so polluted by the nickel smelter emissions that it is virtually devoid of life

Brownfields present both problems and opportunities. Abandoned industrial sites are, at best, eyesores and, at worst, pose real problems with respect to pollution and dispersion of toxic substances to groundwater and surrounding areas. Abandoned brownfield areas and the structures in them often become haunts for squatters, illegal dump sites, and subjects of arson. However, they are usually centrally located near population centers with good access to rail lines, highways, and utilities, thereby providing excellent potential development opportunities. In recognition of these factors, in 1995, the U.S. EPA initiated a brownfield program to encourage cleanup and conversion to beneficial purposes of abandoned contaminated sites. This program provides funds for cleanup and limits liability, particularly in the event of discovery of additional hazardous contamination during cleanup. The program was strengthened by legislation in 2002 that provided additional funds for cleanup and further limited liability.

We recycle water, paper, metals, and a number of other materials; why not recycle land that may have been damaged or contaminated?[3] Restoration of brownfields may entail physical remediation, as well as treatment or removal of chemical contamination. Often, particularly where they are located at abandoned mine sites, brownfields are afflicted by subsidence. In some cases, concrete and stone salvaged from old structures and their foundations can be used as fill to treat subsidence.

Some kinds of chemical contamination can be treated in place. If excavation and disposal of contaminated soil is required, the costs of brownfield restoration may increase dramatically.

In recent years, a strong driving force behind brownfields development has been greatly increased real estate prices. Therefore, it has become attractive to construct dwellings and commercial developments on brownfield sites. Housing developments for private homes have been constructed on renovated brownfields, although such sites are usually more attractive for condominiums and apartment complexes. Many people are willing to trade the open spaces and greenery of suburbs for the convenience and much shorter commutes offered by more concentrated developments closer to urban centers. Ideally, such developments should conform to the standards of **smart growth** defined by the Urban Land Institute as development that is environmentally sensitive, economically viable, community-oriented, and sustainable. For residential developments, smart growth conforms to several criteria. Such a development should be in an identifiable area with a distinct center. A scale that encourages walking as major means of transportation is desirable with wide sidewalks, shade, and benches where pedestrians may gather and rest. Free trolley service may be considered as an additional amenity. It is desirable to have several kinds of structures including dwellings, commercial buildings, and public buildings. Provision should be made for centrally located civic buildings, including municipal offices, library, and post office.

Although much attention has been given to overpopulation and crowded cities, in fact, there is a problem with "shrinking cities" in some parts of the world. According to United Nations demographic studies, for every three cities that are growing in population, two are shrinking. Cities with a loss of at least a third of population in the last generation include St. Louis in the U.S., Phnom Penh in Cambodia and Johannesburg, South Africa. The population of Manchester, England, has fallen by almost half since 1931. Urban areas of the former East Germany lost large numbers of people with the fall of the Berlin Wall and the reunification of Germany. As central cities lose population, buildings and other facilities are abandoned. A major challenge is to find uses for abandoned facilities of shrinking cities, large areas of which have essentially become brownfields.

LITERATURE CITED

1. Bell, Fred G., *Engineering Geology and Construction*, Spon Press (Taylor & Francis), London, 2004.

2. Duff, Craig, Old Ways of Life are Fading as the Arctic Thaws, *New York Times*, October 20, 2005, p. 1.

3. Wall, Roland, Brownfields — The Fine Art of Recycling Land, http://www.acnatsci.org/education/kye/pp/kye5a.00.html.

SUPPLEMENTARY REFERENCES

Bardet, Jean-Pierre, Ed., *Landslide Tsunamis: Recent Findings and Research Directions*, Birkhäuser Verlag, Basel, Switzerland, 2003.

Bolt, Bruce A., *Earthquakes*, 5th ed., W. H. Freeman, New York, 2003.

Bryant, Edward, *Tsunami: The Underrated Hazard*, Cambridge University Press, New York, 2001.

Cornforth, Derek H., *Landslides in Practice: Investigation, Analysis, and Remedial/Preventive Options in Soils*, John Wiley & Sons, Hoboken, NJ, 2005.

Das, Braja M., *Principles of Geotechnical Engineering*, 5th ed., Brooks Cole/Thompson Learning, Pacific Grove, CA, 2002.

Duany, Andres, Elizabeth Plater-Zyberk, and Jeff Speck, *Smart Growth Manual*, McGraw-Hill Professional, New York, 2005.

Evans, Stephen G. and Jerome V. DeGraff, *Catastrophic Landslides: Effects, Occurrence, and Mechanisms*, Geological Society of America, Boulder, CO, 2002.

Francis, Peter and Clive Oppenheimer, *Volcanoes*, 2nd ed., Oxford University Press, New York, 2004.

Greenstein, Rosalind and Yesim Sungu-Eryilmaz, Eds., *Recycling the City: The Use and Reuse of Urban Land*, Lincoln Institute of Land Policy, Cambridge, MA, 2004.

Hsai-Yang Fang and John Daniels, *Introductory Geotechnical Engineering: An Environmental Perspective*, Spon Press, New York. 2005.

Keating, Barbara H., Christopher F. Waythomas, and Alastair Dawson, Eds., *Landslides and Tsunamis*, Birkhauser Verlag, Basel, Switzerland, 2000.

Kehew, Alan E., *Geology for Engineers and Environmental Scientists*, 3rd ed., Pearson Prentice Hall, Upper Saddle River, NJ, 2006.

Keller, Edward A. and Nicholas Pinter, *Active Tectonics: Earthquakes, Uplift, and Landscape*, 2nd ed., Prentice Hall, Upper Saddle River, NJ, 2002.

Kirkwood, Niall, Ed., *Manufactured Sites: Rethinking the Post-Industrial Landscape*, Spon Press, New York, 2001.

Lekkas, E. L., Ed., *Earthquake Geodynamics: Seismic Case Studies*, WIT, Southampton, U.K., 2004.

Leslie, Jacques, *Dams, Development, Disaster: The Epic Struggle Over Dams, Displaced People, and the Environment*, Farrar, Straus and Giroux, New York, 2005.

Martí, Joan and Gerald Ernst, Eds., *Volcanoes and the Environment*, Cambridge University Press, New York, 2005.

Porter, Douglas R., Robert T. Dunphy, and David Salvesen, *Making Smart Growth Work*, Urban Land Institute, Washington, D.C., 2002.

Prager, Ellen, J., *Furious Earth: The Science and Nature of Earthquakes, Volcanoes, and Tsunamis*, McGraw-Hill, New York, 2000.

Robinson, Andrew, *Earthshock: Hurricanes, Volcanoes, Earthquakes, Tornadoes, and other Forces of Nature,* Thames and Hudson, New York, 2002.

Stein, Seth and Michael Wysession, *An Introduction to Seismology, Earthquakes, and Earth Structure*, Blackwell Publishing, Malden, MA, 2003.

Szold, Terry S. and Armando Carbonell, Eds., *Smart Growth: Form and Consequences*, Lincoln Institute of Land Policy, Cambridge, MA, 2002.

Waltham, Tony, Fred Bell, and Martin Culshaw, *Sinkholes and Subsidence: Karst and Cavernous Rocks in Engineering and Construction*, Springer/Praxis, Berlin, 2005.

Waltham, Tony, *Foundations of Engineering Geology*, 2nd ed., Spon Press, New York, 2001.

World Commission on Dams, *Dams and Development: A New Framework for Decision-Making*, Earthscan, London, 2000.

Zeilinga de Boer, Jelle and Donald Theodore Sanders, *Volcanoes in Human History: The Far-Reaching Effects of Major Eruptions,* Princeton University Press, Princeton, NJ, 2002.

QUESTIONS AND PROBLEMS

1. With increasing depth, Earth's solid crust gives way to liquid rock as the subsurface temperature increases. However, Earth's core at even greater depths is solid, despite even more extreme temperatures. Suggest why this is so.

2. Suggest how techniques for monitoring earthquakes may have led to techniques for mapping underground strata useful in areas such as petroleum exploration.

3. Some earthquakes below ocean floors give rise to destructive tsunamis whereas other, equally intense, earthquakes do not. Suggest the reason for this. (It may be helpful to do some research on the Internet in answering this question.)

4. Why is the 79 A.D. eruption of Mount Vesuvius in ancient Rome in a sense the most notable of all volcanic eruptions? A little Internet research on Mount Vesuvius may be appropriate in answering this question.

5. Why do volcanoes have the potential to affect the atmosphere whereas several other destructive geospheric phenomena such as earthquakes do not?

6. What is the difference between weathering and erosion? How are they related? What do they have to do with surface processes on Earth?

7. The year 2004 was particularly bad for mudslides in Central America. Using the Internet or other resources suggest why this was so and estimate the death toll in 2004 because of Central American mudslides.

8. Based on material studied in preceding chapters, suggest a chemical reaction involving groundwater and limestone that can ultimately lead to land subsidence.

9. How does a tsunami differ from a storm surge, although the results are similar? Which gives more warning?

10. How can the construction of large expanses of paved areas near coasts contribute to coastal erosion?

11. Although cement used in construction of dams is often made with so little water that it has to be spread mechanically, then rolled with a roller, cement used in grout is a pumpable liquid. Explain why this is so.

12. Explain why the cryosphere is a very good indicator of, and susceptible to, global warming.

13. See if you can find information regarding conditions at the North Pole during the last summer or two. What does this have to do with the cryosphere and how does it relate to global climate change?

14. Landslides often occur along the sides of reservoirs some time after the reservoir is constructed. Why is this so? Why may faults near the vicinity of a dam be particularly troublesome?

15. What is the role of seismic testing in subsurface excavations?

16. What are some of the advantages of underground storage caverns? Why are they commonly located in limestone formations?

17. What is a salt dome? How are they used for storage facilities? How are storage facilities constructed in salt domes?

18. What determines the useful storage volume of a reservoir constructed by placing a dam across a stream?

19. Look up the U.S. Environmental Protection Agency's brownfields program on the Internet and compile a list of brownfield remediation projects ongoing or completed.

20. List some of the factors involved with the establishment of brownfields that favor their renovation to profitable uses.

13. THE BIOSPHERE: ECOSYSTEMS AND BIOLOGICAL COMMUNITIES

13.1. LIFE AND THE BIOSPHERE

The water-rich boundary region at the interface of Earth's surface with the atmosphere, a paper-thin skin compared to the dimensions of Earth or its atmosphere, is the **biosphere** where life exists. The biosphere includes soil on which plants grow, a small bit of the atmosphere into which trees extend and in which birds fly, the oceans, and various other bodies of water. Although the numbers and kinds of organisms decrease very rapidly with distance above Earth's surface, the atmosphere as a whole, extending many kilometers upward, is essential for life as a source of oxygen, medium for water transport, blanket to retain heat by absorbing outgoing infrared radiation, and protective filter for high-energy ultraviolet radiation. Indeed, were it not for the ultraviolet-absorbing layer of ozone in the stratosphere, life on Earth could not exist in its present form.

This chapter deals with life on Earth. It considers the highly varied locations where life exists and the vastly different conditions of moisture, temperature, sunlight, nutrients, and other factors to which various life-forms adapt. Such conditions may be those of the tropics, with abundant moisture, intense sunlight, high temperatures, and relatively little variations in these and other factors. Or they may be characteristics of inland deserts that are hot during the daytime and cold at night, generally very dry, but subject to occasional torrential rainstorms and flash flooding. Life thrives on land surfaces, in bodies of water, and in sediments in water. The extreme variability of environments in which life exists is matched by the remarkable variety, versatility, and adaptability of the communities of organisms that populate these environments. These range from tropical rain forest communities containing thousands of plant, animal, and microbial species in a small area to austere, exposed mountain rocks subjected to extremes of weather and populated by a thin coating of tenacious lichen, a symbiotic combination of fungi and algae that clings as a thin layer to the rock surface. In addition to dealing with organisms and their environment, this chapter also discusses the intricate relationships among organisms that enable them to coexist with each other and their surroundings.

Understanding life requires defining what it really is. Living organisms are constituted of cells that are bound by a membrane, contain nucleic acid genetic material (DNA), and possess specialized structures that enable the cell to perform its functions. A living organism may consist of only one cell or of billions of cells of many specialized types. All living organisms have two characteristics: (1) they process matter and energy through metabolic processes, and (2) they reproduce. The ability of an organism to process matter and energy is called *metabolism*. Another important characteristic of living organisms is their ability to maintain an internal environment that is favorable to metabolic processes and that may be quite different from the external environment. Warm-blooded animals, for example, maintain internal temperatures that may be much warmer or even cooler than their surroundings. Finally, through succeeding generations, living organisms can undergo fundamental changes in their genetic composition that enable them to adapt better to their environment.

Living species are present in the biosphere because they have evolved with the capability to survive and to reproduce. Every single species in the biosphere has become an expert in these two things; otherwise it would not be here. The key factors for existing — at least long enough to reproduce — are the ability to process energy and to process matter. In so doing, life systems and processes are governed by the principles of thermodynamics and the law of conservation of matter. Organisms handle energy and matter in various ways. Plants, for example, process solar energy by photosynthesis and utilize atmospheric carbon dioxide and other simple inorganic nutrients to make their biomass. Herbivores are animals that eat the matter produced by plants, deriving energy and matter for their own bodies from it. Carnivores in turn feed upon the herbivores.

Life-forms require several things to exist. The appropriate chemical elements must be present and available. Energy for photosynthesis is required in the form of adequate sunlight. Temperatures must stay within a suitable range and preferably should not be subject to large, sudden fluctuations. Liquid water must be available. And, as noted above, a sheltering atmosphere is required. The atmosphere should be relatively free of toxic substances. This is an area in which human influence can be quite damaging, through release of air pollutants that are directly toxic or which react to form toxic products, such as life-damaging ozone produced through the photochemical smog-forming process.

Individual organisms and groups of organisms must maintain a high degree of stability (**homeostasis**, meaning "same status") through a dynamic balance involving inputs of energy and matter and interaction with other organisms and with their surroundings. This requires a high degree of organization and the ability to make continuous compensating adjustments in response to external conditions. For an individual organism, homeostasis means maintaining temperature, levels of water, inputs of nutrients, and other crucial factors at suitable levels. Arguably, the most advantageous evolutionary trait of mammals is their ability to keep their body temperatures within the very narrow limits that are optimum for their biochemical processes. The concept of homeostasis applied to whole ecosystems consisting of groups of organisms and their surroundings is termed **ecosystem stability**. Indeed, homeostasis applies to the entire biosphere. To a large extent, environmental science addresses the homeostasis of the biosphere, and how the critical factors involved in maintaining the dynamic equilibrium of the biosphere are affected by human activities.

The nature of life is determined by the surroundings in which the life-forms must exist. Much of the environment in which organisms live is described by physical factors, including whether or not the surroundings are primarily aquatic or terrestrial. For a terrestrial environment, important physical factors are the nature of accessible soil, availability of water, and availability of nutrients. These are **abiotic factors**. There are also important **biotic factors** relating to the life-forms present, their wastes and decomposition products, their availability as food sources, and their tendencies to be predatory or parasitic.

This chapter discusses life on Earth. To understand the nature of life on Earth, it is important to consider what kinds of life are present, how various species fit into specific habitats, how energy and matter are utilized and cycled, and how various species interact with each other and with their environment. These factors are covered by the science of **ecology**, which addresses how organisms interact with their environment and with each other.

13.2. ORGANISMS AND GREEN SCIENCE AND TECHNOLOGY

There is an extremely strong connection between organisms and green science and technology. The most fundamental reason to practice green science and technology is to maintain an environment on Earth that is hospitable to life. Humans, of course, have a vital self-interest in this endeavor. In general, an environment that is conducive to life and a high diversity of life-forms is, by nature, sustainable. Green science and technology attempt to avoid extremes, such as those of temperature. Conditions that are conducive to life are automatically "green" conditions. Some of the main aspects in which green science and technology are related to organisms and life are the following:

- Conditions under which life thrives are generally, by nature, mild and consistent with green science and technology.

- One of the main characteristics of green science and technology is the absence of toxic substances that harm or kill organisms. Unhealthy or dying organisms are indicative of unsustainable conditions.

- Living organisms and their ecosystems provide models for sustainable anthrospheric systems. Sustainable systems of industrial ecology (Chapter 17) can be largely modeled on natural ecosystems.

- Green science and technology conserve matter and energy to a maximum extent and are characterized by a high degree of recycling. Organisms and their ecosystems have evolved to a maximum degree of efficient energy and matter utilization and provide excellent models for anthrospheric systems.

- Through photosynthesis, plants are outstanding sources of renewable materials in the form of cellulose, lignocellulose, wood, and other materials. Therefore, plants are important and growing sources of materials and fuels for anthrospheric systems.

- Organisms have sophisticated enzyme systems that can perform chemical syntheses and transitions that are either impossible in anthrospheric systems or possible only under extreme conditions.

13.3. ECOLOGY AND LIFE SYSTEMS

To consider the biosphere and its ecology in their entirety, it is necessary to look at several levels in which life exists. The unimaginably huge numbers of individual organisms in the biosphere belong to **species** (kinds of organisms). Groups of organisms of the same species living together and occupying a specified area over a particular period of time constitute a **population**; and that part of Earth on which they dwell is their **habitat**. In turn, various populations coexist in **biological communities**. Members of a biological community interact with each other and with their atmospheric, aquatic, and terrestrial environments to constitute an **ecosystem**. An ecosystem describes the complex manner in which energy and matter are taken in, cycled, and utilized; the foundation on which an ecosystem rests is the production of organic matter by photosynthesis. Assemblies of organisms living in generally similar surroundings over a large geographic area constitute a **biome**. Each biome may contain many ecosystems. The following are examples of important kinds of biomes:

- **Tropical rain forests** characterized by warm temperatures throughout the year and having most of their nutrients contained in the organisms populating the forest

- **Warm-climate evergreen forests** found in the southeastern U.S.

- **Coniferous forests** in temperate climates that have distinct summer and winter seasons and that are populated by cone-bearing trees with needles, such as cedar, hemlock, and pine

- **Temperate deciduous forests** growing in regions with hot, wet summers and cold winters populated by trees that grow new leaves and shed them annually

- **Grasslands** in which grass is anchored in a tough, dense mass of grass roots, and soil called **sod**

- **Hot deserts** populated by cacti, creosote bush, yucca, and other species adapted to high temperatures and sparse moisture

- **Cold deserts** in which tough perennial plants, such as sagebrush and some grasses, survive under cool, dry conditions

- **Tundra** found in arctic regions or at high altitudes. Tundra regions have no trees, cold winters with frost possible even in the summer, permanently frozen subsurface soil called **permafrost**, low productivity, relatively few species, and high vulnerability to environmental insult.

In order to sustain life, an ecosystem must provide energy and nutrients. Energy enters an ecosystem as sunlight. Part of the solar energy is captured by photosynthesis, and part is absorbed to keep organisms warm, which enables their metabolic processes to occur faster. In addition to capturing energy, an ecosystem must provide for recycling essential nutrients, including carbon, oxygen, phosphorus, sulfur, and trace-level metal nutrients, such as iron.

Much of the organization of ecosystems has to do with the acquisition of food by the organisms in it. Virtually all food on which organisms depend is produced by the fixation of carbon from carbon dioxide and energy from light in the form of energy-rich, carbon-rich biomass through the process of photosynthesis. Photosynthesis can be represented by

$$CO_2 + H_2O + h\nu \rightarrow \{CH_2O\} + O_2 \tag{13.3.1}$$

where $h\nu$ represents light energy absorbed in photosynthesis and $\{CH_2O\}$ represents biomass. Thus, the photosynthetic plants in the biosphere are the basic **producers** on which all other members of the community depend for food and for their existence. The rate of biomass production is called **productivity**. It is conventionally expressed as energy or quantity of biomass per unit area per unit time. The food manufactured by producers is utilized by other organisms generally classified as **consumers**.

The sequence of food utilization, starting with biomass synthesized by photosynthetic producers is called the **food chain**. Numerous food chains exist in ecosystems, and there is crossover and overlap between them. Therefore, food chains are interconnected to form intricate relationships called **food webs**. An example of a food chain would be one in which biomass is produced by unicellular algae in a lake and consumed by small aquatic organisms (copepods), which are eaten by small fish which are eaten, in turn, by large fish. Finally, a bald eagle atop the food chain may consume the large fish. As shown by this example, there are several levels of consumption in a food chain called **trophic levels**. In going up the chain, the first through fourth trophic levels are (1) producers, (2) primary consumers (herbivores),

(3) secondary consumers (carnivores), and (4) tertiary consumers, sometimes called "top carnivores." In addition to herbivores and carnivores, there are several other classifications of consumers. **Omnivores** eat both plant matter and flesh. **Parasites** draw their nourishment from a living host. **Scavengers**, such as beetles, flies, and vultures, feed on dead animals and plants. **Detritovores**, such as crabs, earthworms, and some kinds of beetles, feed on **detritus** composed of fragments of dead organisms and undigested wastes in feces. Ultimately, fungi and microorganisms complete the degradation of food matter to simple inorganic forms that can be recycled through the ecosystem, a process called **mineralization**.

Food webs can be divided into two main categories broadly based upon whether the food is harvested from living populations or from the remains of dead organisms. In a **grazing food web** food, along with the energy and nutrient minerals that it contains, is transferred from plants to herbivores and on to carnivores. Dead organisms become part of a **detritus food web** in which various levels of scavengers degrade the organic matter from the dead organisms.

The energy content of food passing through each trophic level may be consumed by respiratory processes to maintain the metabolic activity and movement of the organisms at that level, eventually to be dissipated as heat. Energy incorporated into the bodies of organisms or excreted from them as waste products can become part of the detritus food web. A relatively small, but very important fraction of the energy is passed on to the next trophic level when the organisms are consumed by predators.

It is instructive to consider the flow of energy utilization through various trophic levels. Typically, based upon 100% of the energy from primary producers, the percentages utilized by various trophic levels are decomposers 25%, herbivores 15%, primary carnivores 1.5%, and top carnivores less than 0.1%.

13.4. WHAT IS A BIOLOGICAL COMMUNITY?

A **biological community** consists of an assembly of organisms that occupy a defined space in the environment. The nature of such a community depends upon the physical and chemical characteristics that influence the life-forms in it and on the interactions of the organisms in the community. A biological community is the biological component of an ecosystem, which includes the organisms and their physical environment. The community functions in a manner such that it tends to utilize and convert energy and materials in the most efficient manner that will enable the organisms in it to reproduce and thrive. Therefore, the exchange of matter and energy among the organisms and with their physical environment is a key aspect of an ecosystem. The study of biological communities is called **community ecology**.

There are many interactions of organisms in a community. Many of these interactions are mutually advantageous. For example, grazing animals derive their food from grass growing on grasslands and return nutrients to the grass as manure, urine, and mineralized phosphorus, nitrogen, and potassium from their own bodies after death. Fierce competition exists between species in biological communities. Species compete for the same nutrients, food sources, space, and sunlight. Coexistence often depends upon species finding their own niches. For example, tall trees utilize the sunlight falling directly on a forest canopy for photosynthesis, whereas smaller plants can exist on the more meager light that filters through from above.

Biological communities are subject to constant change. Some of these changes are relatively short term and cyclical, following daily and seasonal patterns. Some desert plants, for example, may lie largely dormant during prolonged periods of dry weather, only to start growing fiercely and burst into bloom when a rare rainstorm occurs. Other changes occur as a habitat undergoes long-term transitions. These may occur, for example, after a forest fire abruptly kills all the tall trees in a community, and the plants and the animals in the community undergo several successions until tall trees are eventually reestablished. Many transitions are the result of human activities, such as those that take place when agricultural land is taken out of production for a number of years. In the past, major transitions in biological communities have occurred with changes in climate, such as those that happened with the retreat of the glaciers after the last Ice Age. Some evidence suggests that global greenhouse warming will cause long-term changes in many biological communities during the next several decades.

Stable biological communities are characterized by a high degree of order and often complex organization. The organisms in a community undergo constant opposing and compensating readjustment of their behavior, feeding, and reproduction in response to each other and to their surroundings. Therefore, such communities are in a state of **homeostasis**. An established, stable biological community as a whole is **homeostatic**.

13.5. PHYSICAL CHARACTERISTICS AND CONDITIONS

Biological communities largely develop in response to their physical environment, which includes the three most critical factors of temperature, moisture, and light, as well as other factors, such as quality of soil and nutrient supply. In the case of an aquatic environment, mixing and circulation, which bring nutrients up from sediment deposits are important. Another factor that must be considered is **variability** of the physical environment. Changes in temperature and moisture level can have particularly strong effects on biological communities, forcing them to change in response to the new conditions. Whereas organisms have adapted well to seasonal differences in their environment, sudden and drastic variations can put a lot of stress on a community. With these factors in mind, it is understandable that some of the most thriving, diverse, productive biological communities are found in the tropics where elevated temperatures favor high metabolic activity, intense light enables a high level of photosynthesis, abundant water favors plant growth, and relatively constant climatic conditions make it unnecessary for organisms to waste much of their effort adapting to seasonal change.

Consideration of the crucial factors in maintaining biological communities leads to the following important concepts:

- **Limiting factors** determine whether a species can exist in a biological community and, if so, its abundance, growth, and distribution.

- **Tolerance limits** describe the lower and higher values of limiting factors below and above which a species cannot live.

- The **critical factor** is the one that is closest to a tolerance limit.

Consider a species of plant that under normal circumstances might be the primary producer of biomass in a biological community. The most prominent limiting factors that determine its survival are light, temperature, and water. Typically, during the late summer months, light is abundant and the temperature remains within a range conducive to plant growth, but the supply of moisture becomes marginal because of drought conditions. Therefore, water is the critical factor and, in a very severe drought, the water supply may fall below the lower tolerance limit such that plants of the particular species wither and die, thus affecting the whole food chain and supply for the biological community.

Topography can have an important influence on a biological community. The population of a community on a south-facing slope (in the Northern Hemisphere) is influenced by the generally warmer temperatures and greater intensity of sunlight compared to that of a north-facing slope. Marked contrasts in communities may exist on either side of a mountain range over which moisture-laden air flows, as illustrated in Figure 13.1. As warm moist air flows over the upwind slope, it is forced to rise causing it to cool and release precipitation. On the downwind slope, the cool air sinks, becomes warmer, and has a drying effect on the vegetation and terrain that it contacts. The absence of rain in this region is called a "rain shadow."

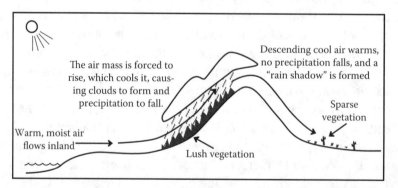

Figure 13.1.   Illustration of the effects of topography on biological communities. A mass of warm moist air forced to rise over a mountain range deposits rain and snow as it does so, enabling a productive community to thrive. On the other side of the range, a "rain shadow" creates drought conditions in which the productivity is low, and a much different community exists.

13.6. EFFECTS OF CLIMATE

Climate refers to weather conditions that exist over the long term, whereas **weather** describes short-term conditions of temperature, precipitation, humidity, wind, and cloud cover. Organisms and their biological communities are largely determined by climate. This section briefly addresses the effects of various aspects of climate on organisms.

The productivity of ecosystems is a function of climate. The most productive ecosystems are those in the tropics. Productivity is lower in the temperate zones and much less near the poles. Seasonal cold has a tremendous effect on plants and hence on productivity. Annual plants are killed by freezing, whereas deciduous trees lose their leaves and stop photosynthesis when it freezes. Thus, cold weather and freezing temperatures tremendously decrease primary productivity.

Specific species of plants and animals thrive only within certain temperature ranges. Animals are either **poikilothermic**, such that their temperatures remain close to those of their surroundings, or **homeothermic**, using intricate physiological processes to maintain their internal temperatures within a narrow range. There are several important ways in which animals maintain desired temperatures. One of the most interesting of these is through circulatory system "heat exchangers" in which blood flowing in veins from the extremities is warmed by outgoing arterial blood. Fur, feathers, layers of fat, and perspiration are all employed to maintain relatively constant temperatures in warm-blooded animals. Some animals hibernate and thus maintain reduced but adequate body temperatures at a low level of metabolism.

The temperatures of plants are largely those of the surroundings. (An interesting exception is the "warm-blooded" skunk cabbage plant that uses respiratory processes to metabolize energy-producing materials stored in its large root to keep its temperature as much as 30°C higher than the surrounding atmosphere on cold spring days.) However, for most plants, temperature extremes cause enzymes to become inactivated and proteins to become denatured. Plant membranes are injured by very high or very low temperatures. Freezing can be very damaging to plants, and the first "killing frost" of autumn signals the demise of many plants that have been thriving during the summer. Ice crystals that form when plants freeze destroy tissue so that it can no longer function. Plants do have some mechanisms to cope with freezing. If the temperature decrease is slow enough, plant tissues can lose water so that they become dehydrated, and ice crystals form on the plant surfaces. Resistance to freezing is also accomplished by the presence of sugars, sugar alcohols, and amino acids in plant fluids. These substances act as a kind of natural antifreeze. Plants also have mechanisms to resist high temperatures. One of these is for plant leaves to turn at an angle such that they do not receive the full force of midday sun.

All life-forms require water. Undesirable levels of moisture, usually moisture deficiency, constitute one of the most common environmental stresses faced by organisms. Drought conditions can be very harmful to biological communities. For a terrestrial community, a lack of water combined with high temperatures may be devastating to plants and animals. During winter or early spring, total levels of water may be adequate, but the water is frozen and not available.

Organisms have various ways to respond to moisture deficiency. Some plants get around the problem by being ephemeral in that their seeds germinate and the plants grow and bloom only when precipitation levels are adequate. Other plant mechanisms for dealing with moisture deficiency include shedding leaves and reducing water loss to the atmosphere by transpiration. This is accomplished by closing stomata and with structures such as thicker leaves or waxy surfaces that slow moisture loss. Succulent plants store large quantities of water in cells, whereas others have extremely deep root systems that reach underground water levels.

Animals have evolved that can tolerate some dehydration. Nocturnal animals avoid the extreme drying conditions of hot desert days. Water generated during metabolic oxidation of food substances is retained in the body. One interesting mechanism of coping with short water supplies is that of producing a very concentrated urine. Such urine, which usually has a very pungent odor, is characteristic of the cat family, which is largely adapted to dry conditions.

13.7. SPECIES

The kinds of species present in a biological community are largely determined by their environment and, in turn, influence their surroundings. The roles played by different species can be categorized in several different ways. Before considering populations of organisms, it is instructive to consider the categories and functions of the species that constitute such populations.

It should be noted that individuals in species are not identical, but exhibit slight differences resulting in **genetic diversity**. Living organisms undergo **mutations** arising from changes in their DNA and through recombinations of their genes that occur with reproduction. Mutations that give rise to favorable traits in individuals give them natural advantages so that they are relatively more likely to survive and reproduce. This **natural selection** allows species to better adapt to their environment and, over many generations, results in permanent changes in genetic composition. This is the process of **evolution**, through which new species are formed. Natural selection and evolution are very important in determining the occupants of biological communities. Organisms evolve that are best adapted to climatic conditions, particularly those of temperature and moisture, and to the presence or deficiency of nutrients. They also evolve in reponse to competition from other species, diseases caused by bacteria and viruses, parasites, and predation by other species.

A diverse biological community cannot exist in which all of the species present are competing for exactly the same limited resources. Therefore, in order to thrive, a species must be specialized and have its own **ecological niche**, which basically describes where the organism lives and how it functions in its environment.

The ecological niche of a species defines the habitat that it occupies, the food and nutrients that it utilizes, its predators, its interaction with the other biotic and abiotic components of the ecosystem, and the seasons and times that it does various things. To coexist with the limited resources available, the members of a biological community must **partition resources**. Organisms vary in the degree to which they require specialized niches.

Niche specialists occupy very well-defined niches, tending to require specific kinds of climatic conditions, habitats, and kinds of foods. Specialists are vulnerable to destruction or disturbance of their niches. Such a specialist is the spotted owl, which requires old-growth timber in the U.S. Northwest for its habitat. Tropical rain forests are particularly notable for the wide variety of niche specialists that occupy them. So many niches are possible because of the stratification of the forests and their long-term environmental and biological stability. That is one reason that damage to or destruction of tropical rain forests poses a particular threat of species extinction.

Niche generalists are organisms that are much more adaptable to different environments. Some niche generalists have adapted so well to habitats modified by humans that they have become nuisances. An example is the coyote, which coexists with humans in some suburban areas, occasionally killing an unwary pet cat. Another example is the Canadian goose, which has taken a strong liking to country club lakes and suburban lawns, and may even aggressively object to sharing these niches with the humans that they encounter. The human species itself is an excellent, resourceful niche generalist.

It is useful to define several other categories of species in discussing ecosystems. **Native species** are those that are naturally present in an ecosystem, whereas those that have been introduced, usually by human intervention, are called **immigrant species** or **alien species**. In some cases immigrant species have been beneficial whereas, in other cases, they have displaced native species with catastrophic results. Some species have such important functions in an ecosystem that they are called **keystone species**. Keystone species include some predators that keep numbers of otherwise destructive species in check, insects that pollinate flowers, and species that modify habitat to provide resources and dwelling space for other species. An example of the last of these is the American alligator, which digs depressions in swampland that provide water habitat for numerous species during droughts.

In considering pollution and damage to ecosystems, **indicator species** are particularly important. These are species whose numbers decline or exhibit symptoms of malaise as a reflection of habitat damage before other major symptoms are observed. The decline of hawks, eagles, and other predatory birds at the top of the food chain was indicative of widespread DDT pesticide pollution during the 1950s. The current loss of frogs and other amphibians worldwide is viewed by some experts as an indication of widespread environmental deterioration and pollution.

13.8. POPULATIONS

Recall that populations consist of groups of the same species of organisms living together and occupying a specified area over a particular period of time. Populations have numerous characteristics, including numbers, genetic composition, birth and death rates, and age and sex distribution. This section addresses some of the important factors involved with populations in biological communities.

All existing species are present on Earth because they have developed exceptional abilities to reproduce and to survive long enough to do so. The potential for reproduction always greatly exceeds the ultimate capacity of an environment to support a population. This capacity is known as the **carrying capacity**. Introduction of members of a population into an area that is amenable to their growth causes their numbers to increase very rapidly, often at a rate approaching exponential growth. Eventually, numbers reach a level around the carrying capacity at which point rapid growth ceases abruptly because of some limiting factor, such as limited food, nutrients, water, air, or shelter, or because of stress from crowding. After the very rapid growth phase ceases, the numbers of organisms tend to fluctuate somewhat around the carrying capacity as shown in Figure 13.2.

In some cases, a large growth in population can cause it to temporarily exceed the carrying capacity by a significant margin, or the carrying capacity can be suddenly reduced by circumstances such as droughts. This can result in a temporary overpopulation until reduced reproduction rates and increased death rates can adjust the population to accord with the carrying capacity. The result can be an abrupt decrease in population, or **population crash**. If the carrying capacity has been altered by overpopulation, such as destruction of grassland by overgrazing, or if it has been reduced by external factors, the new population will stabilize at a figure that reflects the new carrying capacity.

The physical and chemical conditions of a habitat largely determine the organisms that dwell in it. The biological history of a community is also an important

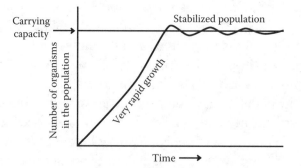

Figure 13.2.   Very rapid growth of a population newly introduced into an environment suitable for its survival, followed by stabilization of numbers around the carrying capacity. This is an idealized picture subject to perturbation by many factors, such as predation, disruption of habitat, or disease.

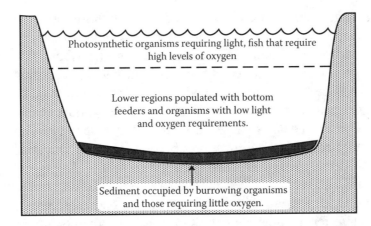

Figure 13.3.   Vertical stratification of a biological community is exhibited in bodies of water, which are often physically stratified because of the temperature/density behavior of water.

factor. Species may dominate simply because they have been there for quite some time, and chance may have played a role in their establishment and survival. The presence and numbers of a species can also be strongly influenced by the other species present. The degree of productivity and the kind of plant biomass produced can largely determine the kinds of organisms in a community. For example, large ruminant animals, such as American Bison, can thrive in areas that are highly productive of prairie grass, and squirrels may thrive in wooded areas that produce large quantities of nuts. Species may modify the environment in ways that assist other species. Damming of streams by beavers, which are keystone species in some habitats, can create pools that provide habitat for certain kinds of plants and other kinds of animals. Predation by other species is also important in determining which species exist in a biological community.

Species divide resources for most effective utilization according to time and space. Some animal species hunt and gather food during the day and others do so at night, thus avoiding needless competition. Some varieties of plants thrive and reproduce in the spring when moisture is abundant, whereas others are adapted to hotter, drier conditions and become dominant during the mid- and late-summer months. One of the major adaptations based upon physical space is that of **vertical stratification**, which is seen, for example, in forests where tall trees get light for photosynthesis from the top layers of the forests and ferns and mosses use the much less intense, diffused light at ground level. As shown in Figure 13.3, vertical stratification is particularly important in bodies of water that are stratified because of the temperature/density behavior of water.

13.9. SURVIVAL OF LIFE SYSTEMS, PRODUCTIVITY, DIVERSITY, AND RESILIENCE

Biological communities are subject to various stressful changes that affect their populations. These can be divided between the two main categories of catastrophic vs. gradual changes, each of which is divided between those caused by natural factors and those resulting from human activities. Examples of catastrophic natural changes include fire, wind damage, flooding, drought, landslide, or volcanic eruption. Catastrophic changes caused by humans include deforestation, cultivation, disruption of surface by strip mining, and some forms of severe pollution. Gradual natural changes include those because of adaptation, evolution, changes in climate, and disease. Gradual changes caused by humans include exploitation of natural, soil, and wildlife resources, salinization of soil because of irrigation, addition or elimination of species, and destruction of habitat.

Ecosystems and biological communities that are suffering from stress exhibit warning signs. One of the most common of these is decreased productivity; if the basic food source for a whole community diminishes, the community as a whole must suffer. Commonly associated with decreased productivity is a loss in nutrients. A well-balanced ecosystem recycles and retains nutrients such as nitrogen. Under stress, such nutrients may be permanently lost from the whole system, whereas levels of pollutants may increase as the ecosystem suffers damage. The loss of indicator species is indicative of problems in an ecosystem, as is a decline in the diversity of species present. Increases in the numbers of some species, particularly predatory insects and disease-causing organisms indicate ecosystem damage.

There are several key parameters that are used to describe the ability of biological communities to survive and thrive. These are the following:

- **Productivity** is the rate at which a biological community generates biomass, almost always by photosynthesis, required to sustain the organisms in it.

- **Diversity** describes how many different species are present and their relative abundances.

- **Inertia** describes the resistance of a community to being altered, damaged, or destroyed.

- **Constancy** is the ability to maintain numbers within the optimum limits that can be supported.

- **Resilience** is the ability to recover from perturbations, which may be of a catastrophic nature and kill large numbers of organisms in the community.

The productivity of a biological community, by the plant producers in it, is influenced by many factors. One factor is the nature of the geosphere to which a terrestrial community is anchored. A key geospheric component is soil; the quality of soil and its nutrient contents largely determine biomass yield. In some communities soil is virtually absent, and what little productivity there is comes from lichen growing on rock surfaces and hardy plants that anchor their roots between rocks.

Topography can influence productivity to an extent. Soil on steep slopes, for example, may be subject to erosion and hence less productive than soil on gently sloping terrain. In general, higher temperatures and higher precipitation favor productivity. Therefore, rain forests tend to be the most productive biological communities, and dry, cold deserts have very little productivity.

Overall, ecosystems in the world's oceans have lower productivity than do those on land. However, biological communities along ocean shores are usually very productive, and estuaries where freshwater and saltwater merge contain the most productive of all biological communities. Warm seawater in tropical regions provides habitat for the most productive marine-based biological communities. On the high seas, cooler temperatures, limited light penetration, and the lack of availability of nutrients tends to keep marine productivity low. Dead biomass and fecal pellets from marine organisms commonly sink to the ocean floor where they are not readily available for recycle of the nutrients, such as phosphorus, that they contain. Areas of particularly high marine productivity occur when nutrient-rich sediments are brought to the surface by convection currents, a phenomenon called **upwelling**.

Species diversity describes an important characteristic of a biological community based upon how many different species are present and their relative abundances. Relatively productive biological communities often show a high level of diversity. Thus, a tropical rain forest provides habitats for a large variety of plants, mammals, reptiles, and lizards; whereas an arctic biological community may have comparatively few. Related to species diversity is the observation that similar biological communities develop under similar conditions in distant geographical locations. **Ecological equivalents** is a term used to describe species that are not closely related to each other genetically, but have evolved similar characteristics and behavior patterns that enable them to occupy similar niches in widely different locations. An example is provided by the African jackal and the North American coyote, which, though entirely different species, fill largely identical kinds of ecological niches on their respective continents and exhibit some similarities in their traits and appearance.

High productivity and species diversity can help a biological community to be resilient and stable. A resilient community subjected to stress, such as from a drought or disease affecting one or more of its members, has the ability to deal with and compensate for change; whereas stress applied to a community with only a few members that is marginally productive may be catastrophic for the community as a whole. One that has a large number of species and a cushion of high productivity will be in a much better position to recover from damaging events and circumstances.

13.10. RELATIONSHIPS AMONG SPECIES

The interactions of species in a biological community have a strong influence on determining its nature. The crucial aspects of such interactions are the following:

- Biomass photosynthesized by plants almost always provides the basic food source on which the rest of the community depends. Most of this food is usually provided by a **dominant plant species**, such as algae growing in a pond.

- The physical nature of the environment is modified by the species in it, particularly the dominant plant species. For example, sagebrush on arid lands holds soil in place and provides shelter for small animals.

- **Competition** exists between different species for food, sunlight, and space.

- **Beneficial** and **antagonistic** relationships may exist between species.

- **Predation** occurs when an organism feeds upon another.

- **Parasitism** describes the activities of organisms that feed off other live **host organisms**, usually without causing death to the host. **Pathogens** are disease-causing organisms, usually microbial bacteria, protozoa, fungi, or viruses.

As shown in Figure 13.4, a dominant plant species anchors the community as its major producer of biomass. In addition to providing most of the food through photosynthesis that the rest of the community uses, the dominant plant species often acts to modify the physical environment of the community in ways that enable the other species to exist in it. For example, the trees in a forest community provide the physical habitat in which birds can nest, relatively safe from predators. In addition, the trees provide shade that significantly modifies the habitat at ground level and prevents the growth of most kinds of low-growing plants, though providing a hospitable habitat for low-growing plants that require relatively low levels of sunlight.

Competition exists between organisms for energy, matter, and space. Members of the same species may compete, as may members of different species. Adult plants may shade the ground such that seedlings of the same or other plant species cannot

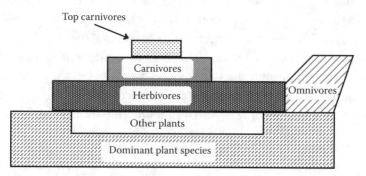

Figure 13.4.   A dominant plant species typically provides most of the food for a biological community. The biomass that it produces is consumed primarily by herbivores, which are fed on by carnivores, of which there may be more than one level. Omnivores feed on both plants and animals.

get started. Competition can be more overt, such as occurs when plants secrete substances through their leaves or roots that deter the growth of other plants. Animals exhibit **territoriality** in which they define areas from which they try to exclude competing animals, including those of their own species that are not one of a mating pair or part of a larger family unit. When two or more similar species exist in the same biological community on a steady basis, it is because they have evolved in ways that reduce competition for food and space; that is, they do not occupy exactly the same ecological niche. This phenomenon is called the **principle of competitive exclusion**.

Symbiotic relationships are those in which species live in close association with each other. Symbiosis is extremely common and is closely related to the other aspects of biological communities, such as the ways in which organisms utilize energy and nutrients. The classic example of symbiosis is that of lichen in which a species of algae exists within the matrix of filaments produced by a fungus. The fungus provides a hospitable physical environment that is anchored to rock, retaining moisture and extracting nutrients from the rock. The algae fix carbon as biomass that is utilized by the fungi. Other examples include **mycorrhizae**, in which fungal hyphae associate with plant roots, enabling the roots to take up adequate quantities of water and nutrients, and nitrogen-fixing bacteria that exist as nodules on the roots of leguminous plants.

The possible beneficial relationships among organisms can be divided into the three classifications as shown in Figure 13.5. **Mutualism** occurs when two species must have each other to exist; **commensalism** when one species requires the other, but the reverse is not true; and **protocooperation** when two species benefit from each other, but can exist independently.

In the broadest definition of the term, **predators** consume the biomass of other organisms. Meat-eating carnivores that pursue and kill their prey are obviously predators. However, so are herbivores that graze on grass, consuming part of the host plant for food, but not killing it. Humans are omnivores that will eat both meat and vegetable matter.

Predation in a biological community can be quite complex. The predators have highly specialized means for getting food, and their potential prey have evolved

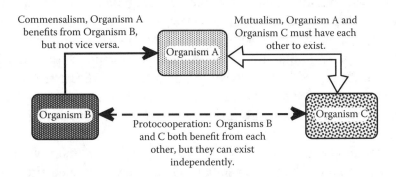

Figure 13.5.   Types of beneficial relationships among organisms.

ways to prevent this from happening. In some cases, predation is of benefit to the prey. Through coevolutionary processes predators and prey have evolved such that both may survive. Stages of the life cycle are very important for both predators and prey. Most very young predators are not sufficiently developed to feed on the prey of adults, who care for them until they can feed themselves. Tadpoles are herbivores, but turn into carnivorous frogs. Very young plants provide excellent food sources for small predators, such as cutworms, whereas the fully grown plants are prey for larger animals, such as ruminant animals.

Different levels of feeding exist in a biological community; these are called **trophic levels**. Trophic levels are illustrated clearly in some aquatic ecosystems. In such systems, the most abundant primary food source constituting the base of the grazing food web may consist of very small, free-floating photosynthesizing organisms called **phytoplankton**. These are fed upon by very small, free-floating, single-celled, and invertebrate animals called **zooplankton**. (The larvae and eggs of many somewhat larger aquatic animals, such as crustaceans, are planktonic.) Small aquatic creatures, such as small fish, feed on the free-floating plankton. Organisms called **filter feeders** efficiently strain water through special structures to remove plankton for food. Organisms that feed on plankton are in turn consumed by larger creatures. This phenomenon gives rise to several levels of carnivores. It can result in the concentration of some kinds of aquatic pollutants that accumulate in progressively higher levels up the food chain, a phenomenon called **biomagnification**.

13.11. CHANGING COMMUNITIES

The nature of biological communities changes over both space and time. The borderline between communities is often a region of unique character and activity. Communities begin, develop, and end, sometimes over very long periods of time, sometimes abruptly. Human intervention is one of the major ways in which biological communities are forced to change.

This section outlines the general way in which a community changes. It should be emphasized, however, that such changes are not as predictable as was once believed. It cannot be known for certain that the succession will proceed in a specific direction, or that it will end up with a readily predicted population of organisms. Random chance plays a part in such uncertainty. Furthermore, human effects, which may be indirect and the result of human activities remote from the community, add a note of uncertainty. However, human intervention intelligently applied can actually hasten the succession processes in biological communities and direct them in beneficial ways. Such measures include, as examples, replanting of forest trees after clear cutting or deliberate introduction of prairie grass species on worn-out agricultural land that is being restored to a native state.

The most fundamental way in which a biological community can change is when a new community becomes established on a newly created site where life of the sort that eventually populates the community has been essentially absent. The process of establishing life-forms in such an area is called **primary succession**. On land, primary succession occurs when a community develops in a location where soil is initially absent. This occurs, for example, with the establishment of a community on a new lava flow or newly exposed rock or gravel removed as overburden in strip mining. Primary succession is one phase of the more general process of **ecological succession**, a term used to describe the changes in species abundance as a result of changes in the environment of a biological community. The initial colonization of such a site is accomplished by a **pioneer species** that is hardy enough to exist under the harsh conditions often presented by a new site. Freshly exposed rock is often colonized by colonies of lichen, described above as symbiotic combinations of fungi and algae. The chemical action of lichen dissolves nutrients from the rock and initiates chemical weathering, which is the first step in soil formation. As soil is formed and crevices develop in the rocks, other organisms begin to populate the community, and eventually displace the pioneer species. Newly established aquatic ecosystems may eventually fill in with debris, and a succession of plants follows that can eventually change the system from an aquatic one to a swamp, then to a grassland, and eventually to a forest.

A mature biological community that lasts for very long time periods is called a **climax community**. (Although climax community is used here, some authorities prefer other terms, such as mature community or relatively stable community, which imply a less permanent state.) Consider the development of a climax community on land. Long before a mature community develops, productivity is low, but goes up gradually as the community acquires increased ability to utilize the resources in it. Soil forms, becomes deeper, and increases in organic content. In general, as the community develops, it becomes richer in organic matter, both living and dead. The number of species usually increases as the total number of organisms rises. The growth of trees and other taller plants enhances the three-dimensional character of the community. The number and variety of habitats become larger. Shading and shelter from larger plants give rise to small microhabitats that are protected from wind, sun, and temperature extremes, thus enabling colonization by small, specialized organisms that thrive in specialized microhabitats. As conditions change, successions of populations of various organisms occupy niches in the community. There is a tendency for successive populations to be both larger and longer lived, such as is the case with trees in an old-growth forest. Finally, a climax community is achieved that has reached a steady-state condition with its terrestrial, atmospheric, and aquatic environment and is in a homeostatic state. It usually has a relatively large amount of organic matter; compare, for example, the high organic content of a rich soil with the negligible amount of organic matter on rock colonized by lichen. A climax community does not exhibit sudden onset of populations of different species or loss of species. The species in the community go through their life cycles in ways that maintain community stability and homeostasis. The following are some important characteristics of a climax community:

- High diversity of life-forms

- A high degree of order

- Narrow specialization of species in their appropriate niches with a high degree of stratification

- Conservation and recycling of nutrients

Secondary succession occurs when a less drastic, but still major change has occurred in a biological community. An example of secondary succession would be the establishment of a highly modified community on forest land that has just been subjected to clear-cutting. Annual plants are the first to colonize newly exposed soil. These are followed by small perennials, such as grass, and successively larger perennials, such as shrubs, bushes, small trees, and finally large trees.

A special kind of community is that which is adapted to periodic and dramatic alteration, especially from fire. **Fire-climax communities** are populated by species that quickly become reestablished following a fire. Some species of trees even require fire to reproduce.

The borders of different biological communities have special characteristics called **edge effects**. A shoreline where grassland borders on a body of water is such an edge. Such areas provide special habitats for various kinds of plants and animals that can take advantage of both kinds of habitat. Edges often show a high degree of biological diversity. Somewhat similar to edges are **patches** that are present within a community because of some discontinuity in the physical structure of the community. A rock outcropping in the middle of a hilly grassland is an example of a patch. Patches may show a different distribution of species from those present in the main community.

13.12. HUMAN EFFECTS

Human intervention has a large potential effect on biological communities. Sometimes such intervention involves drastic physical alteration of the community, such as by plowing grasslands or cutting forests. Other types of human intervention may be less subtle, but may still have drastic effects.

An important way in which humans may influence biological communities is through the introduction of new species. If it is successful in a community, a new species affects those already there and may significantly modify the physical nature of the habitat. New species may prey upon those already present, or serve as prey that attracts predatory species from outside the community. When forests are cut and grasslands established, larger numbers of herbivores and representatives of species not previously present are attracted. These animals in turn attract carnivores that feed on them. Parasites usually accompany newly introduced species that can serve as their hosts.

Some introduced species are particularly destructive to biological communities and habitat. One of the worst of these is the goat, which has a well-earned reputation for indiscriminate consumption of vegetation, destruction of plant life, and damage to sod with its hard hooves. Rats introduced onto islands have wreaked havoc with indigenous species. Domestic house cats reverting to a wild state have wiped out whole populations of birds. Aggressive bird species, particularly house sparrows and starlings, have displaced more desirable native species.

The human species has become inextricably linked with technology such that in a sense *Homo sapiens* are not "natural animals." Much of what is known about the effects of humans on biological communities is negative — destruction of habitat, emission of pollutants to the environment, a potential permanent change in climate from greenhouse gases. These kinds of influences are unfortunate and very harmful to biological communities. However, humans are linked to technology irreversibly, and it will be necessary for humans to adapt themselves and their technologies to the biological communities on which humankind ultimately depends for its existence.

13.13. HUMAN ACTIONS TO PRESERVE AND IMPROVE LIFE ON EARTH

Technology can be harnessed to preserve and improve the condition of life on Earth. A prime example of how that is done is provided by agriculture. With the application of plant genetics, herbicides, fertilizers, and advanced cultivation and harvesting techniques, agricultural interests can vastly increase the productivity of a plot of soil. By building terraces and waterways planted to grass that forms a tough, erosion-resistant sod, the productivity of land may be increased while erosion is slowed to a negligible level.

Human intervention can be used to create and enhance habitats that are not maintained for agricultural production. Although not all reservoirs formed by damming streams are desirable, many provide a welcome variety of habitat for species that live in or around bodies of water. Impounding water can cause suspended material to settle from streams, thus improving stream quality below the dams. In a very limited, but encouraging number of cases, human intervention is being applied to reverse damage done to habitats by human activities in the past. Once meandering streams straightened and turned into ugly, erosive ditches by channelization have been restored in some cases to provide the bending channels that make the stream hospitable to life. Productive wetlands are being restored, or even constructed where none existed before, usually as a means to aid wastewater treatment. Badly used, eroded farmland has been converted to forests and grassland. Special structures can be made and sunk in shallow coastal areas to provide shelter and habitat for marine life; even old ship hulls and airplane fuselages have been used for this purpose.

The restoration of ecosystems by human intervention is called **restoration ecology**. Restoration ecology has become a significant area of human endeavor, and it may be hoped that it will increase in importance as technology is used increasingly to benefit the natural environment. The restoration ecologist needs to be familiar with basic ecology, as well as with the kinds of technology used to rebuild ecosystems. A knowledge of related areas, such as geology, hydrology, limnology, and soil science is also required. After catastrophic floods along the Missouri and Mississippi rivers in the U.S. in 1993, the deliberate decision was made to forego reconstruction of some river dikes destroyed by the flood allowing flooding of surrounding land during periods of high rainfall and reversion of the land to a more natural state. Restoration ecology was applied to some limited areas to restore wetlands and river bottom lands for wildlife habitat.

Much of the work that has been done to preserve wildlife and to restore ecosystems in which wild species exist has been the result of efforts to maintain and increase numbers of game animals. Enlightened hunting and fishing laws have reduced the harvest of many species to sustainable levels. In some cases, these have brought species back from very low numbers or even the brink of extinction. Important examples in the U.S. are American bison, wood ducks, wild turkeys, snowy egrets, and white-tailed deer. In addition to hunting and fishing restrictions, habitat restoration has been very important in increasing numbers of game animals. Restoration of wetland breeding areas has enabled significant increases in numbers of waterfowl.

Information is essential in order to understand, preserve, and enhance biological systems. The capabilities of technology to gather and process information are enormous. Sophisticated chemical analysis techniques provide detailed profiles of the chemical characteristics of the environment in which organisms live, including both nutrients and pollutants. Sensors for temperature, wind, moisture, and sunlight can be used to give a continuous picture of the physical environment. This and other information can be subjected to sophisticated computer analysis to provide a profile of the life system and to direct human intervention in constructive ways.

One of the more useful relatively recent technologies used to study life systems consists of satellite images of Earth, such as those provided by the Landsat satellite. Such images can be gathered by infrared measurements, digitized, and processed by computer to provide profiles of geological features, water on Earth's surface, and vegetation. By remotely sensing the absorption of electromagnetic radiation at specific wavelengths, instruments mounted on satellites can monitor gases or reactive chemical species in the atmosphere. One example of the latter is ClO, a reactive intermediate produced during the photochemical processes that occur as part of ozone depletion from stratospheric chlorofluorocarbons. The levels of greenhouse gases, including carbon dioxide, nitrous oxide, and methane can also be monitored. This information can be used to predict the effects of atmospheric species on life-forms that may occur from global warming or ozone depletion.

Populations of organisms can be enhanced significantly by **habitat restoration** that restores habitats that have been destroyed by human activities. Even more drastic enhancement of populations can be achieved by construction of **artificial habitats**. As examples, a reservoir created by damming a stream is an artificial habitat conducive to the existence of species that thrive in relatively deep, still water, whereas removal of a dam to restore a flowing stream hospitable to species that live in flowing water is an example of habitat restoration.

To a limited degree, plants can be preserved artificially by seed banks in which seeds are stored for long periods of time under appropriate conditions for their preservation. Botanical gardens and arboreta enable growth of plants under artificial conditions that can prevent at least some species from becoming extinct.

The number of animal species that can be maintained in zoos is limited, but this is still of some use for protecting various kinds of animals from extinction. Zoos are being used to a greater degree for wildlife preservation, in some cases with the goal of introducing animal species back into the wild. **Captive breeding** programs have been established to salvage individuals of endangered species from the wild, increase their population by breeding in captivity, and reintroduce them into the wild state. Populations of endangered bird species have been increased by taking eggs from nests of birds in the wild and hatching them in captivity, sometimes with surrogate parents from other bird species. On a much larger scale, fish hatcheries have been in use for many decades to ensure a steady supply of fingerlings, particularly of trout and salmon species. There have been some tentative successes in captive breeding programs to restore species to the wild. In the U.S., captive peregrine falcon and blackfooted ferret have been reintroduced to some areas. The Arabian oryx (a large species of antelope) has been restored to some of its former habitats in the Middle East. Golden lion tamarins have been reintroduced to rain forests in Brazil. The widely publicized reintroduction of the California condor, a large carrion-eating bird, from individuals bred in captivity has been difficult because of the deaths of many of the specimens released and the tendencies of some individuals to prefer feeding by humans rather than scavenging for roadkill carrion. As of 2006, zoos in the U.S. and China were achieving significant success with a mini population explosion of baby giant pandas.

A major problem with captive breeding programs has been the vulnerability of limited numbers of any species population to loss. When only a few individuals remain, the sudden onset of disease can be devastating. Not the least of the problems is the limited genetic diversity of a small population and the adverse effects of inbreeding.

13.14. LAWS AND REGULATIONS

Numerous laws and regulations have been applied to the preservation of wildlife and ecosystems. Gaming laws and the regulations fostered by them were mentioned in the preceding section. In the U.S. such laws have been in effect since the latter 1800s. For the most part these laws restricted the hunting of game and trade in game products, such as pelts.

In 1973, landmark wildlife legislation was passed in the U.S. in the form of the Endangered Species Act. This act was designed to prevent extinction of both animals and plants belonging to threatened species. Because of conflicts of interests, such as that between the lumber industry and groups attempting to preserve Spotted Owl habitat in northwestern U.S. old-growth forests, the enforcement of this act has been controversial. The basic provisions of the Endangered Species Act are the following:

- Requires that a list be compiled of species that are in danger of extinction, thus subject to regulation by the act

- Prohibits hunting or capture of such species

- Forbids trade in the products of listed species, such as pelts or feathers

- Prohibits projects, such as dams, that would threaten listed species

- Provides for protection of habitat

- Requires that plans be formulated to restore species listed under the act

Extinct species are those that no longer exist; examples are the carrier pigeon, the relic leopard frog, and the spiderflower. According to the provisions of the Endangered Species Act, an **endangered species** is one for which the danger of becoming extinct is very high. Over 1000 species are so designated in the U.S., of which almost 40% are mammals. **Threatened species**, such as the gray wolf and grizzly bear, have suffered marked declines in total numbers and extinction or threat thereof in certain localities.

Species recovery plans are mandated under the Endangered Species Act to enable recovery of species to the point that they can be delisted. In a few cases, numbers of some species have increased to a sufficient extent, and conditions for them have improved enough that they have been delisted. One example of such a species is the American alligator, which rebounded from perilously low populations in the late 1960s to relative abundance at present such that by 2006, in some areas, aggressive alligators were threatening pets in some residential areas of Florida.

13.15. ORGANISMS INTERACTING WITH FOREIGN CHEMICALS

The water and soil environments receive a variety of **xenobiotic compounds** defined as those that are foreign to living systems. Microorganisms in water and soil act upon these compounds in four different ways: (1) by absorbing them from the surroundings or by ingestion, a process called bioaccumulation, (2) by using them directly as substrates for energy and biomass production, (3) through cometabolism along with primary metabolic processes, or (4) by joining them chemically with other chemical species present in the organism through the process of conjugation. Of these processes, biodegradation, the metabolic breakdown of substances by microorganisms, is the most important.

Bioaccumulation is the uptake and concentration of environmental chemicals by living systems. In a general sense, the term refers to the process by which substances dissolved and suspended in water or contained in sediments, soil, food, or drinking water are taken into an organism by diffusion from aqueous solution and by ingestion. Although the term applies especially to aquatic organisms, particularly fish, it may be extended to whole series of organisms in food chains. Uptake of environmental chemicals through food chains can result in much higher levels of the chemicals in organisms than would be expected from simple bioaccumulation, thereby resulting in **biomagnification**. Biomagnification can occur, for example, in a succession of organisms starting with herbivores (which live on plant material), progressing through detritovores (which feed on residues from the herbivores), and terminating with carnivores. Loss of a substance from an organism back to the surroundings, such as occurs when a contaminated fish is placed in fresh water, is called **depuration**.

The tendency of a chemical to leave aqueous solution and enter a food chain is important in determining its environmental effects and is expressed through the concept of bioconcentration. **Bioconcentration** (Figure 13.6) may be viewed as a special case of bioaccumulation in which a dissolved substance is selectively taken up from water solution and concentrated in tissue by nondietary routes. As illustrated in Figure 13.6, the model of bioconcentration is based upon a process by which contaminants in water traverse fish gill epithelium and are transported

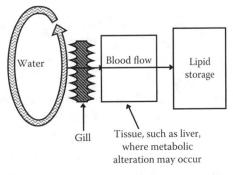

Figure 13.6.   Overall pathway of bioconcentration.

by the blood through highly vascularized tissues to lipid tissue, which serves as a storage sink for hydrophobic substances. Transport through the blood is affected by several factors, including rate of blood flow and degree and strength of binding to blood plasma protein. Prior to reaching the lipid tissue sink, some of the compound may be metabolized to different forms. The concept of bioconcentration is most applicable under the following conditions:

- The substance is taken up and eliminated *via* passive transport processes.

- The substance is metabolized slowly.

- The substance has a relatively low water solubility.

- The substance has a relatively high lipid solubility.

Substances that undergo bioconcentration are hydrophobic and tend to undergo transfer from water media to fish lipid tissue. The simplest model of bioconcentration

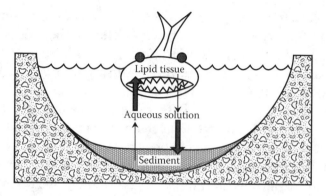

Figure 13.7.   Partitioning of a hydrophobic chemical species among sediment, water, and lipid tissue. Heavier arrows denote the preference of the chemical for sediment and lipid tissue compared to aqueous solution.

views the phenomenon on the basis of the physical properties of the contaminant and does not account for physiologic variables (such as variable blood flow) or metabolism of the substance. Such a simple model forms the basis of the **hydrophobicity model** of bioconcentration in which bioconcentration is regarded from the viewpoint of a dynamic equilibrium between the substance dissolved in aqueous solution and the same substance dissolved in lipid tissue.

Because of the strong attraction of hydrophobic species for insoluble materials such as humic matter, many organic pollutants in the aquatic environment are held by sediments in bodies of water. Bioaccumulation of these materials must, therefore, consider transfer from sediment to water to organism as illustrated in Figure 13.7.

Quantitatively, the **bioconcentration factor**, which is the ratio of the concentration of a specified substance in lipid tissue of an organism to its concentration in water, expresses the degree to which foreign chemicals are taken up by an organism. Bioconcentration factors can range up to approximately 1 million meaning that some organic compounds are a million times more soluble in lipid tissue than in water. The long-chain alcohol, n-octanol, has been used as a surrogate for fish lipid tissue. In such a case, the ratio of the concentration of a lipid-soluble chemical in n-octanol to that in water with which the n-octanol is in contact, the **octanol–water partition coefficient**, is used as an estimate of the uptake of a chemical by fish lipid tissue. Values typically range from 10 to 10^7. For land animals, a **biotransfer factor** can be used to express the relative concentration of a substance in animal tissue to its daily intake in the diet of an animal.

Detoxication refers to the biological conversion of a toxic substance to a less toxic species, which may still be a relatively complex, or even more complex material. An example of detoxication is illustrated below for the enzymatic conversion of paraoxon (a highly toxic organophosphate insecticide) to p-nitrophenol, which has only about 1/200 the toxicity of the parent compound,

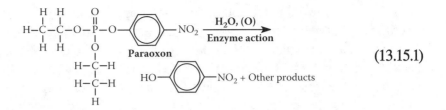

(13.15.1)

13.16. BIODEGRADATION

An important aspect of the interaction of organisms with xenobiotic compounds is **biodegradation,** the alteration of chemical species by biochemical processes. In nature, biodegradation is usually carried out by the enzyme systems of microorganisms. Biodegradation may involve relatively small changes in the parent molecule, such as substitution or modification of a functional group. Biodegradation is the most important means by which xenobiotic compounds are eliminated as environmental pollutants. In the most favorable cases, the compound is completely destroyed such that the end result is conversion of relatively complex organic compounds to CO_2, H_2O, and inorganic salts, a process called **mineralization**. Usually, the products of biodegradation are molecular forms that tend to occur in nature compared to their synthetic precursors.

Several terms should be distinguished in considering the biochemical aspects of biodegradation. Biotransformation is what happens to any substance that is metabolized by the biochemical processes in an organism and is altered by these processes. Metabolism is divided into the two general categories of catabolism, which is the breaking down of more complex molecules, and anabolism, which is the building up of life molecules from simpler materials. The substances subjected to biotransformation may be naturally occurring or anthropogenic (made by human activities). They may consist of xenobiotic molecules that are foreign to living systems.

Biodegradation of an organic compound occurs in a stepwise fashion and is rarely the result of the activity of a single specific organism. Usually several strains of microorganisms, often existing synergistically, are involved. These may utilize different metabolic pathways and a variety of enzyme systems. Most important processes involved in the breakdown of chemical species in the water and soil environments are enzymatic biodegradation of organic matter by microorganisms in the aquatic and terrestrial environments. It occurs by way of a number of stepwise, microbially catalyzed reactions.

Although biodegradation is normally regarded as degradation to simple inorganic species such as carbon dioxide, water, sulfates, and phosphates, the possibility must always be considered of forming more complex or more hazardous chemical species. An example of the latter is the production of volatile, soluble, toxic methylated forms of arsenic and mercury from inorganic species of these elements by bacteria under anaerobic conditions.

It is well known that microbial communities develop the ability to break down xenobiotic compounds metabolically when exposed to them in the environment. This has become particularly obvious from studies of biocidal compounds (those that kill organisms) in the environment. In general, such compounds are readily degraded by bacteria that have been in contact with the compounds for prolonged periods, but not by bacteria from unexposed sites. The development of microbial cultures with the ability to degrade materials to which they are exposed is described as **metabolic adaptation**. In rapidly multiplying microbial cultures, enough generations are involved so that metabolic adaptation can include genetic changes that favor microorganisms that have developed the ability to degrade a specific pollutant. Metabolic adaptation may also include increased numbers of microorganisms capable of degrading the substrate in question and enzyme induction.

Xenobiotic compounds are usually attacked by enzymes whose prime function is to react with other compounds, a process that provides neither carbon nor energy called **cometabolism**. Cometabolism usually involves relatively small modifications of the substance that is cometabolized (the secondary substrate) relative to the primary substrate. Other organisms act on the cometabolized products to complete the biodegradation process.

The rates and efficacy of biodegradation of organic substances depend upon several obvious factors. These include the concentration of the substrate compound; whether or not molecular oxygen, O_2, is available; presence of phosphorus and nitrogen nutrients; availability of trace element nutrients; the presence of a suitable organism; absence of toxic substances; and the presence of appropriate physical conditions (temperature, growth matrix). In addition to their biochemical properties, the physical properties of compounds, including volatility, water-solubility, organophilicity, tendency to be sorbed by solids, and charge play a role in determining the biodegradability of organic compounds. Trace amounts of micronutrients, such as calcium, iron, and zinc are needed to support biological processes and as constituents of enzymes.

The amenability of a compound to biochemical attack by microorganisms is expressed as its **biodegradability**. The biodegradability of a compound is influenced by its physical characteristics, such as solubility in water and vapor pressure, and by its chemical properties, including molecular mass, molecular structure, and presence of various kinds of functional groups, some of which provide a "biochemical handle" for the initiation of biodegradation. Biodegradability is an important consideration in green chemistry in the case of products that are dispersed to the environment. For example, biodegradable plastics, such as polylactic acid, are desirable because they are amenable to biodegradation.

Recalcitrant or **biorefractory** substances are those that resist biodegradation and tend to persist and accumulate in the environment. Such materials are not necessarily toxic to organisms, but simply resist their metabolic attack. In general, compounds of biological origin readily undergo biodegradation. Thus, proteins and carbohydrates are readily metabolized by organisms. Lipids, including fats and oils, are slower to undergo biodegradation because of their generally low solubilities. Initiation of the biodegradation of hydrocarbons tends to be slow. However, if an alcohol (–OH) or carboxylic acid (–CO$_2$H) group is attached to a hydrocarbon structure, biodegradation is much faster. Hydrolysis, the splitting of a molecule with the addition of H$_2$O, generally occurs very readily, so that compounds with ester or amide groups tend to break down quickly. "Green" chemical species of biological origin are often preferred when such materials are likely to enter the water or soil environments because of their relatively facile biodegradability.

SUPPLEMENTARY REFERENCES

Allsopp, Dennis, Kenneth J. Seal, and Christine C. Gaylarde, *Introduction to Biodeterioration*, Cambridge University Press, New York, 2004.

Atlas, Ronald M. and Jim Philp, *Bioremediation: Applied Microbial Solutions for Real-World Environmental Cleanup*, ASM press, Washington, D.C., 2005.

Begon, Michael, Colin R. Townsend, and John L. Harper, *Ecology: From Individuals to Ecosystems*, 4th ed., Blackwell, Malden, MA, 2006.

Caswell, H., Ed., *Food Webs: From Connectivity to Energetics*, Elsevier Academic Press, San Diego, CA, 2005.

Coker, Paddy and Paul Ganderton, *Ecological Biogeography*, Prentice Hall, Upper Saddle River, NJ, 2005.

Crawford, Ronald L. and Don L. Crawford, Eds., *Bioremediation: Principles and Applications*, Cambridge University Press, Cambridge, U.K., 2005.

Falk, Donald A., Margaret A. Palmer, and Joy B. Zedler, Eds., *Foundations of Restoration Ecology*, Island Press, Washington, D.C., 2006.

Fingerman, Milton and Rachakonda Nagabhushanam, Eds., *Bioremediation of Aquatic and Terrestrial Ecosystems*, Science Publishers, Enfield, NH, 2005.

Hindmarsh, Richard and Geoffrey Lawrence, Eds., *Recoding Nature: Critical Perspectives on Genetic Engineering*, UNSW Press, Sydney, 2004.

Kumar, Anil, *Genetic Engineering*, Nova Science Publishers, New York, 2006.

Lovelock, James, *Gaia: A New Look at Life on Earth*, Oxford University Press, New York, 2000.

Mayhew, Peter J., *Discovering Evolutionary Ecology: Bringing Together Ecology and Evolution*, Oxford University Press, New York, 2006.

Miller, G. Tyler, *Essentials of Ecology*, Thompson, Brooks/Cole, Pacific Grove, CA, 2005.

Samson, Paul R. and David Pitt, Eds., *The Biosphere and Noosphere Reader: Global Environment, Society, and Change*, Routledge, London, 1999.

Scragg, Alan, *Environmental Biotechnology*, 2nd ed., Oxford University Press, New York, 2004.

Singh, Ved Pal and Raymond D. Stapleton, Eds., *Biotransformations: Bioremediation Technology for Health and Environmental Protection*, Elsevier Science, Amsterdam, 2002.

Smith, Thomas M. and Robert Leo Smith, *Elements of Ecology*, 6th ed., Benjamin Cummings, San Francisco, CA, 2006.

Smil, Vaclav, *The Earth's Biosphere: Evolution, Dynamics, and Change*, MIT Press, Cambridge, MA, 2002.

Talley, Jeffrey, Ed., *Bioremediation of Recalcitrant Compounds*, Taylor & Francis, Boca Raton, FL, 2005.

Torr, James, Ed., *Genetic Engineering*, Greenhaven Press, San Diego, CA, 2006.

Van Andel, Jelte and James Aronson, Eds., *Restoration Ecology: The New Frontier*, Blackwell Publishing, Malden, MA. 2006.

Wackett, Lawrence P. and C. Douglas Hershberger, *Biocatalysis and Biodegradation: Microbial Transformation of Organic Compounds*, ASM Press, Washington, D.C., 2001.

Warren, A. and J. R. French, Eds., *Habitat Conservation: Managing the Physical Environment*, John Wiley & Sons, New York, 2001.

QUESTIONS AND PROBLEMS

1. In what sense is the biosphere a particularly thin layer on Earth?

2. Give two characteristics of all living organisms.

3. Define homeostasis and explain why it is important for life.

4. Explain the sense in which a biological community is a subcategory of an ecosystem.

5. What is a food chain and what is the special role of producers in a food chain?

6. As they pertain to biological communities, define and relate limiting factors, tolerance limits, and the critical factor.

7. Based on a climatological/topographical factor, explain why there may be marked differences in the kind and quantity of vegetation on one side of a mountain range as compared to the opposite side.

8. How do plants cope with dry conditions and with subfreezing temperatures?

9. What is an ecological niche? Do you occupy an ecological niche? If so, define and explain it.

10. Explain why niche specialists are more likely than niche generalists to become endangered species.

11. Relate the plot shown in Figure 13.2 to what is happening to Earth's human population. Discuss the ramifications of a stabilized human population and of a possible population crash.

12. How do populations tend to avoid destructive competition in terms of both space and time?

13. Explain how proper balances of the following may enable human populations to exist and thrive: productivity, diversity, inertia, constancy, and resilience.

14. Explain how organisms may be ecological equivalents, even though they are not closely related genetically.

15. What is the special role of a dominant plant species in a biological community?

16. Distinguish among mutualism, commensalism, and protocooperation. Can you cite specific examples of these kinds of relationships from your own life?

17. Relate the phenomenon of biomagnification to the various classes of organisms shown in Figure 13.4.

18. Explain how a pioneer species and a climax community are on opposite ends of the spectrum of geological succession.

19. What are the major characteristics of a climax community?

20. Explain how human actions can be employed to preserve and improve life on Earth?

21. Define detoxication and cite an example of the phenomenon.

22. Distinguish among the following: bioaccumulation, biomagnification, and bioconcentration.

23. What is the octanol-water partition coefficient? What is it used to predict?

24. How is cometabolism involved in the biodegradation of xenobiotic compounds?

25. In what sense are living organisms indicators of optimum conditions for green science and technology?

26. What is the distinction between parasites and scavengers?

27. Suggest an essential ecological role played by scavengers and detritovores.

28. Although predatory carnivores that catch up with individual prey are definitely bad news for individual prey, how might they benefit a prey population as a whole?

29. Look up overturn as the term applies to stratified bodies of water. What role might be played by overturn in aquatic ecology?

30. How may loss of water help enable plants to cope with cold temperatures?

31. Describe the ways in which human beings deserve the title of niche generalists?

32. Terrestrial ecosystems largely rest upon, and depend upon, soil. How, in turn, does soil depend on terrestrial ecosystems for its quality?

33. In what respects do humans resemble other animals in respect to territoriality?

34. Compare natural ecosystems with industrial systems with respect to the production of waste products. Can you name any true waste products in a natural ecosystem? What about the argument that fossil fuels, such as coal, are true waste products of natural ecosystems?

14. TOXIC EFFECTS ON ORGANISMS AND TOXICOLOGICAL CHEMISTRY

14.1. TOXIC SUBSTANCES AND GREEN SCIENCE AND TECHNOLOGY

Ultimately, most pollutants and hazardous substances are of concern because of their toxic effects. The general aspects of these effects and the toxicological chemistry of specific classes of chemical substances are addressed in this chapter. In order to understand toxicological chemistry, it is essential to have some understanding of biochemistry, the science that deals with chemical processes and materials in living systems. Biochemistry was summarized in Chapter 3.

It is very important to consider the toxic effects of substances when dealing with green science and technology. That is because one of strongest driving forces behind green science and technology is avoidance of the use of toxic substances and reduction of exposure to such substances. In this endeavor, it is essential to know which substances are toxic, their chemical nature, and the ways in which they act to produce toxic effects.

Toxicology

A **poison**, or **toxicant**, is a substance that is harmful to living organisms because of its detrimental effects on tissues, organs, or biological processes. **Toxicology** is the science of poisons. These definitions are subject to a number of qualifications. Whether a substance is poisonous depends upon the type of organism exposed, the amount of the substance, and the route of exposure. In the case of human exposure, the degree of harm done by a poison can depend strongly upon whether the exposure is to the skin, by inhalation, or through ingestion.

Toxicants to which subjects are exposed in the environment or occupationally may be in several different physical forms. This may be illustrated for toxicants that are inhaled. **Gases** are substances such as carbon monoxide in air that are normally in the gaseous state under ambient conditions of temperature and pressure. **Vapors** are gas-phase materials that have evaporated or sublimed from liquids or solids.

Dusts are respirable solid particles produced by grinding bulk solids, whereas **fumes** are solid particles from the condensation of vapors, often metals or metal oxides. **Mists** are liquid droplets.

Often, a toxic substance is in solution or mixed with other substances. A substance with which the toxicant is associated (the solvent in which it is dissolved or the solid medium in which it is dispersed) is called the **matrix**. The matrix may have a strong effect upon the toxicity of the toxicant.

There are numerous variables related to the ways in which organisms are exposed to toxic substances. One of the most crucial of these, **dose,** is discussed in Section 14.2. Another important factor is the **toxicant concentration**, which may range from the pure substance (100%) down to a very dilute solution of a highly potent poison. Both the **duration** of exposure per exposure incident and the **frequency** of exposure are important. The **rate** of exposure and the total time period over which the organism is exposed are both important situational variables. The exposure **site** and **route** also affect toxicity.

It is possible to classify exposures on the basis of acute vs. chronic and local vs. systemic exposure, giving four general categories. **Acute local** exposure occurs at a specific location over a time period of a few seconds to a few hours and may affect the exposure site, particularly the skin, eyes, or mucus membranes. The same parts of the body can be affected by **chronic local** exposure, for which the time span may be as long as several years. **Acute systemic** exposure is a brief exposure or exposure to a single dose and occurs with toxicants that can enter the body, such as by inhalation or ingestion, and affect organs such as the liver that are remote from the entry site. **Chronic systemic** exposure differs in that the exposure occurs over a prolonged time period.

In discussing exposure sites for toxicants, it is useful to consider the major routes and sites of exposure, distribution, and elimination of toxicants in the body as shown in Figure 14.1. The major routes of accidental or intentional exposure to toxicants by humans and other animals are the skin (percutaneous route), the lungs (inhalation, respiration, and pulmonary route), and the mouth (oral route); minor routes of exposure are rectal, vaginal, and parenteral (intravenous or intramuscular, a common means for the administration of drugs or toxic substances in test subjects). The way that a toxic substance is introduced into the complex system of an organism is strongly dependent upon the physical and chemical properties of the substance. The pulmonary system is most likely to take in toxic gases or very fine, respirable solid or liquid particles. In other than a respirable form, a solid usually enters the body orally. Absorption through the skin is most likely for liquids, solutes in solution, and semisolids, such as sludges.

The defensive barriers that a toxicant may encounter vary with the route of exposure. For example, toxic elemental mercury is readily absorbed through the alveoli in the lungs, much more readily than through the skin or gastrointestinal tract. Most test exposures to animals are through ingestion or gavage (introduction into the stomach through a tube). Pulmonary exposure is often favored with subjects

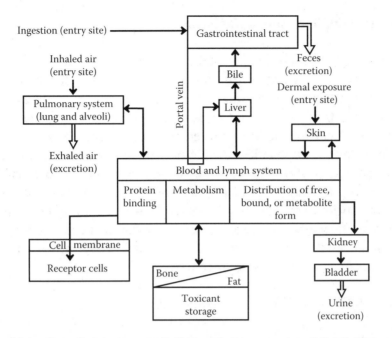

Figure 14.1.   Major sites of exposure, metabolism, and storage, routes of distribution, and elimination of toxic substances in the body.

that may exhibit refractory behavior when noxious chemicals are administered by means requiring a degree of cooperation from the subject. Intravenous injection may be chosen for deliberate exposure when it is necessary to know the concentration and effect of a xenobiotic substance in the blood. However, pathways used experimentally that are almost certain not to be significant in accidental exposures can give misleading results when they avoid the body's natural defense mechanisms.

An interesting historical example of the importance of the route of exposure to toxicants is provided by cancer caused by contact of coal tar with skin. The major barrier to dermal absorption of toxicants is the **stratum corneum** or horny layer. The permeability of skin is inversely proportional to the thickness of this layer, which varies by location on the body in the order soles and palms > abdomen, back, legs, and arms > genital (perineal) area. Evidence of the susceptibility of the genital area to absorption of toxic substances is to be found in accounts of the high incidence of cancer of the scrotum among chimney sweeps in London described by Sir Percival Pott, Surgeon General of Britain during the reign of King George III. The cancer-causing agent was coal tar condensed in chimneys. This material was more readily absorbed through the skin in the genital areas than elsewhere leading to a high incidence of scrotal cancer (a condition aggravated by a lack of appreciation of basic hygienic practices, such as bathing and regular changes of underclothing).

Organisms can serve as indicators of various kinds of pollutants. In this application, organisms are known as **biomonitors**. Higher plants, fungi, lichens, and mosses can be useful biomonitors for heavy metal pollutants in the environment.

Synergism, Potentiation, and Antagonism

The biological effects of two or more toxic substances can be different in kind and degree from those of one of the substances alone. One of the ways in which this can occur is when one substance affects the way in which another undergoes any of the steps in the kinetic phase as discussed in Section 14.7 and illustrated in Figure 14.9. Chemical interaction between substances may affect their toxicities. Both substances may act upon the same physiologic function, or two substances may compete for binding to the same receptor (molecule or other entity acted upon by a toxicant). When both substances have the same physiologic function, their effects may be simply **additive** or they may be **synergistic** (the total effect is greater than the sum of the effects of each separately). **Potentiation** occurs when an inactive substance enhances the action of an active one and **antagonism** when an active substance decreases the effect of another active one.

14.2. DOSE–RESPONSE RELATIONSHIPS

Toxicants have widely varying effects upon organisms. Quantitatively, these variations include minimum levels at which the onset of an effect is observed, the sensitivity of the organism to small increments of toxicant, and levels at which the ultimate effect (particularly death) occurs in most exposed organisms. Some essential substances, such as nutrient minerals, have optimum ranges above and below which detrimental effects are observed (see Section 14.5 and Figure 14.4).

Factors such as those just outlined are taken into account by the **dose–response** relationship, which is one of the key concepts of toxicology. **Dose** is the amount, usually per unit body mass, of a toxicant to which an organism is exposed. **Response** is the effect upon an organism resulting from exposure to a toxicant. In order to define a dose–response relationship, it is necessary to specify a particular response, such as death of the organism, as well as the conditions under which the response is obtained, such as the length of time from administration of the dose. Consider a specific response for a population of the same kinds of organisms. At relatively low doses, none of the organisms exhibit the response (e.g., all live) whereas at higher doses, all of the organisms exhibit the response (e.g., all die). In between, there is a range of doses over which some of the organisms respond in the specified manner and others do not, thereby defining a dose–response curve. Dose–response relationships differ among different kinds and strains of organisms, types of tissues, and populations of cells.

Figure 14.2 shows a generalized dose–response curve. Such a plot may be obtained, for example, by administering different doses of a poison in a uniform manner to a homogeneous population of test animals and plotting the cumulative percentage of deaths as a function of the log of the dose. The dose corresponding to the mid-point (inflection point) of the resulting S-shaped curve is the statistical estimate of the dose that would kill 50% of the subjects and is designated as LD_{50}.

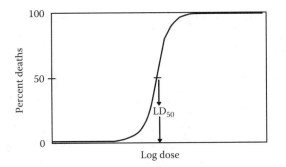

Figure 14.2.   Illustration of a dose–response curve in which the response is the death of the organism. The cumulative percentage of deaths of organisms is plotted on the Y axis.

The estimated doses at which 5% (LD_5) and 95% (LD_{95}) of the test subjects die are obtained from the graph by reading the dose levels for 5% and 95% fatalities, respectively. A relatively small difference between LD_5 and LD_{95} is reflected by a steeper S-shaped curve and *vice versa*. Statistically, 68% of all values on a dose–response curve fall within ±1 standard deviation of the mean at LD_{50} and encompass the range from LD_{16} to LD_{84}.

14.3. RELATIVE TOXICITIES

Table 14.1 illustrates standard **toxicity ratings** that are used to describe estimated toxicities of various substances to humans. For an adult human of average size, a "taste" of a super toxic substances (just a few drops or less) is fatal. In June of 1997, a research chemist died as the result of exposure to a minute amount of dimethyl mercury. The exposure had occurred some months before as the result of a few drops of the compound absorbed through the skin after being spilled on latex gloves used by the chemist. Later tests showed that dimethyl mercury penetrates such gloves in a matter of seconds. A teaspoonful of a very toxic substance could be fatal to a human. However, as much as a liter of a slightly toxic substance might be required to kill an adult human.

When there is a substantial difference between LD_{50} values of two different substances, the one with the lower value is said to be the more **potent**. Such a comparison must assume that the dose–response curves for the two substances being compared have similar slopes.

So far, toxicities have been described primarily in terms of the ultimate effect, that is, deaths of organisms, or lethality. This is obviously an irreversible consequence of exposure. In many, and perhaps most cases, **sublethal** and **reversible** effects are of greater importance. This is obviously true of drugs, where death from exposure to a registered therapeutic agent is rare, but other effects, both detrimental and beneficial, are usually observed. By their very nature, drugs alter biological processes; therefore, the potential for harm is almost always present. The major consideration in establishing drug dose is to find a dose that has an adequate therapeutic

Table 14.1. Toxicity Scale with Example Substances[a]

Substance	Approsimate LD_{50}	Toxicity Rating
	-10^5	1. Practically nontoxic
DEHP[b] ⟶	$-$	$> 1.5 \times 10^4$ mg/kg
Ethanol ⟶	-10^4	2. Slightly toxic, 5×10^3
Sodium chloride ⟶	$-$	to 1.5×10^4 mg/kg
Malathion ⟶	-10^3	3. Moderately toxic,
Chlordane ⟶	$-$	500 to 5000 mg/kg
Heptachlor ⟶	-10^2	
	$-$	4. Very toxic, 50 to
Parathion ⟶	-10	500 mg/kg
	$-$	5. Extremely toxic,
TEPP[c] ⟶	-1	5 to 50 mg/kg
	$-$	6. Supertoxic,
Tetrodotoxin[d] ⟶	-10^{-1}	<5 mg/kg
	$-$	
	-10^{-2}	
	$-$	
TCDD[e] ⟶	-10^{-3}	
	$-$	
	-10^{-4}	
	$-$	
Botulinus toxin ⟶	-10^{-5}	

[a] Doses are in units of mg of toxicant per kg of body mass. Toxicity ratings on the right are given as numbers ranging from 1 (practically nontoxic) through 6 (supertoxic) along with estimated lethal oral doses for humans in mg/kg. Estimated LD_{50} values for substances on the left have been measured in test animals, usually rats, and apply to oral doses.

[b] Bis(2-ethylhexyl)phthalate

[c] Tetraethylpyrophosphate

[d] Toxin from pufferfish

[e] TCDD represents 2,3,7,8,-tetrachlorodibenzodioxin, commonly called "dioxin."

effect without undesirable side effects. A dose–response curve can be established for a drug that progresses from noneffective levels through effective, harmful, and even lethal levels. A low slope for this curve indicates a wide range of effective dose and a wide **margin of safety** (see Figure 14.3). This term applies to other substances, such as pesticides, for which a large difference between the dose that kills a target organism and one that harms a desirable species is needed.

14.4. REVERSIBILITY AND SENSITIVITY

Sublethal doses of most toxic substances are eventually eliminated from an organism's system. If there is no lasting effect from the exposure, it is said to be **reversible**. However, if the effect is permanent, it is termed **irreversible**. Irreversible effects of exposure remain after the toxic substance is eliminated from the organism. Figure 14.3 illustrates these two kinds of effects. For various chemicals and different subjects, toxic effects may range from the totally reversible to the totally irreversible.

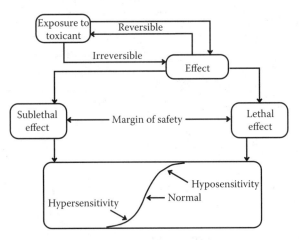

Figure 14.3.   Effects of and responses to toxic substances.

Hypersensitivity and Hyposensitivity

Examination of the dose–response curve shown in Figure 14.2 reveals that some subjects are very sensitive to a particular poison (e.g., those killed at a dose corresponding to LD_5), whereas others are very resistant to the same substance (e.g., those surviving a dose corresponding to LD_{95}). These two kinds of responses illustrate **hypersensitivity** and **hyposensitivity**, respectively; subjects in the midrange of the dose–response curve are termed **normals**. These variations in response tend to complicate toxicology in that there is not a specific dose guaranteed to yield a particular response, even in a homogeneous population.

In some cases hypersensitivity is induced. After one or more doses of a chemical, a subject may develop an extreme reaction to it. This occurs with penicillin, for example, in cases where people develop such a severe allergic response to the antibiotic that exposure is fatal if countermeasures are not taken.

14.5. XENOBIOTIC AND ENDOGENOUS SUBSTANCES

Xenobiotic substances are those that are foreign to a living system, whereas those that occur naturally in a biologic system are termed **endogenous**. The levels of an endogenous substance must usually fall within a particular concentration range for metabolic processes to occur normally. Levels below a normal range may result in a deficiency response or even death, and the same effects may occur above the normal range. This kind of response is illustrated in Figure 14.4.

Examples of endogenous substances in organisms include various hormones, blood glucose, and some essential metal ions, including Ca^{2+}, K^+, and Na^+. The optimum level of calcium in human blood serum covers a narrow range of 9 to 9.5 mg/dl. Below these values a deficiency response known as hypocalcemia occurs, manifested by muscle cramping, and above about 10.5 mg/dl hypercalcemia occurs, the major effect of which is kidney malfunction.

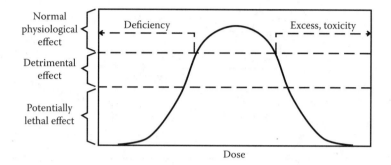

Figure 14.4.   Biological effect of an endogenous substance in an organism showing optimum level, deficiency, and excess.

14.6. TOXICOLOGICAL CHEMISTRY

Toxicological chemistry is the science that deals with the chemical nature and reactions of toxic substances, including their origins, uses, and chemical aspects of exposure, fates, and disposal. Toxicological chemistry addresses the relationships between the chemical properties and molecular structures of molecules and their toxicological effects. Figure 14.5 outlines the terms discussed above and the relationships among them.

The processes by which organisms metabolize xenobiotic species are enzyme-catalyzed Phase I and Phase II reactions, which are described briefly here.

Lipophilic xenobiotic species in the body tend to undergo **Phase I reactions** that make them more water-soluble and reactive by the attachment of polar functional groups, such as –OH (Figure 14.6). Most Phase I processes are "microsomal mixed-function oxidase" reactions catalyzed by the cytochrome P-450 enzyme system associated with the **endoplasmic reticulum** of the cell and occurring most abundantly in the liver of vertebrates.

The polar functional groups attached to a xenobiotic compound in a Phase I reaction provide reaction sites for **Phase II reactions**. Phase II reactions are **conjugation reactions** in which enzymes attach **conjugating agents** to xenobiotics, their Phase I reaction products, and nonxenobiotic compounds (Figure 14.7). The **conjugation product** of such a reaction is usually less toxic than the original xenobiotic compound, less lipid-soluble, more water-soluble, and more readily eliminated from

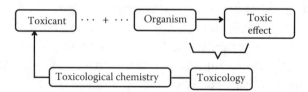

Figure 14.5.   Toxicology is the science of poisons. Toxicological chemistry relates toxicology to the chemical nature of toxicants.

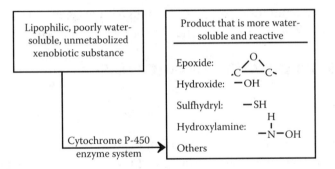

Figure 14.6.   Illustration of Phase I reactions.

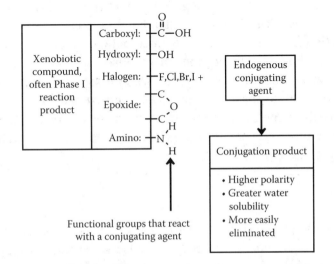

Figure 14.7.   Illustration of Phase II reactions.

the body. The major conjugating agents and the enzymes that catalyze their Phase II reactions are glucuronide (UDP glucuronyltransferase enzyme), glutathione (glutathionetransferase enzyme), sulfate (sulfotransferase enzyme), and acetyl (acetylation by acetyl transferase enzymes). The most abundant conjugation products are glucuronides. A glucuronide conjugate is illustrated in Figure 14.8, where –X–R represents a xenobiotic species conjugated to glucuronide, and R is an organic moiety.

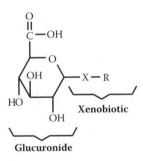

Figure 14.8.   Glucuronide conjugate formed from a xenobiotic, HX-R.

For example, if the xenobiotic compound conjugated is phenol, HXR represents the HOC_6H_5 molecule, X is the O atom, and R represents the phenyl group, C_6H_5.

14.7. KINETIC PHASE AND DYNAMIC PHASE

Kinetic Phase

The major routes and sites of absorption, metabolism, binding, and excretion of toxic substances in the body are illustrated in Figure 14.1. Toxicants in the body are metabolized, transported, and excreted, they have adverse biochemical effects, and they cause manifestations of poisoning. It is convenient to divide these processes into two major phases, a kinetic phase and a dynamic phase.

In the **kinetic phase**, a toxicant or the metabolic precursor of a toxic substance (**protoxicant**) may undergo absorption, metabolism, temporary storage, distribution, and excretion, as illustrated in Figure 14.9. A toxicant that is absorbed may be passed through the kinetic phase unchanged as an **active parent compound**, metabolized to a **detoxified metabolite** that is excreted or converted to a toxic **active metabolite**. These processes occur through Phase I and Phase II reactions discussed in the preceding section.

Dynamic Phase

In the **dynamic phase** (Figure 14.10), a toxicant or toxic metabolite interacts with cells, tissues, or organs in the body to cause some toxic response. The three major subdivisions of the dynamic phase are the following:

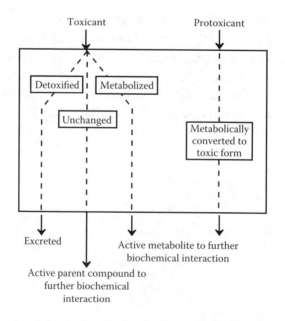

Figure 14.9.   Processes involving toxicants or protoxicants in the kinetic phase.

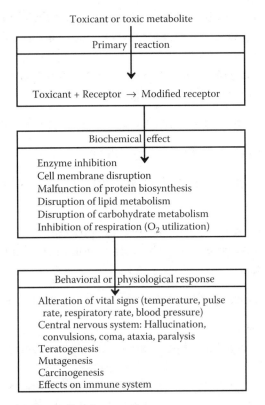

Figure 14.10.   The dynamic phase of toxicant action.

- **Primary reaction** with a receptor or target organ

- A biochemical response

- Observable effects

A primary reaction in the dynamic phase occurs when a toxicant or an active metabolite reacts with a receptor. The process leading to a toxic response is initiated when such a reaction occurs. A typical example is when benzene epoxide produced by the metabolic oxidation of benzene,

$$\bigcirc \xrightarrow{\text{Metabolic oxidation}} \bigcirc\!\!\!\!\diagup_O \quad \text{Benzene epoxide} \qquad (14.7.1)$$

forms an adduct with a nucleic acid unit in DNA (receptor) resulting in alteration of the DNA. This reaction is an **irreversible** reaction between a toxicant and a receptor. A **reversible** reaction that can result in a toxic response is illustrated by the binding between carbon monoxide and oxygen-transporting hemoglobin (Hb) in blood:

$$O_2Hb + CO \leftrightarrow COHb + O_2 \qquad (14.7.2)$$

The binding of a toxicant to a receptor may result in some kind of biochemical effect in the dynamic phase. The major ones of these are the following:

- Impairment of enzyme function by binding to the enzyme, coenzymes, metal activators of enzymes, or enzyme substrates

- Alteration of cell membrane or carriers in cell membranes

- Interference with carbohydrate metabolism

- Interference with lipid metabolism resulting in excess lipid accumulation ("fatty liver")

- Interference with respiration, the overall process by which electrons are transferred to molecular oxygen in the biological oxidation of energy-yielding substrates

- Stopping or interfering with protein biosynthesis by the action of toxicants on DNA

- Interference with regulatory processes mediated by hormones or enzymes

There are many kinds of responses to toxicants. Among the more immediate and readily observed manifestations of poisoning are alterations in the **vital signs** of **temperature, pulse rate, respiratory rate**, and **blood pressure**. Poisoning by some substances may cause an abnormal skin color (jaundiced yellow skin from CCl_4 poisoning) or excessively moist or dry skin. Toxic levels of some materials or their metabolites cause the body to have unnatural **odors**, such as the bitter almond odor of HCN in tissues of victims of cyanide poisoning. Symptoms of poisoning manifested in the eye include **miosis** (excessive or prolonged contraction of the eye pupil), **mydriasis** (excessive pupil dilation), **conjunctivitis** (inflammation of the mucus membrane that covers the front part of the eyeball and the inner lining of the eye lids), and **nystagmus** (involuntary movement of the eyeballs). Some poisons cause a moist condition of the mouth, whereas others cause a dry mouth. Gastro-intestinal tract effects including pain, vomiting, or paralytic ileus (stoppage of the normal peristalsis movement of the intestines) occur as a result of poisoning by a number of toxic substances.

Central nervous system poisoning may cause **convulsions, paralysis, hallu-cinations**, and **ataxia** (lack of coordination of voluntary movements of the body), as well as abnormal behavior, including agitation, hyperactivity, disorientation, and delirium. Severe poisoning by some substances, including organophosphates and car-bamates causes **coma**, the term used to describe a lowered level of consciousness.

Chronic responses to toxicant exposure include mutations, cancer, birth defects, and effects on the immune system. Other observable effects, some of which may occur soon after exposure, include gastrointestinal illness, cardiovascular disease, hepatic (liver) disease, renal (kidney) malfunction, neurologic symptoms (central and peripheral nervous systems), and skin abnormalities (rash, dermatitis).

Often the effects of toxicant exposure are subclinical in nature. These include some kinds of damage to immune system, chromosomal abnormalities, modification of functions of liver enzymes, and slowing of conduction of nerve impulses.

14.8. TERATOGENESIS, MUTAGENESIS, CARCINOGENESIS, IMMUNE SYSTEM EFFECTS, AND REPRODUCTIVE EFFECTS

Teratogens are chemical species that cause birth defects, the process of **teratogenesis**. These usually arise from damage to embryonic or fetal cells. However, mutations in germ cells (egg or sperm cells) may cause birth defects, such as Down's syndrome.

The biochemical mechanisms of teratogenesis are varied. These include enzyme inhibition by xenobiotics, deprivation of the fetus of essential substrates, such as vitamins, interference with energy supply, or alteration of the permeability of the placental membrane.

Mutagens alter DNA to produce inheritable traits in the process of **mutagenesis**. Although mutation is a natural process that occurs even in the absence of xenobiotic substances, most mutations are harmful. The mechanisms of mutagenicity are similar to those of carcinogenicity, and mutagens often cause birth defects as well. Therefore, mutagenic hazardous substances are of major toxicological concern.

To understand the biochemistry of mutagenesis, it is important to recall from Chapter 3 that DNA contains the nitrogenous bases adenine, guanine, cytosine, and thymine. The order in which these bases occur in DNA determines the nature and structure of newly produced RNA, a substance produced as a step in the synthesis of new proteins and enzymes in cells. Exchange, addition, or deletion of any of the nitrogenous bases in DNA alters the nature of RNA produced and can change vital life processes, such as the synthesis of an important enzyme. This phenomenon, which can be caused by xenobiotic compounds, is a mutation that can be passed on to progeny, usually with detrimental results.

There are several ways in which xenobiotic species may cause mutations. It is beyond the scope of this work to discuss these mechanisms in detail. For the most part, however, mutations due to xenobiotic substances are the result of chemical alterations of DNA, such as those discussed in the two examples below.

Nitrous acid, HNO_2, is an example of a chemical mutagen that is often used to cause mutations in bacteria. To understand the mutagenic activity of nitrous acid it should be noted that three of the nitrogenous bases — adenine, guanine, and cytosine — contain the amino group, $-NH_2$. The action of nitrous acid is to replace amino groups with a hydroxy group. When this occurs, the DNA may not function in the intended manner, causing a mutation to occur.

Alkylation consisting of the attachment of a small alkyl group, such as $-CH_3$ or $-C_2H_5$, to an N atom on one of the nitrogenous bases in DNA is one of the

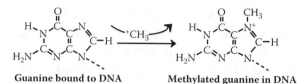

Guanine bound to DNA   Methylated guanine in DNA

Figure 14.11.   Alkylation of guanine in DNA.

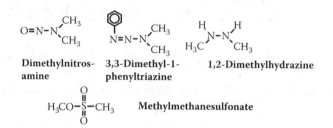

Dimethylnitros-   3,3-Dimethyl-1-   1,2-Dimethylhydrazine
amine   phenyltriazine

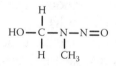 Methylmethanesulfonate

Figure 14.12.   Examples of simple alkylating agents capable of causing mutations.

most common mechanisms leading to mutation. The methylation of "7" nitrogen in guanine in DNA to form *N*-Methylguanine is shown in Figure 14.11. *O*-alkylation may also occur by attachment of a methyl or other alkyl group to guanine's oxygen atom.

A number of mutagenic substances act as alkylating agents, which usually function by attaching the methyl group to one of the nitrogenous bases in DNA. Prominent among these are the compounds shown in Figure 14.12.

Alkylation occurs by way of generation of positively charged electrophilic species that bond to electron-rich nitrogen or oxygen atoms on the nitrogenous bases in DNA. The generation of such species usually occurs by way of biochemical and chemical processes. For example, dimethylnitrosamine (structure in Figure 14.12) is activated by oxidation through cellular NADPH to produce the following highly reactive intermediate:

$$
\begin{array}{c}
\text{H} \\
| \\
\text{HO}-\text{C}-\text{N}-\text{N}{=}\text{O} \\
|\quad\ \ | \\
\text{H}\quad \text{CH}_3
\end{array}
$$

This product undergoes several nonenzymatic transitions to generate the positively charged methyl carbocation, $^{+}CH_3$. This species can bind to the nitrogenous bases on DNA leading to mutations and causing the uncontrolled cell replication characteristic of cancer.

One of the more notable mutagens is tris(2,3-dibromopropyl)phosphate, commonly called "tris," that was used as a flame retardant in children's sleepwear. Tris was found to be mutagenic in experimental animals and its metabolites were found in children wearing the treated sleepwear. This strongly suggested that tris is absorbed through the skin and its uses were discontinued.

Carcinogenesis

Cancer is a condition characterized by the uncontrolled replication and growth of the body's own cells (somatic cells). **Carcinogenic agents** may be categorized as follows:

- Chemical agents, such as nitrosamines and polycyclic aromatic hydrocarbons

- Biological agents, such as hepadnaviruses or retroviruses

- Ionizing radiation, such as x-rays

- Genetic factors, such as selective breeding

Clearly, in some cases, cancer is the result of the action of synthetic and naturally occurring chemicals. The role of xenobiotic chemicals in causing cancer is called **chemical carcinogenesis**. It is often regarded as the single most important facet of toxicology and clearly the one that receives the most publicity.

Chemical carcinogenesis has a long history. As noted earlier in this chapter, in 1775, Sir Percivall Pott, Surgeon General serving under King George III of England, observed that chimney sweeps in London had a very high incidence of cancer of the scrotum, which he related to their exposure to soot and tar from the burning of bituminous coal. Around 1900, a German surgeon, Ludwig Rehn, reported elevated incidences of bladder cancer in dye workers exposed to chemicals extracted from coal tar; 2-naphthylamine,

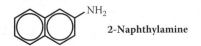

2-Naphthylamine

was shown to be largely responsible. Other historical examples of carcinogenesis include observations of cancer from tobacco juice (1915), oral exposure to radium from painting luminescent watch dials (1929), tobacco smoke (1939), and asbestos (1960).

Large expenditures of time and money on the subject in recent years have yielded a much better understanding of the **biochemical bases of chemical carcinogenesis**. The overall processes for the induction of cancer may be quite complex, involving numerous steps. However, it is generally recognized that there are two major steps in carcinogenesis: an **initiation stage** followed by a **promotional stage**. These steps are further subdivided as shown in Figure 14.13.

Initiation of carcinogenesis may occur by reaction of a **DNA-reactive species** with DNA or by the action of an **epigenetic carcinogen** that does not react with DNA and is carcinogenic by some other mechanism. Most DNA-reactive species are **genotoxic carcinogens** because they are also mutagens. These substances react

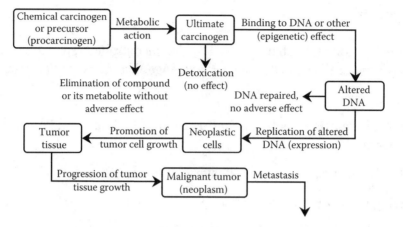

Figure 14.13.   Outline of the process by which a carcinogen or procarcinogen may cause cancer.

irreversibly with DNA. They are either electrophilic or, more commonly, metabolically activated to form electrophilic species, as is the case with electrophilic $^+CH_3$ generated from dimethylnitrosamine, discussed under mutagenesis above. Cancer-causing substances that require metabolic activation are called **procarcinogens**. The metabolic species actually responsible for carcinogenesis is termed an **ultimate carcinogen**. Some species that are intermediate metabolites between procarcinogens and ultimate carcinogens are called **proximate carcinogens**. Carcinogens that do not require biochemical activation are categorized as **primary** or **direct-acting carcinogens**. Some example procarcinogens and primary carcinogens are shown in Figure 14.14.

Most substances classified as epigenetic carcinogens are **promoters** that act after initiation. Manifestations of promotion include increased numbers of tumor cells and decreased length of time for tumors to develop (shortened latency period). Promoters do not initiate cancer, are not electrophilic, and do not bind with DNA. The classic example of a promoter is a substance known chemically as decanoylphorbol acetate or phorbolmyristate acetate, a substance extracted from croton oil.

Chemical carcinogens usually have the ability to form covalent bonds with macromolecular life molecules. Such covalent bonds can form with proteins, peptides, RNA, and DNA. Although most binding is with other kinds of molecules, which are more abundant, the DNA adducts are the significant ones in initiating cancer. Prominent among the species that bond to DNA in carcinogenesis are the alkylating agents which attach alkyl groups — such as methyl (CH_3) or ethyl (C_2H_5) — to DNA. A similar type of compound, **arylating agents**, act to attach aryl moieties, such as the phenyl group

Naturally occurring carcinogens that require bioactivation

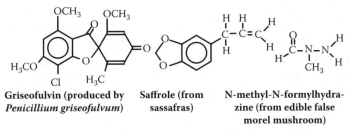

Griseofulvin (produced by   Saffrole (from   N-methyl-N-formylhydra-
Penicillium griseofulvum)   sassafras)   zine (from edible false
morel mushroom)

Synthetic carcinogens that require bioactivation

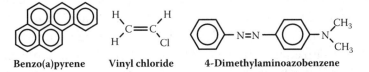

Benzo(a)pyrene   Vinyl chloride   4-Dimethylaminoazobenzene

Primary carcinogens that do not require bioactivation

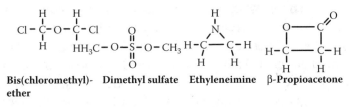

Bis(chloromethyl)-   Dimethyl sulfate   Ethyleneimine   β-Propioacetone
ether

Figure 14.14.   Examples of the major classes of naturally occurring and synthetic carcinogens, some of which require bioactivation, and others of which act directly.

Methyl groups attached to
N (left) or O (right) in
guanine contained in DNA

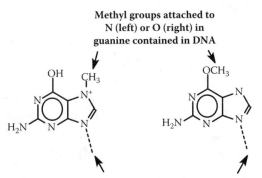

Attachment to the remainder of the DNA molecule

Figure 14.15.   Alkylated (methylated) forms of the nitrogenous base guanine.

to DNA. As shown by the examples in Figure 14.15, the alkyl and aryl groups become attached to N and O atoms in the nitrogenous bases that compose DNA. This alteration in the DNA can initiate the sequence of events that results in the growth and replication of neoplastic (cancerous) cells. The reactive species that donate alkyl groups in alkylation are usually formed by metabolic activation as shown for dimethylnitrosamine in the discussion of mutagenesis earlier in this section.

Testing for Carcinogens

Only a few chemicals have definitely been established as human carcinogens. A well-documented example is vinyl chloride, $CH_2=CHCl$, which is known to have caused a rare form of liver cancer (angiosarcoma) in individuals who cleaned autoclaves in the poly(vinyl chloride) fabrication industry. In some cases, chemicals are known to be carcinogens from epidemiological studies of exposed humans. Animals are used to test for carcinogenicity, and the results can be extrapolated, although with much uncertainty, to humans.

Mutation inferring carcinogenicity is the basis of the **Bruce Ames** test, in which observations are made of the reversion of mutant histidine-requiring *Salmonella* bacteria back to a form that can synthesize its own histidine. The test makes use of enzymes in homogenized liver tissue to convert potential procarcinogens to ultimate carcinogens. Histidine-requiring *Salmonella* bacteria are inoculated onto a medium that does not contain histidine, and those that mutate back to a form that can synthesize histidine establish visible colonies that are assayed to indicate mutagenicity.

Animal tests for carcinogens that make use of massive doses of chemicals may give results that cannot be accurately extrapolated to assess cancer risks from smaller doses of chemicals. This is because the huge doses of chemicals used kill large numbers of cells, which the organism's body attempts to replace with new cells. Rapidly dividing cells greatly increase the likelihood of mutations that result in cancer simply as the result of rapid cell proliferation, not genotoxicity.

Immune System Response

The **immune system** acts as the body's natural defense system to protect it from xenobiotic chemicals, infectious agents, such as viruses or bacteria, and neoplastic cells, which give rise to cancerous tissue. Adverse effects on the body's immune system are being increasingly recognized as important consequences of exposure to hazardous substances. Toxicants can cause **immunosuppression**, which is the impairment of the body's natural defense mechanisms. Xenobiotics can also cause the immune system to lose its ability to control cell proliferation, resulting in leukemia or lymphoma.

Another major toxic response of the immune system is **allergy** or **hypersensitivity**. This kind of condition results when the immune system overreacts to the presence of a foreign agent or its metabolites in a self-destructive manner. Among the xenobiotic materials that can cause such reactions are beryllium, chromium, nickel, formaldehyde, pesticides, resins, and plasticizers.

14.9. HEALTH HAZARDS

In recent years, attention in toxicology has shifted away from readily recognized, usually severe, acute maladies that developed on a short time scale as a result of brief, intense exposure to toxicants, toward delayed, chronic, often less severe illnesses caused by long-term exposure to low levels of toxicants. Although the total impact of the latter kinds of health effects may be substantial, their assessment is very difficult because of factors such as uncertainties in exposure, low occurrence above background levels of disease, and long latency periods.

Assessment of Potential Exposure

A critical step in assessing exposure to toxic substances, such as those from hazardous waste sites is evaluation of potentially exposed populations. The most direct approach to this is to determine chemicals or their metabolic products in organisms. For inorganic species, this is most readily done for heavy metals, radionuclides, and some minerals, such as asbestos. Symptoms associated with exposure to particular chemicals may also be evaluated. Examples of such effects include skin rashes or subclinical effects, such as chromosomal damage.

Epidemiological Evidence

Epidemiological studies applied to toxic environmental pollutants, such as those from hazardous wastes, attempt to correlate observations of particular illnesses with probable exposure to such wastes. There are two major approaches to such studies. One approach is to look for diseases known to be caused by particular agents in areas where exposure is likely from such agents in hazardous wastes. A second approach is to look for **clusters** consisting of an abnormally large number of cases of a particular disease in a limited geographic area, then attempt to locate sources of exposure to hazardous wastes that may be responsible. The most common types of maladies observed in clusters are spontaneous abortions, birth defects, and particular types of cancer.

Epidemiologic studies are complicated by long latency periods from exposure to onset of disease (which, in the case of cancer can be 20 years or more), lack of specificity in the correlation between exposure to a particular waste, pollutant, or substance to which exposure has taken place in the workplace, and the occurrence of a disease, and background levels of a disease in the absence of exposure to a hazardous waste capable of causing the disease.

An important part of estimating the risks of adverse health effects from exposure to toxicants involves extrapolation from experimentally observable data. Usually, the end result needed is an estimate of a low occurrence of a disease in humans after a long latency period resulting from low-level exposure to a toxicant for a long period of time. The data available are almost always taken from animals exposed to high levels of the substance for a relatively short period of time. Extrapolation is then made using linear or curvilinear projections to estimate the risk to human populations. There are, of course, very substantial uncertainties in this kind of approach.

Toxicological considerations are very important in estimating potential dangers of pollutants and hazardous waste chemicals. One of the major ways in which toxicology interfaces with the area of hazardous wastes is in **health risk assessment**, providing guidance for risk management, cleanup, or regulation needed at a hazardous waste site based upon knowledge about the site and the chemical and toxicological properties of wastes in it. Risk assessment includes the factors of site characteristics, substances present (including indicator species), potential receptors, potential exposure pathways, and uncertainty analysis. It may be divided into the following major components:

- Identification of hazard

- Dose–response assessment

- Exposure assessment

- Risk characterization

SUPPLEMENTARY REFERENCES

Baselt, Randall C., *Disposition of Toxic Drugs and Chemicals in Man*, 6th ed., Biomedical Publications, Foster City, CA, 2002.

Benigni, Romualdo, Ed., *Quantitative Structure-Activity Relationship (QSAR) Models of Mutagens and Carcinogens*, CRC Press, Boca Raton, FL, 2003.

Bingham, Eula, Barbara Cohrssen, and Charles H. Powell, *Patty's Toxicology*, 5th ed., John Wiley & Sons, New York, 2001.

Clayson, David Barringer, *Toxicological Carcinogenesis,* CRC Press/Lewis Publishers, Boca Raton, FL, 2001.

Dart, Richard C., *Medical Toxicology*, 3rd ed., Lippincott, Williams & Wilkins, Philadelphia, 2003.

Fenton, John, *Toxicology: A Case-Oriented Approach*, CRC Press, Boca Raton, FL, 2002.

Hodgson, Ernest, *A Textbook of Modern Toxicology*, 3rd ed., John Wiley & Sons, New York, 2004.

Hodgson, Ernest and Joyce A. Goldstein, Metabolism of toxicants: phase I reactions and pharmacogenetics, in *Introduction to Biochemical Toxicology*, 3rd ed., Ernest Hodgson and Robert C. Smart, Eds., Wiley-Interscience, New York, 2001, pp. 67–113, chap. 5.

Hoffman, David J., Barnett A. Rattner, G. Allen Burton, Jr., and John Cairns, Jr., *Handbook of Ecotoxicology*, 2nd ed., Lewis Publishers/CRC Press, Boca Raton, FL, 2002.

Ioannides, Costas, *Enzyme Systems That Metabolise Drugs and Other Xenobiotics*, John Wiley & Sons, New York, 2002.

Klaassen, Curtis D. and John B. Watkins III, Eds., *Casarett and Doull's Essentials of Toxicology*, McGraw-Hill Medical, New York, 2003.

Klaassen, Curtis D., Ed., *Casarett and Doull's Toxicology: The Basic Science of Poisons*, 6th ed., McGraw-Hill Medical, New York, 2001.

Landis, Wayne G. and Ming-Ho Yu, *Introduction to Environmental Toxicology: Impacts of Chemicals upon Ecological Systems*, 3rd ed., CRC Press, Boca Raton, FL, 2004.

Leonard, Barry, Leonard, Ed., *Report on Carcinogens: Carcinogen Profiles*, 10th ed., Collingdale, PA, 2002.

Mommsen, T.P. and T.W. Moon, *Environmental Toxicology*, Elsevier, Boston, MA, 2005.

Parvez, S.H., Ed., *Molecular Responses to Xenobiotics*, Elsevier, Amsterdam, 2001.

Pohanish, Richard P. and Marshall Sittig, *Sittig's Handbook of Toxic and Hazardous Chemicals and Carcinogens*, Knovel Corporation, Norwich, New York, 2002.

Wilson, Samuel H. and William A. Suk, *Biomarkers of Environmentally Associated Disease*, Boca Raton, FL, 2002.

Yu, Ming-Ho, *Environmental Toxicology: Biological and Health Effects Pollutants*, 2nd ed., CRC Press, Boca Raton, FL, 2005.

QUESTIONS AND PROBLEMS

1. How are conjugating agents and Phase II reactions involved with some toxicants?

2. What are Phase I reactions? What enzyme system carries them out? Where is this enzyme system located in the cell?

3. Name and describe the science that deals with the chemical nature and reactions of toxic substances, including their origins, uses, and chemical aspects of exposure, fates, and disposal.

4. What is a dose–response curve?

5. What is meant by a toxicity rating of 6?

6. What are the three major subdivisions of the *dynamic phase* of toxicity, and what happens in each?

7. Characterize the toxic effect of carbon monoxide in the body. Is its effect reversible or irreversible? Does it act on an enzyme system?

8. Of the following, choose the one that is **not** a biochemical effect of a toxic substance: (1) impairment of enzyme function by binding to the enzyme, (2) alteration of cell membrane or carriers in cell membranes, (3) change in vital signs, (4) interference with lipid metabolism, and (5) interference with respiration.

9. Distinguish among teratogenesis, mutagenesis, carcinogenesis, and immune system effects. Are there ways in which they are related?

10. As far as environmental toxicants are concerned, compare the relative importance of acute and chronic toxic effects and discuss the difficulties and uncertainties involved in studying each.

11. What are some of the factors that complicate epidemiologic studies of toxicants?

12. Match the type of exposure on the left, below, with its description or example from the right.

a. Acute local	i. Cancer of the mouth at age 31 by a person who has been using snuff (chewing tobacco) since age 9
b. Chronic systemic	
c. Acute systemic	ii. Bladder cancer developed by coal tar dye workers after many years of work
d. Chronic local	iii. Fatal dose of cyanide
	iv. Occurs at a specific location over a time period of a few seconds to a few hours and may affect the exposure site, particularly the skin, eyes or mucous membranes

13. A distinction was made between the kinetic phase and dynamic phase of toxicology. Of the following, the untrue statement pertaining to these is:
   a. In the dynamic phase, the toxicant reacts with a receptor or target organ in the primary reaction step.

   b. The kinetic phase involves absorption, metabolism, temporary storage, distribution, and, to a certain extent, excretion of the toxicant or its precursor compound called the protoxicant.

   c. In the dynamic phase, there is a biochemical response.

d. Following the biochemical response, physiological and/or behavioral manifestations of the effect of the toxicant may occur.

e. Phase 2 reactions commonly occur in the dynamic phase.

14. Pertaining to the major sites of exposure, metabolism, and storage, routes of distribution and elimination of toxic substances from the body as shown in Figure 14.1, the true statement of the following is:

a. The most likely place to find an unmetabolized, water-insoluble xenobiotic compound several weeks after it was absorbed through the skin is in the liver.

b. The most likely place for loss from the body of a nonvolatile, Phase 2 metabolite of a xenobiotic compound is through exhaled air.

c. There are no barriers that could prevent a cancer-causing metabolite of a procarcinogen from reaching DNA that it might affect.

d. An entry site that presents the fewest barriers, either from a screening organ or from barriers inherent to the entry site, itself, is the lung.

e. Systemic poisons are most likely to affect the skin at the point where they are absorbed.

15. Why were the genital areas of chimney sweeps in England during the 1700s particularly susceptible to cancer caused by coal tar?

16. Substance "A" and substance "B" taken separately have only minimal toxic effects, whereas taken together at similar dose levels, severe adverse effects occur. What sort of phenomenon is illustrated by this example in relation to toxic substances?

17. In what sense are immunosuppression and hypersensitivity opposites?

18. What is the distinction between epigenetic carcinogens and genotoxic carcinogens?

19. Toxicologically, what is the distinction between a detoxified metabolite and an active metabolite?

20. What are some examples of materials that have been shown directly to produce cancer in humans?

15. BIOPRODUCTIVITY FOR A GREENER FUTURE

15.1. FROM BIOMATERIALS TO PETROLEUM AND BACK AGAIN

This chapter is about the production of materials from living sources, especially plants. There is nothing new about this source of materials. Throughout virtually all of humankind's brief stay on Earth, wood and fiber were the primary raw materials used by humans to make dwellings, as fuel, and in many other applications. Animals feeding on plant biomass also provided key raw materials, such as leather for shoes, fur wraps to protect against Ice Age cold, wool from sheep, and silk from the cocoons of insects.

Around 1900, chemists began to develop means to make a variety of materials from cheap and abundant petroleum. This was especially so with the development of synthetic polymers made from small petroleum molecules. These polymers could be used to produce plastics, synthetic fibers, and synthetic rubber. Usually making these materials required significant modification of petroleum molecules by breaking larger molecules of petroleum liquids down to smaller molecules of gaseous materials, rearranging molecular structures, and attaching oxygen (partial oxidation) or nitrogen. During the 1900s, such efforts were very successful and hundreds of different petrochemical-based materials became available. A vast petrochemicals industry developed in which textiles, plastics, rubber, and a variety of other useful materials were synthesized chemically from liquid petroleum and gaseous hydrocarbon feedstocks. By 2000, with the exception of metals and wood employed as a building material, most of the materials used in the anthrosphere came from petroleum, and plant biomass was largely supplanted as a raw material.

But, by the early 2000s, a major problem had developed because it had become evident that petroleum was reaching its peak as a source of organic raw materials. True, petroleum supplies may last longer than most pessimists believe. New sources may still be found, especially in extreme and remote locations, such as below deep ocean floors. Somewhat less than half of the petroleum in an oilfield is removed by conventional means, and advanced recovery techniques have the potential to extract as much as has been taken previously. Recycling of materials and conservation of petroleum can reduce consumption. But, inevitably, sooner or later, petroleum will become very scarce and largely unavailable.

And so it is back to substances produced by living organisms as a source of raw materials and feedstocks for industry. These materials are produced originally by **photosynthesis** by green plants and some microorganisms, especially cyanobacteria, a complex process represented by the simple equation

$$CO_2 + H_2O + h\nu \rightarrow \{CH_2O\} + O_2 \tag{15.1.1}$$

in which $h\nu$ represents solar energy and $\{CH_2O\}$ is a simplified formula for biomass. Virtually all organisms depend upon photosynthesis for food, either directly or somewhat higher on the food chain. The food supply chain is so short, that one year of a major global shortfall in photosynthesis, such as might happen in the case of an asteroid impact or massive volcanic eruption, would result in the starvation of millions of animals including humans. Photosynthesis removes excess carbon dioxide from the atmosphere and restores elemental oxygen.

15.2. TYPES OF BIOMATERIALS

As noted above, biomass from plants will have to become the most abundantly used replacement for petroleum-based materials. There are several important categories of biomass raw materials, which are discussed in this section.

By photosynthetic processes, plants generate **carbohydrates**, for which the approximate simple formula is CH_2O. The most fundamental of the carbohydrates, and the one generated directly by photosynthesis, is **glucose** (chemical formula $C_6H_{12}O_6$), known as a simple sugar or monosaccharide. Two molecules of simple sugars with chemical formulas of $C_6H_{12}O_6$ can bond together chemically with the elimination of a molecule of H_2O to produce a disaccharide (chemical formula $C_{12}H_{22}O_{11}$). Sucrose used as a sweetener in foods and drinks is a disaccharide. Many molecules of simple sugars can bond together with the elimination of a molecule of H_2O for each one bonded to produce polysaccharides consisting of huge macromolecules. Starch and cellulose are the most prominent such polysaccharides.

As discussed in Section 15.8, glucose is being studied intensely as a feedstock for making a variety of substances. It is most readily obtained by breaking down starch produced by grains. Cornstarch is the most abundant source of glucose. Cellulose found in woody plant material and in substances such as cotton is a very abundant potential source of glucose, although it is more difficult to extract glucose from cellulose than from starch. Sucrose squeezed as sap from sugarcane is a ready source of glucose and of the similar sugar fructose. Starch is an excellent raw material for many applications, as is cellulose. Chemically modified starch and chemically altered cellulose both have a variety of applications.

Lignin is a biological polymer with a complex structure, which is associated with cellulose in woody parts of plants, binding fibers of cellulose together in lignocellulose. In a number of applications, most prominently paper making, it is necessary to separate the lignin from the cellulose. Unfortunately, lignin is a refractory material of rather variable molecular composition that has few uses. It would be of great use if plants could be developed that produced copious quantities of cellulose without lignin. The cotton plant does this, producing a pure form of cellulose in the cotton fiber, but the small ball of cotton is only a small fraction of the cotton plant biomass.

Many useful **lipid oils** are extracted from a variety of plant seeds including soybeans, sunflowers, and corn. In addition to their food uses, these oils are used in a large variety of applications including raw materials for making other chemical products, lubricants, and as biodiesel fuels. Part of the usefulness of lipid oils in many applications is because of their similarity to petroleum hydrocarbons. Volatile solvents, most commonly the 6-carbon straight-chain alkane *n*-hexane, C_6H_{14}, are used to extract oils from plant sources. In this process, the solvents are distilled off from the extract and recirculated through the process.

Some plants produce hydrocarbons directly in the form of **terpenes**. Most of the plants that produce terpenes are conifers (evergreen trees and shrubs such as pine and cypress), plants of the genus *Myrtus*, and trees and shrubs of the genus *Citrus*. One of the most common terpenes emitted by trees is α-pinene, a principal component of turpentine, a material formerly widely used in paint formulations because it reacts with atmospheric oxygen to form a peroxide, and then a hard resin that produces a durable painted surface. The only commercial source of natural rubber, at the moment, is the Brazilian rubber tree *Hevea brasiliensis*. The hydrocarbon terpenes that occur in rubber trees can be tapped from the trees as a latex suspension in tree sap. Steam treatment and distillation can be employed to extract terpenes from sources such as pine or citrus tree biomass.

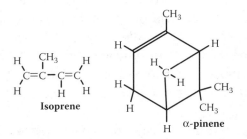

Isoprene   α-pinene

Grain seeds are rich sources of protein, almost always used for food, but potentially useful as chemical feedstocks for specialty applications. An exciting possibility just now coming to fruition in a practical sense is to transplant genes into plants so that they will make specialty proteins, such as medicinal agents.

Biological materials used as sources of feedstocks are usually complex mixtures, which makes separation of desired materials difficult. However, some compensation is made for that disadvantage in that in some biological starting materials nature has done much of the synthesis of the final product. Most biomass materials are partially oxidized as is the case with carbohydrates, which contain approximately one oxygen atom per carbon atom (compared to petroleum hydrocarbons which have no oxygen). This can avoid expensive, sometimes difficult oxidation steps, which may involve potentially hazardous reagents and conditions. The complexity of biomass sources can make the separation and isolation of desired constituents relatively difficult.

There are several main pathways by which feedstocks can be obtained from biomass. The most straightforward of these is a simple physical separation of biological materials, such as squeezing oil from oil-bearing biomass or tapping latex from rubber trees. Only slightly more drastic treatment consists of extraction of oils by organic solvents. Physical and chemical processes can be employed to remove useful biomass from the structural materials of plants, which consist of lignocellulose composed of cellulose bound together by lignin "glue."

15.3. PLANT PRODUCTIVITY

Somewhat less than 0.5% of the energy of sunlight falling on green plants is captured and converted to chemical energy in biomass. Despite this extremely low conversion efficiency, enormous quantities of biomass are generated each year. In the U.S. alone, approximately 700 billion pounds of corn, soybeans, wheat, rice, and other small-grain cereals are produced each year. But this is only about half of the total biomass potentially available for use as raw material. The by-product biomass left in the field from crop production is around 2600 pounds/acre on a dry-weight basis. In the case of corn, sorghum, soybeans, and wheat, alone, the U.S. agricultural system generates around 520 billion pounds per year of dry crop by-product biomass. This does not consider wood, grass, and other nongrain plants, which likewise produce huge quantities of biomass.

Plant productivity has increased significantly in recent decades. For production of grains alone, dwarf varieties of wheat, rice, and barley during the 1950s (the "green revolution") have increased yields around 3-fold. Some noncereal crops cultivated for their biomass can boost production greatly. Genetic engineering (see Section 15.4) has enormous potential to increase biomass production. Especially great increases could occur if plants could be developed with relatively higher photosynthesis efficiencies.

For food production alone, to say nothing of biomaterials production, it is essential that conditions be maintained that are conducive to maximum plant productivity. One of the major factors in maintaining bioproductivity is climate. Global warming with accompanying drought and other effects detrimental to plant growth is one of the major factors threatening bioproductivity. Another threat is the loss of productive land to development, paving it over to construct highways, parking lots, and buildings. Land use is crucial, and the tendency in some societies of many individuals to want to live in buildings far from urban centers and surrounded by large expanses of land devoted to the growing of ornamental grass and other plants of little use for the production of food or biomaterials is deplorable.

Not to be overlooked in the production of biomass are organisms that do not grow on land. Especially promising is the possibility of growing photosynthetic marine algae in shallow ponds of saltwater from oceans or from underground brackish water sources located on desert lands unsuitable for other plants. The oldest and one of the most productive photosynthetic organisms are actually bacteria, the **cyanobacteria**. Despite their long-standing and misleading name of "blue-green algae," these prokaryotic organisms may have yellow, brown, red, black, or green casts in addition to appearing blue-green. They are the oldest known organisms appearing as fossils in rock formations dating back 3.5 billion years. As testimony to their photosynthetic prowess, they were responsible for generating Earth's atmospheric oxygen and are believed to be responsible for majority of the biomass that later developed into petroleum. They later took up residence in certain eukaryotic cells, a hospitable atmosphere in return for which they generated food for their hosts and developed into the plant chloroplasts where plant photosynthesis occurs. Cyanobacteria thrive in water contaminated by sewage and have the potential to make a significant contribution to biomaterials if procedures can be developed to harvest them efficiently.

15.4. GENETIC MATERIAL AND ITS MANIPULATION

One of the more promising developments for the production of biomaterials is offered by genetic engineering, the ability to manipulate genetic material in organisms to provide a variety of effects on plants and the materials generated by them.

Recall from Chapter 3 that nucleic acids are huge biopolymer molecules that are basically the code for living organisms. Deoxyribonucleic acid (DNA), was noted as the macromolecule that stores and passes on genetic information that organisms need to reproduce and synthesize proteins. Recall that DNA is composed of repeating units called *nucleotides* each consisting of a molecule of the sugar 2-deoxy-β-D-ribofuranose, a phosphate ion, and one of the four nitrogen-containing bases, adenine, cytosine, guanine, and thymine (conventionally represented by the letters A, C, G, and T, respectively). DNA is one of two nucleic acids, the other one of which is ribonucleic acid (RNA). Like DNA, RNA consists of repeating nucleotides, but the sugar in RNA is β-D-ribofuranose and it contains uracil instead of thymine in its bases. The structural formulas of segments of DNA are shown in Figure 15.1; those of RNA are very similar except for the different sugar and uracil instead of thymine among the four nitrogenous bases.

The structure of DNA is a key aspect of its function and its elucidation by Watson and Crick in 1953 was a scientific insight that set off a revolution in biology that is going on to this day. The huge DNA molecules consist of two strands counterwound with each other and held together by hydrogen bonds (discussed in Chapter 4, Section 4.1 and illustrated for water molecules in Figure 4.1). A representation of this structure is shown in Figure 15.2. In this structure, the hydrogen bonds connecting complementary bases on the two strands are represented by dashed lines. Because

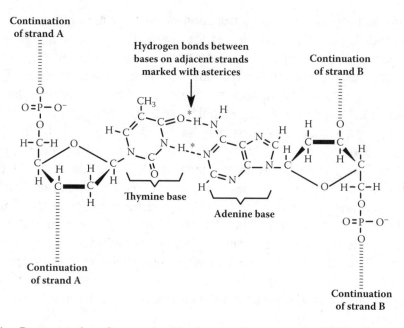

Figure 15.1.   Representation of two nucleotides in two adjacent strands of DNA showing hydrogen bonding between the bases thymine and adenine. These two bases bonded together by hydrogen bonds constitute a base pair.

Figure 15.2.   Representation of the double helix structure of DNA. Hydrogen bonds between complementary bases on the two strands are shown by dashed lines.

of their structures that make hydrogen bonding possible, adenine on one strand is always hydrogen-bonded to thymine on the opposite strand and guanine to cytosine. During cell division, the two strands of DNA unwind and each generates a complementary strand for the DNA of each new cell. Figure 15.1 shows a representation of two complementary strands of DNA in which thymine and adenine from the two strands are hydrogen-bonded together. When guanine is opposite cytosine, the two bases are bonded by three hydrogen bonds.

In organisms that are more complex than bacteria and those which have eukaryotic cells, DNA is divided into units associated with protein molecules called **chromosomes**. The number of these varies with the organism; humans have 23 pairs of chromosomes, a total of 46. The strands of DNA in chromosomes, in turn, are divided into sequences of nucleotides, each distinguished by the nitrogen-containing base in it. These sequences of nucleotides give directions for the synthesis of a specific kind of protein or polypeptide. (Polypeptide is a general term for polymers of amino acids; proteins are the relatively long-chain polypeptides.) These specific groups of nucleotides, each of which has a specific function, are called **genes**. When a particular protein is made, DNA produces a nucleic acid segment designated *m*RNA, which goes out into the cell and causes the protein to be formed through a process called **transcription** and **translation** (the gene is said to be **expressed**). Whereas the number of chromosomes in an organism is now easy to determine, the number of genes is a matter of debate and is still not known for most organisms.

Proteins are the biological molecules that make up much of the structure of cells and that perform most of the key functions of living organisms. Proteins are made according to directions provided by cellular DNA. The steps in protein synthesis are the following:

1. The DNA in a gene that is specific for a particular protein transfers information for the protein synthesis to RNA.

2. The RNA links with a cell ribosome, which is the protein-synthesizing entity of the cell.

3. Using directions provided by the RNA, the ribosome assembles amino acids into a protein.

4. The protein performs the function for which it is designed in the organism; for example, it may function as an enzyme to carry out metabolic processes.

As the entities that give the directions for protein synthesis, genes are obviously of the utmost importance in living organisms. As discussed later in this section, genes can now be transferred between different kinds of organisms and will direct the synthesis of the protein for which they are designed in the recipient organism. It is now known that a number of human diseases are the result of defective genes, and there is a genetic tendency toward getting other kinds of diseases. For example, certain gene characteristics are involved in susceptibility to breast cancer.

Genome Sequencing and Green Science and Technology

The Human Genome Project and related genome sequencing of other organisms have a number of implications for green science and technology. One of the key goals of green chemistry is to use chemicals that have maximum effectiveness for their stated purpose with minimum side effects. This certainly applies to pharmaceuticals in which knowledge of the human genome may enable development of drugs that do exactly what they are supposed to do without affecting nontarget systems. This means that drugs can be made very efficiently with little waste material.

Some of the most important effects of DNA sequencing as it relates to green science and technology has to do with a wide variety of organisms other than humans. With an exact knowledge of DNA and the genes that it contains, it is possible to deal with organisms on a highly scientific basis in areas such as pest control and the biosynthesis of raw materials. An accurate map of the genetic makeup of insects, for example, should result in the synthesis of precisely targeted insecticides, which kill target pests without affecting other organisms. Such insecticides should be effective at very low doses, thus minimizing the amount of insecticide that has to be synthesized and applied, consistent with the goals of green technology.

An exact knowledge of the genomes of organisms is extremely helpful in the practice of genetic engineering in which genes are transferred between species to enable production of desired proteins and to give organisms desirable characteristics, such as pest resistance. A number of medically useful proteins and polypeptides are now produced by genetically engineered microorganisms, most commonly genetically modified *Escherichia coli* bacteria. Perhaps the greatest success with this technology has been the biosynthesis of human insulin, a lack of which causes diabetes in humans. Two genes are required to make this relatively short polypeptide which consists of only 51 amino acids. Other medically useful substances produced by genetically engineered organisms include human growth hormone, tissue plasmogen activator that dissolves blood clots formed in heart attacks and strokes, and various vaccine proteins to inoculate against diseases such as meningitis, hepatitis B, and influenza.

15.5. GENETIC ENGINEERING

Ever since humans started raising crops (and later animals) for food and fiber, they have modified the genetic makeup of the organisms that they use. This is particularly evident in the cultivation of domestic corn, which is physically not at all like its wild ancestors. Until now, breeding has been a slow process. Starting with domestication of wild species, selection and controlled breeding have been used to provide desired properties, such as higher yield, heat and drought tolerance, cold resistance, and resistance to microbial or insect pests. For some domesticated species, these changes have occurred over thousands of years. During the 1900s, increased understanding of genetics greatly accelerated the process of breeding different varieties.

The development of high-yielding varieties of wheat and rice during the "green revo-lution" of the 1950s has prevented (or at least postponed) starvation of millions of people. A technology that enabled a quantum leap in productivity of domestic crops was the development of **hydrids** from crossing of two distinct lines of the same crop, dating in a practical sense from the mid-1900s.

This section discusses the genetic modification of organisms to enhance their value. It addresses plants primarily because more effort has been made and more things have been accomplished in plant breeding than with other kinds of organisms. However, the general principles discussed apply to animals and other kinds of organisms as well.

Traditional breeding normally takes a long time and depends largely upon random mutations to generate desirable characteristics. One of its greatest limitations has been that it is essentially confined to the same species, whereas more often than not, desired characteristics occur in species other than those being bred. Since about the 1970s, however, the possibility has arisen of using **transgenic technology** to transfer genes from one organism to an entirely different kind. This has raised a vast array of possibilities for greatly modified species that could be applied to many different purposes. And it has led as well to a number of concerns regarding unintended consequences of the technology. Ideally, transgenic technology can be used beneficially in plant breeding to increase tolerance to stress, increase yield, enhance the value of the end product by enriching it in desired biochemicals such as essential amino acids, and otherwise make plants more useful.

Transgenic technology is possible because of the existence of DNA within cells. This long-chain biological polymer directs cell reproduction and metabolism as discussed in Section 15.4. Transgenic technology is possible because a gene in DNA will make the protein for which it is designed in an organism that is quite different from the one in which the gene originated. So a gene transferred from one organism to another as a segment of DNA will often perform the function for which it was developed in the recipient organism. The details of how segments of DNA are transferred between organisms are beyond the scope of this work. Enzymes are used in the process, with restriction enzymes cutting out desired regions of DNA and ligase enzymes joining the ends of DNA together. Enzymes are used to further manipulate and amplify the DNA.

Perhaps the most difficult aspect of transgenic technology is identifying the genes responsible for desired characteristics and locating them among the millions of repeating units comprising the DNA strand. In addition to identifying specific genes, it is necessary to learn how they interact with other genes and the mechanisms by which they are regulated and expressed, the process by which a gene generates a specific protein.

After a specific gene is isolated, it is cloned by insertion into a bacterium, which reproduces the gene many times. In order for a gene to generate a desired protein at the appropriate time and location in a plant, a **promoter** must be added that functions as a switch. The easiest promoter to use is a **constitutive promoter** that causes the gene to be expressed in most of the plant's tissues and throughout its lifetime. The most successful promoter for this purpose is designated CaMV35S, which is isolated from the cauliflower mosaic virus. Other more specific promoters have also been used, such as those that are induced by light and function during photosynthetic processes. Much of the current effort in transgenic technology is devoted to the use of specific promoters that cause the gene to be expressed only where and when its protein product is needed.

So far, two major methods have been used to insert genes into a plant cell. The **gene gun** uses a very small projectile to literally shoot genetic information into cells. This method has been used with monocot ("grassy") species including corn and wheat. It suffers from a low percentage of "hits."

The most widely employed method of gene insertion is the *Agrobacterium* method widely used on dicot (broad-leafed) species, such as potatoes and soybeans, and more recently adapted to monocots as well. This method uses a bacterium that thrives in soil called *Agrobacterium tumafaciens* (the cause of crown gall disease in plants), which infects plants, using the plants' metabolic processes for its own reproduction. The mechanism by which *Agrobacterium* is used to insert genetic information into plant cells is complicated and not completely understood. The bacterium enters a plant through a wound in the plant stem or leaves. Somehow the DNA incorporated into the bacterium is transferred through plant cell protoplasm to the plant DNA. This process may occur when the plant DNA becomes uncoiled during cell reproduction.

Only a few percent of plant cells targeted for gene insertion actually incorporate and express the gene. Therefore, it is necessary to have some means of knowing if the gene insertion has been successful. This is accomplished by the insertion of **marker genes** that make plant cells resistant to herbicidal compounds or antibiotics that kill normal plant cells. Plant cells are placed in media containing the toxic materials, and those cells that reproduce are the ones into which the desired genes have been successfully inserted. Following selection of the viable cells that presumably contain the desired transplanted genes, the cells are grown in tissue cultures in the presence of growth-promoting hormones and nutrients required for growth. This leads to the production of whole plants that produce seeds. Additional plants are grown from these seeds and evaluated for the desired characteristics.

Once plants containing desired transgenes have been produced, an exhaustive evaluation process occurs. This process has several objectives. The most obvious of these is an evaluation of the transplanted gene's activity to see if it produces adequate quantities of the protein for which it is designed. Another important characteristic is whether or not the gene is passed on reliably to the plant's progeny through successive generations. It is also important to determine whether the modified plant grows and yields well and if the quality of its products is high.

Only a few strains of plants are amenable to the insertion of transgenes and, normally, their direct descendants do not have desired productivity or other characteristics required for a commercial crop. Therefore, transgenic crops are crossbred with high-yielding varieties. The objective is to develop a cross that retains the transgene while having desired characteristics of a commercially viable crop. The improved variety is subjected to exhaustive performance tests in greenhouses and fields for several years and in a number of locations. Finally, large numbers of genetically identical plants are grown to produce seed for commercial use.

15.6. BIOMATERIALS AND THEIR PROCESSING

Biomaterials, sometimes referred to as plant/crop-based resources, is a term used here to describe a broad range of materials produced by plants including those grown as crops. As supplies of petroleum become exhausted, biomaterials will have to serve as raw materials for polymers, plastics, rubber, adhesives, and the vast variety of other feedstocks now provided by petroleum. This will require a complete overhaul of production systems that have developed around petroleum as a feedstock. Whereas petroleum consists of hydrocarbons, the high-volume biomaterials are carbohydrates, lignins, and plant oils. Chemically, a major distinguishing feature between petroleum materials and biomaterials is that the latter contain oxygen. This can actually be an advantage in that many of the materials now made from petroleum are synthesized by partial oxidation of the hydrocarbon feedstocks. Whereas with petroleum it is necessary to deal with whatever material comes from the ground, the raw material from plants can be manipulated by growing different varieties of plants and now by genetically engineering plants to produce desired materials. Ideally, plants could be used as bioreactors to synthesize desired intermediates or even end products.

Currently, the major sources of biomaterials are wood and agricultural crops and the processing streams of these biomaterials. A biomaterial-based economy would undoubtedly have crops dedicated to the production of specific kinds of materials. Aside from food and animal feed, a variety of materials are now made from biomaterials. In volume, by far the greatest source of biomaterials is wood, which is used directly for lumber and is processed to make paper, paperboard, and wood-based composites. Oils and gums with a variety of applications are extracted from wood. Non-food starch is used to make polymers, resins, and adhesives. Vegetable oils from sources such as soybeans are used to make resins, paints, surfactants, and inks. Formerly the source of all rubber, natural rubber is still employed for tires and other rubber goods. Cellulose is processed to make polymers and textile fibers. Small amounts of lignin are used to make vanillin, adhesives, and chemicals used in tanning.

It is useful to consider the sources of biomaterials in relation to the processes to which they may be subjected as shown in Figure 15.3. In this figure, each numbered square indicates a specific class of biomaterials acted on by processing of a specific level of sophistication. Of these, the least sophisticated and productive is represented by square 4A in which by-products and wastes are processed by current means, including those used in the petrochemical industry. At present, a current

	A. Byproducts and waste materials	B. Existing crops and trees	C. Crops dedicated to specific applications	D. Genetically engineered plants
1. Processing unique molecules	1A	1B	1C	1D
2. Biochemical processing	2A	2B	2C	2D
3. Adapted chemical processing	3A	3B	3C	3D
4. Current processing	4A	4B	4C	4D

Figure 15.3.   Illustrations of the options available for the use of a variety of biomaterials through different levels of processing sophistication.

area of high activity is represented by square 2B in which existing biomaterials are subjected to biochemical processing or 3C in which dedicated crops are subjected to adapted chemical processing. The ultimate in sophistication is represented by 1D in which processes designed for unique specialty molecules are applied to such molecules produced from genetically engineered plants. In general, therefore, technology needs to move as rapidly as possible from 4A to 1D to realize the maximum potential of biomaterials. Specific operations that are carried out or that potentially will be carried out in this matrix are discussed below.

There are many current processes for utilization of crop and forestry by-products (4A in Figure 15.3). In processing forestry products, spent pulping liquors are used to make lignosulfonate surfactants, and tannin is produced from bark. Grains processed for food also yield starch-based adhesives, citrates, amino acids, and furfural, a solvent and chemical raw material. Adapted chemical processing (3A) may enable extraction of relatively more valuable materials from by-products, such as sugars that can be fermented to alcohols. A great deal of effort is being made in biochemical processing of by-products (2A) using microorganisms or enzymes to break down the biomaterials to a variety of molecules that can have value.

Current processing with existing chemistry, largely based upon the petrochemical industry, is not very attractive for existing crops and trees (4B). Adapted chemical processing (3B) may enable better utilization of cellulosic materials from the woody parts of existing crop plants. Biochemical processing of existing crops, especially grains, (2B) is in fact widely practiced today. The most commonly cited process is the production of billions of kilograms of high-fructose corn syrup from corn grain. Glucose obtained biochemically from cornstarch is a valuable raw material and can be used to make a variety of chemicals.

Until now, the kinds of crops grown are largely determined by demand for food products and animal feed. However, when the use of biomaterials as chemical feedstocks is taken into consideration, the mix of existing crops and particular strains of such crops could change significantly. For example, it is now generally desirable to grow varieties of corn that produce maximum amounts of grain with minimum quantities of cornstalk and leaf material. With adapted chemical processing (3C) or biochemical processing (2C), it might be possible to make greater use of the non-grain portion of the corn plant leading to greater plantings of hybrids with larger stalks. Greater demand for plant oils in chemical synthesis combined with improved methods of extracting and processing such oils could lead to a shift toward the growing of oil-producing plants.

Though somewhat controversial and not without risks, the potential of genetically modified plants to produce valuable biomaterials will lead to much greater production of such plants in the future. Genetic engineering of plants provides great possibilities for providing an entirely new raw materials base for making all of the products currently based on the petrochemical industry. The potential exists to generate a wide variety of materials with many uses. Plants may be developed that provide large quantities of specific chemicals, thereby enabling industry to bypass some of the synthesis steps. An intriguing possibility is to develop varieties with higher efficiencies of photosynthesis, a process that currently converts only a very small fraction of incident sunlight to biochemical energy. Doubling current rates of photosynthesis would vastly increase the quantities of biomaterials produced.

Transgenic technologies have already been applied to the modification of plants and are in various stages of development from evaluation through commercial production. These technologies have been applied to the alteration of carbohydrates (sugars, starch, and cellulose). The distribution of oils and fatty acids has been altered to give more desirable mixes of these materials. An important alteration has been in the relative quantities of types of proteins generated by plants. Many plant sources of protein are deficient in one or more amino acids making these plant materials relatively less complete sources of protein for a balanced diet. Genetic engineering has been used to add essential amino acids to plant protein sources. The characteristics of fibers produced by plants have been modified transgenically, making plants sources of more desirable fibers. Antibodies, industrial enzymes, and desirable polymers can be produced by transgenic plants. An important accomplishment of transgenic plants is the production of specialty carotenoids, sterols, and other valuable secondary compounds.

In addition to producing unique materials in large quantities through genetically engineered plants, genetic engineering can be applied as well to processes by which biomaterials are converted to finished products. Genetically engineered microorganisms can be expected to produce specialty chemicals that are very difficult or impossible to make by conventional chemical means. Very efficient genetically engineered microorganisms and enzyme systems derived from them can be expected to generate enormous quantities of commodity chemical feedstocks from highly productive genetically engineered plants.

The uses of microorganisms operating in fermentation processes to generate commodity chemicals were discussed above. Plants are the other kind of organism that can be used for producing chemicals. Indeed, the nutrients used for fermentation processes come originally from plants. Fermentation is, in a sense, not a very efficient means of producing chemicals because of the consumption of nutrients to support the microorganisms and their reproduction and because of the generation of large quantities of by-products. Plants, which generate their own biomass from atmospheric carbon dioxide and water are very efficient producers of materials. Wood and the cellulose extracted from it are prime examples of such materials.

In addition to their efficient production of biomass, plants offer distinct advantages in their production and harvesting. Genetics determine the materials that a plant makes, and once a crop is growing in a field, the products it is programmed for will be produced without fear of contamination by other organisms, which is always a consideration in fermentation. Plants can be grown by relatively untrained personnel using well-known agricultural practices. Plant matter is relatively easy to harvest in the form of grains, stalks, and leaves, which can be taken to a specialized facility to extract needed materials.

The production of feedstocks and other chemical commodities from plants has been limited by the genetic restrictions inherent to plants. Now, however, transgenic plants can be bred to produce a variety of materials directed by genes transplanted from other kinds of organisms. For example, as discussed in Section 15.12, plants have even been developed to synthesize plastics. Another limitation of the production of materials by plants has been the mixture of these materials with other matter generated by plants. The intimate mixture of useful wood cellulose with relatively useless lignin is a prime example of this problem. Again, transgenic technology can be expected to be helpful in developing plants that produce a relatively pure product (such as the almost-pure cellulose in cotton).

The potential of plants to produce useful products has been greatly increased by the development of hybrid plants with spectacular abilities to generate biomass by photosynthesis. Corn is one of the more productive field crops, and hybrid varieties produce large quantities of grain and plant biomass. Sugarcane is noted for its ability to produce biomass, some in the form of sugar, much more in the cane stalk biomass, which has relatively few uses, other than for fuel. One of the more exciting developments of productive hybrid plants is the hybrid poplar tree which, nourished by minimal amounts of fertilizer and watered by economical trickle irrigation systems, grows within a few years to a harvestable size for the production of wood pulp and wood for plywood. The ability of these trees to generate cellulose that can be converted to glucose means that they may serve as the basis of an entire plant-based chemicals industry. The possibility exists that they can be genetically engineered to produce other chemicals as well.

One of the more promising areas of materials production by genetically engineered plants is their ability to generate specialized proteins in large quantities. Ventria Bioscience is a company that specializes in the production of specialty proteins important in human and animal health biosynthesized by genetically engineered plants. The advantage of plants in this application is that they have the potential for generating very large ton quantities of proteins compared to bacteria, yeast, or fungi, which are traditionally used. Among the first proteins produced by genetically engineered plants are lactoferrin and lysozyme.

Lactoferrin is an iron-binding protein encountered in breast milk and epithelial surface secretions including saliva and tears. It is a natural antimicrobial agent against bacteria, fungi, and viruses that acts in part by depriving microorganisms of iron. It is an antioxidant. Lactoferrin can be used to treat acute diarrhea, a leading cause of death among infants and children in less developed countries. It is also effective in the treatment of topical (skin surface) infections, including fungal infections and inflammations.

Lysozyme protein is produced in the body along with the same secretions that contain lactoferrin. Lysozyme acts as a protective barrier against microorganisms thereby preventing infections. It is effective against bacteria, viruses, and fungi. It is used to treat the same conditions as lactoferrin.

15.7. FEEDSTOCKS

Feedstocks are the main ingredients that go into the production of chemical products. Reagents act upon feedstocks and often the two are not readily distinguished. Feedstock selection largely dictates the reactions and conditions that will be employed in a chemical synthesis and is, therefore, of utmost importance in the practice of green chemistry. A feedstock should be as safe as possible. The source of a feedstock can largely determine its environmental impact, and the acquisition of the feedstock should not strain Earth's resources. The process of isolating and concentrating a feedstock can add to the potential harm of otherwise safe materials. This is true of some metal ores in which corrosive and toxic reagents (in the case of gold, cyanide) are used to isolate the desired material.

As a general rule, it is best if feedstocks come from renewable sources rather than depletable resources. A biomass feedstock, for example, can be obtained as a renewable resource grown by plants on land, whereas a petroleum-based feedstock is obtained from depletable crude oil resources. However, the environmental tradeoffs between these two sources may be more complex than first appears in that the petroleum feedstock may simply be pumped from a few wells in Saudi Arabia, whereas the biomass may require large areas of land, significant quantities of fertilizer, and large volumes of irrigation water for its production. However, as discussed above, dwindling supplies of petroleum will dictate that in the future most organic feedstocks will have to come from biological sources.

Much of the challenge and potential environmental harm in obtaining feedstocks is in separating the feedstock from other materials. This is certainly true with petroleum, which consists of many different hydrocarbons, only one of which may be needed as the raw material for a particular kind of product. Cellulose from wood, which can be converted to paper and a variety of chemicals, is mixed intimately with lignin, from which it is separated only with difficulty.

In evaluating the suitability of a feedstock, it is not sufficient to consider just the hazards attributable to the feedstock itself and its acquisition. That is because different feedstocks require different processing and synthetic operations downstream that may add to their hazards. If feedstock A requires use of a particularly hazardous material to convert it to product, whereas feedstock B can be processed by relatively benign processes, feedstock B should be chosen. This kind of consideration points to the importance of considering the whole life cycle of materials rather than just one aspect.

15.8. GLUCOSE FEEDSTOCK

The glucose molecule provides a promising platform for a number of different organic syntheses. In addition to being produced in abundance by plants, glucose is a partially oxidized material, advantageous where a partially oxidized product is made. It also contains hydroxyl groups (–OH) around the molecule, which act as sites for the attachment of various functionalities. Glucose is metabolized by essentially all organisms, so it serves as an excellent starting point for biosynthesis reactions using enzymes. Glucose and many of its products are biodegradable, adding to their environmental acceptability.

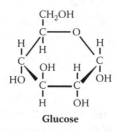

Glucose

Glucose can be obtained by enzyme-catalyzed processes from other sugars, including sucrose and fructose. Virtually all of the glucose that is now used is obtained from the enzymatic hydrolysis of cornstarch. It is also possible to obtain glucose with some difficulty by the enzymatic hydrolysis of cellulose. The difficulty in doing this is compensated by the enormous quantities of cellulose available in wood and other biomass sources. The greatest use of glucose for synthesis is by fermentation with yeasts to produce ethanol,

$$H-\overset{\overset{\displaystyle H}{|}}{\underset{\underset{\displaystyle H}{|}}{C}}-\overset{\overset{\displaystyle H}{|}}{\underset{\underset{\displaystyle H}{|}}{C}}-OH$$

Ethanol

an alcohol widely used as a gasoline additive, solvent, and chemical feedstock. A by-product of this fermentation process is carbon dioxide, which when heated to a high enough temperature and pressurized to the point at which the distinction between liquids and gases disappeared, becomes a supercritical fluid with a number of green chemical and technological applications.

Glucose is widely used as a starting material for the biological synthesis of a number of different biochemical compounds. These include ascorbic acid, citric acid, and lactic acid. Several amino acids used as nutritional supplements, including lysine, phenylalanine, threonine, and tryptophan, are biochemically synthesized starting with glucose. The vitamins folic acid, ubiquinone, and enterochelin are also made biochemically from glucose.

In addition to the predominantly biochemical applications of glucose mentioned above, this sugar can be used to make feedstocks for chemical manufacture. The possibilities for so doing are now greatly increased by the availability of genetically engineered microorganisms that can be made to express genes for the biosynthesis of a number of products. Sophisticated genetic engineering is required to make chemical feedstocks because these are materials not ordinarily produced biologically. As an example of the potential of glucose for making important feedstocks, consider the synthesis from glucose of adipic acid,

$$HO-\overset{\overset{\displaystyle O}{\|}}{C}-\overset{\overset{\displaystyle H}{|}}{\underset{\underset{\displaystyle H}{|}}{C}}-\overset{\overset{\displaystyle H}{|}}{\underset{\underset{\displaystyle H}{|}}{C}}-\overset{\overset{\displaystyle H}{|}}{\underset{\underset{\displaystyle H}{|}}{C}}-\overset{\overset{\displaystyle H}{|}}{\underset{\underset{\displaystyle H}{|}}{C}}-\overset{\overset{\displaystyle O}{\|}}{C}-OH$$

Adipic acid

a feedstock consumed in large quantities to make nylon. The conventional synthesis of this compound starts with benzene, a volatile, flammable hydrocarbon that is believed to cause leukemia in humans. The synthesis involves several steps using catalysts at high pressure and corrosive oxidant nitric acid, which releases air pollutant nitrous oxide (N_2O). The first step is the addition to benzene over a Ni/Al_2O_3 catalyst at pressures 25 to 50 times atmospheric pressure of explosive hydrogen gas (H_2)

$$\bigcirc + 3H_2 \rightarrow \bigcirc \qquad (15.8.1)$$

to produce cyclohexane, which is then subjected to oxidation in air at 9 atm pressure over a cobalt catalyst,

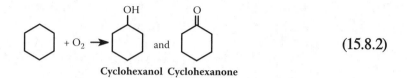

(15.8.2)

Cyclohexanol Cyclohexanone

to produce a mixture of cyclohexanol, a cyclic alcohol, and cyclohexanone, a cyclic ketone. This mixture is then reacted with oxidizing, corrosive, 60% nitric acid over a Ni/Al$_2$O$_3$ catalyst at 25 to 50 atm pressure to give the adipic acid feedstock.

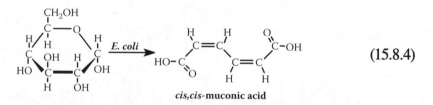

(15.8.3)

Throughout the synthesis process, elevated temperatures of approximately 250°C are employed. The N$_2$O released by the synthesis of adipic acid in the manufacture of nylon accounts for a significant fraction of worldwide N$_2$O releases. The potential dangers and environmental problems with this synthesis are obvious.

As an alternative to the chemical synthesis of adipic acid as discussed in the preceding paragraph, a biological synthesis using genetically modified *Escherichia coli* bacteria and a simple hydrogenation reaction has been devised. The bacteria convert glucose to *cis,cis*-muconic acid:

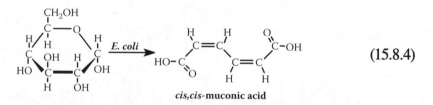

cis,cis-muconic acid

(15.8.4)

The muconic acid is then treated under relatively mild conditions with H$_2$ under 3 atm pressure over a platinum catalyst to give adipic acid.

Another organic chemical that potentially can be produced by the action of transgenic microorganisms on glucose is catechol, used as a feedstock to make flavors, pharmaceuticals, carbofuran pesticide, and other chemicals. About 20 million kilograms per year worldwide of this compound are now manufactured chemically starting with propylene and carcinogenic benzene, both derived from depleting petroleum sources. Toxic phenol is generated as an intermediate, and it is oxidized to catechol with 70% hydrogen peroxide, which at this concentration is a violently reactive, hazardous oxidant. These steps require some rather severe conditions and stringent precautions in handling hydrogen peroxide reagent. *E. coli* bacteria of a genetically modified strain designated AB2834/pKD136/pKD9/069A, produce catechol from glucose and, if yields can be gotten to acceptable levels, biosynthesis could become a major source of this important chemical.

Catechol

Another potentially important organic feedstock that has now been synthesized from glucose using transgenic *E. coli* is 3-dehydroshikimic acid:

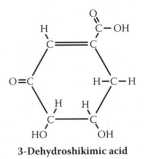

3-Dehydroshikimic acid

This compound is an important intermediate in the production of aromatic amino acids, gallic acid, vanillin, and other chemicals. It also has antioxidant properties. Antioxidants are organic compounds that react with oxygen-containing, reactive free radical species, such as hydroxyl radical, HO·. With their unpaired electrons (which make them free radicals), these species oxidize materials such as oils, fats, and lubricating oils and greases, causing deterioration in quality. By reacting with the free radicals, antioxidants stop their action. An abundant source of 3-dehydroshikimic acid could lead to its much wider application as an antioxidant.

15.9. CELLULOSE FEEDSTOCK

The most abundant natural material produced by organisms is **cellulose** synthesized biologically by the joining of glucose molecules with the loss of one H_2O molecule for each bond formed (see Figure 15.4). This makes the chemical formula of cellulose $(C_6H_{10}O_5)_n$, where n ranges from about 1500 to 6000 or more. Most cellulose is made by plants, with total amounts exceeding 500 *billion* metric tons per year worldwide. Cellulose makes up the sturdy cell walls of plants. Wood is about 40% cellulose, leaf fibers about 70%, and cotton, one of the purest sources of cellulose, about 95%. Cellulose occurs in different forms and is always associated with hemicellulose (a material also composed of carbohydrate polymers) and lignin, a biopolymer of varied composition and bonding composed largely of aromatic units.

The first major step in cellulose utilization, such as extraction of cellulose fibers for making paper, consists of separating the cellulose from its matrix of lignocellulose

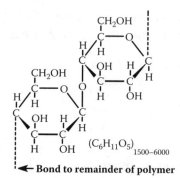

Figure 15.4.   Segment of the cellulose molecule in which from 1500 to several thousand anhydro-glucose units (glucose molecules less H_2O) are bonded together.

(hemicellulose and lignin). This step has been the cause of many problems in utilizing cellulose because of the harsh chemical processing that has been employed. Lignin residues impart color to the cellulose, so wood pulp used in making paper has to be bleached with oxidants that alter the structure of the coloring agents. Bleaching used to be done almost entirely with elemental Cl_2, and salts of hypochlorite ion, ClO^-, which produced chlorinated organic impurities and pollutants. Therefore, ozone and hydrogen peroxide are preferred bleaching agents.

A finely divided form of cellulose called **microcrystalline cellulose** is produced by appropriate physical and chemical processing of cellulose. This material has many uses in foods in which it imparts smoothness, stability, and a quality of thickness and in pharmaceutical preparations and cosmetics. Added to food, indigestible cellulose contributes bulk and retains moisture.

Chemically modified cellulose is used to make a wide variety of materials. Like the glucose that comprises it, cellulose has an abundance of –OH groups to which various other groups can be bonded to impart a variety of properties. One of the oldest synthetic fabrics, rayon, is made by treating cellulose with base and carbon disulfide, CS_2, then extruding the product through fine holes to make thread. In a similar process, chemically treated cellulose is extruded through a long narrow slot to form a sheet of transparent film called cellophane.

As seen by the structure in Figure 15.4, each unit of the cellulose polymer has three –OH groups that are readily attached to other functional groups leading to chemically modified cellulose. One of the most common such products is cellulose acetate, an ester (see Section 5.4 and Reaction 5.4.1) used primarily for apparel and home furnishings fabrics in which most of the –OH groups on cellulose are replaced by acetate groups by reaction with acetic anhydride (see below):

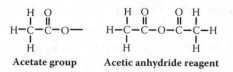

Acetate group   Acetic anhydride reagent

Although the cellulose feedstock for cellulose acetate synthesis is certainly a "green" material, acetic anhydride used to make the acetate is a corrosive, toxic chemical that produces poorly healing wounds on exposed flesh. Furthermore, potentially hazardous solvents, such as dichloromethane, are used in some processes for making cellulose acetate.

Another cellulose ester that has been widely manufactured is cellulose nitrate, a material in which the $-OH$ groups on cellulose are replaced by $-ONO_2$ groups by treating cellulose with a mixture of nitric acid (HNO_3) and sulfuric acid (H_2SO_4). Cellulose nitrate makes transparent film and was used in the early days of moving pictures for movie film. However, one of the other major uses of this material is as an explosive, so cellulose nitrate can burn violently giving off highly toxic fumes of NO_2 gas. In years past, this characteristic has led to several tragic fires involving human fatalities. Its use is now largely restricted to lacquer coatings, explosives, and propellants. Although the cellulose raw material is green, neither the process for making cellulose nitrate involving strong acids, nor the flammable product would qualify as green.

From the preceding discussion, it is apparent that cellulose is an important raw material for the preparation of a number of materials. The reagents and conditions used to convert cellulose to other products are in some cases rather severe. It may be anticipated that advances in the science of transgenic organisms will result in alternative biological technologies that will enable conversion of cellulose to a variety of products under relatively mild conditions.

Feedstocks from Cellulose Wastes

Large quantities of cellulose-rich waste biomass are generated as by-products of crop production in the form of straw remaining from grain harvest, bagasse residue from the extraction of sucrose from sugarcane, and other plant residues representing a large amount of essentially free raw material that could be converted to chemical feedstocks. One way in which this can be done is by the use of enzyme systems to break the cellulose down into glucose sugar, used directly as a feedstock or fermented to produce ethanol. Direct conversion of cellulose wastes to feedstocks is another route. Fortunately, nature has provided efficient microorganisms for this purpose in the form of rumen bacteria that live in the stomachs of cattle and related ruminant animals. It has been found that these bacteria function well in large fermenters from which oxygen is excluded if the plant residues are first treated with lime ($Ca(OH)_2$ and $CaCO_3$), producing short-chain organic acids that exist as their calcium salts in the presence of lime.

The organic acids produced by rumen bacteria in animals are absorbed from the digestive systems of the animals and used as food. The acids produced in digesters are in the form of calcium salts, primarily calcium acetate, calcium propionate, and calcium butyrate. These materials can be processed to produce feedstocks for a variety of organic syntheses. Acidification of the salts yields the corresponding organic acids as shown by the structural formulas below:

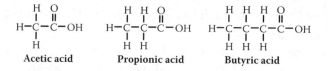

Acetic acid   Propionic acid   Butyric acid

Reaction of these acids with elemental hydrogen (hydrogenation) can be used to convert them to alcohols:

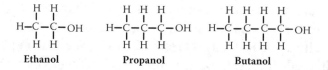

Ethanol   Propanol   Butanol

Heat treatment of the calcium salts of the organic acids at 450°C produces ketones, such as those shown below. These compounds are valuable feedstocks for a number of different chemical synthesis operations.

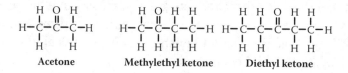

Acetone   Methylethyl ketone   Diethyl ketone

15.10. LIGNIN FEEDSTOCK

Lignin, a chemically complex biopolymer that is associated with cellulose in plants and serves to bind cellulose in the plant structure, ranks second in abundance only to cellulose as a biomass material produced by plants. Lignin is normally regarded as a troublesome waste in the processing and utilization of cellulose. The characteristic that makes lignin so difficult to handle in chemical processing is its inconsistent, widely variable molecular structure as shown by the segment of lignin polymer in Figure 15.5. This structure shows that much of the carbon is present in aromatic rings that are bonded to oxygen-containing groups. Because of this characteristic, one of the potential uses of lignin is to produce phenolic compounds, which have the –OH group bonded to aromatic rings. The abundance of hydroxyl (–OH), methoxyl (–OCH$_3$), and carbonyl (C=O) groups in lignin also suggests potential chemical uses for the substance. A significant characteristic of lignin is its resistance to biological attack. This property, combined with lignin's highly heterogeneous nature makes it a difficult substrate to use for the enzyme-catalyzed reactions favored in the practice of green chemistry to give single pure products useful as chemical feedstocks.

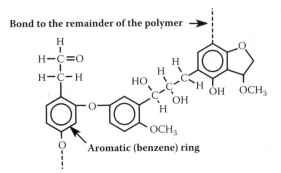

Figure 15.5.   Segment of a lignin polymer molecule showing aromatic character and the disorganized, variable chemical structure that makes lignin a difficult material to use as a feedstock.

Lignin generated as a by-product in the extraction of cellulose from wood is now largely burned for fuel. It has some uses for binders to hold materials together in coherent masses, fillers, resin extenders, and dispersants. There is also some potential to use lignin as a degradation-resistant structural material, such as in circuit boards.

15.11. CHEMICAL PRODUCTION BY BIOSYNTHESIS

Fermentation refers to the action of microorganisms on nutrients under controlled conditions to produce desired products. Fermentation for some products is anoxic (anaerobic, O_2 absent) and for others oxic (aerobic, O_2 is present) fermentation is used. Fermentation processes have been used for thousands of years to produce alcoholic beverages, sauerkraut, vinegar, pickles, cheese, yogurt, and other foods. Ethanol, the alcohol in alcoholic beverages, is the most widely produced chemical made by fermentation. Lactic acid

Lactic acid

has also been produced by fermentation processes for many years. More recently, fermentation has been applied to the production of a wide variety of organic acids, antibiotics, enzymes, and vitamins.

Starting in the 1940s, one of the major products of industrial fermentation has been penicillin, of which there are several forms. Figure 15.6 shows a simplified diagram of a facility for production of this life-saving antibiotic. Following penicillin, fermentation processes were developed for the production of several other significant antibiotics.

Selection of the appropriate microorganism is the most important consideration of a successful fermentation production process. The microorganisms have to have

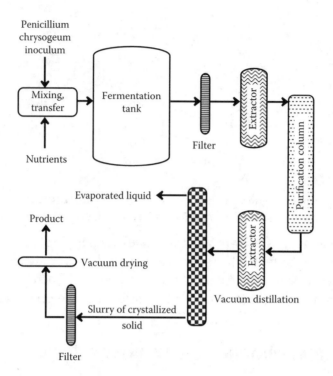

Filter

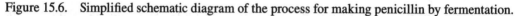

Figure 15.6.   Simplified schematic diagram of the process for making penicillin by fermentation.

the proper nutrients, the choice of which can affect the kind and yield of the product. Sterile conditions must be maintained, and sterilization of equipment and media is accomplished by heating to 125°C to 150°C for appropriate lengths of time. Air entering the fermenter must be filtered and sterilized. The temperature of fermentation is important, with fermentation rates increasing up to an optimum temperature, after which they decrease sharply with increased temperatures as the enzymes used by the microorganisms are destroyed (denatured). This kind of temperature relationship has increased interest in the use of thermophilic microorganisms that exist at boiling water temperatures in hot springs. If such organisms can be engineered to produce desired products, the rate of product generation may increase markedly. Both the levels of oxygen (which must be excluded from anaerobic processes) and pH must be controlled precisely. Modern fermentation processes use a variety of sensors to continuously monitor conditions in the fermentation tank and computerized control to accurately control all the parameters.

Fermentation is undergoing tremendous development with the use of transgenic microorganisms (to which genes have been transferred) to make specific kinds of substances. The most common and valuable substances made by transgenic microorganisms consist of a variety of proteins. These include proteins and smaller molecule polypeptides that are used as pharmaceuticals. The best example of such a substance is human insulin, which is now produced in large quantities by transgenic microorganisms.

Until recently, fermentation has not been widely employed to make commodity chemicals used on a large scale. An exception is the large-scale production of ethanol from the fermentation of glucose sugar by yeasts. Now mandated as a gasoline additive in some parts of the U.S. by law, huge and growing quantities of ethanol are made by fermentation of glucose derived from corn, and ethanol production is an important market for corn. It is not clear that this is a truly green technology, and some authorities believe that the energy consumed and the environmental damage from more intensive cultivation of corn outweigh the benefits of using this grain to produce ethanol fuel. Advances in transgenic microbiology have now raised the possibility of using fermentation for the production of a variety of chemicals and chemical feedstocks, several examples of which are discussed in this chapter.

Bioconversions of Synthetic Chemicals

Most of the biochemical operations described so far in this chapter pertain to natural products which, by their nature, would be expected to be amenable to the action of enzymes. The mild conditions under which enzymes operate, the readily available, safe reagents that they employ (such as molecular O_2 for oxidations), and the high specificity of enzyme catalysts make biocatalyzed reactions attractive for carrying out chemical processes on synthetic chemicals, such as those from petroleum sources. This section discusses two examples of enzyme-catalyzed processes applied to chemical processes on synthetic chemicals that would otherwise have to be performed with chemical reagents under much more severe conditions.

p-Hydroxybenzoic Acid from Toluene

The potential for use of biosynthesis applied to synthetic chemicals can be illustrated by the synthesis of ***p*-hydroxybenzoic acid,**

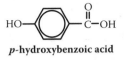

p-hydroxybenzoic acid

an important intermediate used in the synthesis of pharmaceuticals, pesticides, dyes, preservatives, and liquid crystal polymers. It is currently made by reacting potassium phenolate,

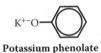

Potassium phenolate

with carbon dioxide under high pressure at 220°C, which converts slightly less than half of the potassium phenolate to the desired product and produces substantial impurities. The process dates back to the early 1860s, around 150 years ago, long before there were any considerations of pollutants and wastes. It requires severe conditions and produces metal and phenol wastes. Reactive alumina powder (Al_2O_3) used to catalyze the process has been implicated in a 1995 explosion at a facility to produce p-hydroxybenzoic acid that killed 4 workers.

A biosynthetic alternative to the synthesis described above has been attempted with *Pseudomonas putida* bacteria genetically engineered to carry out several steps in the synthesis of p-hydroxybenzoic acid starting with toluene. A key to the process is the attachment at the *para* position on toluene of a hydroxyl group by the action of toluene-4-monooxygenase (T4MO) enzyme system transferred to *Pseudomonas putida* from *Pseudomonas mendocina*:

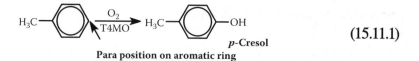

$$\text{(15.11.1)}$$

p-Cresol

Para position on aromatic ring

The next step is carried out by p-cresol methylhydroxylase (PCMH) enzyme from a strain of *Pseudomonas putida* that yields p-hydroxybenzyl alcohol followed by conversion to p-hydroxy-benzaldehyde:

$$\text{(15.11.2)}$$

The last step is carried out by an aromatic aldehyde dehydrogenase enzyme designated PHBZ also obtained from a strain of *Pseudomonas putida* and consists of the conversion of the aldehyde to the p-hydroxybenzoic acid product:

$$\text{(15.11.3)}$$

Through elegant genetic manipulation, the chemical processes described above were achieved leading to the desired product. In addition to providing the enzymes to carry out the desired steps, it was also crucial to block steps that would consume intermediates and give undesired by-products that would consume raw material and require separation from the product. Although it is a long way from showing that the complex biochemical synthesis process actually gives the desired product to the final goal of having a practical process that can be used on a large scale, the results described above certainly show the promise of transgenic organisms in carrying out chemical syntheses.

Production of 5-Cyanovaleramide

The second biocatalyzed process to be considered is the conversion of adiponitrile to 5-cyanovaler-amide. This conversion was required for the synthesis of a new chemical used for crop protection. This process can be carried out chemically with a stoichiometric mixture of adiponitrile with water and a manganese dioxide catalyst under pressure at 130°C as shown by the following reaction:

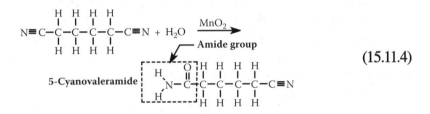

$$(15.11.4)$$

If the reaction is run to 25% completion, an 80% selectivity for the 5-cyanovaleramide is achieved, with the other fraction of the adiponitrile that reacts going to adipamide, in which the second $-C\equiv N$ functional group is converted to an amide group. Carrying the reaction beyond 25% completion resulted in unacceptable levels of conversion to by-product adipamide.

The isolation of the 5-cyanovaleramide product from the chemical synthesis described above entails dissolving the hot reaction mixture in toluene solvent, which is then cooled to precipitate the product. The unreacted adiponitrile remains in toluene solution from which it is recovered to recycle back through the reaction. For each kilogram of 5-cyanovaleramide product isolated, approximately 1.25 kg of MnO_2 required disposal; this is definitely not a green chemical process!

As an alternative to the chemical synthesis described above, a biochemical synthesis was developed using organisms that had nitrile hydratase enzymes to convert the $C\equiv N$ functional group to the amide group. The microorganism chosen for this conversion was designated *Pseudomonas chloroaphis* B23. The cells of this organism were immobilized in beads of calcium alginate, the salt of alginic acid isolated from the cell walls of kelp. It was necessary to run the process at 5°C, above which temperature the enzyme lost its activity. With this restriction, multiple runs were performed to convert adiponitrile to 5-cyanovaleramide. During these runs, 97% of the adiponitrile was reacted, with only 4% of the reaction going to produce by-product adipamide. The water-based reaction mixture was simply separated mechanically from the calcium alginate beads containing the microorganisms, which were then recycled for the next batch of reactant. The water was distilled off the product to leave an oil, from which the 5-cyanovaleramide product was dissolved in methanol, leaving adipamide and other by-products behind. In contrast to the enormous amount of waste catalyst produced in the chemical synthesis of 5-cyanovaleramide, only 0.006 kg of catalyst waste residue was produced per kg of product. And the waste microbial catalyst was 93% water, so its disposal was not a problem.

15.12. DIRECT BIOSYNTHESIS OF POLYMERS

Cellulose in wood and cotton is only one example of the numerous significant polymers that are made biologically by organisms. Other important examples are wool and silk, which are protein polymers. A big advantage of these kinds of polymers from an environmental viewpoint is that polymers made biologically are also the ones that are most likely to be biodegradable. Attempts have been made to synthesize synthetic polymers that are biodegradable; these efforts have centered on those prepared from biodegradable monomers, such as lactic acid.

From the standpoint of green chemistry, it is ideal to have polymers that are made by organisms in a form that is essentially ready to use. Recently, interest has focused on poly(hydroxyalkanoate) compounds, of which the most common are polymers of 3-hydroxybutyric acid:

$$HO-\overset{\overset{\displaystyle O}{\|}}{C}-\overset{\overset{\displaystyle H}{|}}{\underset{\underset{\displaystyle H}{|}}{C}}-\overset{\overset{\displaystyle H}{|}}{\underset{\underset{\displaystyle O}{|}}{C}}-\overset{\overset{\displaystyle H}{|}}{\underset{\underset{\displaystyle H}{|}}{C}}-H$$

3-Hydroxybutyric acid

This compound and related ones have both a carboxylic acid group ($-CO_2H$) and an alcohol ($-OH$) group. A carboxylic acid can bond with an alcohol with the elimination of a molecule of H_2O forming an *ester linkage*, one of which is outlined by the dashed box in the structure below. Because the hydroxyalkanoates have both functional groups, the molecules can bond with each other to form polymer chains:

$$---\overset{\overset{\displaystyle O}{\|}}{C}-\overset{\overset{\displaystyle H}{|}}{\underset{\underset{\displaystyle H}{|}}{C}}-\overset{\overset{\displaystyle H}{|}}{\underset{\underset{\displaystyle CH_3}{|}}{C}}-O-\overset{\overset{\displaystyle O}{\|}}{C}-\overset{\overset{\displaystyle H}{|}}{\underset{\underset{\displaystyle H}{|}}{C}}-\overset{\overset{\displaystyle H}{|}}{\underset{\underset{\displaystyle CH_3}{|}}{C}}-O-\overset{\overset{\displaystyle O}{\|}}{C}-\overset{\overset{\displaystyle H}{|}}{\underset{\underset{\displaystyle H}{|}}{C}}-\overset{\overset{\displaystyle H}{|}}{\underset{\underset{\displaystyle CH_3}{|}}{C}}-O-\overset{\overset{\displaystyle O}{\|}}{C}-\overset{\overset{\displaystyle H}{|}}{\underset{\underset{\displaystyle H}{|}}{C}}-\overset{\overset{\displaystyle H}{|}}{\underset{\underset{\displaystyle CH_3}{|}}{C}}-O---$$

Ester groups are among the most common in a variety of biological compounds, such as fats and oils, and organisms possess enzyme systems that readily attack ester linkages. Therefore, the poly(hydroxy-alkanoate) compounds are amenable to biological attack. Aside from their biodegradability, polymers of 3-hydroxybutyric acid and related organic acids that have $-OH$ groups on their hydrocarbon chains (alkanoates) can be engineered to have a variety of properties ranging from rubber-like to hard solid materials.

It was first shown in 1923 that some kinds of bacteria make and store poly(hydroxyalkanoate) ester polymers as a reserve of food and energy. In the early 1980s it was shown that these materials have thermoplastic properties, meaning that they melt when heated and resolidify when cooled. This kind of plastic can be very useful, and the thermoplastic property is rare in biological materials. One commercial operation was set up for the biological synthesis of a polymer in which 3-hydroxybutyrate groups alternate with 3-hydroxyvalerate groups, where valeric acid has a 5-carbon atom chain. This process uses a bacterium called *Ralstonia eutropia* fed glucose and the sodium salt of propionic acid to make the polymer in fermentation vats. Although the process works, costs are high because of problems common to most microbial fermentation synthesis processes: the bacteria have to be provided with a source of food, yields are relatively low, and it is difficult to isolate the product from the fermentation mixture.

Developments in genetic engineering have raised the possibility of producing poly(hydroxyalkanoate) polymers in plants. The plant *Arabidopsis thaliana* has accepted genes from bacterial *Alcaligenes eutrophus* that have resulted in plant leaves containing as much as 14% poly(hydroxybutyric acid) on a dry weight basis. Transgenic *Arabidopsis thaliana* and *Brassica napus* (canola) have shown production of the copolymer of 3-hydroxybutyrate and 3-hydroxyvalerate. If yields can be raised to acceptable levels, plant-synthesized poly(hydroxyalkanoate) materials would represent a tremendous advance in biosynthesis of polymers because of the ability of photosynthesis to provide the raw materials used to make the polymers.

QUESTIONS AND PROBLEMS

1. Discuss advantages that biological feedstocks have over petroleum. Are there disadvantages?

2. What is a fundamental chemical difference between petroleum and biological feedstocks?

3. Name some characteristics of an ideal feedstock.

4. Name three kinds of reactions used in processing feedstocks. Which is best from the viewpoint of green chemistry?

5. Name several categories of biomass that can be used for feedstocks. Which of these is the least useful?

6. How are oils extracted from plant sources?

7. Use chemical formulas to make the argument that carbohydrates are a more oxidized chemical feedstock than hydrocarbons.

8. What are the two main biological sources of materials?

9. Name some categories of chemicals routinely produced by fermentation.

10. Which pharmaceutical material has been produced by fermentation for many years?

11. What is the first, most important consideration in developing a fermentation process for production of a chemical?

12. What is the significance of temperature in fermentation processes? What happens if temperature is too high?

13. Which chemical is made in largest quantities by fermentation?

14. In which fundamental respect are plants more efficient producers of material than fermentation?

15. Which relatively recent advance in biotechnology has greatly increased the scope of materials potentially produced by plants?

16. Why are hybrid poplar trees particularly important in the production of raw materials?

17. Describe the structural characteristics of glucose and other carbohydrates that make them good platforms for chemical synthesis.

18. Give a disadvantage and an advantage of the use of cellulose as a source of glucose.

19. List some of the hazards associated with the chemical synthesis of adipic acid used to make nylon.

20. Give a major concern with the use of benzene as a feedstock.

21. What is a chemical characteristic of 3-dehydroshikimic acid that could lead to much greater uses for it?

22. Although the chemical formula of glucose is $C_6H_{12}O_6$, that of the cellulose polymer made from glucose is $(C_6H_{10}O_5)_n$ where n is a large number. Since cellulose is made from glucose, why is the cellulose formula not $(C_6H_{12}O_6)_n$?

23. Why is wood pulp consisting mostly of cellulose, treated with oxidants? Which oxidants are preferred, and which has lost favor?

24. Give some examples of useful chemically modified cellulose. Which of these has proven to be rather dangerous?

25. In ruminant animals that have bacteria in their stomachs that digest cellulose, the rumen bacteria and the organic acids they generate are passed on through the digestive tract where the bacterial biomass is dissolved, with the products and the organic acids previously generated absorbed by the animal as food. Suggest why basic limestone is used in the large batch processes that use rumen bacteria in digesters to produce organic acids from cellulose.

26. Why is it difficult to deal with lignin as a chemical feedstock?

27. What is the current main use of waste lignin?

28. Give the main advantage of biopolymers from an environmental viewpoint.

29. Which structural feature of hydroxyalkanoates enables them to make polymeric molecules?

30. What was the original source of poly(hydroxyalkanoate) polymers? How is it now proposed to produce them?

31. Although enzymes have not developed specifically to act upon synthetic compounds, they have some specific advantages that make them attractive for carrying out chemical processes on synthetic compounds. What are some of these advantages?

32. Name two chemicals for which it has been shown that enzymatic processes can actually convert synthetic raw materials to chemical products normally made by nonbiological chemical reactions.

16. THE ANTHROSPHERE AS PART OF THE GLOBAL ENVIRONMENT

16.1. THE EARTH AS MADE BY HUMANS

In Section 2.6, the **anthrosphere** was defined as that part of the environment made or modified by humans and used for their activities, in a sense, the Earth as made by humans. The anthrosphere has become an integral part of Earth's environment strongly influenced by and, in turn, strongly influencing the other four environmental spheres — the geosphere, hydrosphere, biosphere, and atmosphere. Therefore, the remaining chapters in this book discuss the anthrosphere including what constitutes it, what its function is, and the effects that it has. This requires consideration of the technological, engineering, and industrial aspects of human activities carried out in the anthrosphere. It is essential to recognize that humans *will use* technology to provide the food, shelter, and goods that they need for their well-being and survival. The challenge is to interweave technology with considerations of the environment and ecology such that the two are mutually advantageous.

Figure 2.8 in Chapter 2 illustrates the anthrosphere and some of its main constituents. These can be viewed in terms of the following three major aspects:

1. **Anthrospheric constructs**, including dwellings, factories, commercial centers, and mines

2. **Anthrospheric flows** of materials, energy, communications, and people

3. **Anthrospheric conduits** for the transmission of materials, energy, people and communications

These three major aspects of the anthrosphere are summarized in Figure 16.1 and serve as the basis in this chapter for the detailed discussion of the anthrosphere and its constituents. The chapter emphasizes the relationship of the anthrosphere to green science and technology and the key role that its proper management must play in sustainability.

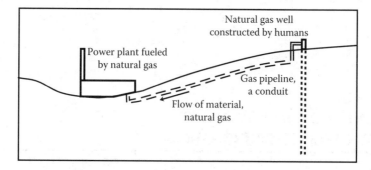

Figure 16.1.   Illustration of the three main aspects of the anthrosphere — constructs, flows, and conduits.

16.2. CONSTRUCTS IN THE ANTHROSPHERE

The dwellings of humans and the other buildings designed to accommodate humans, such as stores and schools, have an enormous influence on human well-being and on the surrounding environment. In relatively affluent societies, the quality of living space has improved dramatically during the last century. Homes have become much more spacious per occupant and are largely immune to the extremes of weather conditions. Such homes are equipped with a huge array of devices, such as indoor plumbing, climate control, communications equipment, and entertainment centers. The comfort factor for occupants has increased enormously. In some areas of the world, however, people are forced to survive with substandard dwellings that are not even safe or healthy. Acts of nature, such as the devastating Indian Ocean tsunami at the end of 2004 or the destructive U.S. Gulf Coast Hurricane Katrina of 2005, have destroyed tens of thousands of dwellings and other buildings that humans use.

The construction and use of modern homes and the other buildings in which people spend most of their time, place tremendous strains on Earth's support systems and cause a great deal of environmental damage. Typically, as part of the siting and construction of new homes, shopping centers, and other buildings, the landscape is rearranged drastically at the whims of developers. Topsoil is removed, hills are cut down, and low places are filled in an attempt to make the surrounding environment fit to a particular architectural scheme. The construction of modern buildings consumes large amounts of resources, such as concrete, steel, plastic, and glass, as well as the energy required to make synthetic building materials. The operation of a modern building requires additional large amounts of energy and of materials, such as water.

The good news is that there is a large potential to design, construct, and operate homes and other buildings in a manner consistent with environmental preservation and improvement. One obvious way in which this can be done is to reevaluate the kinds of materials used in buildings to minimize the use of steel, plastics, and other materials that require large amounts of resources and energy to fabricate. Substitution of renewable materials, such as wood, and nonfabricated materials, such as quarried stone, can save large amounts of energy and reduce environmental impact. In some parts of the world, sun-dried adobe blocks made from soil are practical building materials that require little energy to fabricate.

Recycling of building materials and of whole buildings can save large amounts of materials and minimize environmental damage. At a low level, stone, brick, and concrete can be used as fill material on which new structures may be constructed. Bricks are often recyclable, and recycled used bricks can make useful and quaint materials for walls and patios. Given careful demolition practices, wood can often be recycled. Buildings can be designed with recycling in mind. This means using architectural design conducive to adding stories and annexes and to rearranging existing space. Utilities may be placed in readily accessible passageways rather than being imbedded in structural components to facilitate later changes and additions.

Technological advances can be used to make buildings much more sustainable. Advanced window design using multiple panes and infrared-blocking glass can significantly reduce energy consumption. Modern insulation materials are highly effective. Advanced heating and air-conditioning systems operate with a high degree of efficiency. Automated and computerized control of building utilities, particularly those used for cooling and heating, can significantly reduce energy consumption by regulating temperatures and lighting to the desired levels at specific locations and times in the building.

Advances in making buildings airtight and extremely well insulated can lead to problems with indoor air quality. Carpets, paints, paneling, and other manufactured components of buildings give off organic vapors, such as formaldehyde, solvents, and monomers used to make plastics and fabrics. This can lead to "sick building syndrome" in which the occupants become ill due to exposure to indoor air pollutants. Advanced air purification systems with suitable filters can enable humans to live in airtight, energy-efficient buildings without ill effects.

Other than dwellings, schools, and commercial buildings, there are many other structures that humans make in the anthrosphere that have important effects on anthrospheric sustainability. There is not room here to discuss all of these in detail. However, many factors should be considered in the construction of such facilities. As just one example, consider options for electrical power plant construction. Wind power is certainly renewable, safe and nonpolluting. However, a wind power installation of a size sufficient to provide electricity to a major city could cover many square kilometers of area, although virtually all of the land on which it is located could be used for other purposes. A large nuclear power plant of modern design scores less well in the categories of renewability, safety, and pollution (though certainly not as potentially harmful as claimed by the more strident critics of nuclear energy), but occupies a relatively miniscule area (all of which must be devoted to the power plant and its ancillary facilities).

16.3. ANTHROSPHERIC FLOWS

Throughout the anthrosphere there is a constant flow of materials, energy, information, and people. A very large share of the environmental impact of modern civilization is because of these flows. Raw materials and energy move from mines and other sources to factories in large quantities. Goods produced by factories are distributed throughout the world by ship, rail, and truck. Electricity, natural gas, and water are fed to homes and other buildings. Wastewater and refuse must be transported to recycling centers and treatment and disposal sites. People are constantly moving by private automobiles, airplanes, and trains. The flux of movement of materials, energy, people, and information is arguably the best measure of how advanced a country is by conventional measures of economic development.

Much of the environmental and sustainability impact of modern civilizations is the result of moving materials, energy, and people (especially with modern technologies, movement of information, except printed media, has negligible environmental and resource impact). The most visible impact results from movement by automobile and truck. These conveyances require conversion of vast areas of land to highways and parking facilities that cannot be used for other purposes. Exhaust emissions from automobiles and trucks are the largest sources of air pollutants and carbon dioxide, which is responsible for global warming.

Reduction of flows of materials, energy, and people is one of the best ways to reduce environmental/resource impact and to enhance sustainability. **Dematerialization** (less use of materials), a major objective of sustainability, can significantly reduce material flow. Greater energy use efficiency, such as more energy-efficient homes and more energy-efficient automobiles can significantly reduce the required flows of natural gas, electricity, and gasoline. Reducing the need for humans to commute to work (see discussion of the telecommuter society later in this chapter) can significantly diminish the flow of people, resulting in less congestion and lower energy consumption.

16.4. ANTHROSPHERIC CONDUITS

Anthrospheric conduits is a term used here to refer to the means by which materials, energy, people, and communications are moved. The choice of such means has a very large impact upon the environment and sustainability. Most anthrospheric conduits are parts of the infrastructure discussed below. The highways and streets required to distribute materials by truck cover large areas of land that cannot be used for other purposes and that are paved, preventing water infiltration into groundwater reservoirs. The same flows of material can be accommodated by rail lines that have a much smaller footprint per unit capacity and that do not cover vast areas with water-impregnable concrete or asphalt. Surface transport of fuels requires conduits that disturb Earth's surface, consume space, and generally have major environmental impacts. Pipelines are an alternative that, although disruptive of the surface during installation, function below ground with essentially no environmental impact (except in rare cases when they break, sometimes leading to very damaging releases or even fires). Similarly, distribution of electricity or electronic communications in neighborhoods by wires mounted on poles results in unsightly distribution systems that are prone to damage and disruption by natural phenomena such as ice formation. Buried underground cables and fiber-optic systems are generally immune to disruption by natural phenomena and leave essentially no surface imprint.

16.5. INFRASTRUCTURE

The **infrastructure** refers to the utilities, facilities, and systems used in common by members of a society and on which the society is dependent for its normal function. The infrastructure includes both physical components — roads, bridges, and pipelines — and the instructions — laws, regulations, and operational procedures — under which the physical infrastructure operates. Parts of the infrastructure may be publicly owned, such as the U.S. Interstate Highway system and some European railroads, or privately owned, as is the case with virtually all railroads in the U.S. Some of the major components of the infrastructure of a modern society are the following:

- Transportation systems, including railroads, highways, and air transport systems

- Energy generating and distribution systems

- Buildings

- Telecommunications systems

- Water supply and distribution systems

- Waste treatment and disposal systems, including those for municipal wastewater, municipal solid refuse, and industrial wastes

In general, the infrastructure refers to the facilities that large segments of a population must use in common for a society to function. In a sense, the infrastructure is analogous to the operating system of a computer. A computer operating system determines how individual applications operate and the manner in which they distribute and store the documents, spreadsheets, and illustrations created by the applications. Similarly, the infrastructure is used to move raw materials and power to factories and to distribute and store their output. An outdated, cumbersome computer operating system with a tendency to crash is detrimental to the efficient operation of a computer. In a similar fashion, an outdated, cumbersome, broken-down infrastructure causes society to operate in a very inefficient manner and is subject to catastrophic failure.

For a society to be successful, it is of the utmost importance to maintain a modern, viable infrastructure. Such an infrastructure is consistent with environmental protection and sustainability. Properly designed utilities and other infrastructural elements, such as water supply systems and wastewater treatment systems, minimize pollution and environmental damage. Components of the infrastructure are subject to deterioration. To a large extent, this is because of natural aging processes. Fortunately, many of these processes can be slowed or even reversed. Corrosion of steel structures, such as bridges, is a big problem for infrastructures; however, use of corrosion-resistant materials and maintenance with corrosion-resistant coatings can virtually stop this deterioration process. The infrastructure is subject to human insult, such as vandalism, misuse, and neglect. Often the problem begins with the design and basic concept of a particular component of the infrastructure. For example, many river dikes destroyed by flooding should never have been built because they attempt to thwart to an impossible extent the natural tendency of rivers to flood periodically.

Technology plays a major role in building and maintaining a successful infrastructure. Many of the most notable technological advances applied to the infrastructure were made from 150 to 100 years ago. By 1900, railroads, electric utilities, telephones, and steel building skeletons had been developed. The net effect of most of these technological innovations was to enable humankind to "conquer," or at least temporarily subdue nature. The telephone and telegraph helped to overcome isolation, high-speed rail transport and later air transport conquered distance, and dams were used to control rivers and water flow.

The development of new and improved materials continues to have a significant influence on the infrastructure. From about 1970 to 1985 the strength of steel commonly used in construction nearly doubled. During the latter 1900s, significant advances were made in the properties of structural concrete. Superplasticizers enabled mixing cement with less water, resulting in a much less porous, stronger concrete product. Polymeric and metallic fibers used in concrete made it much stronger. For dams and other applications, in which a material stronger than earth but not as strong as conventional concrete is required, roller-compacted concrete consisting of a mixture of cement with silt or clay has been found to be useful. The silt or clay used is obtained on site with the result that both construction costs and times are lowered.

The major challenge in designing and operating the infrastructure in the future will be to construct and operate it in a manner consistent with maximum sustainability to the benefit of humankind and the Earth support systems on which we all depend. Obvious examples of environmentally friendly infrastructures are state-of-the-art sewage treatment systems, high-speed rail systems that can replace inefficient highway transport, and renewable, environmentally-friendly public power systems. More subtle approaches with a tremendous potential for making the infrastructure more environmentally friendly include employment of workers at computer terminals in their homes so that they do not need to commute, instantaneous electronic mail systems that avoid the necessity for moving letters physically, and solar-electric-powered installations to operate remote signals and relay stations, which avoids having to run electric power lines to them.

Whereas advances in technology and the invention of new machines and devices enabled rapid advances in the development of the infrastructure during the 1800s and early 1900s, it may be anticipated that advances in electronics, computers, nanotechnology, and other developing areas will have a comparable effect in the future. One of the areas in which the influence of modern electronics and computers is most visible is in telecommunications. Dial telephones and mechanical relays were perfectly satisfactory in their time, but have been made totally obsolete by innovations in electronics, computer control, and fiber optics. Air transport controlled by a truly modern, state-of-the-art computerized control system (which, unfortunately, is not yet fully installed in the U.S.) could enable present airports to handle many more airplanes safely and efficiently, thus reducing the need for airport construction. Sensors for monitoring strain, temperature, movement, and other parameters can be imbedded in the structural members of bridges and other structures. Information from these sensors can be processed by computer to warn of failure and to aid in proper maintenance. Many similar examples could be cited.

Although the payoff is relatively long term, intelligent investment in infrastructure pays very high rewards. In addition to the traditional rewards in economics and convenience, properly designed additions and modifications to the infrastructure can pay large returns in environmental improvement and sustainability as well.

16.6. TRANSPORTATION AND THE TELECOMMUTER SOCIETY

Ways of getting around and of moving materials and belongings have always been central to human existence and lifestyle. From the time that humans first started using primitive tools, they have invented devices to increase their ability to move themselves and their burdens. For land transport, the earliest of these were simple poles on which slain animals or other things were carried by two people or dragged by one (devices called the travois or slide car). These developed into skis and sledges. A major advance was the domestication of animals and training them to carry and drag loads. Then came the discovery of the wheel, arguably the greatest single increment in transportation technology of all time.

On water, primitive log dugouts and rafts greatly increased human mobility. These subsequently developed into canoes and boats propelled by human-powered ores and paddles. An advance comparable to the wheel on land was the discovery of the sail and development of means to use it effectively to propel boats and ships on water.

For both land and water transport, the development (around 1800) of steam engines light enough to fit on self-propelled vehicles provided an enormous impetus to transportation. Steam-powered ships and boats freed water transport from the vagaries of the wind and enabled movement of boats upstream, though at a fearful price of death and destruction from boiler explosions and fires. The marriage of the steam engine mounted on a locomotive with steel rails enabled the development of railroads, which totally revolutionized land transport and completely changed human economic and social systems. The next huge advance came with the development of successful internal combustion (gasoline) engines in the late 1800s. These relatively light and compact power plants made the automobile a practical reality and made air transport possible. In more industrialized nations in the modern era, private transport, commuting, and travel over relatively short distances are largely by automobile and goods are largely transported by trucks. The flexibility, convenience, and independence afforded by private automobiles and by trucks have made them extremely popular. Technical advances have greatly improved automobile efficiency, comfort, and safety.

Few aspects of modern industrialized society have had as much influence on the environment and sustainability as developments in transportation. These effects have been both direct and indirect, not the least of which is the heavy toll in lives and injuries from automobile accidents. Many of the direct effects are those resulting from the construction and use of transportation systems. The most obvious example of this is the tremendous effects that the widespread use of automobiles, trucks, and buses have had upon the environment. Entire landscapes have been rearranged to construct highways, interchanges, and parking lots. Emissions from the internal combustion engines used in automobiles are the major source of air pollution in most urban areas. The effects of modern transportation practices on resources have been enormous. The most notable of these has been the exhaustion in little more than a century of vast petroleum reserves that end up being burned as fuel for automobiles and other conveyances. This petroleum, wastefully burned with its combustion products dissipated to the atmosphere, would have served for centuries as raw material for petrochemicals manufacture.

The indirect environmental effects of widespread use of automobiles are enormous. The automobile has made possible the "urban sprawl" that is characteristic of residential and commercial patterns of development in the U.S., and in many other industrialized countries as well. The paving of vast areas of watershed and alteration of runoff patterns have contributed to flooding and water pollution. Discarded, worn-out automobiles have caused significant waste disposal problems. Vast enterprises of manufacturing, mining, and petroleum production and refining required to support the "automobile habit" have been very damaging to the environment.

Air transportation has become the method of choice for long-distance movement of passengers. Movement of people and high-value freight by airplane is now reliable, safe, and relatively inexpensive. Technological advances in aircraft and engine design, as well as operation and control, made possible by better materials and computers, have given a tremendous impetus to air transportation. It is clearly the mode of choice for overseas travel and long-distance travel over land, but definitely needs to be better integrated with rail transport for distances of up to a few hundred kilometers.

Advanced technology is being used, and still has a huge unrealized potential, for the improvement of transportation systems in areas such as speed, convenience, safety, and energy efficiency. High-tech computerized control has enabled automobiles to be much more efficient while emitting far smaller amounts of pollutants. Advanced air-traffic-control systems allow the operation of more aircraft in smaller spaces and at much closer intervals with much greater safety than was previously possible. **Mixed modes** of transport can be very successful; an example is the movement of truck trailers and their contents over long distances by rail followed by final distribution by tractor trailer truck. A systems approach that enables tradeoffs to be made among speed, energy consumption, noise, pollution, convenience, and other factors offers much promise in the transportation area. Applications of advanced engineering and technology to transportation can be of tremendous benefit to the environment. Modern rail and subway transportation systems, concentrated in urban areas and carefully connected to airports for longer distance travel, can enable the movement of people rapidly, conveniently, and safely, with minimum environmental damage. Although pitifully few in number in respect to the need for them in the U.S., examples of such systems are emerging in progressive cities, showing the way to environmentally friendly transportation systems of the future. Many European and Asian countries are relatively far advanced with respect to sustainable transportation systems.

A new development that has significantly reshaped the way humans move, where they live, and how they live, is the growth of a **telecommuter society**, composed of workers who do their work at home and "commute" through their computers, modems, FAX machines, and the internet connected by way of high-speed telephone communication lines. These technologies, along with several other developments in modern society, have made such a work pattern possible and desirable. An increasing fraction of the work force deals with information in their jobs. In principle, information can be handled just as well from a home office as it can from a centralized location, often an hour or more commuting distance from the worker's dwelling.

Actually, home and its immediate surroundings were where most work was done prior to the industrial revolution, whose assembly lines and large centralized factories demanded that workers come to a particular location for their work shifts. This work pattern and the prosperity that it brought with it resulted in the establishment of a huge suburban population that had left the cities and farms. The idyllic dream of suburbia has all too often given way to urban sprawl, traffic congestion, and long, tedious commutes. The associated environmental problems have been enormous as has the pressure put on limited resources.

It is estimated that almost 20% of the U.S. work force, a total of around 30 million people, could work very well from their homes. This tendency has been accelerated in recent years by the downsizing of many U.S. industries and "outsourcing" of work to private parties, sometimes from the companies to the workers that have been laid off. There are disadvantages, of course. People working at home may become isolated, careless in their habits, and "desocialized" from the workplace environment and the "office politics" that can play a role in advancement and training. Indeed, it is possible to visualize another type of home workplace activity in which "telecommuter counselors" advise home workers about how to manage such problems.

The changes that will result from widespread use of telecommuting are many and profound. With properly sited housing, workers can live in a rural environment with minimal disturbance to the surroundings. The flow of information has essentially no environmental costs compared to the movement of people on daily long commutes. The potential benefits to family life are obvious and may play a strong role in reshaping society positively. Whole communities may be transformed by telecommuters. Rather than being populated by people who are gone most of the day to a job far away, telecommuter communities will be occupied by people who are available on flexible schedules during the day and who are not too exhausted by long, tiring commutes to engage in civic activities at night. Telecommuters are likely to be intelligent, ambitious, and self-motivated, with all that these characteristics imply for civic activities. Cultural activities should thrive in environmentally friendly telecommuter communities combining the best of rural, suburban, and urban life.

16.7. THE COMMUNICATIONS REVOLUTION

It has become an overworked cliché that we live in an information age. Nevertheless, the means to acquire, store, and communicate information have expanded at an incredible pace during recent years. This phenomenon is changing society and has the potential to have numerous effects on the environment and sustainability.

The major areas to consider in respect to information are its acquisition, recording, computing, storing, displaying, and communicating. Consider, for example, the detection of a pollutant in a major river. Data pertaining to the nature and concentration of the pollutant may be obtained with a combination gas chromatograph and mass spectrometer. Computation by digital computer is employed to determine the identity and concentration of the pollutant. The data can be stored on a magnetic disk, displayed on a video screen, and communicated instantaneously all over the world by satellite and fiber-optic cable.

The acquisition, manipulation, and storage of information are all crucial parts of an information system. No aspect is more important, however, than the transmission of information, quickly, accurately, and over long distances. This constitutes the area of communications, now performed predominantly by telecommunications. The major constituents of a telecommunications system are illustrated in Figure 16.2. These are (1) a means of converting the information in the form in which it is found to a form that can be transmitted, (2) a means and medium for transmission, and (3) a device or system for receiving the transmitted signal and converting it to sound, video display, computer language, or other form understandable by humans or other

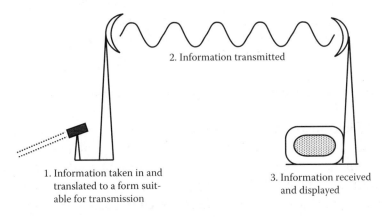
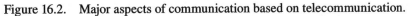

2. Information transmitted

1. Information taken in and
translated to a form suitable for transmission

3. Information received
and displayed

Figure 16.2.   Major aspects of communication based on telecommunication.

machines. These three components may be termed, respectively, the **transmitter**, **medium**, and **receiver**.

For transmission over a long distance, information is **encoded** to produce a form that the medium employed can handle. The first electrical device capable of virtually instantaneous long-distance communication was the telegraph, which encoded information as dots and dashes, that is, in a **discrete** form. The telephone was able to encode sound waves to continuous electrical waves that could be transmitted over wires, a form of signal called an **analog** signal. Interestingly, modern telecommunications systems convert analog information, such as the human voice, to a discrete form for transmission, then put it back into an analog form at the receiver. This is done by a process called **digitization** in which the analog signal is sampled at regular, closely spaced intervals, and its characteristics converted to numbers, which are transmitted. This digital information may then be converted back to analog information.

All the aspects of information and communication listed above have been tremendously augmented by recent technological advances. Perhaps the greatest such advance has been that of silicon integrated circuits, which have had a massive impact on capabilities for recording, computing, storing, displaying, and communicating information in vast quantities. The most rapid means of manipulating massive amounts of information is to combine standard electronics with **photonics**, which deals with information carried and recorded by photons of light. Using photonics, information can be acquired by a video camera, stored on compact video disks, transmitted by fiber optics, and displayed with a cathode-ray tube or light-emitting diode display. Photonics relates to information transfer and storage in two major ways — optical memory and optical communication. Somewhat misnamed, **optical memory** consists of information recorded on microscopic grooves of a rotating disk, the most familiar example of which is the compact disk (CD) for playing music. The information is recorded on the disk digitally in a binary system represented by series of reflective dots (to represent "1") and nonreflective dots (to represent "0").

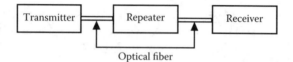

Figure 16.3.   Three major components of an optical communications system.

The amounts of information that can be so stored are staggering — a single compact video disk 9 cm in diameter can store the contents of several long books.

The medium of photonics (see preceding paragraph) has been adapted to communication with remarkable success through **optical communication** using optical fibers to transmit information. The three major components of an optical communication system are shown in Figure 16.3. One of the keys to its success is the very thin, flexible **optical fiber** through which photons of light can be transmitted. Optical fiber consists of extremely pure glass (essential to prevent unacceptable loss of light intensity during transmission) sheathed with a transparent cladding having a lower index of refraction that prevents loss of light from the fiber. The transmitter in the system takes in a signal, amplifies it, converts it to light with a device called a light emitting diode (LED), and sends the optical signal on its way with an emitter. These light signals lose their power and have to be amplified at regular intervals. This is accomplished by a repeater, which detects the light signal with a photodetector and converts it to an electronic signal that is amplified electronically, then changed back to light and sent on through optical fibers with an emitter. At the receiver, the light signal is again converted to an electronic signal by a photodetector, amplified, and put out in the needed form, such as sound in a telephone headset.

The central characteristic of communication in the modern age is the combination of telecommunications with computers called **telematics**. Automatic teller machines use telematics to make cash available to users at locations far from the customer's bank. Information used for banking, for business transactions, and in the media depends upon telematics.

There exists a tremendous potential in the application of the "information revolution" to environmental improvement. An important advantage is the ability to acquire, analyze, and communicate information about the environment. For example, such a capability enables detection of perturbations in environmental systems, analysis of the data to determine the nature and severity of the pollution problems causing such perturbations, and rapid communication of the findings to all interested parties.

16.8. TECHNOLOGY AND ENGINEERING

Technology

Much of the remainder of this chapter is dedicated to a discussion of the ways in which humans use the anthrosphere to provide for their needs. For the most part, these needs are **goods and services**. Goods can include the most fundamental things such as food and shelter. Services include transportation and communication discussed above. For the most part, goods and services are provided by technology, which is discussed in this section. The provision of goods and services has enormous implications for sustainability and the environment.

Technology refers to the ways in which humans do and make things with materials and energy. In the modern era, technology is to a large extent the product of **engineering** based on scientific principles. Science deals with the discovery, explanation, and development of theories pertaining to interrelated natural phenomena of energy, matter, time, and space. Based on the fundamental knowledge of science, engineering provides the plans and means to achieve specific practical objectives. Technology uses these plans to carry out the desired objectives. Technology makes use of tools, techniques, and systems for carrying out specific objectives, such as manufacturing consumer goods; moving people, manufactured goods, and materials; or minimizing environmental impact of human activities. At an accelerated pace during the last several years, technology has made use of computers to accomplish its objectives.

Technology has a long history, and, indeed, goes back into prehistory to times when humans used primitive tools made from stone, wood, and bone. As humans settled in cities, human and material resources became concentrated and focused, such that technology began to develop at an accelerating pace. Technology got a tremendous boost from the discovery that metals could be worked and shaped, giving rise to the pursuit of **metallurgy**. Metallurgy probably began with processing of native elemental copper around 4000 BC, followed by the widespread use of bronze, an alloy of tin and copper. The Bronze Age lasted until around 1200 BC, when iron replaced bronze for tools and weapons. Other early technological innovations predating the rise of Greek and Roman civilizations included domestication of the horse, discovery of the wheel, architecture to enable construction of substantial buildings, control of water for canals and irrigation, and writing for communication.

A major advance during the Greek and Roman eras was the development of **machines**, including the windlass, pulley, inclined plane, screw, catapult for throwing missiles in warfare, and water screw for moving water. Later, the waterwheel was developed to harness power from falling water. The generated power was transmitted by wooden gears. Many technological innovations came from China, sometimes several hundred or even a thousand years before they appeared in Europe. As examples, China had rotating fans for ventilation by around AD 200, printing with wood blocks by around AD 740, and gunpowder about a century later. The 1800s saw an explosion in technology. Among the major advances during this century were widespread use of steam power, steam-powered railroads, the telegraph, telephone, electricity as a power source, textiles, use of iron and steel in building and bridge construction, cement, photography, and invention of the internal combustion engine, which revolutionized transportation in the following century. It may be argued that in a relative sense, advances in technology during the 1800s were at least as great as those that have occurred since 1900, and certainly laid the groundwork for the vast advances that have taken place during the last 100 years.

Since about 1900, advancing technology has been characterized by vastly increased uses of energy; greatly increased speed in manufacturing processes, information transfer, computation, transportation, and communication; automated control; a huge new variety of chemicals; new and improved materials for new applications; and, more recently, the widespread application of computers to manufacturing, communication, and transportation. Arguably, the greatest impact during the last 100 years has been the application of electronics to technology. Electronics, as it is now understood, began with the discovery of the radio by Guglielmo Marconi in 1896. Electronics got an enormous boost in the early 1900s with the development of vacuum tubes that had the ability to control and amplify electronic signals and to make radio transmission possible over vast distances. A special vacuum tube, the cathode-ray tube, made television and radar possible. Solid-state devices that are remarkably small and fast have supplanted vacuum tubes in virtually all applications and have made modern computers possible.

The development of electronics probably illustrates better than anything else the revolution in technology resulting from applications of basic and applied science. In modern times, science and technology are inseparable and synergistic. The principles of science are applied to make technological advances possible. For example, radio communication and associated electronics were made possible by understanding of electromagnetic waves.

In transportation, the development of passenger-carrying airplanes has resulted in an astounding change in the ways in which people get around. In addition, large amounts of high-priority freight are now moved by air.

The technological advances of the present century are largely attributable to improved materials. For example, since the 1930s, shortly before World War II, airliners have been made of special strong alloys of aluminum; these are being supplanted by even more advanced composites that are lighter and stronger than metals, thereby reducing aircraft mass and increasing fuel efficiency. Synthetic materials with a significant impact on modern technology include plastics, fiber reinforced materials, composites, and ceramics.

Several aspects of technology should be emphasized here. It is important to realize that technology has a tremendous ability to increase productivity. Therefore, improved technology in areas such as manufacturing and agriculture has been largely responsible for the accelerating production of goods and services that has been characteristic of the last two centuries. In so doing, technology has resulted in the displacement of entire occupations and the creation of entirely new ones. An examination of technology and all of the aspects of the modern world that it influences clearly illustrates its role in the **interconnectedness** of important global and societal issues, such as productivity, resource utilization, environmental impact, employment, and overpopulation.

Until very recently, technological advances were made largely without heed to environmental impacts. Now, however, the greatest technological challenge is to reconcile technology with environmental sustainability consequences. The survival of humankind and of the planet that supports it now require that the established two-way interaction between science and technology become a three-way relationship including sustainability through the application of green science and technology. That is, of course, a major theme of this book.

Engineering

Engineering uses fundamental knowledge acquired through science to provide the plans and means to achieve specific objectives in areas such as manufacturing, communication, and transportation. At one time engineering could be divided conveniently between military and civil engineering. With increasing sophistication, civil engineering evolved into even more specialized areas, such as mechanical engineering, chemical engineering, electrical engineering, and environmental engineering. Other engineering specialties include aerospace engineering, agricultural engineering, biomedical engineering, CAD/CAM (computer-aided design and computer-aided manufacturing engineering, which are discussed in the following paragraphs), ceramic engineering, industrial engineering, materials engineering, metallurgical engineering, mining engineering, plastics engineering, and petroleum engineering.

Mechanical engineering is the branch of engineering that deals with machines and the manner in which they handle forces, motion, and power. This discipline arose as a separate area with the development of the enormous capabilities of the steam engine in the early 1800s. The major objective of mechanical engineering is to develop and improve machines that produce goods and services. The scientific principles on which engineering is based include consideration of the laws that govern forces and motion (dynamics); the thermodynamic laws that govern energy, power, and heat; transfer of materials and fluids; and other factors, including vibration control, materials properties, and wear minimization through proper lubrication. In addition to mechanical components, other machine components with which mechanical engineering must deal are electric, electronic, hydraulic, and fluidic components.

There are several objectives of machine design. Machines must produce whatever is needed in high quality to minimize expensive rejects. Speed of production is of utmost importance. Finally, costs must be minimized to enable adequate return on capital. An important factor involved in this endeavor is that of materials that can maintain exacting tolerances with minimum wear under, sometimes, severe conditions. Control of machines is particularly important. This is especially true with greatly increased use of automation and robotics. The availability of inexpensive, fast, sophisticated computers is revolutionizing control of machines and the processes that they carry out.

In the past, many of the machines and processes developed through mechanical engineering have contributed to environmental degradation and pollution. The availability of gargantuan earth-moving equipment has enabled strip-mining, destruction of wildlife habitat, and damming of natural streams. Efficient machine-equipped factories have often produced massive amounts of pollution and have provided noisy and dangerous conditions for workers. As with other branches of engineering, a major emphasis has now been placed on mechanical engineering designed to maximize sustainability and to improve environmental quality. Examples of this include machinery designed to minimize noise, much improved energy efficiency in machines, and the uses of earth-moving equipment for environmentally beneficial purposes, such as restoration of strip-mined lands and construction of wetlands.

Electrical engineering grew from the rapidly developing electrical power industry starting in the late 1800s. The theory of electrical engineering is largely based on the mathematical formulation of the laws of electricity by the Scotsman James Clerk Maxwell in 1864, though the profession began in a primitive manner with the invention of the telegraph in 1837. Electrical engineering is concerned with the generation, transmission, and utilization of electrical energy. Of particular importance in this area has been the early utilization of alternating current, which is more complex to use, but more efficient in some applications than direct current. Proper application of electrical engineering can be very helpful in the environmental arena. Efficient generation, distribution, and utilization of electrical energy constitute one of the most promising avenues of endeavor leading to environmental improvement. The modern practice of burying electrical transmission lines in urban areas has minimized visual pollution from unsightly overhead power lines and made the power distribution system less vulnerable to interruptions from weather events.

Electronics engineering deals with phenomena based on the behavior of electrons in vacuum tubes and other devices. As such, it is very much concerned with electromagnetic radiation ("radio waves" and microwaves) and transmission of radiofrequency signals through the air and space. Much of the early work in electronics engineering was based on fundamental studies of the behavior of electrons in vacuum tubes dating from the first decades of the 1900s. Though slow, prone to failure (because of the tendency of their hot filaments to burn out), and power consumptive by modern standards, these devices enabled development of practical radios, television, radar, and chemical instrumentation.

The fact that modern times are largely an "electronic era" is because of the invention of the solid-state electronic device known as the **transistor** in 1948. This enabled a truly remarkable development that has changed society enormously in the latter part of this century — the **silicon integrated circuit**. Integrated circuits are made by depositing and etching microscopic electronic circuits containing many transistors, capacitors, and resistors on single silicon chips as small as a pinhead and rarely larger than a dime. Although the cost of integrated circuit chips has remained about constant since 1960, the number of transistors possible on each chip has doubled regularly and may eventually reach a billion or so. This phenomenon has led to the vast array of consumer electronic devices, computers, and control systems that are a fact of modern life — and this is probably only the beginning.

The principles of electronics applied through electronic engineering for sustainability and environmental improvement are enormous. Automated factories can turn out goods with lowest possible consumption of energy and materials, while minimizing air and water pollutants and production of hazardous wastes. During the last two decades, electronic control of electrical power production, distribution, and utilization has enabled greater production of light and usable energy without the construction of massive numbers of new power plants. Sophisticated electronic control and the development of electronically based photovoltaic cells are enabling practical utilization of solar energy. Nuclear power generation, which can certainly play a major role in generating electrical energy without production of greenhouse gases, can be made virtually fail-safe by electronic systems that do not doze, daydream, or have bad habits that are always potential problems with systems that rely primarily on the performance of fallible humans.

Chemical engineering uses the principles of chemical science, physics, and mathematics to design and operate processes that generate products and materials through controlled chemical reactions. Historically, a key concept of chemical engineering has been that of **unit processes**, such as mixing, distillation, filtration, and heat exchange that are combined in a variety of ways to carry out chemical processes. These in turn are based on laws of thermodynamics, chemical kinetics, mass and heat transfer, and fluid flow.

Thermodynamics, the science of heat and energy phenomena, transfers, and conversions, is of crucial importance in chemical engineering. Thermodynamics enables computation of heat required and produced in chemical production, deals with the feasibility of partition of materials between phases (see mass transfer in the following paragraph), and provides information regarding the effects of temperature on chemical equilibrium (degree to which a reaction goes to completion). Whereas thermodynamics deals with equilibrium situations, **chemical kinetics** deals with speeds of chemical reactions. It explains whether or not a reaction proceeds at a sufficient rate to be practical or so rapidly as to be hazardous, and how long reactants must remain together for a desired reaction to occur. A key aspect of chemical kinetics is **catalysis**, in which catalysts, which enable chemical reactions to occur, are either mixed with reactant mixtures (homogeneous catalysis) or held on solid surfaces (heterogeneous catalysis).

Mass transfer involves separation of phases of materials and transfer of materials between phases. The most straightforward example of mass transfer is distillation in which one component of a liquid mixture is vaporized and condensed to a pure form. Sorption of pollutant organic matter from aqueous solution onto activated carbon, precipitation and filtration of solids, and extraction of materials from aqueous solution into an organic solvent are all unit operations involving mass transfer.

Much of chemical engineering deals with **heat transfer**, which in chemical plants is accomplished by conduction, convection, radiation, and as latent heat (condensation/evaporation, particularly of water).

Unlike other manufacturing and assembly operations that deal with the manipulation of objects and solids, virtually all of the material transferred in a chemical plant is in the form of fluids, so **fluid flow** becomes a major consideration. Consideration must be given to whether or not a fluid can be induced to flow at a sufficiently rapid rate to be practical, as well as what may be done to make a fluid flow adequately. Another major consideration is the loss of energy as fluids flow through pipes, over solid catalysts, and through chemical reactors.

Control of the processes described above is of particular importance in chemical engineering; a poorly regulated process can very rapidly get out of control and ruin the product or cause a fire or explosion. The transfers of fluids and energy that occur in chemical plants are particularly amenable to automated and computerized control. Application of the principles of green engineering can make chemical engineering much more safe, sustainable, and efficient.

16.9. ACQUISITION OF RAW MATERIALS

Manufacturing involves the processing of a wide variety of materials. Some materials are used with relatively little processing, whereas others are the result of highly sophisticated manufacturing processes. The acquisition and processing of materials has a number of environmental implications. There are numerous potential environmental effects of mining. These include removal and distribution of overburden in strip mining, disturbance of watersheds and aquifers, and release of pollutants to waterways. On the other hand, the development of sophisticated materials through the application of materials science has the potential to be very beneficial to the environment. As an example, underground fuel storage tanks made of polymer-reinforced fiberglass are corrosion proof and will not leak unless subjected to drastic physical insult. Their use virtually eliminates leakage, which was a significant problem with older steel tanks and makes it unnecessary to use anticorrosive coatings, which themselves pose a potential for soil and water pollution.

Raw Materials

Raw materials consist of the substances required for manufacturing processes. Raw materials may be minerals, hydrocarbons from petroleum, wood, plant fibers, cellulose and other materials. Such materials can be obtained from either **extractive** (nonrenewable) or from **renewable** sources.

The extractive industries are those which take irreplaceable resources from the Earth's crust. This is normally done by mining, but may also include pumping petroleum and withdrawal of natural gas. The raw materials extracted may be divided broadly into the categories of inorganic minerals, such as iron ore, clay used for firebrick, and gravel, and materials of organic origin, such as coal, lignite, or petroleum. Inorganic minerals that are mined include aluminum (from bauxite), antimony, asbestos, barite, beryl, bismuth, cadmium, chromite, cobalt, columbium-tantalum, copper, gem and industrial diamond, feldspar, fluorspar, gold, gypsum, iron ore, lead, magnesium, manganese, mercury, molybdenum, nickel, phosphate rock (hydroxyapatite and fluorapatite), platinum group metals, potash, selenium, silver, tellurium, tin, titanium, tungsten, uranium oxide, vanadium, and zinc.

Renewable raw materials from plant sources are discussed in some detail in Chapter 15. The prime example of a renewable material is wood. From an environmental viewpoint wood is an ideal resource. Forests conserve soil by preventing soil erosion and provide recreational areas and watersheds. The removal of carbon dioxide from air by photosynthesis carried out by trees is a significant mechanism for the removal of greenhouse-gas carbon dioxide from the atmosphere. Natural rubber and cotton are two other examples of renewable resources. The use of these substances saves irreplaceable petroleum that otherwise would be consumed in manufacturing synthetic rubber and synthetic fabrics.

Unfortunately, many resources are simply not renewable. It is impossible to grow aluminum, nickel, phosphorus, sulfur, or any of the other essential elemental resources. For these irreplaceable resources, use minimization, substitution by more abundant alternate resources, conservation, and recycling are of utmost importance in maintaining sustainability.

Materials from Earth's Crust: Mining

Geological and geochemical factors are crucial in mining, particularly in locating ore deposits. Deposits of metals often occur in masses of igneous rock that have been forced in a solid or molten state into the surrounding rock strata; such masses are called *batholiths*. Other geological factors to consider include age of rock, fault zones, and rock fractures. The crucial step of finding ore deposits falls in the category of mining geology.

Surface mining, commonly called strip mining, is used to extract minerals that occur near the surface. A common example of surface mining is quarrying of rock. In some cases where stone is mined for construction, quarrying is done by cutting the stone into slabs and blocks. Most rock is loosened by blasting before being removed as irregular chunks. Although quarry blasting used to be done by expensive and potentially hazardous dynamite or gunpowder, it is now accomplished primarily by granular ammonium nitrate soaked with fuel oil. Though powerful, this explosive is particularly safe for the user because it requires a booster in addition to a simple blasting cap for detonation. Sand and gravel are sometimes surface-mined by simply digging the material from deposits. Coal is commonly strip-mined with giant shovels that are employed primarily for removing overburden, leaving the thinner coal seam to be loaded with smaller equipment.

Gravel, sand, and some other minerals, such as gold, often occur in so-called placer deposits to which they have been carried and deposited by running water. In such cases, mining can be accomplished by dredging from a boom-equipped barge. Another approach that can be used is hydraulic mining with large streams of water. One interesting approach for more coherent deposits is to cut the ore with intense water jets, then suck up the resulting small particles with a pumping system.

For many minerals, underground mining is the only practical means of extraction. An underground mine can be very complex and sophisticated. The structure of the mine depends upon the nature of the deposit. It is, of course, necessary to have a shaft that reaches to the ore deposit. This is normally a vertical shaft, but for inclined beds it can be inclined. Horizontal tunnels extend out into the deposit, and provision must be made for sumps to remove water and for ventilation. In designing an underground mine, many factors must be considered by the mine engineer. These include the depth, shape, and orientation of the ore body, as well as the nature and strength of the rock in and around it; thickness of overburden; and depth below the surface. Of crucial importance is support of the mine roof. In some cases roof support is accomplished with a room and pillar technique in which pillars of ore are left to support the roof. Bolts may have to be put into the roof to secure it, and posts may have to be installed to hold up the roof. In stabilizing mine structures, it is important to apply rock mechanics to determine the behavior of rock under stress and how such behavior is altered by mining. The compressive and tensile strengths of the rock must be considered, along with the fact that rock behaves elastically when a load is placed on it.

Usually, significant amounts of processing are required before a mined product is used or even moved from the mine site. Even rock to be used for aggregate and for road construction must be crushed and sized. Crushing is a necessary first step for further processing of ores. Some minerals occur to an extent of a few percent or even less in the rock taken from the mine and must be concentrated on site so that the residue does not have to be hauled far. For metals mining, these processes, as well as roasting, extraction, and similar operations are covered under the category of **extractive metallurgy**.

Manufactured Materials

To an increasing degree, manufactured materials with special properties are being used for structures, machines, and other applications. The use of such materials has both positive and negative aspects from the environmental viewpoint. For example, most such manufactured materials require significant amounts of energy and may use irreplaceable petroleum as a raw material. On the other hand, some such materials are much more enduring and suitable than those which they replace. Manufactured materials are discussed in more detail in Section 16.12.

16.10. AGRICULTURE: THE MOST BASIC INDUSTRY

The most basic human need is that for food. Without adequate supplies of food, the most pristine and beautiful environment becomes a hostile place for human life. The industry that provides food is **agriculture**, an enterprise concerned primarily with growing crops and livestock. Agriculture is absolutely essential to the maintenance of the huge human populations now on Earth. Soil on which food is grown and the agricultural enterprise by which food is produced are discussed in Chapter 11. The most productive sections of Earth's surface are dedicated to food production making the agricultural enterprise the largest part of the anthrosphere. In this section, the provision of food and its effect on the anthrosphere and other environmental spheres are discussed briefly.

Agriculture can be divided into the two main categories of **crop farming**, in which plant photosynthesis is used to produce grain, fruit, and fiber, and **livestock farming**, in which domesticated animals are grown for meat, milk, and other animal products. The major divisions of crop farming include production of cereals, such as wheat, corn, or rice; animal fodder, such as hay; fruit; vegetables; and specialty crops, such as sugarcane, sugar beets, tea, coffee, tobacco, cotton, and cacao. Livestock farming involves raising cattle, sheep, goats, swine, asses, mules, camels, buffalo, and various kinds of poultry. In addition to meat, livestock produce dairy products, eggs, wool, and hides. Freshwater fish and even crayfish are raised on "fish farms." Beekeeping provides honey and the bees provide an essential function in pollinating fruit trees and other flowering crops.

The crops that provide for most of human caloric food intake, as well as much food for animals, are **cereals**, which are harvested for their starch-rich seeds. In addition to corn, wheat used for making bread and related foods, and rice consumed directly, other major cereal crops include barley, oats, rye, sorghum, and millets.

As applied to agriculture and food, **vegetables** are plants or their products that can be eaten directly by humans. A large variety of different parts of plants are consumed as vegetables. These include leaves (lettuce), stems (asparagus), roots (carrots), tubers (potato), bulb (onion), immature flower (broccoli), immature fruit (cucumber), mature fruit (tomato), and seeds (pea). According to this system, fruits, which are bodies of plant tissue containing the seed, are a subclassification of vegetables. Many kinds of vegetables are grown and consumed. Some others in addition to those mentioned above include beets, cabbage, celery, leek, pepper, pumpkin, spinach, squash, and watermelon.

For the most part, though not invariably, fruits and nuts (which are fruits, botanically) grow on trees. The ubiquitous peanut is actually a legume. Common fruits include apple, peach, apricot, citrus (orange, lemon, lime, and grapefruit), banana, cherry, and various kinds of berries. Many kinds of nuts are grown. Among the exotic kinds of nuts are the bambarra groundnut from tropical Africa, the Chile hazel from Chile, the dika nut from West Africa, the pili nut from the Pacific tropics, and the yeheb nut from Somalia. In addition to consumption as food, nuts are also used for oil, ink, varnish, spices, ornamentals, soap substitutes, polishing (ground walnut shells), tanning, and poisons. With diminishing supplies of petroleum raw material used to make some of these substances, it is likely that nuts will come to play a stronger role as sustainable sources of some important products.

Agriculture is based on domestic plants engineered by early farmers from their wild plant ancestors. Without perhaps much of an awareness of what they were doing, early farmers selected plants with desired characteristics for the production of food. This selection of plants for domestic use brought about a very rapid evolutionary change, so profound that the products often barely resemble their wild ancestors. Plant breeding based on scientific principles of heredity is a very recent development dating from around 1900. One of the major objectives of plant breeding has been to increase yield. An example of success in this area is the selection of dwarf varieties of rice, which yield much better and mature faster than the varieties that they replaced. Such rice along with dwarf varieties of wheat, barley, and other grains were largely responsible for the "green revolution" dating from about the 1950s. Yields can also be increased by selecting plants for resistance to insects, drought, and cold. In some cases the goal is to increase nutritional value, such as in the development of corn high in lysine, an amino acid essential for human nutrition.

The development of hybrids has vastly increased yields and other desired characteristics of a number of important crops. Basically, **hybrids** are the offspring of crosses between two different **true-breeding** strains. Often quite different from either parent strain, hybrids tend to exhibit "hybrid vigor" resulting in significantly higher yields of food and biomass. The most success with hybrid crops has been obtained with corn (maize). Corn is one of the easiest plants to hybridize because of the physical separation of the male flowers, which grow as tassels on top of the corn plant, from female flowers, which are attached to incipient ears on the side of the plant. Despite great success with conventional breeding techniques, application of recombinant DNA technology is well on its way to overshadowing all the advances in plant breeding made to date.

In addition to plant strains and varieties, numerous other factors are involved in crop production. Weather is an obvious factor, and shortages of water, chronic in many areas of the world, are mitigated by irrigation. Here automated techniques and computer control are beginning to play an important, often more environmentally friendly, role by minimizing the quantities of water required. The application of chemical fertilizer has vastly increased crop yields. The judicious application of pesticides, especially herbicides, but including insecticides and fungicides as well, has increased crop yields and reduced losses greatly. Use of herbicides has had an environmental benefit in reducing the degree of mechanical cultivation of soil required. Indeed, conservation tillage agriculture (see Chapter 11, Section 11.10), which requires judicious application of herbicides, is now widely practiced on some crops.

Agriculture has enormous effects on the geosphere and other environmental spheres. The displacement of native plants, destruction of wildlife habitat, erosion, pesticide pollution, and other environmental aspects of agriculture have enormous potential for environmental damage, and it is essential that agricultural practice be as environmentally friendly as possible. On the other hand, growth of domestic crops removes (at least temporarily) greenhouse gas carbon dioxide from the atmosphere and provides renewable sources of energy and fiber that can substitute for petroleum-derived fuels and materials.

One of the most rapid and profound changes in the environment that has ever taken place was the conversion of vast areas of the North American continent from forests and grasslands to cropland. Throughout most of the continental U.S., this conversion took place predominantly during the 1800s. The effects of it were enormous. Huge acreages of forest lands that had been stable since the last Ice Age were suddenly deprived of stabilizing tree cover and subjected to water erosion. Prairie lands put to the plow were destabilized and subjected to extremes of heat, drought, and wind that resulted in the blowing away of topsoil, culminating in the Dust Bowl of the 1930s.

Now, valuable farmland faces a new threat posed by the urbanization of rural areas. Prime agricultural land has been turned into subdivisions and paved over to create parking lots and streets. Increasing urban sprawl has led to the need for more highways. In a vicious continuing circle, the availability of new highway systems has enabled even more development. The ultimate result of this pattern of development has been the removal of once productive farmland from agricultural use.

In addition to the destruction of farmland to build factories, roads, housing, and other parts of the infrastructure associated with industrialization, there are other factors that tend to decrease grain production as economic activity increases. One of the major ones of these is emission of air pollutants that are toxic to plants. Water pollution can seriously curtail fish harvests. Intensive agriculture uses large quantities of water for irrigation. If groundwater is used for irrigation, aquifers may become rapidly depleted.

Food production and consumption are closely linked with industrialization and the growth of technology. It is an interesting observation that those countries that develop high population densities prior to major industrial development experience two major changes that strongly impact food production and consumption:

1. Cropland is lost as a result of industrialization; if the industrialization is rapid, increases in grain crop productivity cannot compensate fast enough for the loss of cropland to prevent a significant fall in production.

2. As industrialization raises incomes, the consumption of livestock products increases, such that demand for grain to produce more meat, milk, and eggs rises significantly.

Three countries that have experienced rapid industrialization after achieving a high population density are Japan, Taiwan, and South Korea; now China, vastly larger and more populous is undergoing the same changes. In each case, starting as countries that were largely self sufficient in grain supplies, Japan, Taiwan, and South Korea lost 20 to 30% of their grain production and became heavy grain importers over an approximately three-decade time period. The effects of these changes on global grain supplies and prices was relatively small because of the limited population of these countries — the largest, Japan, had a population of only about 100 million. Since about 1990, however, China has been experiencing economic growth at a rate of about 10% per year. With a population in 2006 of 1.3 billion people, China's economic activity has an enormous effect on global markets. It may be anticipated that this economic growth, coupled with increase in population, will result in a demand for grain and other food supplies that will cause disruptive food shortages and dramatic price increases.

On a positive note, agriculture has been a sector in which sustainability has seen some notable advances during the last 50 to 75 years. This has occurred largely under the umbrella of soil conservation. The need for soil conservation became particularly obvious during the Dust Bowl years of the 1930s, when it appeared that much of the agricultural production capacity of the U.S. would be swept away from drought-stricken soil by erosive winds. In those times and areas in which wind erosion was not a problem, water erosion took its toll. Ambitious programs of soil conservation have largely alleviated these problems. Wind erosion has been minimized by practices such as low-tillage agriculture, strip cropping in which crops are grown in strips alternating with strips of summer-fallowed crop stubble, and conversion of marginal cultivated land to pasture or forests. The application of low-tillage agriculture and the installation of terraces and grass waterways have greatly reduced water erosion.

As discussed in detail in Chapter 15, a huge challenge facing agriculture in the future will be the need to produce biomaterials by photosynthesis. There are several ways of meeting this challenge. Much of the biomaterial needed in the form of stalks, leaves, hulls, and other by-products can be produced along with cereal grain, sugar, nuts, and other food products. The production of animal protein requires ten times or more grain than does the direct provision of protein from plant sources. Furthermore, reduced production of livestock to free up land for growth of biomaterials could be accomplished without any loss of nutrition (probably leading to healthier, more vegetarian-oriented diets). Conversion of marginal cropland from the growth of annual grain crops to perennial crops grown for biomaterials could be better for the land.

As of 2006, the most well-developed source of nonfood biomaterials was ethanol, an alcohol made by fermentation of sugar derived from corn grain. Fuel ethanol production from corn has become a major economic factor in the U.S. corn belt. At the beginning of 2006, the state of Iowa had 19 ethanol plants with eight more scheduled to go into production by the end of the year. Within approximately a

1-year period, the price of ethanol fuel had increased from \$1.00/gal to \$1.75/gal. The prospect of a strong market for corn to produce ethanol was having a noted effect on farmers' plans to allot farmland areas devoted to corn vs. other crops. Some authorities have expressed concern that demand for fuel ethanol could lead to food shortages. However, protein by-product of fermentation of corn-derived sugars largely conserves the value of corn for animal feed. Advocates of the production of diesel fuel from soybean oil likewise contend that about 80% of the soybean remains available for animal feed.

16.11. INDUSTRIES

An **industry** is an enterprise that makes a kind of good or provides a particular service needed by humans for their existence and well-being. Various industries have developed because specialization of human activities makes for the greatest efficiency in providing goods and services. The kinds of industries that a country or region has depend upon the availability of needed attributes, such as raw materials, human resources, or availability of transport.

Classification of Industries

Industries fall into various classes. In an early stage of development, **basic-need industries** providing essentials of food, clothing, fuel, and shelter are emphasized, but become less important relative to more discretionary industries as wealth increases. The most common example of this is the high percentage of people engaged in agriculture at early stages of development, which dwindles to a very low figure (currently less than 3% in the U.S.) as the industrial and economic base becomes more developed. Somewhat arbitrarily, industries may be divided among the following categories:

- **Food production** Agriculture and fishing

- **Extractive mineral industries**: those involved with the mining of minerals, such as those used as sources of metals (energy sources are addressed in separate categories here)

- **Renewable resource industries**: Forestry, production of nonfood crops, such as cotton

- **Renewable energy industry**: a small, but of necessity, growing industry dealing with the utilization of renewable energy resources, such as solar energy, wind power, and biomass energy

- **Extractive energy industry**: coal mining, uranium ore mining, petroleum, and natural gas

- **Manufacturing**: Conversion of raw materials or articles to higher-value goods

- **Construction**: Building and erection of dwellings, buildings, railroads, highways, and other components of the infrastructure

- **Utilities**: Electricity distribution systems, natural gas

- **Communications**: Telecommunications, media communications

- **Transportation**: Rail, highway, air, barge, ship

- **Wholesale and retail trade**: provides the interface between the production of goods and their sale for consumption and use

- **Finance**: Banks and other entities that provide the financial resources and transactions required for industries and trade

- **Services**: Law, medicine, motels, recreation, and many others

- **Government**: National, regional (state, provincial), city, and local entities that provide needed services and regulation

Industrial growth in developing societies can be described by the sequence of (1) traditional society devoted largely to the most fundamental economic needs of food and shelter; (2) preconditions for takeoff; (3) takeoff, a growth period characterized by heavy investment in industrial development; (4) drive to maturity in which the industrial base becomes mature and well diversified, and (5) age of high mass consumption, which emphasizes high consumption of consumer goods and a large service industry. It may be hoped that a sixth stage will become dominant in which societies achieve equilibrium with the environment and the resources that must sustain it. Such a stage would be characterized by limited consumption of disposable consumer goods, highly efficient utilization of energy, land use consistent with environmental harmony, zero population growth, and a high quality of life.

Manufacturing

Once a device or product is designed and developed, it must be made — synthesized or manufactured. This may consist of the synthesis of a chemical from raw materials, casting of metal or plastic parts, assembly of parts into a device or product, or any of the other things that go into producing a product that is needed in the marketplace (Figure 16.4).

Manufacturing activities have a tremendous influence on the environment. Energy, petroleum to make petrochemicals, and ores to make metals must be dug from, pumped from, or grown on the ground to provide essential raw materials. The potential for environmental pollution from mining, petroleum production, and

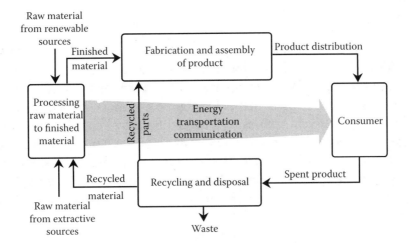

Figure 16.4.   General outline of an industrial manufacturing process.

intensive cultivation of soil is enormous. Huge land-disrupting factories and roads must be built to transport raw materials and manufactured products. The manufacture of goods carries with it the potential to cause significant air and water pollution and production of hazardous wastes. The earlier in the design and development process that environmental considerations are taken into account, the more "environmentally friendly" a manufacturing process will be.

Automation, Robotics, and Computers in Manufacturing

Three relatively new developments that have revolutionized manufacturing and that continue to do so are automation, robotics, and computers. These topics are discussed at greater lengths later in this chapter. All make use of machines and electronic instruments to do tasks formerly done by humans. Automation does repetitive tasks automatically according to directions programmed into a machine. Robotics uses machines to mimic the actions of humans, often carrying out tasks that require a number of steps and may require modification of actions based upon circumstances. Computers are superfast calculating machines that can process information very rapidly and direct manufacturing processes.

16.12. MATERIALS SCIENCE

Materials science deals with the composition, properties, and applications of substances used to make devices, machines, and structures. All kinds of materials may be used for various purposes, including metals, plastics, wood, glass, concrete, and ceramics. A vast variety of characteristics of materials must be considered, including strength, weight, fire resistance, and costs. A particularly important consideration in materials science has become consideration of sustainability of sources of materials, the application of green science and technology to materials science.

Polymers

To a very large extent, materials consist of polymers made up of smaller monomer molecules (see Chapter 3, Section 3.11). There are many kinds of polymers, both natural and synthetic. Among the important natural polymers classified as materials are cellulose and lignin in plant matter. Synthetic polymers include plastics (such as polyethylene), fibers used in textiles, synthetic rubber, and epoxy resins (widely used in composites, see below). One special class of polymers consists of **elastomers** that readily flex and stretch, like rubber. Many polymers are used to make fibers, which can be woven into fabrics. Premium plastics are polymers that have especially outstanding properties, such as high strength or resistance to heat, abrasion, or chemical attack.

Ceramics

One of the fastest growing areas of materials science has been ceramics. **Ceramics** are inorganic substances that usually involve ionically bound constituents, silicon, and oxygen, and that are usually formed by high temperature processes. Ceramics may contain aluminum as well as carbides, nitrides, and borides. Ceramics tend to be hard, rigid, and resistant to high temperatures and chemical attack. Because of their desirable properties, they have found increasing application in a number of areas. The major raw materials used to make ceramics are clays, of which the main types are montmorillonite ($Al_2(OH)_2Si_4O_{10}$), illite ($K_{0-2}Al_4(Si_{8-6}Al_{0-2})O_{20}(OH)_4$), and kaolinite ($Al_2Si_2O_5(OH)_4$); silica ($SiO_2$); feldspar, such as potassium feldspar, $KAlSi_3O_8$; and various synthetic chemicals.

Ceramics are used for many purposes, including refractory bricks and linings, crucibles, furnace tubes, and as composites with metals. One of the major applications of ceramics is as electrical materials. Some ceramics are excellent insulators, and maintain their insulating properties even at high temperatures. Other ceramics, such as silicon carbide, are semiconductors that are used in transistors, rectifiers, photocells, thermistors, and electrical heating elements. Other materials can be blended with ceramics to give composites that retain the desirable properties of ceramics, such as their thermal resistance, while improving other characteristics, such as resistance to mechanical stress.

For the most part, ceramics are made from elements that are so abundant in Earth's crust that they may be regarded as almost renewable. The substitution of ceramics for scarce materials, such as metals, is very much in keeping with the practice of green technology.

Composites

An important class of materials with growing uses consists of **composites** composed of two or more materials to give a material that has better overall properties of strength, density, cost, or electrical resistance/conductivity than their individual constituents. The materials in composites may be organic polymers, inorganic substances, metals, or metal alloys shaped as fibers, rods, particles, porous solids, sheets, or other forms. Composites are usually composed of high-strength reinforcing materials imbedded in some sort of matrix. The matrix can be a metal, plastic, ceramic, or other moldable material. Among the reinforcing materials used are fibers composed of substances such as glass, carbon, or boron nitride, or larger filaments of metal wires, silicon carbide, or other materials.

Nature provides a variety of fibers that can serve as renewable fiber sources in composites. These include both plant and animal products. Keratin protein fibers from bird feathers may have excellent strengths as composite components. Silk and wool are examples of natural animal fibers. One of the more intriguing possibilities has been to use spider web material as a natural fiber. It is one of the strongest materials known. Efforts to cultivate spider web production in a manner analogous to production of silk from silkworm farms have been stymied by the natural tendency of spiders to not get along with each other, and various genetic engineering approaches are being tried to get other organisms to produce the material.

Various lignocellulosic materials from plants are used as fibers in composites. One such material dating from thousands of years ago is straw mixed with clay or mud to make sun-baked bricks. Bamboo is another source of fiber long used by humans, especially in the Orient. Jute and hemp are strong natural fibers. One of the most productive sources of natural plant fiber is kenaf (a member of the hibiscus family) that thrives in warm climates with long growing seasons. Up to three times as productive of biomass as common fast-growing trees (southern pine), kenaf plants may reach heights of 5 meters in a single growing season. Kenaf contains long, stringy, fibrous bast. Bast is typically isolated from the kenaf plant material by what is called a *retting process* in which microorganisms degrade the light, spongy material in which the bast is contained, and the bast is removed mechanically from the residue. The kenaf plant has been advocated by some as an alternative to wood pulp for paper manufacture and some paper has been produced from kenaf.

16.13. AUTOMATION

Automation uses machines working automatically and repetitively in tasks repeated multiple times and employing mechanical and electrical devices integrated into systems to replace or extend human physical and mental activities. Automation is most widely employed on assembly lines. Primitive forms of automation were known in ancient times; an early example consists of float devices used to control water levels in Roman plumbing systems. An automated operation is directed by a **control system**, which provides the desired response as a function of time or location. In a relatively simple **open-loop system**, information regarding the desired output is fed to a controller (control actuator) that directs a process to provide the

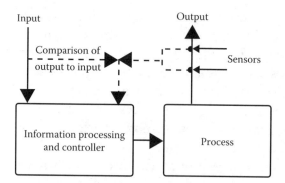

Figure 16.5.   Major components of a feedback control system.

desired response. In a more sophisticated **closed-loop system** the output is measured and compared to the desired response for feedback.

As shown in Figure 16.5, modern automated systems make use of **feedback control** systems containing the following components: (1) input in the form of a **reference value** or **set point** with which output is to be matched, for example, a pH needed for a wastewater stream fed into a bioreactor designed to degrade the wastes in the stream; (2) the process being controlled, for example, the addition of acid or base as needed for pH adjustment; (3) output, such as the actual pH of the wastewater stream; (4) sensing element, such as a glass electrode that produces an electrical potential indicative of the pH of the stream in which it is immersed; and (5) control mechanism, such as valves attached to tanks of acid and base solution and actuated by small electric motors and servomechanisms.

The simplest level of automation is **mechanization**, in which a machine is designed to increase the strength, speed, or precision of human activities. A backhoe for dirt excavation is an example of mechanization. **Open-loop, multifunctional** devices perform tasks according to preset instructions, but without any feedback regarding whether or how the task was done. **Closed-loop, multifunctional** devices use process feedback information to adjust the process on a continuous basis. The highest level of automation is **artificial intelligence** in which information is combined with simulated reasoning to arrive at a solution to a new problem or perturbation that may arise in the process.

In all but the simplest modern automated systems, computers are used for control. This enables complex sets of instructions to be given and simultaneous control of a number of interrelated factors. An important aspect of more sophisticated control is the capability for **decision making** based upon computer processing of multiple kinds of information. This can serve a number of purposes, including correction of errors, maintenance of safe conditions, and process optimization. The decision-making process enables human interaction as well by sensing undesirable changes and correcting for them, a process termed *error detection and recovery.*

Automation has been most widely applied in manufacturing and assembly. In this application, there are two major approaches. The first of these is **fixed automation** in which the mechanical design of the equipment determines the motions by means of hardware (cams and gears) and wiring, semipermanent features of the equipment that cannot be changed except by rebuilding the device or replacing parts on it. Fixed automation is most applicable to large numbers of repetitive processes that take place over a long time period. **Programmable automation** is used to give a machine instructions that can be changed for different purposes. An advanced form of this type of automation is **flexible automation**, in which the machine is automatically reprogrammed for different tasks on demand, such that it can, for example, do different assembly tasks as different items reach it on an assembly line.

Automation has found a large number of applications in industry. The first of these was in machining, the operations by which metal is shaped by cutting, grinding, and drilling. Chemical synthesis and production is especially amenable to automation because of the relative ease by which flows of materials and heat and parameters such as pressure or pH are controlled or sensed. Electronics manufacturing in its modern form has always taken advantage of automation in functions, such as parts placement on circuit boards or wire wrapping, the process in which terminal pins are connected by wires.

Automation has been used for decades in the communications industry, which is now fully automated. This was done through telephone dialing systems, now replaced by "touch tone" telephone systems. Once dominated by somewhat cumbersome, though generally effective, electromechanical switching and control systems, telephone communication is now controlled by sophisticated computer systems (swift, powerful, generally quite reliable, but capable of causing widespread consternation when catastrophic crashes occur).

Transportation makes use of a variety of automation. Simple examples are cruise control and antilock brakes on automobiles. Much more sophisticated automated functions in transportation are automatic pilots for aircraft and complex integrated systems for urban rail and subway systems. A broad range of financial transactions in banking and retail trades are now handled automatically. Another service industry in which automation is playing a major role is in health care, such as in sensing blood glucose levels in diabetics and injecting appropriate amounts of insulin to regulate blood glucose.

Not all of the effects of automation on society and on the environment are necessarily good. One obvious problem is increased unemployment and attendant social unrest resulting from displaced workers. Another is the ability that automation provides to enormously increase the output of consumer goods at more affordable prices. This capability greatly increases demands for raw materials and energy, putting additional strain on the environment. To attempt to address such concerns by cutting back on automation is unrealistic, so societies must learn to live with it and to use it in beneficial ways. There are many beneficial applications of automation. Automated processes can result in much more efficient utilization of energy and materials for production, transportation, and other human needs. A prime example is the greatly increased gasoline mileage achieved during the last approximately 20 years by the application of computerized, automated control of automobile engines and transmissions. Automation in manufacturing and chemical synthesis is used to produce maximum product from minimum raw material. Production of air and water pollutants and of hazardous wastes can be minimized by the application of automated processes. By replacing workers in dangerous locations, automation can contribute significantly to worker health and well-being.

16.14. ROBOTICS

Robotics is an extension of automation in which a machine mimics human activities through physical movement. The most common type of robotic device is the mechanical arm that can pick up objects, move them, and reorient them, or perform other activities, such as welding on an assembly line. A **robot** is a multifunctional device that can perform a variety of tasks by manipulating tools and other objects according to a preprogrammed set of instructions that can be changed as needed. Objects are manipulated by means of mechanical "arms" called **mechanical manipulators** consisting of links fastened by flexible joints that can extend (like a piston in a cylinder), twist, or rotate (Figure 16.6). Specialized end effectors are attached to the ends of robot arms to accomplish specific functions. The most common such device is a **gripper** used like a hand to grasp objects. A crucial component of robots consists of **servomechanisms** in which low-energy signals from electronic control devices are used to direct the actions of a relatively large and powerful mechanical system.

The "brains" of a robot consist of a computer that tells the robot what to do, that is, the motion sequence that it is required to go through. These instructions may be entered into the computer by a skilled programmer using computer language. A more intuitive approach is to actually lead the robot through the desired motion sequence and have it transfer the information to a computer for later repetition. Efforts are now under way to develop robots that can respond to voice instructions. The robot's computer is used for more than just directing its motions, and includes functions such as data processing, communication with other devices and with humans, decision making, and response to sensors. The last of these is a rapidly advancing area as robots become equipped with larger numbers of increasingly sensitive and sophisticated sensors for sensing motion, pressure, temperature, and light. A form of vision

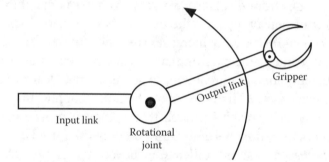

Figure 16.6.   Representation of a robot manipulator and gripper with a rotational joint.

that enables robots to recognize shapes, colors, textures, and sizes and locations of objects will become more important in the future.

The industrial applications of robots can be divided into at least four major areas. The first of these consists of **moving materials and objects**, such as transferring parts from a conveyer to pallets. An important special aspect of moving objects is placing parts in machines in the correct way for further processing. A second major area in which industrial robots are used is in **processing operations**, that is, in performing particular operations or steps in manufacturing. The most common of these is welding metals, such as in automobile assembly; other examples include grinding, smoothing, polishing, and spray painting. Robots are especially useful in **assembly**, where their reliability, speed, and accuracy are of utmost value. Finally, **inspection** for quality control may be accomplished by robots.

Robots have enormous, largely unrealized, potential in green technology. This is because robots can work continuously in environments that are unhealthy or dangerous to humans. Robots also excel at repetitive motions that, performed by humans over long periods of time, can lead to repetitive motion disorders with damage to nerves, muscles, and joints. As an example of the use of robots in dangerous situations, robots have been developed to approach and disarm suspicious packages that may contain bombs. Robots can perform in high-radiation environments such as may be generated in nuclear reactor accidents. Another area in which robots can be used is in mining activities where rockfalls, poisonous gases, and explosion hazards may endanger miners. With robots, it may be possible to develop previously inaccessible mineral deposits, such as those at depths beyond which humans cannot function safely.

16.15. COMPUTERS AND TECHNOLOGY

In modern times, nobody needs to be told of the importance of computers. Dating from just a handful of huge, cumbersome devices that first appeared in the mid-1900s, computers have advanced from machines that do calculations to those that are everywhere and that provide directions for most human activities. They are found in all kinds of apparatus and are the inseparable companions to hundreds of millions of people who use them for word processing, record keeping, communication, commerce, and virtually every other activity at work or in the home.

Computer-aided design (CAD) is employed to convert an idea to a manufactured product. Whereas innumerable sketches, engineering drawings, and physical mockups used to be required to bring this transition about, computer graphics are now used. Thus computers can function to provide a realistic visual picture of a product, to analyze its characteristics and performance, and to redesign it based upon the results of computer analysis. The design can be subjected to analyses, such as those dealing with its heat-transfer or mechanical-strain characteristics. The capabilities of computers in this respect are enormous. Closely linked to CAD is **computer-aided manufacturing (CAM)** which employs computers to plan and control manufacturing operations, for quality control, and to manage entire manufacturing plants.

A major application of computers in industry is the combination of computer-aided design and computer-aided manufacturing or so-called CAD/CAM. The tremendous data acquisition and processing capabilities of a computer are used to create and analyze a design, then modify and optimize it. In CAM, the computer can be used directly to process data regarding production, assembly, and other aspects of manufacturing, and to make adjustments to optimize the manufacturing process.

Applications of CAD/CAM continue to undergo explosive growth. Until the time that it was produced, the largest project to date using CAD/CAM was Boeing Aircraft's new 777 jetliner, the first production model plane to be produced ready for flight testing without prior production of a mockup. First flown in 1994, the product was within 0.5 mm of perfect alignment, compared to variances of more than 10 mm for previous models. More recently, Boeing Aircraft's 787 "Dreamliner," scheduled for commercial use in 2008, was designed by CAD/CAM. Described by the manufacturer as a "super-efficient" aircraft, the 787 is designed with passenger capacities of 210 to 330 and ranges up to 16,000 km (depending upon the version). It is reputed to use 20% less fuel than comparable aircraft. It also has built-in self-monitoring systems that report problems and needed maintenance to ground-based computers. Much of the improved performance of the aircraft is because of much greater use of composite materials in the fuselage and wing compared to older aircraft, a substitution that was very much aided by CAD/CAM.

CAD/CAM has taken as much as 2 years off of the period from concept to production of new automobiles. This has aided greatly in the ability of automobile manufacturers to respond to changing markets and changing conditions, such as limitations in fuel supply. The development of relatively new hybrid automobiles has relied heavily on CAD/CAM.

CAD/CAM offers a number of other advantages in addition to providing better products produced more quickly. It can show, for example, if parts are likely to interfere with each other, which offers obvious safety advantages in areas such as aircraft operation. The interaction of people with controls and machines can be simulated and parts optimized for best interaction and reduction of repetitive motion injury. CAD/CAM has the potential to significantly reduce amounts of materials used and, particularly, wastage in cutting objects. Therefore, it provides significant environmental benefits.

Computers have found wide application in **modeling**, the construction of mathematically oriented theoretical schemes that mimic complex systems, such as manufacturing processes, ecosystems, and climate. One caution — not always observed as it should be — is that models are only as good as the information that goes into them and should always be correlated and verified as closely as possible with what actually happens in the systems modeled.

The application of computers has had a profound influence on environmental concerns. One example is the improved accuracy of weather forecasting that has resulted from sophisticated and powerful computer programs and hardware. Related to this are the uses of weather satellites, which could not be placed in orbit or operated without computers. Satellites operated by computer control are used to monitor pollutants and map their patterns of dispersion. Computers are widely used in modeling complex ecosystems, climate, and other environmentally relevant systems.

Computers are essential to the modern practice of green technology. Computerized control of manufacturing and commercial processes can enable the most efficient possible production and delivery of goods and services with minimum consumption of materials and energy. Safety can be greatly improved by computers that do not have to rely on human judgment and are not subject to human misjudgment.

Computers and their networks are susceptible to mischief and sabotage by outsiders. The exploits of "computer hackers" in breaking into government and private sector computers have been well documented. Important information has been stolen and the operation of computers has been seriously disrupted by hackers with malicious intent. Most computer operations are connected with others through the Internet, enabling communication with employees at remote locations and instant contact with suppliers and customers. Numerous kinds of protection are available for computer installations. Such protection comes in the form of both software and hardware. Special encryption software can be used to put computer messages in code that is hard to break. Hardware and software barriers to unauthorized corporate computer access — firewalls — continue to become more sophisticated and effective.

16.16. THINKING SMALL: MICROMACHINES AND NANOTECHNOLOGY

Micromachines

Micromachines consisting of working devices of the order of a millimeter in size have been constructed and may have a number of useful applications in the future. As an example of the capability of micromachining, a model of an automobile with 24 moving parts capable of propelling itself has been built by a subsidiary of Toyota Motor Corporation in Japan (see Figure 16.7). Although there might not be many practical uses for a model of a 1936 Toyota, about as large as a grain of rice that can traverse the length of a matchstick in about 1 sec, there are numerous

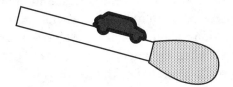

Figure 16.7.   Relative sizes of a working miniaturized model of Toyota's first automobile, a 1936 model, compared to a match. The model has 24 parts, including an electric motor that can propel the miniature car at a speed of up to 50 mm/sec.

potential uses for micromachines. An example of such a device would be a monitor for vital signs, such as blood pressure, attached to a wristwatch and capable of injecting medication if an acute health emergency develops. It may be possible to develop diagnostic devices small enough to propel themselves through blood vessels and relay pictures outside the body. Microcamera systems now exist that are swallowed and broadcast a record of digestive system conditions as they go through the stomach and intestines. Micro devices might even be equipped to perform micro-surgical procedures. Devices to inspect pipes internally are also feasible and would be very useful.

A special category of micromachines consists of those fabricated on semiconductors using technology well developed in making integrated circuits (microchips) for computer and electronics applications. These kinds of devices have been called *microelectromechanical systems*, or MEMS. An example of such a MEMS device is a miniature pressure sensor containing a silicon diaphragm that flexes in response to pressure.

Nanotechnology

Nanotechnology refers to devices the size of a nanometer (nm) and **nanoscience** is the science of the extremely small. A nanometer is very small, about the length of 10 hydrogen atoms laid in a row, 1/1000 the size of a typical bacterial cell, or 1 millionth the size of a pinhead. With printing the size of nanometers, it would be possible to store the contents of the entire Library of Congress within the volume of a sugar cube. Although named after the nanometer, nanotechnology is a term that is applied to devices up to about 1,000 times larger, the size of a micrometer. More specifically, the **nanorealm** is now understood to deal with materials and systems in which at least one dimension is 1 to 100 nm, are governed by the fundamental physical and chemical properties of molecules on a molecular size scale, and that can be connected to form larger entities.

Nanoscience and nanotechnology deal with a realm that borders that in which the behavior of atoms and molecules are governed by the laws of quantum mechanics and the macrorealm where more classical laws for the behavior of matter hold true. The bottom range of the nanorealm is limited by the size of atoms, which is the lower limit for dimensions of devices. Nanoscience has been made possible by several inventions that enable observation and manipulation at atomic levels. The scanning tunneling microscope and the atomic force microscope, the tip of which can apply accurately chosen forces to individual atoms and molecules, have enabled observation and manipulation of individual atoms. The capabilities of such devices were demonstrated in 1989 when Donald M. Eigler wrote the letters IBM with individual xenon atoms and made a picture of the result.

There are two general approaches to making nanostructures. The classical approach is "top–down" in which macro materials are etched and materials are added to bulk material. A later, very promising, approach is bottom–up in which atoms and molecules undergo self-assembly under the appropriate conditions.

One of the most interesting and useful materials for making nanodevices are carbon nanotubes. Discovered in 1991, these are hollow tubes of a diameter only about 1/10,000 that of a human hair (only about 10 carbon atoms across) composed of rolled-up sheets of carbon hexagonal structures produced by electrical discharges between carbon electrodes.

Much of the interest in nanotechnology lies in electronics. Nanoscale molecular structures exhibit unique electron transport characteristics that make them particularly useful in electronics. Microchips have now been prepared with circuit lines only 100 nm wide. Nanoscale transistors have been demonstrated, and molecular switches have been made from single molecules. In 2001, IBM scientists constructed the first array of transistors from carbon nanotubes. In early 2006, it was announced that Columbia University researchers had made the first transistor that was a hybrid of a carbon nanotube with a molecule capable of acting as a switch. The carbon nanotube was cut leaving a space the size of the molecule, which was inserted between the two nanotube segments. The ends of the nanotubes into which the molecular switch was inserted were prepared for bonding to the molecule by an oxidation process.

Several products of nanotechnology have already been commercialized. Ultrasensitive sensors for disc drives have been made with nonmagnetic layers of nanodimensions sandwiched between magnetic layers, which allows construction of very sensitive disc drives with much increased capacities. A chemotherapeutic agent effective against AIDS-related Kaposi's sarcoma has been encapsulated in liposomes, spheres of lipids (see Chapter 3, Section 3.16) about 100 nm in diameter to deliver the drug efficiently. Zeolites with pores less than a nanometer in size, have been developed as very efficient catalysts for breaking large hydrocarbon molecules from crude oil down into smaller molecules useful in gasoline. Nanocrystalline particles have been used to make catalysts, colorless sunblocks, and stronger ceramics.

There is a strong potential connection between nanoscience and biological science. Indeed, the ultimate nanodevice is the living cell which, in some cases smaller than a micrometer in size, functions as a highly sophisticated biochemical factory capable of replication. At an even smaller level are viruses, segments of RNA genetic material (see Chapter 3, Section 3.16) that very cleverly hijack cell metabolism and replicate at a very rapid rate using the host's biochemical machinery. There is much interest in (and concern about) making artificial cell-like structures to carry out various functions.

Much of the interest in nanomaterials and nanodevices stems from their unique and potentially superior mechanical, electrical, optical, or chemical properties. Nanotechnology has considerable potential for the development of green science and technology. One of the most active areas in green chemistry is the development of more effective, efficient, and specific catalysts for carrying out desired chemical processes; catalysts based upon nanotechnology may be very effective in such applications. Nanomembranes designed for specific porosity, surface characteristics, and surface chemistry can enable much more efficient separation processes and serve as highly effective catalyst supports. Nanomaterials may be used to bind with and remove water pollutants. Nanostructured titanium dioxide activated with solar radiation has the potential to detoxify water and air pollutants. Miniaturized devices making use of nanotechnology can enable much lower material requirements (dematerialization), one of the basic objectives of green technology. Similarly, because of its inherently lower scale, processes based on nanotechnology will produce less waste. Very small computers operating at very high speeds based on nanotechnology can be useful in achieving the objectives of green technology.

16.17. HIGH TECHNOLOGY

What is high technology, "high tech"? Some have said, "To a caveman, it is the wheel." Certainly, in their day, the telegraph and the steam locomotive were high technology, as was the biplane aircraft of World War I. In present times, however, there is a group of technologies that can be labeled as high-tech from the perspective of the modern era. In speaking of high-tech, a number of terms come to mind, including the following: voice synthesis, bionics, plasma systems, artificial intelligence, computerized language translation, remote sensing, cryogenics, photovoltaics, sonar, computer-controlled prosthetic devices, lasers, composite materials, ceramics, light-emitting diodes, fiber optics, genetic engineering, robotics, nanotechnology, cryptography, CAT scanners, MRI imaging, digital audio, digital video, as well as computer-aided design (CAD), instruction (CAI), graphics (CAG), and manufacturing (CAM). Several important aspects of high technology are summarized briefly in this section.

Several areas of high-tech have already been addressed in this chapter, including computers, robotics, certain aspects of materials science, and nanotechnology. Central to these and all high-tech areas are computers and the microchip, integrated circuits, and microprocessors that make them possible. Any other field of modern high technology is dependent upon computers. Computers are central to modern telecommunications, they make robots possible, they are essential to the development of new formulations for exotic materials, and make possible the exacting conditions under which such materials must be made.

Endeavors in **space** certainly can be classified as high-tech. The realities of high costs and other more pressing priorities have prevented the development of "factories on the moon" or permanently populated orbiting space stations that many predicted when the "Space Age" became reality in the late 1950s. The exotic space combat visualized in "Star Wars" concepts has given way, to a large extent, to the decline of major nuclear powers and to the realities of grubby little wars fought by more conventional means including even suicide bombers. The greatest practical success in space so far has been in the launching of telecommunications satellites that have revolutionized global communications. Space technology is used very successfully in weather forecasting and in studying threats to the global climate, particularly greenhouse warming and atmospheric ozone depletion. Additional uses of space technology will undoubtedly develop in the future.

Lasers, devices that generate and transmit a condensed and directed beam of light, probably symbolize high-tech more than any device other than the computer. Laser technology is an area that has come of age in areas such as laser printers, laser scanning of purchases, and laser drilling of diamonds. Lasers have also found significant uses in medicine.

Advanced **biotechnology** uses biochemical processes to perform tasks and to make products that would otherwise be impossible to make. Biotechnology directed through bioengineering makes use of enzymes, recombinant DNA, gene splicing, cloning, and other biological phenomena. The most exciting area of biotechnology in recent years is the one dealing with gene splicing or recombinant DNA wherein DNA material from one organism is inserted into another to give an organism with desired characteristics, such as the ability to make a specific protein. Several significant products have been produced by gene splicing. One such product is Humulin, a form of insulin identical to human insulin; another is human growth hormone. There is a high level of activity in biotechnology as applied to agriculture. Particularly promising are prospects to develop plants that are resistant to insects or to herbicides applied to competing plants, plants that can be made to fix nitrogen (through symbiotic bacteria growing on their roots), and growth promoters for plants and perhaps animals.

High-tech is finding increasing uses in medicine and in pharmaceuticals. Examples of high-tech medicine include arthroscopic surgery that greatly reduces the invasiveness of surgical procedures, real-time sensors for the control of such things as blood sugar in diabetics, and magnetic resonance imaging techniques. Prosthetic devices controlled by computers have significant promise in aiding the physically handicapped to lead more normal lives. Among the high-tech medical devices in common use are CAT scanners, defibrillators, pacemakers to regulate heartbeat, and heart-lung machines. Medical sonographs are devices that image internal organs to show abnormalities, such as tumor masses, using reflected high frequency sound signals processed by computer. Common organs that are imaged by this technique include the brain (using echoencephalogy) and the heart (using echocardiography). A refined version called doppler sonography is employed to show blood flow. Dialysis machines are devices through which a patient's blood is passed for removal of impurities across a semipermeable membrane as a substitute for nonfunctional kidneys. Another medical area that depends upon high-tech equipment and techniques is nuclear medicine, which uses radioactive substances that emit gamma rays to image body organs to diagnose abnormalities, particularly tumors.

High-tech pharmaceuticals use sophisticated techniques to design and synthesize new drugs that are extremely potent and targeted against specific maladies with minimum side effects. High-tech medicine and pharmaceuticals both make considerable use of biotechnology. Nanomaterials for the delivery of pharmaceuticals are now being developed.

SUPPLEMENTARY REFERENCES

Allenby, Braden, *Reconstructing Earth: Technology and Environment in the Age of Humans*, Island Press, Washington, D.C., 2005.

Amirouche, Farid, *Principles of Computer-Aided Design and Manufacturing*, Pearson Prentice Hall, Upper Saddle River, NJ, 2004.

Asthana, Rajiv, Ashok Kumar, and Narendra B. Dahotre, *Materials Processing and Manufacturing Science*, Elsevier Academic Press, Burlington, MA, 2005.

Azapagic, Adisa, Slobodan Perdan, and Roland Clift, Eds., *Sustainable Development in Practice: Case Studies for Engineers and Scientists*, John Wiley & Sons, Hoboken, NJ, 2004.

Baker, Susan, *Sustainable Development*, Routledge, New York, 2006.

Banister, David, *Unsustainable Transport*, Spon Press, New York, 2005.

Barrow, C.J., *Environmental Management for Sustainable Development*, 2nd ed., Routledge, New York, 2006.

Beloff, Beth, Marianne Lines, and Dicksen Tanzil, Eds., *Transforming Sustainability Strategy into Action: The Chemical Industry*, Wiley-Interscience, Hoboken, NJ, 2005.

Benhabib, Beno, *Manufacturing: Design, Production, Automation and Integration*, Marcel Dekker, New York, 2003.

Bigg, Tom, Ed., *Survival for a Small Planet: The Sustainable Development Agenda*, Earthscan Publications, London, 2004.

Biondo, Ronald J. and Jasper S. Lee, *Introduction to Plant and Soil Science and Technology*, 2nd ed., Interstate Publishers, Danville, IL, 2003.

Booker, Richard and Earl Boysen, *Nanotechnology for Dummies*, John Wiley & Sons, Hoboken, NJ, 2005.

Brandon, Peter S. and Patrizia Lombardi, *Evaluating Sustainable Development in the Built Environment*, Blackwell Science, Malden, MA, 2005.

Brand, Ralf, *Synchronizing Science and Technology with Human Behaviour*, Earthscan Publications, Sterling, VA, 2005.

Burton, L. DeVere and Elmer L. Cooper, *Agriscience: Fundamentals and Applications*, 4th ed., Thomson/Delmar Learning, Clifton Park, New York, 2005.

Cahn, R.W., P. Haasen, and E.J. Kramer, Eds., *Materials Science and Technology: A Comprehensive Treatment*, Wiley-VCH, Weinheim, Germany, 2005.

Canniffe, Eamonn, *Urban Ethic: Design in the Contemporary City*, Routledge, New York, 2006.

Corey, Kenneth E. and Mark I. Wilson, *Urban and Regional Technology Planning: Planning Practice in the Global Knowledge Economy,* Routledge, New York, 2006.

Dewulf, Jo and Herman Van Langenhove, *Renewables-Based Technology: Sustainability Assessment,* John Wiley & Sons, Hoboken, NJ, 2006.

Elizabeth, Lynne and Cassandra Adams, *Alternative Construction: Contemporary Natural Building Methods*, John Wiley & Sons, Hoboken, NJ, 2005.

Gad-el-Hak, Mohamed, Ed., *MEMS: Applications*, 2nd ed., CRC Press/Taylor & Francis, Boca Raton, FL, 2006.

Gad-el-Hak, Mohamed, Ed., *MEMS: Background and Fundamentals*, CRC Press/Taylor & Francis Group, 2005.

Graedel, Thomas E. and Jennifer A. Howard-Grenville, *Greening the Industrial Facility: Perspectives, Approaches, and Tools*, Springer-Verlag, New York, 2005.

Hall, J. Storrs, *Nanofuture: What's Next for Nanotechnology*, Prometheus Books, Amherst, New York, 2005.

Hayes, Brian, Infrastructure: A Field Guide to the Industrial Landscape, W.W. Norton, New York, 2005.

Hitchcock, Darcy and Marsha, Willard, *The Business Guide to Sustainability*, Earthscan, London, 2006.

Howell, Paul R., *Earth, Air, Fire and Water: Elements of Materials Science*, 2nd ed., Pearson Custom Publishing, Boston, MA, 2005.

Hummel, Rolf E., *Understanding Materials Science: History, Properties, Applications,* 2nd ed., Springer-Verlag, New York, 2004.

Ikerd, John E., *Sustainable Capitalism: A Matter of Common Sense*, Kumarian Press, Bloomfield, CT, 2005.

Jeong, Howon, *Globalization and the Physical Environment*, Chelsea House Publishers, Philadelphia, 2006.

Jha, Raghbendra and Bhanu Murthy, *Environmental Sustainability: A Consumption Approach*, Routledge, New York, 2006.

Kalpakjian, Serope and Steven Schmid, *Manufacturing, Engineering and Technology*, 5th ed., Prentice Hall, Upper Saddle River, NJ, 2005.

Kelsall, Robert W, Ian W Hamley, and Mark Geoghegan, Eds., *Nanoscale Science and Technology*, John Wiley & Sons, Hoboken, NJ, 2005.

Krishnamoorthy, C.S. and S. Rajeev, *Computer Aided Design: Software and Analytical Tools*, Alpha Science International, Harrow, U.K., 2005.

Liu, Chang, *Foundations of MEMS*, Upper Saddle River, Pearson Prentice Hall, Upper Saddle River, NJ, 2005.

Low, Nicholas, *The Green City: Sustainable Homes, Sustainable Suburbs,* Routledge, New York, 2005.

Lyshevski, Sergey, Edward, *Nano- and Micro-Electromechanical Systems: Fundamentals of Nano- and Microengineering,* 2nd ed., CRC Press, Boca Raton, FL, 2005.

Maluf, Nadim and Kirt Williams, *Introduction to Microelectromechanical Systems Engineering*, 2nd ed., Artech House, Boston, MA, 2004.

Meadows, Donella, Jørgen Randers, and Dennis Meadows, *The Limits to Growth: The 30-Year Update*, Chelsea Green Publishing Company, White River Junction, VT, 2004.

Miller, Donald and Gert de Roo, Eds., *Integrating City Planning and Environmental Improvement: Practicable Strategies for Sustainable Urban Development*, 2nd ed., Ashgate, Brookfield, VT, 2004.

Munier, Nolberto, *Introduction to Sustainability: Road to a Better Future,* Springer-Verlag, New York, 2005.

Olson, Robert and David Rejeski, Eds., *Environmentalism and the Technologies of Tomorrow: Shaping the Next Industrial Revolution,* Island Press, Washington, D.C., 2005.

Ooi, Giok Ling, *Sustainability and Cities: Concept and Assessment*, World Scientific Publishing, New York, 2005.

Purvis, Martin and Alan Grainger, Eds., *Exploring Sustainable Development: Geographical Perspectives*, Earthscan Publications, Sterling, VA, 2004.

Robinson, Nicholas A., Ed., *Strategies Toward Sustainable Development: Implementing Agenda 21*, Oceana Publications, Dobbs Ferry, New York, 2004.

Ruth, Matthias, Ed., *Smart Growth and Climate Change: Regional Development, Infrastructure, and Adaptation,* Edward Elgar Publishing, Northhampton, MA, 2006.

Saxena, Anupam and Birendra Sahay, *Computer Aided Engineering Design*, Springer-Verlag, New York, 2005.

Schulte, Jürgen, *Nanotechnology: Global Strategies, Industry Trends and Applications*, John Wiley & Sons, Hoboken, NJ, 2005.

Seidel, Peter and Ervin Laszlo, Eds., *Global Survival: The Challenge and Its Implications for Thinking and Acting*, SelectBooks, New York, 2006.

Sikdar, Subhas K., Peter Glavic, and Ravi Jain, Eds., *Technological Choices for Sustainability*, Springer-Verlag, Berlin, 2004.

Simpson, R. David, Michael A. Toman, and Robert U. Ayres, Eds., *Scarcity and Growth Revisited: Natural Resources and the Environment in the New Millennium*, Resources for the Future, Washington, D.C., 2005.

Smith, William F. and Javad Hashemi, *Foundations of Materials Science and Engineering*, 4th ed., McGraw-Hill, Boston, MA, 2005.

Stenerson, Jon, *Industrial Automation and Process Control*, Prentice Hall, Upper Saddle River, NJ, 2003.

Theodore, Louis and Robert G. Kunz, *Nanotechnology: Environmental Implications and Solutions*, John Wiley & Sons, Hoboken, NJ, 2005.

Todd, Nancy Jack, *A Safe and Sustainable World: The Promise of Ecological Design*, Island Press, Washington, D.C., 2005.

Uldrich, Jack, *Investing in Nanotechnology: Think Small, Win Big*, Adams Media, Avon, MA, 2006.

Wallace, Bill, *Becoming Part of the Solution: The Engineer's Guide to Sustainable Development*, American Council of Engineering Companies, Washington, D.C., 2005.

Woodin, Michael and Caroline Lucas, *Green Alternatives to Globalisation: A Manifesto*, Pluto Press, London, 2004.

Wright, R. Thomas and Ryan A. Brown, *Technology: Design and Applications*, Goodheart-Wilcox Co., Tinley Park, IL, 2003.

QUESTIONS AND PROBLEMS

1. Classify each of the following as (A) an anthrospheric construct, (B) anthrospheric flow, (C) anthrospheric conduit: (1) Water in a hydroelectric turbine, (2) a hydroelectric turbine, (3) airport runway, (4) workers entering a factory, (5) a house, (6) hallway in a school building.

2. Outline the life cycle analysis of an automobile with the view of minimizing the environmental impact of its manufacture, use, and disposal.

3. It has been stated that many modern dwellings are constructed and operated "out of the context" of their surrounding. What do you think this means? Suggest how dwellings might be more "in context" of their surroundings.

4. What is the distinction between extractive and renewable sources of raw materials?

5. Explain the relationship of geology and geochemistry to the location and utilization of extractive sources of raw materials.

6. What is extractive metallurgy?

7. What are polymers? How are they important in materials science?

8. What are ceramics? Why are they of growing importance in materials science?

9. What are the advantages of composites as materials?

10. What is the distinction between an open-loop and closed-loop system in automation?

11. What is the function of a set point in a feedback control system used in automation?

12. What is meant by robotics? What is the distinction between robotics and automation?

13. What are the "brains" and "hands," respectively, of a robot?

14. What is meant by CAD/CAM? What advantages does it offer?

15. What are mixed modes of transport? What advantages do they offer?

16. What is the distinction between photonics and electronics?

17. What did the first successful long-distance communication device, the telegraph, have in common with modern digital computers?

18. What are the key components of an optical communication system?

19. What would have been considered "high-tech" devices in the late 1800s?

20. In what respects might it be argued that the biological sciences are the most "high-tech" of all modern areas?

21. List and explain the direct and indirect effects on the environment from a highway-based transportation system.

22. In Section 16.7 are discussed some major aspects of information and its communication. Discuss how these areas may pertain to weather forecasting.

23. Discuss some of the adverse effects on the environment from the development of agriculture in the U.S., during the 1800s. In what sense is agriculture, itself, now being threatened?

24. What are the consequences of a country undergoing rapid industrial development after it has achieved a high population density.

25. What is meant by the infrastructure? List the specific parts of the infrastructure on which you depend.

26. Compare the ways in which the development and use of computers that is now ongoing are analogous to the development and application of electricity in the late 1800s and early 1900s.

27. How is engineering related to science, and how are both related to technology?

28. In what sense may it be argued that relative advances in technology were as great during the 1800s as they have been during the 1900s?

29. Which invention in the early 1900s enabled rapid developments in electronics?

30. What device from the late 1940s has enabled the explosive growth in electronics that has occurred since then?

31. Distinguish between extractive and renewable energy resources.

32. In what sense is wood an ideal raw material environmentally?

33. Define what is meant by an industry.

34. What is automation? How do computers contribute to modern automation?

35. Distinguish between automation and robotics. How are the two related?

36. List some specific ways in which CAD/CAM is contributing to a "new industrial revolution."

37. Look up carbon nanotubes on the Internet. How do they relate to nanotechnology?

38. In what senses is a living cell the ultilmate nanodevice?

39. Suggest technologies that qualified as "high-tech" in 1900.

40. Look up self-assembly and nanotechnology on the Internet. What does it mean? What are the advantages and potential of self-assembly in nanotechnology?

17. INDUSTRIAL ECOLOGY FOR SUSTAINABLE RESOURCE UTILIZATION

17.1. INTRODUCTION

In Chapter 1 it was noted that sustainability depends on the availability of energy and materials that are constantly replenished and inexhaustible. Such a utopian goal collides with the reality of the demands that each person places on resources and energy and increasing numbers of people. As noted in Chapter 1, one of the strongest warnings of the problems of increasing population and the demands it places on resources was posted by Paul Ehrlich, the Stanford University biologist who wrote the 1968 book, *The Population Bomb*.[1] Ehrlich predicted a grim future as resources and energy were exhausted and populations crashed from starvation. A rebuttal was posted in several works including *Hoodwinking the Nation*[2] by University of Maryland economist Julian Simon. He ridiculed Ehrlich's pessimistic outlook, placing his faith in free market economics and human ingenuity. Simon said these would always find a way to overcome shortages and that there was no need to worry about running out of materials, food, and energy.

For the next 40 years following the publication of Ehrlich's book, it appeared that Simon was right, leading people and their governments to a false sense of security. Although food shortages led to malnutrition, especially in parts of Africa, and some people did starve, these problems were attributable more to wars and inadequate, corrupt political systems than to shortages of material goods. There were petroleum shortages in the 1970s, but these were also due to political disputes rather than real shortages of oil. High prices of fuel accompanying these disruptions were followed by years of the lowest prices for petroleum products (in inflation-adjusted terms) since the first production of crude oil in the mid 1860s. Efforts to conserve petroleum in the United States, such as federally mandated automobile fuel economy standards, were neglected as huge, fuel-wasting vehicles became more and more popular.

By the early 2000s, however, it was becoming obvious that shortages of fuel and materials were real and would grow in importance as factors in modern society. Petroleum prices reached record levels in 2005/2006 and all indications were that such high prices would continue and would rise in the future. Commodities such as copper, zinc, and silver increased dramatically in price. As is the nature of pricing such commodities, price spikes were followed by rollbacks from their peaks, but prices still remained at painfully high levels. A new factor that until the end of the 1900s had not had much effect was the rapid development of China and India as economic powers. With huge populations, these countries are rapidly undergoing massive changes in their social and economic systems as their people emerge from social and economic systems of extreme poverty that placed relatively little demand per person on resources. As middle classes have grown in these and other formerly impoverished countries, legitimate quests for "the good life" have demanded more energy, more and better food, and more materials for housing and other needs.

Common sense tells us that we live on a planet that has limited resources and limited natural capital to support populations growing in numbers and in affluence. Nor is it simply a matter of shortages of material and energy. Modern civilizations have damaged Earth's support systems, and the harm continues with greenhouse gases released to the atmosphere causing global warming, productive farmland covered with highways, buildings, and other structures, and degraded air and water quality. The central challenge to modern societies is to reverse these unsustainable trends and to find ways for humankind to live within Earth's carrying capacity. It will not be easy and success is by no means assured. Unfortunately, failure is an option, but one that would inflict enormous misery upon Earth's populations.

It is now obvious that resources and energy have become the limiting factors in the development of modern industrialized societies. This occurs at a time when two countries with enormous populations — China and India — are rapidly becoming industrialized and are demanding their share of increasingly scarce minerals, energy, food, fiber, and the other things provided by Earth to keep modern industrialized societies going. Although it has long been noted that the United States with 5% of Earth's population is the largest consumer of Earth's resources, the U.S. is losing that dubious distinction, having now been passed by China in consumption of the key commodities of grain, meat, coal, and steel. If China continues its 2006 growth rate of about 8% by year, its consumption of resources per person will pass the U.S. in the early 2030s when its population will exceed 1.4 billion people. Then it would consume two thirds of the amount of grain produced worldwide in 2006, double the amount of the current world production of paper, and essentially 100 million barrels of oil per day, exceeding current world production of 84 million barrels per day. With these kinds of consumption figures, it is simply impossible to keep going with a "business as usual" approach. Either advanced societal systems will have to find a way to live within the carrying capacity of Earth, or they will collapse. Finding a way to live within our means is a central challenge to humankind today.

There are two key aspects of sustainability that — with luck — will enable modern societies to exist in relative comfort and bring less affluent societies to acceptable living standards in a sustainable matter. The two essential attributes of sustainable industrialized societies are the practice of industrial ecology in which industrial systems operate in ways analogous to sustainable natural ecosystems discussed in this chapter and the provision of adequate amounts of sustainable energy, the topic of Chapter 18.

It is now obvious that resources and energy have become the limiting factors in the development of modern industrialized societies. This occurs at a time when two countries with enormous populations — China and India — are rapidly becoming industrialized and are demanding their share of increasingly scarce minerals, energy, food, fiber, and the other things provided by Earth to keep modern industrialized societies going. It is simply impossible to keep going with a "business as usual" approach. Either advanced societal systems will have to find a way to live within the carrying capacity of Earth, or they will collapse. Finding a way to live within our means is a central challenge to humankind today.

There are two key aspects of sustainability that — with luck — will enable modern societies to exist in relative comfort and bring less affluent societies to acceptable living standards in a sustainable matter. The two essential attributes of sustainable industrialized societies are the practice of industrial ecology in which industrial systems operate in ways analogous to sustainable natural ecosystems discussed in this chapter and the provision of adequate amounts of sustainable energy, the topic of Chapter 18.

17.2. THE OLD, UNSUSTAINABLE WAY

Figure 17.1 illustrates the old way of industrial production that prevailed in the earlier days of the industrial revolution. Raw materials and energy were taken from nonrenewable sources, including wood from old growth timber, and products were made without consideration of wastes or pollutants. Air and water pollutants

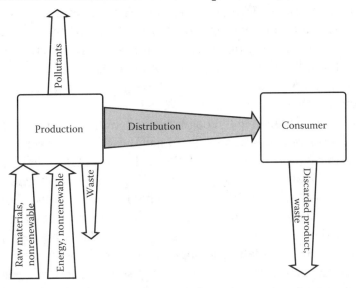

Figure 17.1.   The old model of production without regard to consumption of nonrenewable energy and resources or consideration of pollutant release or waste disposal.

were simply released to the atmosphere and waterways, respectively, and by-products — solid, semisolid, and liquid wastes — were placed in the geosphere without consideration of their effects. After products were used, they were simply discarded to the environment without consideration of environmental effects. The idea of sustainability was a foreign concept. As a result, air and water became polluted, and the land was littered with waste dumps and discarded consumer products.

Eventually, and at an increasing pace during the latter 1900s, the undesirable features of the model outlined above became increasingly obvious. Greater efficiencies were achieved in the utilization of both energy and materials. The release of pollutants was increasingly limited by regulations, and the disposal of wastes to the geosphere became regulated. Pollution prevention became an important consideration with redesign of systems to avoid pollutant release. But there remained resistance to such measures, especially when higher costs were entailed. It became obvious that a better way was needed, which turned out to be the application of industrial ecology as discussed below.

17.3. EARTH SYSTEMS ENGINEERING AND MANAGEMENT

There are two extremes with respect to humans' relationship to Earth and its resources. The first of these, as outlined above, is to proceed to exploit Earth's support systems with increasing populations without regard to environmental consequences and sustainability. The second is to restore Earth to its natural state supporting a very limited population. The problem is that the first of these is unsustainable and the second simply will not happen. To successfully deal with the limitations of Earth's resources and the pressures on it, it is useful to consider **earth systems engineering and management** defined as the capacity to rationally engineer and manage human technology systems and related elements of natural systems in such a way as to provide the requisite functionality while facilitating the active management of strongly coupled natural systems. Earth systems management recognizes the overwhelming influence of human activities on Earth and its systems such as atmospheric and oceanic systems as well as the hydrologic, carbon, nitrogen, sulfur, and phosphorus cycles.

Earth systems management recognizes that human activities are massively impacting the Earth and its systems and, in fact, have done so ever since substantial human populations have existed on the planet. Human history has been characterized by a series of industrial revolutions including the establishment of agriculture several millennia ago, the development of heavy industries in the 1800s, the explosive growth of production and resource exploitation in the 1900s, and the current high-tech era of the early 2000s. As a result, systems of technology have evolved, and economic systems, population demographics, and human cultures have changed greatly. As a consequence, major natural systems on Earth have been increasingly and overwhelmingly dominated by humans and their activities.

Earth systems engineering and management recognizes the reality of the pervasive influence of humans on Earth, such that Earth as it is now known is largely a product of human design, and it advocates managing human activities in a responsible and ethical way that makes human impacts positive rather than detrimental. It is, therefore, an integral part of any viable system of industrial ecology (discussed in the following paragraphs). The current practice of engineering and management is based upon the premise that essentially all the important things are known about systems being managed and studied and that there are definite objectives to be accomplished within a particular time span. This approach does not work for Earth systems engineering and management, because Earth and its human denizens constitute a system that is far too complex and variable for such a simple view. Therefore, Earth systems engineering and management must continually evolve through the interaction of Earth systems and those who would manage them.

The successful application of Earth systems engineering and management requires thorough knowledge of environmental sciences, ecology, cycles of matter, and other aspects of Earth science. It also demands a knowledge of the nature and operation of the anthrosphere, technology, and the science of industrial ecology. It requires constant monitoring of results and readjustment of measures applied, that is, an adaptive approach. An analysis is required of the current status quo, what is right, what is wrong, and what needs to be changed. Goals need to be set and measures put in place to accomplish those goals. A continuous process is required of monitoring results and of modifying approaches and goals based upon the results of monitoring.

The most commonly cited example of a system that has been subjected to an Earth systems engineering and management plan is the Florida Everglades. This is a large subtropical freshwater marsh that supports a variety of flora (notably, tall-growing saw grass) and fauna (notably alligators). The Everglades region has been greatly impacted by humans, with 6 million residents dwelling between its eastern shore and Florida's Atlantic coast. Much of the impact has been the result of agricultural development involving drainage and diversion of water to support agriculture, especially sugarcane and citrus. Runoff from agricultural developments has raised fertilizer nitrogen and phosphate levels in Everglades water. The normal flow of water in the Everglades has been impacted by numerous canals and levees and some elevated highways. These intrusions had a marked detrimental effect upon the Everglades, especially with greatly reduced water flow and high levels of water pollution. Fish populations declined significantly and numbers of wading birds were reduced by 90 to 95%.

In 1993, the U.S. Army Corps of Engineers undertook an analysis of the Everglades problem that concluded that much of the remedy lay in increasing the residence time of freshwater in the ecosystem with fewer canals and levees diverting water rapidly to tidal areas. According to the plan put forth in 2004, the water flow through the Everglades will be corrected and storage facilities constructed to enable more even flow of water and avert water shortages. It is hoped that these measures will restore the capacity of the Everglades to support its normal plant and animal life while providing adequate fresh water supplies for neighboring populations of people.

Ecological Engineering

Ecological engineering seeks to integrate the anthrosphere and its activities with natural ecosystems to mutual advantage. Combining systems ecology with engineering, ecological engineering is a very important area of Earth systems engineering and management. It develops eco-technologies through designing, constructing, and managing systems of natural ecology integrated with the anthrosphere.

The greatest success to date from the application of ecological engineering is the construction and operation of wetlands to treat wastewater. Constructed wetlands do not necessarily duplicate the living species and other aspects of natural wetlands and may even be located in areas where wetlands have not previously existed. They do provide the essential ingredients — water, sunlight, nutrients, and confined conditions — that enable wetland-based ecosystems to develop and thrive. Other endeavors involving ecological engineering include restoration ecology of areas damaged by development, phytoremediation using plants to remove pollutants from polluted areas, stream restoration, and soil bioengineering that uses ecosystems and the plants in them to reduce soil erosion and increase agricultural productivity in a sustainable manner.

Somewhat related to ecological engineering and applicable to industrial ecology, **biomimicry** is the use of natural systems as the basis for design of anthropogenic systems. Areas in which humans have learned from, or are in the process of learning from, nature include design of strong, lightweight structures based upon shells of organisms; nontoxic materials, such as dyes; surfaces with special properties; and highly selective catalysts that carry out chemical reactions under relatively mild conditions.

17.4. THE EMERGENCE OF INDUSTRIAL ECOLOGY

An alternative to the old, wasteful ways of production and consumption outlined in Section 17.2 and an important tool in Earth systems engineering and management was formulated in the late 1990s and early 2000s with the emergence of the new concept of industrial ecology. **Industrial ecology** is simply a comprehensive approach to production, distribution, utilization, and termination of goods and services in a manner that maximizes mutually beneficial utilization of materials and energy among enterprises thereby minimizing consumption of nonrenewable raw materials and energy while preventing the production of wastes and pollutants. The practice of industrial ecology involves optimization of materials utilization starting with raw material and progressing through finished material, to component, to product and, finally, to the final fate of obsolete product and its components. In addition to materials and resources, industrial ecology considers energy and capital. Industrial ecology is analogous to ecology in nature in which organisms create intricate interdependent webs whereby individual kinds of organisms utilize waste products from others for their own needs. This results in extremely high efficiencies of resource utilization in nature with essentially zero wastes. Organisms operate in natural ecosystems. Similarly, enterprises that practice industrial ecology operate in **industrial ecosystems**.

Several important analogies exist between industrial ecology and natural ecology. One of these is the analogy behind the evolution of organisms and their ecosystems through natural selection in which those organisms most suited to particular environments evolve through genetic processes. Similarly, systems of industrial ecology have evolved through natural selection processes. If a waste product becomes available, an enterprise is likely to develop to utilize the product. If a more efficient enterprise for utilizing the waste appears, it will become dominant. It should be kept in mind, however, that something like "intelligent design" has the potential to be applied to industrial ecosystems. Based upon knowledge of industrial enterprises, their needs, and their markets, and utilizing the enormous computing powers now available to planners, systems of industrial ecology that are functional and efficient can be planned, constructed, and operated.

Just as natural ecosystems are not static and undergo ecological succession, such as grasslands $\rightarrow$ shrub lands $\rightarrow$ forests, industrial ecosystems undergo a form of ecological succession. Rapid developments in manufacturing techniques, changes in sources of raw materials or energy, and shifting markets will inevitably change the mix of enterprise in an industrial ecosystem and the ways in which they interact. Therefore, a form of natural selection must operate to provide the most efficient possible systems of industrial ecology.

Figure 17.2 shows industrial ecology operating in an idealized manner. This figure illustrates the minimal inputs of raw materials and energy, minimum production of wastes, and maximum circulation of materials within the systems that are characteristic of ideal systems.

17.5. THE FIVE MAJOR COMPONENTS OF AN INDUSTRIAL ECOSYSTEM

Industrial ecosystems can be broadly defined to include all types of production, processing, and consumption. It is useful to define five major components of an industrial ecosystem. These are (1) a primary materials producer, (2) a source or sources of energy, (3) a materials-processing and manufacturing sector, (4) a waste-processing sector, and (5) a consumer sector. In such an idealized system operating with the best practice of industrial ecology, the flow of materials among the major hubs is very high. Each constituent of the system evolves in a manner that maximizes the efficiency with which the system utilizes materials and energy.

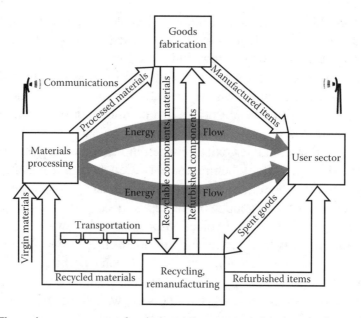

Figure 17.2.   The major components of an industrial ecosystem showing maximum flow of materials within the system.

It is convenient to consider the **primary materials producers** and the **energy generators** together because both materials and energy are required for the industrial ecosystem to operate. The primary materials producer or producers may consist of one or several enterprises devoted to providing the basic materials that sustain the industrial ecosystem. Most generally, in any realistic industrial ecosystem, a significant fraction of the material processed by the system consists of virgin materials. In a number of cases, and increasingly so as pressures build to recycle materials, significant amounts of the materials come from recycling sources.

The processes that virgin materials entering the system are subjected to vary with the kind of material, but can generally be divided into several major steps. Typically, the first step is extraction, designed to remove the desired substance as completely as possible from the other substances with which it occurs. This stage of materials processing can produce large quantities of waste material requiring disposal, as is the case with some metal ores in which the metal makes up a small percentage of the ore that is mined. In other cases, such as corn grain providing the basis of a corn products industry, the "waste" — in this specific example, the cornstalks associated with the grain — can be left on the soil to add humus and improve soil quality. A concentration step may follow extraction to put the desired material into a purer form. After concentration, the material may be put through additional refining steps that may involve separations. Following these steps, the material is usually subjected to additional processing and preparation leading to the finished materials. Throughout the various steps of extraction, concentration, separation, refining, processing, preparation, and finishing, various physical and chemical operations are used, and wastes requiring disposal may be produced. Recycled materials may be introduced at various parts of the process, although they are usually introduced into the system following the concentration step.

The extraction and preparation of energy sources can follow many of the steps outlined above for the extraction and preparation of materials. For example, the processes involved in extracting uranium from ore, enriching it in the fissionable uranium-235 isotope, and casting it into fuel rods for nuclear fission power production include all of those outlined above for materials. On the other hand, some rich sources of coal are essentially scooped from a coal seam and sent to a power plant for power generation with only minimal processing, such as sorting and grinding.

Recycled materials added to the system at the primary materials, and energy production phase may be from both pre- and postconsumer sources. As examples, recycled paper may be macerated and added at the pulping stage of paper manufacture. Recycled aluminum may be added at the molten metal stage of aluminum metal production.

Finished materials from primary materials producers are fabricated to make products in the **goods fabrication** and **manufacturing sector**, which is often a very complex system. For example, the manufacture of an automobile requires steel for the frame, plastic for various components, rubber in tires, lead in the battery, and copper in the wiring, along with a large number of other materials. Typically, the first step in materials manufacturing and processing is a forming operation. For example, sheet steel suitable for making automobile frames may be cut, pressed, and welded into the configuration needed to make a frame. At this step some wastes may be produced that require disposal. An example of such wastes consists of carbon fiber/epoxy composites left over from forming parts such as jet aircraft engine housings. Finished components from the forming step are fabricated into finished products that are ready for the consumer market.

The materials processing and manufacturing sector presents several opportunities for recycling. At this point it may be useful to define two different streams of recycled materials:

- **Process recycle streams** consisting of materials recycled in the manufacturing operation itself

- **External recycle streams** consisting of materials recycled from other manufacturers or from postconsumer products

Materials suitable for recycling can vary significantly. Generally, materials from the process recycle streams are quite recyclable because they are the same materials used in the manufacturing operation. Recycled materials from the outside, especially those from postconsumer sources, may be quite variable in their characteristics because of the lack of effective controls over recycled postconsumer materials. Therefore, manufacturers may be reluctant to use such substances.

In the **consumer sector**, products are sold or leased to the consumers who use them. The duration and intensity of use vary widely with the product; paper towels are used only once, whereas an automobile may be used thousands of times over many years. In all cases, however, the end of the useful lifetime of the product is reached and it is either (1) discarded or (2) recycled. The success of a total industrial ecology system may be measured largely by the degree to which recycling predominates over disposal.

Recycling has become so widely practiced that an entirely separate **waste-processing sector** of an economic system may now be defined. This sector consists of enterprises that deal specifically with the collection, separation, and processing of recyclable materials and their distribution to end users. Such operations may be entirely private or they may involve cooperative efforts with governmental sectors. They are often driven by laws and regulations that provide penalties against simply discarding used items and materials, as well as positive economic and regulatory incentives for their recycle.

17.6. INDUSTRIAL METABOLISM

Industrial metabolism refers to the processes to which materials and components are subjected in industrial ecosystems. It is analogous to the metabolic processes that occur with food and nutrients in biological systems. Like biological metabolism, industrial metabolism may be addressed at several levels. A level of industrial metabolism at which green chemistry, especially, comes into play is at the molecular level where substances are changed chemically to give desired materials or to generate energy. Industrial metabolism can be addressed within individual unit processes in a factory, at the factory level, at the industrial ecosystem level, and even globally.

A significant difference between industrial metabolism as it is now practiced and natural metabolic processes relates to the wastes that these systems generate. Natural ecosystems have developed such that true wastes are virtually nonexistent. For example, even those parts of plants that remain after biodegradation of plant materials form soil humus (see Chapter 11) that improves the conditions of soil on which plants grow. Anthropogenic industrial systems, however, have developed in ways that generate large quantities of wastes, where a waste may be defined as *dissipative use of natural resources*. Furthermore, human use of materials has a tendency to dilute and dissipate materials and disperse them to the environment. Materials may end up in a physical or chemical form from which reclamation becomes impractical because of the energy and effort required. A successful industrial ecosystem overcomes such tendencies.

Organisms performing their metabolic processes degrade materials to extract energy (catabolism) and synthesize new substances (anabolism). Industrial ecosystems perform analogous functions. The objective of industrial metabolism in a successful industrial ecosystem is to make desired goods with the least amount of by-product and waste. This can pose a significant challenge. For example, to produce lead from lead ore for the large electrical battery market requires mining large quantities of ore, extracting the relatively small fraction of the ore consisting of lead sulfide mineral, and roasting and reducing the mineral to get lead metal. The whole process generates large quantities of lead-contaminated tailings left over from mineral extraction and significant quantities of the by-product sulfur dioxide, which must be reclaimed to make sulfuric acid and not released to the environment. The recycling pathway, by way of contrast, takes essentially pure lead from recycled batteries and simply melts it down to produce lead for new batteries; the advantages of recycling in this case are obvious. Industrial metabolic processes that emphasize recycling are desirable because recycling gives essentially constant reservoirs of materials in the recycling loop.

Living organisms have elaborate systems of control. Considering the metabolism that occurs in an entire natural ecosystem, it is seen to be **self-regulating**. If herbivores that consume plant biomass become too abundant and diminish the stock of the biomass, their numbers cannot be sustained, the population dies back, and their food source rebounds. The most successful ecosystems are those in which this self-regulating mechanism operates continuously without wide variations in populations. Industrial systems do not inherently operate in a self-regulating manner that is advantageous to their surroundings, or even to themselves in the long run. Examples of the failure of self-regulation of industrial systems abound in which enterprises have wastefully produced large quantities of goods of marginal value, running through limited resources in a short time, and dissipating materials to their surroundings, polluting the environment in the process. Despite these bad experiences, within a proper framework of laws and regulations designed to avoid wastes and excess, industrial ecosystems can be designed to operate in a self-regulating manner. Such self-regulation operates best under conditions of maximum recycling in which the system is not dependent upon a depleting resource of raw materials or energy.

17.7. MATERIALS FLOW AND RECYCLING IN AN INDUSTRIAL ECOSYSTEM

Figure 17.3 provides an overview of materials flow in an industrial ecosystem. There are several important aspects of a complete industrial ecosystem. One of these is that (as discussed in the preceding section) there are several points at which materials may be recycled in the system. A second aspect is that there are several points at which wastes are produced. The potential for the greatest production of waste lies in the earlier stages of the cycle in which large quantities of materials with essentially no use associated with the raw material, such as ore tailings, may require disposal. In many cases, little if anything of value can be obtained from such wastes. The best thing to do with them is to return them to their source (usually a mine), if possible. Another big source of potential wastes, and often the one that causes the

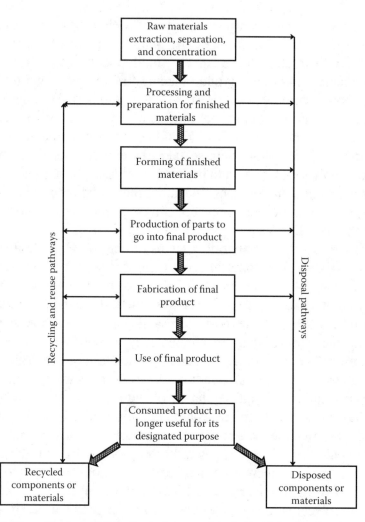

Figure 17.3.   Outline of materials flow through a complete industrial ecosystem.

most problems, consists of postconsumer wastes generated when a product's life cycle is finished. With a properly designed industrial ecology cycle, such wastes can be minimized and, ideally, totally eliminated.

In general, the amount of waste per unit output decreases in going through the industrial ecology cycle from virgin raw material to final consumer product. Also, the amount of energy expended in dealing with waste or in recycling decreases farther into the cycle. For example, waste iron from the milling and forming of automobile parts may be recycled from a manufacturer to the primary producer of iron as scrap steel. To be used, such steel must be remelted and run through the steel manufacturing process again, with a considerable consumption of energy. However, a postconsumer item, such as an engine block, may be refurbished and recycled to the market with relatively less expenditure of energy.

17.8. THE KALUNDBORG INDUSTRIAL ECOSYSTEM

The most often cited example of a functional industrial ecosystem is that of Kalundborg, Denmark. The various components of the Kalundborg industrial ecosystem are shown in Figure 17.4. To a degree, the Kalundborg system developed spontaneously, without being specifically planned as an industrial ecosystem. It is based upon two major energy suppliers, the 1500-MW ASNAES coal-fired electrical power plant and the 4 to 5 million tons/year Statoil petroleum refining complex, each the largest of its kind in Denmark. The electric power plant sells process steam to the oil refinery, from which it receives fuel gas and cooling water. Sulfur removed from the petroleum goes to the Kemira sulfuric acid plant. By-product heat from the two energy generators is used for district heating of homes and commercial establishments, as well as to heat greenhouses and a fish farming operation. Steam from the electrical power plant is used by the $2 billion/year Novo Nordisk pharmaceutical plant, a firm that produces industrial enzymes and 40% of the world's supply of insulin. This plant generates a biological sludge that is used by area farms for fertilizer. Calcium sulfate produced as a by-product of sulfur removal by lime scrubbing from the electrical plant is used by the Gyproc company to make wallboard. The wallboard manufacturer also uses clean-burning gas from the petroleum refinery as fuel. Fly ash generated from coal combustion goes into cement and roadbed fill. Lake Tisso serves as a freshwater source. Other examples of efficient materials utilization associated with Kalundborg include use of sludge from the plant that treats water

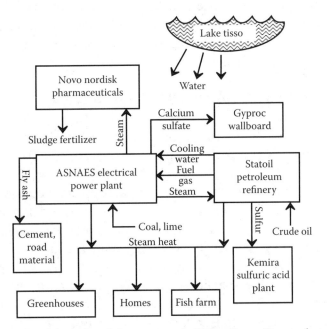

Figure 17.4.   Schematic of the industrial ecosystem in Kalundborg, Denmark.

and wastes from the fish farm's processing plant for fertilizer, and blending of excess yeast from Novo Nordisk's insulin production as a supplement to swine feed.

The development of the Kalundborg complex occurred over a long period of time, beginning in the 1960s, and provides some guidelines for the way in which an industrial ecosystem can evolve naturally. The first of many synergistic (mutually advantageous) arrangements was cogeneration of usable steam along with electricity by the ASNAES electrical power plant. The steam was first sold to the Statoil petroleum refinery, then, as the advantages of large-scale, centralized production of steam became apparent, steam was also provided to homes, greenhouses, the pharmaceutical plant, and the fish farm. The need to produce electricity more cleanly than was possible simply by burning high sulfur coal resulted in two more synergies. Operation of a lime-scrubbing unit for sulfur removal on the power plant stack produced large quantities of calcium sulfate, which found a market in the manufacture of gypsum wallboard. It was also found that a clean-burning gas by-product of the petroleum refining operation could be substituted in part for the coal burned in the power plant, further reducing pollution.

The implementation of the Kalundborg ecosystem occurred largely because of the close personal contact among the managers of the various facilities in a relatively close social and professional network over a long period of time. All the contracts have been based upon sound business fundamentals and have been bilateral. Each company has acted upon its perceived self-interest, and there has been no master plan for the system as a whole. The regulatory agencies have been cooperative, but not coercive in promoting the system. The industries involved in the agreements have fit well, with the needs of one matching the capabilities of the other in each of the bilateral agreements. The physical distances involved have been small and manageable; it is not economically feasible to ship commodities such as steam or fertilizer sludge for long distances.

17.9. ENVIRONMENTAL IMPACTS AND WASTES

By its nature, industrial production has an impact upon the environment. Whenever raw materials are extracted, processed, used, and eventually discarded, some environmental impacts will occur. In designing an industrial ecological system, several major kinds of environmental impacts must be considered in order to minimize them and keep them within acceptable limits. These impacts and the measures taken to alleviate them are discussed in subsequent paragraphs.

For most industrial processes, the first environmental impact is that of extracting raw materials. This can be a straightforward case of mineral extraction, or it can be less direct, such as utilization of biomass grown on forest or crop land. A basic decision, therefore, is the choice of the kind of material to be used. Wherever possible, materials should be chosen that are not likely to be in short supply in the foreseeable future. As an example, the silica used to make the lines employed for fiber-optics communication is in unlimited supply and a much better choice for communication lines than copper wire made from limited resources of copper ore.

Air Pollutants

Industrial ecology systems should be designed to reduce or even totally eliminate air pollutant emissions. Among the most notable recent progress in that area has been the marked reduction and even total elimination of solvent vapor emissions (volatile organic carbon, VOC), particularly those from organochlorine solvents. Some progress in this area has been made with more effective trapping of solvent vapors. In other cases, the use of the solvents has been totally eliminated. This is the case for chlorofluorocarbons (CFCs), which are no longer used in plastic foam blowing and parts cleaning because of their potential to affect stratospheric ozone. Other air pollutant emissions that should be eliminated are hydrocarbon vapors, including those of methane, CH_4, and oxides of nitrogen or sulfur.

Water Pollutants

Discharges of water pollutants should be entirely eliminated wherever possible. For many decades, efficient and effective water treatment systems have been employed that minimize water pollution. However, these are "end of pipe" measures, and it is much more desirable to design industrial systems such that potential water pollutants are not even generated.

Wastes and Hazardous Wastes

Industrial ecology systems should be designed to prevent production of liquid wastes that may have to be sent to a waste processor. Such wastes fall into the two broad categories of water-based wastes and those contained in organic liquids. Under current conditions the largest single constituent of so-called "hazardous wastes" is water. Elimination of water from the waste stream automatically prevents pollution and reduces amounts of wastes requiring disposal. The solvents in organic wastes largely represent potentially recyclable or combustible constituents. A properly designed industrial ecosystem does not allow such wastes to be generated or to leave the factory site.

In addition to liquid wastes, many solid wastes must be considered in an industrial ecosystem. The most troublesome are toxic solids that must be placed in a secure hazardous waste landfill. The problem has become especially acute in some industrialized nations in which the availability of landfill space is severely limited. In a general sense, solid wastes are simply resources that have not been properly utilized. Closer cooperation among suppliers, manufacturers, consumers, regulators, and recyclers can minimize quantities and hazards of solid wastes.

Although ideal working systems of industrial ecology do not produce any hazardous wastes, such wastes are, in fact, generated in modern industrial systems. **Hazardous wastes** are waste materials that may pose dangers to humans or the environment including the anthrosphere. These include a vast variety of materials including explosives, compressed gases, flammable liquids, flammable solids, oxidizing materials that provide oxygen for combustion, corrosive materials such as acids, poisonous materials, disease-causing agents, and radioactive materials. For regulatory purposes, hazardous wastes are classified according to (1) **characteristics** of **ignitability, corrosivity, reactivity, toxicity** and (2) **listed wastes** consisting of specific substances or classes of substances assigned specific hazardous waste numbers.

An important aspect of the management of wastes is **waste segregation**. Keeping wastes separated makes the utilization, treatment, and disposal of wastes much easier. For example, hydrocarbon solvents used for degreasing parts not mixed with other wastes can be reclaimed by distillation or used as fuel. If such solvents are mixed with organohalide compounds, purification becomes more complicated and use as fuel much more difficult because of the production of acidic hydrogen chloride (HCl) in the combustion products. Some of the most difficult wastes to deal with are those consisting of mixtures of hazardous organic and inorganic materials including water and in the form of a sludge.

Hazardous waste management refers to the processes to which wastes are handled from their production, through treatment, utilization, and final disposal. In descending order of desirability, hazardous waste management attempts to accomplish the following:

- Do not produce it

- If making it cannot be avoided, produce only minimum quantities

- Recycle it

- If it is produced and cannot be recycled, treat it, preferably in a way that makes it nonhazardous

- If it cannot be rendered nonhazardous, dispose of it in a safe manner

- Once it is disposed, monitor it for leaching and other adverse effects

Two important aspects of hazardous waste management are **waste reduction** (cutting down quantities of wastes from their sources) and **waste minimization** (utilization of treatment processes that reduce the quantities of wastes requiring ultimate disposal). Recycling wastes for useful purposes is highly desirable and is one of the basic processes of industrial ecology.

An important aspect of hazardous waste management is **waste treatment** designed to convert wastes to nonhazardous forms. A number of physical and chemical processes are used to reduce the hazards of wastes. Some of these are separation procedures. For example, many hazardous waste materials are composed largely of water, and separation of relatively small quantities of the toxic or otherwise hazardous constituents from relatively large quantities of water that can safely be discharged to a water treatment facility can greatly reduce the quantities of wastes that must be handled. Processes that reduce the mobility of wastes by decreasing their solubilities or volatilities can be an important waste treatment procedure. Chemically, measures may be taken, such as neutralization of waste acids with base, neutralization of bases with acids, reaction of oxidants with reducing agents, or the precipitation of soluble metal ions as insoluble salts, such as the conversion of soluble lead ion (Pb^{2+}) to virtually insoluble lead sulfide (PbS).

If no use can be found for a hazardous waste material, some means of disposal must be employed. With present practice, this is done for the most part in **secure chemical landfills** designed with features such as impermeable polymer liners and layers of impermeable clay to prevent release of hazardous constituents. One of the main considerations is the prevention of leaching of water-soluble hazardous constituents into groundwater permeating the landfill. Gas-impermeable caps on landfills are used to prevent releases of toxic or pollutant vapors to the atmosphere. A problem with secure landfills is that none can be made completely secure. Furthermore, unlike radioactive wastes that do eventually decay to nonradioactive forms, some chemical wastes never degrade such that release of hazardous constituents — perhaps centuries later — will eventually occur. The best practice of industrial ecology calls for never placing in the ground materials that have the potential to pose significant hazards if they are later released, exposed to groundwater, or exposed to the atmosphere.

17.10. THREE KEY ATTRIBUTES: ENERGY, MATERIALS, DIVERSITY

By analogy with biological ecosystems, a successful industrial ecosystem should have (1) renewable energy, (2) complete recyling of materials, and (3) species diversity for resistance to external shocks. These three key characteristics of industrial ecosystems are addressed here.

Unlimited Energy

Energy is obviously a key ingredient of an industrial ecosystem. Unlike materials, the flow of energy in even a well-balanced closed industrial ecosystem is essentially one-way in that energy enters in a concentrated, valuable form, such as chemical energy in natural gas, and leaves in a dilute, disperse form as waste heat. An exception is the energy that is stored in materials. This can be in the form of energy that can be obtained from materials, such as by burning rubber tires, or it can be in the form of an **energy credit**, which means that by using a material in its refined form, energy is not consumed in making the material from its raw material precursors. A prime example of this is the energy credit in metals, such as that in aluminum metal, which can be refined into new aluminum objects requiring only a fraction of the energy consumed to refine the metal from aluminum ore. On the other hand, recycling and reclaiming some materials can require a lot of energy, and the energy consumption of a good closed industrial ecosystem can be rather high.

Given the needed elements, any material can be made if a sufficient amount of energy is available. The key energy requirement is a source that is abundant and of high quality, that can be used efficiently, and that does not produce unacceptable by-products.

Although energy is ultimately dissipated from an industrial ecosystem, it may go through two or more levels of use before it is wasted. An example of this would be energy from natural gas burned in a turbine linked to a generator, the exhaust gases used to raise steam in a power plant to run a steam turbine, and the relatively cool steam from the turbine used to heat buildings.

Natural ecosystems run on unlimited, renewable energy from the sun or, in some specialized cases, from geochemical sources. Successful industrial ecosystems must also have abundant, sustainable sources of energy to be sustained for an indefinite period of time. As of the present, industrial systems run primarily on fossil fuels, but that cannot continue indefinitely because of diminishing supplies of fossil fuel and the problem with global warming from release of carbon dioxide during combustion. The question of energy supplies and utilization is so great that all of Chapter 18 is devoted to it.

Efficient Material Utilization

A system of industrial ecology is successful if it reduces demand for materials from virgin sources. Strategies for reduced material use may be driven by technology, by economics, or by regulation. The four major ways in which material consumption may be reduced are (1) using less of a material for a specific application, an approach called **dematerialization**; (2) **substitution** of a relatively more abundant and safe material for one that is scarce and/or toxic; (3) **recycling**, broadly defined; and (4) extraction of useful materials from wastes, sometimes called **waste mining**. These four facets of efficient materials utilization are outlined in this section.

Dematerialization is a term given to the use of less material. There are numerous recent examples of reduced uses of materials for specific applications. One example is the transmission of greater electrical power loads with less copper wire by using higher voltages on long-distance transmission lines. Copper is also used much more efficiently for communications transmission than it was in the early days of telegraphy and telephone communication. Amounts of silver used per roll of photographic film have decreased significantly in recent years and, with the advent of digital photography, the use of silver in imaging is decreasing rapidly. The layer of tin plated onto the surface of a "tin can" used for food preservation and storage is much lower now. In response to the need for greater fuel economy, the quantities of materials used in automobiles have decreased significantly over the last two decades, a trend reversed, unfortunately, by the more recent increased demand for large sport-utility vehicles (SUVs). Automobile storage batteries now use much less lead for the same amount of capacity than they did in former years. The switch from 6-V to 12-V automotive batteries in the 1950s enabled use of lighter wires, such as those from the battery to the electrical starter. Somewhat later, the change to steel-belted radial tires enabled use of lighter tires and resulted in greatly increased tire lifetimes so that less rubber was used for tires.

One of the most commonly cited examples of dematerialization is that resulting from the change from vacuum tubes to solid-state circuit devices in radios, television sets, and other electronic equipment. Actually, this conversion should be regarded as material substitution as transistors replaced vacuum tubes, followed by spectacular mass reductions as solid-state circuit technology advanced.

Dematerialization can be expected to continue as technical advances, some rapid and spectacular, others slow and incremental, continue to be made. Some industries lead the way out of necessity. Aircraft weight has always played a crucial role in determining performance, so the aircraft manufacturing sector is one of the leaders in dematerialization.

Substitution of more sustainable and less expensive materials is complementary to dematerialization in reducing materials use. The substitution of solid-state components for electronic vacuum tubes and the accompanying reduction in material quantities has already been cited. The substitution of polyvinylchloride (PVC) siding in place of wood on houses has resulted in dematerialization over the long term because the plastic siding does not require paint. With the major shift from photographic film to digital cameras in the early 2000s, silver in the film was replaced by other materials.

Technology and economics combined have been leading factors in materials substitution. For example, the technology to make PVC pipe for water and drain lines has enabled its use in place of more expensive cast iron, copper, and even lead pipe (in the last case, toxicity from lead contamination of water is also a factor to be considered).

A very significant substitution that has taken place over recent decades is that of aluminum for copper and other substances. Copper, although not a strategically short metal resource, nevertheless is not one of the more abundant metals in relation to the demand for it. Considering its abundance in the geosphere and in sources such as coal ash, aluminum is a very abundant metal. Now aluminum is used in place of copper in many high-voltage electrical transmission applications. Aluminum is also used in place of brass, a copper-containing alloy, in a number of applications. Aluminum roofing substitutes for copper in building construction. Aluminum cans are used for beverages in place of tin-plated steel cans.

There have been a number of subsitutions of chemicals in recent years, many of them driven by environmental concerns and regulations resulting from those concerns. One of the greater of these has been the substitution of hydrochlorofluorocarbons (HCFCs) and hydrofluorocarbons (HFCs) for chlorofluorocarbons (freons or CFCs) driven by concerns over stratospheric ozone depletion. Substitutions of nonhalogenated solvents, supercritical fluid carbon dioxide, and even water with appropriate additives for chlorinated hydrocarbon solvents will continue as environmental concerns over these solvents increase.

Substitutions for metal-containing chemicals promise to reduce costs and toxicities. One such substitution that has greatly reduced the possibilities for lead poisoning is the use of titanium-based pigments in place of lead for white paints. In addition to lead, cadmium, chromium, and zinc are also used in pigments, and substitution of organic pigments for these metals in paints has reduced toxicity risks. Copper, chromium (chromate), and arsenic were used for many years to treat wood (CCA lumber). Because of the toxicity of arsenic, particularly, substitutes are being developed for these metals in wood. It should be pointed out, however, that the production of practically indestructible CCA lumber has resulted in much less use of wood and has saved the materials and energy required to replace wood that has rotted or been damaged by termites.

Recycling at levels approaching 100% of many kinds of materials must occur for a true and complete industrial ecosystem to be realized. In principle, given a finite supply of all the required elements and abundant energy, essentially complete recycling can be achieved. A central goal of industrial ecology is to develop efficient technologies for recycling that reduce the need for virgin materials to the lowest possible levels. Another goal must be to implement process changes that eliminate dissipative uses of toxic substances, such as heavy metals, that are not biodegradable and pose a threat to the environment when they are discarded.

For consideration of recycling, matter can be put into four separate categories. The first of these consists of elements that occur abundantly and naturally in essentially unlimited quantities in consumable products. Food is the ultimate consumable product. Soap is consumed for cleaning purposes, discarded down the drain, precipitated as its insoluble calcium salt, then finally biodegraded. Materials in this category of recyclables are discharged into the environment and recycled through natural processes or for very low-value applications, such as sewage sludge used as fertilizer on soil.

A second category of recyclable materials consists of elements that are not in short supply, but are in a form that is especially amenable to recycling. Wood is one such commodity. At least a portion of wood taken from buildings that are being razed can and should be recycled. The best example of a kind of commodity in this class is paper. Paper fibers can be recycled up to five times, and the nature of paper is such that it is readily recycled. More than ⅓ of world paper production is currently from recycled sources, and that fraction should exceed 50% within the next several decades. The major impetus for paper recycling is not a shortage of wood to make virgin paper, but rather a shortage of landfill space for waste paper.

A third category of recyclables consists of those elements, mostly metals, for which world resources are low. Chromium and the platinum group of precious metals are examples of such elements. Given maximum incentives to recycle, especially through the mechanism of higher prices, it is likely that virgin sources of these metals can make up any shortfall not met by recycling in the foreseeable future.

A fourth category of materials to consider for recycling consists of parts and apparatus, such as automotive parts discussed previously. In many cases, such parts can be refurbished and reused. Even when this is not the case, substantial deposits at the time of purchase can provide incentives for recycling. In order for components to be recycled efficiently, they must be designed with reuse in mind in aspects such as facile disassembly. Such an approach has been called *design for environment* (DFE) and is discussed in more detail in Section 17.14.

Combustion to produce energy can be a form of recycling. For some kinds of materials, combustion in a power plant is the most cost-effective and environmentally safe way of dealing with materials. This is true, for example, of municipal refuse that contains a significant energy value because of combustible materials in it, as well as a variety of items that potentially could be recycled for the materials in them. However, once such items become mixed in municipal refuse and contaminated with impurities, the best means of dealing with them is simply combustion.

It should be noted that recycling comes with its own set of environmental concerns. One of the greatest of these is contamination of recycled materials with toxic substances. In some cases, motor oil, especially that collected from the individual consumer sector, can be contaminated with organohalide solvents and other troublesome impurities. Food containers pick up an array of contaminants and, as a consequence, recycled plastic is not generally regarded as a good material for food applications. Substances may become so mixed with use that recycling is not practical. This occurs particularly with synthetic fibers, but it may be a problem with plastics, glass, and other kinds of recyclable materials.

Waste mining is a term applied to the extraction of useful materials from wastes that has significant, largely unrealized potential for the reduction in use of virgin materials. Waste mining can often take advantage of the costs that must necessarily be incurred in treating wastes, such as flue gases. Sulfur is one of the best examples of a material that is now commonly recovered from wastes. Sulfur is a constituent of all coal and can be recovered from flue gas produced by coal combustion. It would not be cost-effective to use flue gas simply as a source of sulfur. However, because removal of sulfur dioxide from flue gas is now required by regulation, the incremental cost of recovering sulfur as a commodity, rather than simply discarding it, can make sulfur recovery economically feasible.

There are several advantages to recovering a useful resource from wastes. One of these is the reduced need to extract the resource from a primary source. Therefore, every kilogram of sulfur recovered from flue gas means one less kilogram of sulfur that must be extracted from sulfur ore sources. By using waste sources, the primary source is preserved for future use. Another advantage is that extraction of a resource from a waste stream can reduce the toxicity or potential environmental harm from the waste stream. As noted previously, arsenic is a by-product of the refining of some other metals. The removal of arsenic from the residues of refining such metals significantly reduces the toxicities and potential environmental harm by the wastes. Coal ash, the residue remaining after the combustion of coal for power generation, has a significant potential as a source of iron (ferrosilicon), silicon, and aluminum, and perhaps several other elements as well. An advantage of using coal ash in such applications is its physical form. For most power applications, the feed coal is finely ground, so that the ash is in the form of a powder. This means that coal ash is already in the physical form most amenable to processing for recovery of by-products. For a particular coal feedstock, coal ash is homogeneous, which offers some definite advantages in processing and resource recovery. A third advantage of coal ash is that it is dry, so no additional energy needs to be expended in removing water from an ore.

Diversity and Robust Character of Industrial Ecosystems

Successful natural ecosystems are highly diverse, as a consequence of which they are also very robust. **Robustness** means that if one part of the system is perturbed, there are others that can take its place. Consider what happens if the numbers of a top predator at the top of a food chain in a natural ecosystem are severely reduced because of disease. If the system is well balanced, another top predator is available to take its place.

The energy sector of industrial ecosystems often suffers from a lack of robustness. Examples of energy vulnerability have become obvious with several "energy crises" during recent history. Another requirement of a healthy industrial ecology system that is vulnerable in some societies is water. In some regions of the world, both the quantity and quality of water are severely limited. A lack of self-sufficiency in food is a third example of vulnerability. Vulnerability in food and water are both strongly dependent upon climate, which, in turn, is tied to environmental concerns as a whole.

17.11. LIFE CYCLES: EXPANDING AND CLOSING THE MATERIALS LOOP

In a general sense, the traditional view of product utilization is the one-way process of extraction → production → consumption → disposal shown in the upper portion of Figure 17.5. Materials that are extracted and refined are incorporated into the production of useful items, usually by processes that produce large quantities of waste by-products. After the products are worn out, they are discarded. This essentially one-way path results in a relatively large exploitation of resources, such as metal ores, and a constant accumulation of wastes. As shown at the bottom of Figure 17.5, however, the one-way path outlined above can become a cycle in which manufactured goods are used, then recycled at the end of their life spans. As one aspect of such a cyclic system, it is often useful for manufacturers to assume responsibility for their products, to maintain "stewardship." Ideally, in such a system, a

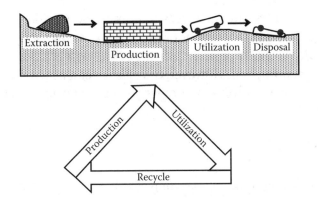

Figure 17.5. The one-way path of conventional utilization of resources to make manufactured goods followed by disposal of the materials and goods at the end consumes large quantities of materials and makes large quantities of wastes (top). In an ideal industrial ecosystem (bottom), the loop is closed and spent products are recycled to the production phase.

product and/or the material in it would have a never-ending life cycle; when its useful lifetime is exhausted, it is either refurbished or converted into another product.

From the discussion in the preceding paragraphs and in the rest of this chapter, it may be concluded that industrial ecology is all about *cyclization of materials*. This approach is summarized in a statement attributed to Kumar Patel of the University of California at Los Angeles, "The goal is *cradle to reincarnation*, since if one is practicing industrial ecology correctly there is no grave." For the practice of industrial ecology to be as efficient as possible, cyclization of materials should occur at the highest possible level of material purity and stage of product development as discussed under embedded utility in the following subsection.

In considering life cycles, it is important to note that commerce can be divided into the two broad categories of **products** and **services**. Whereas most commercial activity used to be concentrated on providing large quantities of goods and products, demand has been largely satisfied for some segments of the population, and the wealthier economies are moving more to a service-based system. Much of the commerce required for a modern society consists of a mixture of services and goods. The trend toward a service economy offers two major advantages with respect to waste minimization. Obviously, a pure service involves little material. Secondly, a service provider is in a much better position to control materials to ensure that they are recycled and to control wastes, ensuring their proper disposal. A commonly cited example is that of photocopy machines. They provide a service, and a heavily used copy machine requires frequent maintenance and cleaning. The parts of such a machine and the consumables, such as toner cartridges, consist of materials that eventually will have to be discarded or recycled. In this case, it is often reasonable for the provider to lease the machine to users, taking responsibility for its maintenance and ultimate fate. The idea could even be expanded to include recycling of the paper processed by the copier, with the provider taking responsibility for recyclable paper printed by the machine.

In many cases, to be practical, recycling must be practiced on a larger scale than simply that of a single industry or product. For example, recycling plastics used in soft drink bottles to make new soft drink bottles is not allowed because of the possibilities for contamination. However, the plastics can be used as raw material for automotive parts. Usually, different companies are involved in making automotive parts and soft drink bottles.

Product Stewardship

The degree to which products are recycled is strongly affected by the custody of the products. For example, batteries containing cadmium or mercury pose significant pollution problems when they are purchased by the public; used in a variety of devices, such as calculators and cameras; then discarded through a number of channels, including municipal refuse. However, when such batteries are used within a single organization, it is possible to ensure that almost all of them are returned for recycling. In cases such as this, systems of stewardship can be devised in which marketers and manufacturers exercise a high degree of control of the product. This can be done through several means. One is for the manufacturer to retain ownership of the product, as is commonly practiced with photocopy machines. Another mechanism is one in which a significant part of the purchase price is refunded for trade-in of a spent item. This approach could work very well with batteries containing cadmium or mercury. The normal purchase price could be doubled, then discounted to half with the trade-in of a spent battery.

Embedded Utility

Figure 17.6 is an "energy/materials pyramid" showing that the amounts of energy and materials involved decrease going from the raw material to the finished product. The implication of this diagram is that significantly less energy, and certainly no more materials, are involved when recycling is performed near the top of the materials flow chain rather than near the bottom.

In low-level recycling, a material or component is taken back to near the beginning of the steps through which it is made. For example, an automobile engine block might be melted down to produce molten metal from which new blocks are then cast. With high-level recycling, the item or material is recycled as close to the final product as possible. In the case of the automobile engine block, it may be cleaned, the cylinder walls rehoned, the flat surfaces replaned, and the block used as the platform for assembling a rebuilt engine. In this example and many others that can be cited, high-level recycling uses much less energy and materials and is inherently more efficient.

The greater usability and lower energy requirements for recycling products higher in the order of material flow are called **embedded utility**. One of the major

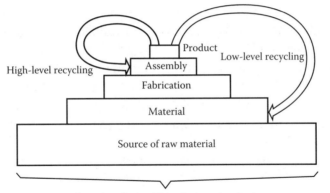

Figure 17.6.   A material flow chain or energy/materials pyramid. Less energy and materials are involved when recycling is done near the end of the flow chain, thus retaining embedded utility.

objectives of a system of industrial ecology and, therefore, one of the main reasons for performing life-cycle assessments is to retain the embedded utility in products by measures such as recycling as near to the end of the material flow as possible, and replacing only those components of systems that are worn out or obsolete. An example of the latter occurred during the 1960s when efficient and safe turboprop engines were retrofitted to still serviceable commercial aircraft airframes to replace relatively complex piston engines, thus extending the lifetime of the aircraft by a decade or more.

17.12. LIFE-CYCLE ASSESSMENT AND SCOPING

From the beginning, industrial ecology must consider process/product design in the management of materials, including the ultimate fates of materials when they are discarded. The product and materials in it should be subjected to an entire **life-cycle assessment** or analysis. A life-cycle assessment applies to products, processes, and services through their entire life cycles from extraction of raw materials — through manufacturing, distribution, and use — to their final fates from the viewpoint of determining, quantifying, and ultimately minimizing their environmental impacts. It takes account of manufacturing, distribution, use, recycling, and disposal. Life-cycle assessment is particularly beneficial in determining the relative environmental merits of alternative products and services. At the consumer level, this could consist of an evaluation of paper vs. Styrofoam drinking cups. On an industrial scale, life-cycle assessment could involve evaluation of nuclear vs. fossil energy-based electrical power plants.

A basic step in life-cycle analysis is **inventory analysis**, which provides qualitative and quantitative information regarding consumption of material and energy resources (at the beginning of the cycle) and releases to the anthrosphere, hydrosphere, geosphere, and atmosphere (during or at the end of the cycle). It is based upon various materials cycles and budgets, and it quantifies materials and energy required as input and the benefits and liabilities posed by products. The related area of **impact analysis** provides information about the kind and degree of environmental impacts resulting from a complete life cycle of a product or activity. Once the environmental and resource impacts have been evaluated, it is possible to do an **improvement analysis** to determine measures that can be taken to reduce impacts on the environment or resources.

In making a life-cycle analysis the following must be considered:

- If there is a choice, selection of the kinds of materials that will minimize waste

- Kinds of materials that can be reused or recycled

- Components that can be recycled

- Alternate pathways for the manufacturing process or for its various parts

Although a complete life-cycle analysis is expensive and time-consuming, it can yield significant returns in lowering environmental impacts, conserving resources, and reducing costs. This is especially true if the analysis is performed at an early stage in the development of a product or service. Improved computerized techniques are making significant advances in the ease and efficacy of life-cycle analyses. Until now, life-cycle assessments have been largely confined to simple materials and products such as reusable cloth vs. disposable paper diapers. A major challenge now is to expand these efforts to more complex products and systems such as aircraft or electronics products.

A crucial early step in life-cycle assessment is **scoping** — the process of determining the boundaries of time, space, materials, processes, and products to be considered. Consider, as an example, the manufacture of parts that are rinsed with an organochloride solvent in which some solvent is lost by evaporation to the atmosphere, by staying on the parts, during the distillation and purification process by which the solvent is made suitable for recycling, and by disposal of waste solvent that cannot be repurified. The scope of the life-cycle assessment could be made very narrow by confining it to the process as it exists. An assessment could be made of the solvent losses, the impacts of these losses, and means for reducing the losses, such as reducing solvent emissions to the atmosphere by installation of activated carbon air filters or reducing losses during purification by employing more efficient distillation processes. A more broadly scoped life-cycle assessment would be to consider alternatives to the organochloride solvent. An even broader scope would consider whether the parts even need to be manufactured; are there alternatives to their use?

17.13. CONSUMABLE, RECYCLABLE, AND SERVICE (DURABLE) PRODUCTS

In industrial ecology, most treatments of life-cycle analysis make the distinction between **consumable products**, which are essentially used up and dispersed to the environment during their life cycle and **service or durable products**, which essentially remain in their original form after use. Gasoline is clearly a consumable product, whereas the automobile in which it is burned is a service product. It is useful, however, to define a third category of products that clearly become "worn-out" when employed for their intended purpose, but which remain largely undispersed to the environment. The motor oil used in an automobile is such a substance in that most of the original material remains after use. Such a category of material may be called a **recyclable commodity**.

Desirable Characteristics of Consumables

Consumable products include laundry detergents, hand soaps, cosmetics, windshield washer fluids, fertilizers, pesticides, laser printer toners, and all other materials that are impossible to reclaim after they are used. The environmental implications of the use of consumables are many and profound. In the late 1960s and early 1970s, for example, nondegradable surfactants in detergents caused severe foaming and esthetic problems at water treatment plants and sewage outflows, and the phosphate builders in the detergents promoted excessive algal growth in receiving waters, resulting in a condition known as eutrophication. Lead in consumable leaded gasoline was widely dispersed to the environment when the gasoline was burned. These problems have now been remedied with the adoption of phosphate-free detergents with biodegradable surfactants and mandatory use of unleaded gasoline.

Because they are destined to be dispersed into the environment, consumables should meet several "environmentally friendly" criteria, including the following:

- **Degradability**: This usually means biodegradability, such as that of household detergent constituents that occurs in waste treatment plants and in the environment. Chemical degradation may also occur.

- **Nonbioaccumulative**: Lipid-soluble, poorly biodegradable substances, such as DDT and PCBs tend to accumulate in organisms and to be magnified through the food chain. This characteristic should be avoided in consumable substances.

- **Nontoxic**: To the extent possible, consumables should not be toxic in the concentrations that organisms are likely to be exposed to them. In addition to their not being acutely toxic, consumables should not be mutagenic, carcinogenic, or teratogenic (cause birth defects).

Desirable Characteristics of Recyclables

Recyclables is used here to describe materials that are not used up, in the sense that laundry detergents or photocopier toners are consumed, but are not durable items. In this context, recyclables can be understood to be chemical substances and formulations. The HCFCs used as refrigerant fluids fall into this category, as does ethylene glycol mixed with water in automobile engine antifreeze/antiboil formulations (although rarely recycled in practice).

Insofar as possible, recyclables should be minimally hazardous with respect to toxicity, flammability, and other hazards. For example, both volatile hydrocarbon solvents and organochloride (chlorinated hydrocarbon) solvents are recyclable after use for parts degreasing and other applications requiring a good solvent for organic materials. The hydrocarbon solvents have relatively low toxicities but may present flammability hazards during use and reclamation for recycling. The organochloride solvents are less flammable but may present a greater toxicity hazard. An example of such a solvent is carbon tetrachloride, which is so nonflammable that it was once used in fire extinguishers, but the current applications of which are highly constrained because of its high toxicity.

An obviously important characteristic of recyclables is that they should be designed and formulated to be amenable to recycling. In some cases, there is little leeway in formulating potentially recyclable materials; motor oil, for example, must meet certain performance criteria, including the ability to lubricate, stand up to high temperatures, and other attributes, regardless of its ultimate fate. In other cases, formulations can be modified to enhance recyclability. For example, the use of bleachable or removable ink in newspapers enhances the recyclability of the newsprint, enabling it to be restored to an acceptable level of brightness.

For some commodities, the potential for recycling is enormous. This can be exemplified by lubricating oils. The volume of motor oil sold in the U.S. each year for gasoline engines is about 2.5 billion liters, a figure that is doubled if all lubricating oils are considered. A particularly important aspect of utilizing recyclables is their collection. In the case of motor oil, collection rates are low from consumers who change their own oil, and they are responsible for the dispersion of large amounts of waste oil to the environment.

Desirable Characteristics of Service Products

Because, in principle at least, service products are destined for recycling, they have comparatively lower constraints on materials and higher constraints on their ultimate disposal. A major impediment to the recycling of service products is the lack of convenient channels through which they can be put into the recycling loop. Television sets and major appliances such as washing machines or ovens have many recyclable components, but often end up in landfills and waste dumps simply because there is no handy means for getting them from the user and into the recycling loop.

In such cases, government intervention may be necessary to provide appropriate channels. One partial remedy to the disposal/recycling problem consists of leasing arrangements or payment of deposits on items such as batteries to ensure their return to a recycler. The terms "de-shopping" or "reverse shopping" describe a process by which service commodities would be returned to a location such as a parking lot where they could be collected for recycling. According to this scenario, the analogy to a supermarket would be a facility in which service products are disassembled for recycling.

Much can be done in the design of service products to facilitate their recycle. One of the main characteristics of recyclable service products must be ease of disassembly so that remanufacturable components and recyclable materials, such as copper wire, can be readily removed and separated for recycling.

17.14. DESIGN FOR ENVIRONMENT

Design for environment is the term given to the approach of designing and engineering products, processes, and facilities in a manner that minimizes their adverse environmental impacts and, where possible, maximizes their beneficial environmental effects. In modern industrial operations, design for environment is part of a larger scheme termed "design for X," where "X" can be any one of a number of characteristics such as assembly, manufacturability, reliability, and serviceability. In making such a design, numerous desired characteristics of the product must be considered, including ultimate use, properties, costs, and appearance. Design for environment requires that the designs of the product, the process by which it is made, and the facilities involved in making it conform to appropriate environmental goals and limitations imposed by the need to maintain environmental quality. It must also consider the ultimate fate of the product, particularly whether or not it can be recycled at the end of its normal life span.

Products, Processes, and Facilities

In discussing design for environment, the distinctions among products, processes, and facilities must be kept in clear perspective. **Products** — automobile tires, laundry detergents, and refrigerators — are items sold to consumers. **Processes** are the means of producing products and services. For example, tires are made by a process in which hydrocarbon monomers are polymerized to produce rubber molded in the shape of a tire with a carcass reinforced by synthetic fibers and steel wires. A **facility** is where processes are carried out to produce or deliver products or services. In cases where services are regarded as products, the distinction between products and processes becomes blurred. For example, a lawn care service delivers products in the forms of fertilizers, pesticides, and grass seeds, but also delivers pure services including mowing, edging, and sod aeration.

Although products tend to get the most public attention in consideration of environmental matters, processes often have more environmental impact. Successful process designs tend to stay in service for many years and be used to make a wide range of products. Whereas the product of a process may have minimal environmental impact, the process by which the product is made may have marked environmental effects. An example is the manufacture of paper. The environmental impact of paper as a product, even when improperly discarded, is not terribly great, whereas the process by which it is made involves harvesting wood from forests, high use of water, potential emission of a wide range of air pollutants, and other factors with profound environmental implications.

Processes develop symbiotic relationships when one provides a product or service utilized in another. An example of such a relationship is the one between steel making and the process for the production of oxygen (required in the basic oxygen process) by which carbon and silicon impurities are oxidized from molten iron to produce steel. The long lifetimes and widespread applicability of popular processes make their design for environment of utmost importance.

The nature of a properly functioning system of industrial ecology is such that processes are even more interwoven than would otherwise be the case, because by-products from some processes are used by other processes. Therefore, the processes employed in such a system and the interrelationships and interdependencies among them are particularly important. A major change in one process may have a "domino effect" on the others.

Key Factors in Design for Environment

Two key choices that must be made in design for environment are those involving materials and energy. The choices of materials in an automobile illustrate some of the possible tradeoffs. Steel as a component of automobile bodies requires relatively large amounts of energy and involves significant environmental disruption in the mining and processing of iron ore. Steel is a relatively heavy material, so more energy is involved in moving automobiles made of steel. However, steel is durable, has a high rate of recycling, and is produced initially from abundant sources of iron ore. Aluminum is much lighter than steel and quite durable. It is one of the most commonly recycled commodities. Good primary sources of aluminum, bauxite ores, are not as abundant as iron ores, and large amounts of energy are required in the primary production of aluminum. Plastics are another source of automotive components. The light weight of plastic reduces automotive fuel consumption, plastics with desired properties are readily made, and molding and shaping plastic parts is a straightforward process. However, plastic automobile components have a low rate of recycling.

Three related characteristics of a product that should be considered in design for environment are durability, repairability, and recyclability. **Durability** simply refers to how well the product lasts and resists breakdown in normal use. Some products are notable for their durability; ancient two-cylinder John Deere farm tractors from the 1930s and 1940s are legendary in farming circles for their durability, enhanced by the affection engendered in their owners who tend to preserve them. **Repairability** is a measure of how easy and inexpensive it is to repair a product. A product that can be repaired is less likely to be discarded when it ceases to function for some reason. **Recyclability** refers to the degree and ease with which a product or components of it may be recycled. An important aspect of recyclability is the ease with which a product can be disassembled into constituents consisting of a single material that can be recycled. It also considers whether the components are made of materials that can be recycled.

Hazardous Materials in Design for Environment

A key consideration in the practice of design for environment is the reduction of the dispersal of hazardous materials and pollutants. This can entail the reduction or elimination of hazardous materials in manufacture, an example of which was the replacement of stratospheric ozone-depleting chlorofluorocarbons (CFCs) in foam blowing of plastics. If appropriate substitutes can be found, somewhat toxic and persistent chlorinated solvents should not be used in manufacturing applications such as parts washing. The use of hazardous materials in the product — such as batteries containing toxic cadmium, mercury, and lead — should be eliminated or minimized. Pigments containing heavy metal cadmium or lead should not be used if there are any possible substitutes. The substitution of hydrochlorofluorocarbons and hydrofluorocarbons for ozone-depleting CFCs in products (refrigerators and air conditioners) is an example of a major reduction in environmentally damaging materials in products. The elimination of extremely persistent polychlorinated biphenyls (PCBs) from electrical transformers removed a major hazardous waste problem because of the use of a common product.

17.15. INHERENT SAFETY

The use of nitroglycerin in the construction of the Central Pacific Railroad in California during 1860s provides a lesson in safe handling of materials. On April 3, 1866, 70 crates of nitroglycerin exploded aboard the California-bound steamship European being unloaded on the Caribbean coast of Panama killing 50 people and causing substantial damage. Only two weeks later, two crates of nitroglycerin, which had been refused delivery because of their condition, exploded in an office of the Wells Fargo Company in San Francisco killing 15 people and causing massive destruction, athough instantaneously resolving the question regarding what to do with the damaged goods. Only two days later, six workers on the Central Pacific Railroad in the Sierra Nevada mountains were killed by an explosion of nitroglycerin that was being used in place of much less effective black gunpowder, for blasting on the railroad construction. As a consequence of these accidents, California legislators quickly passed legislation forbidding the transportation of nitroglycerin through San Francisco and Sacramento.

It appeared that the measures taken by the California authorities would prevent the use of the powerfully explosive nitroglycerin in the construction of the Central Pacific causing a huge delay of the project. However, James Howden, a British chemist, won contracts to make nitroglycerin on site for the Central Pacific construction and he produced up to 100 lb/d of the explosive as needed with no further fatalities resulting from nitroglycerin explosions on the massive construction project. This was an early example of safe handling of material. The glycerin and sulfuric and nitric acids needed to make nitroglycerin could be transported to the site in relative safety and with no explosion hazard. The truly hazardous nitroglycerin was made only as needed so that at no time was there a large amount available in one place to cause an explosion. Finally, the nitroglycerin was only transported short distances greatly minimizing the possibility of transportation accidents with this notoriously sensitive explosive.

A chemical process is said to be inherently safe when permanent measures have been integrated into the process to reduce or eliminate specific hazards. Five approaches by which this may be done are the following:

1. Use only minimum quantities of hazardous substances. In the example of nitroglycerin, discussed above, relatively small quantities of the explosive were made at any one time.

2. Use a less hazardous substance. In 1867, Alfred Nobel invented dynamite, which consisted of nitroglycerin absorbed onto a carrier such as sawdust and which was much safer to handle than pure nitroglycerin but almost as effective as an explosive.

3. Use safer conditions. In a chemical process, this might consist of carrying out reactions at lower temperatures and pressures over a catalyst so that if a malfunction occurs, the results are much less catastrophic.

4. Simplify. As a general rule, simpler processes are safer processes. Added steps add possibilities for things to go wrong.

5. Maximize continuous steady-state operation. Just as most aircraft accidents occur during takeoff and landing, plant malfunctions are more likely during startup and shutdown. Therefore, continuous processes are much more desirable than batch processes and should be operated as long as possible without interruption.

Increased Safety with Smaller Size

Safety can often be increased significantly by the simple expedient of downsizing operations and quantities of materials, a strategy of **minimization**. A common example of minimization is the substitution of small continuous flow reactors for large batch processes. In scaling up from batch laboratory processes to commercial

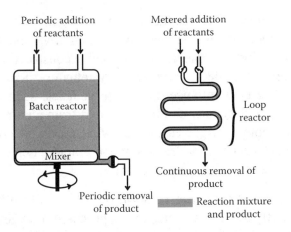

Figure 17.7.   Batch reactor (left) and continuous flow reactor (right) for chemical synthesis. Note the much smaller volume (typically by a factor of 1/100) of reaction mixture and product in the continuous flow reactor.

production, large batch reactors often have been employed. Such a reactor containing large amounts of material can be very problematic if something goes wrong, such as an uncontrolled runaway reaction. Such reactors often have been used because of slow mixing and heating that required reactants to be in contact for long periods of time, whereas the actual reaction is often very quick once the reactants are in contact. It is often desirable to substitute a very small loop reactor with efficient mixing and rapid energy transfer so that the quantities of reactants in the reaction process are very much reduced (Figure 17.7). In such a case, only much smaller quantities are involved and, if anything goes wrong, the amounts of material that must be dealt with are much smaller.

17.16. TWELVE PRINCIPLES OF GREEN ENGINEERING

Green Engineering is an approach to engineering, design, and production based upon sustainability, minimum environmental impact, and favorable economic and social aspects. Green engineering may be conducted within a framework of 12 principles, each of which is discussed here.[3]

1. **Inherent sustainability and reduction of hazard.** The sustainability of a process can be enhanced and its hazard reduced by **circumstantial control** in which the conditions under which the process is carried out are carefully regulated. For example, the environmental and safety problems posed by large SUVs can be reduced by equipping them with emissions control devices and requiring drivers to obey speed limits and other traffic laws. Such circumstantial controls are prone to failure; emissions control devices wear out and fail and drivers routinely break speed limits and other regulations designed to reduce their hazard to themselves and others.

Inherency in contrast involves the design of systems that by their nature are safer and more sustainable and environmentally friendly. In the case of transportation, for example, this may involve construction of rail transportation systems run electrically. Such systems are demonstrably safer and less polluting than reliance on private vehicles, such as SUVs. Furthermore, because electricity can be generated from a number of sources, including renewable alternatives such as wind power, electrified rail systems of transportation are not dependent upon rapidly depleting petroleum resources and are indefinitely sustainable.

2. **Prevention vs. treatment and remediation.** Until recently it was often cheaper (in strictly monetary terms) to use processes that generated wastes and simply disposed of them to the environment. Now, as the costs of treating and remediating wastes are taken into full account, it has become much more cost-effective to prevent the generation of wastes rather than making or treating or remediating them. An important principle of waste prevention is to use processes in which maximum fractions of material put into the process end up in the final product.

3. **Facilitate separation.** The recovery, recycle, and reuse of materials and components are often made much more complicated and expensive — sometimes impossible — by difficulty in separating them. At the component level, careful consideration should be given to facilitating separation of reusable components. For example, from the standpoint of later recovery, it is much more desirable to fasten components together with screws or other mechanical fasteners rather than with adhesive.

4. **Maximize efficiency in the utilization of materials, energy, space, and time.** This aspect of green engineering means that operations are performed with minimum consumption of materials and energy as rapidly as possible and in as little space as possible. These goals may often be accomplished by process intensification, for example, in chemical synthesis, using a small reactor operated under intense conditions of pressure, temperature, and agitation thus maximizing throughput while minimizing space and time required. These conditions must be balanced with the need to avoid hazards, such as increased danger of explosion, that comes with process intensification.

5. **Driving processes by pulling output.** Well known as a basic rule of chemistry, Le Châtelier's principle states that a system placed under stress will respond in a manner that relieves the stress. In the case of a chemical synthesis, the time to produce a product can be lowered, and the quantity of product produced may be increased by increasing quantities of one of the reagents required. This "input-pushing" of a reaction can result in large quantities of wastes from the reagent used in excess. The same end

can be achieved in a manner more consistent with green engineering by an "output-pulled" approach. For example, a chemical reaction can be driven more strongly to completion by removing a product, such as the one desired, or a by-product such as water. "Just-in-time" manufacturing in which product is made as needed by the consumer is an example of output-pushed processing.

6. **Conservation of complexity.** Much of the cost, including environmental cost, of making a product lies in fabricating relatively complex components or assemblies. Relatively higher efficiencies can be attained by retaining as much as is practical of the complexity of the product. This normally means favoring reuse of a product over recycling the materials in it. For example, an automobile cylinder head is a moderately complex engine component requiring careful casting, honing, and assembly. If a worn cylinder head is simply melted down and recycled for its iron, all of the complexity in it is lost. However, if it is replaned and refurbished, the complexity is retained for much greater efficiency.

7. **Durability, but not immortality.** Durability of a product and the materials in it means that it will last through its useful lifetime. However, a product should be degradable and not environmentally "immortal." An example is that of biodegradable polymers, such as polylactic acid, which are durable in the applications for which they are used but that do eventually biodegrade when discarded.

8. **Meet the need without excess.** Overdesign can be costly in terms of resources and even environmental damage. A prime example is the large vehicles favored by some people, often traveling almost empty while consuming far more fuel than necessary to haul their passengers to their destinations. CFCs (freons) and PCBs were much more degradation resistant than their uses required. The former ended up depleting stratospheric ozone whereas the latter ended up as very persistent environmental pollutants. Excess can be avoided by high specificity. Finely targeted insecticides that kill specific insects while not harming other organisms are examples of high specificity.

9. **Use of fewer kinds of materials.** Recycling is greatly complicated by the use of a wide variety of materials that make it difficult to isolate pure products for recycling. For example, the diverse nature of plastics in automobiles can make it very difficult to recover plastic materials in a form suitable for recycling. This difficulty can be alleviated by using only a few kinds of readily identified plastics.

10. **Utilization of available materials and energy.** Efficiencies can be obtained by utilizing locally available sources of materials and energy which are cheaper and require less transportation. Utilization of locally

available construction materials, such as wood in lumber-producing regions, can enable less costly and more environmentally-friendly construction. In chemical production, heat generated by exothermic processes can be captured to provide energy for other processes that require heat. Rather than dissipating energy from braking as useless heat, hybrid electric vehicles use the electric motor for braking and produce electricity that is stored in a battery for later use.

11. **Design for later use.** Some efficiencies can be obtained by designing components for commercial "afterlife." A commendable trend in urban redevelopment has been to convert old factories and warehouses into restaurants, microbreweries, and even apartment housing. Such a conversion can be facilitated by designing the buildings in the beginning with an "afterlife" in mind. Computers often have short replacement cycles, but are perfectly suitable for many more years of use by people who find the capabilities of computers that are several years old perfectly adequate for their needs.

12. **Maximize use of renewable materials and energy rather than depleting resources.** The most common examples of renewable resources are those made by biological processes. However, recyclable materials that originally came from depleting resources — aluminum, for example — can be considered at least partially renewable. The practice of green engineering requires maximum use of renewable materials and minimization of depleting resources.

LITERATURE CITED

1. Ehrlich, Paul, R., *The Population Bomb*, Ballantine Books, New York, 1968.

2. Simon, Julian, *Hoodwinking the Nation,* Transaction Publishers, Somerset, NJ, 1999.

3. Anastas, Paul T. and Julie B. Zimmerman, Design through the twelve principles of green engineering, *Environmental Science and Technology*, 37, 95A–101A, 2003.

SUPPLEMENTARY REFERENCES

Allenby, Braden, R*econstructing Earth: Technology and Environment in the Age of Humans*, Island Press, Washington, D.C., 2005.

Ayres, Robert U. and Leslie W. Ayres, Eds., *A Handbook of Industrial Ecology*, Edward Elgar Publishing, Cheltenham, U.K., 2002.

Barr-Kumar, Raj, *Green Architecture: Strategies for Sustainable Design*, Barr International, Washington, D.C., 2003.

Beer, Tom and Alik Ismail-Zadeh, *Risk Science and Sustainability: Science for Reduction of Risk and Sustainable Development for Society,* Kluwer Academic, Boston, MA, 2003.

Booth, Douglas E., *Hooked on Growth: Economic Addictions and the Environment*, Rowman and Littlefield Publishers, Lanham, MD, 2004.

Ehrenfeld, John R., Industrial ecology: coming of age, *Environmental Science and Technology,* 36, 281A–285A, 2002.

Graedel, Thomas E., The evolution of industrial ecology, *Environmental Science and Technology*, 34, 28A–31A, 2000.

Graedel, Thomas E. and B.R. Allenby, *Industrial Ecology*, 2nd ed., Prentice Hall, Upper Saddle River, NJ, 2003.

Graedel, Thomas E. and Jennifer A. Howard-Grenville, *Greening the Industrial Facility: Perspectives, Approaches, and Tools*, Springer-Verlag, New York, 2005.

Guinée, Jeroen B., Ed., *Handbook on Life Cycle Assessment: Operational Guide to the ISO Standards,* Kluwer Academic, Boston, MA, 2002.

Hendrickson, Chris T., *Environmental Life Cycle Assessment Using Economic Input-Output Analysis*, Resources for the Future, Washington, D.C., 2006.

Levett, Roger, A Better Choice of Choice: Quality of Life, Consumption and Economic Growth, Fabian Society, London, 2003.

Lifset, Reid and Thomas E. Graedel, Industrial ecology: goals and definitions, in *A Handbook of Industrial Ecology*, Robert U. Ayres and L. Ayres, Eds., Edward Elgar, Cheltenham, U.K., 2002.

Lutz, Wolfgang and Warren Sanderson, Eds., *The End of World Population Growth: Human Capital and Sustainable Development in the 21st Century,* Earthscan, Sterling, VA, 2004.

Madu, Christian N., *Handbook of Environmentally Conscious Manufacturing*, Kluwer Academic, Boston, MA, 2001.

McDonough, William and Michael Braungart, *Cradle to Cradle: Remaking the Way We Make Things,* North Point Press, New York, 2002.

Nemerow, Nelson L., Zero *Pollution for Industry: Waste Minimization Through Industrial Complexes,* John Wiley & Sons, New York, 1995.

Seuler-Hausmann, Jan-Dirk, Christa Liedke, and Ernst Ulrich, Eds., *Eco-Efficiency and Beyond — Towards the Sustainable Enterprise*, Greenleaf Publishing, Sheffield, U.K., 2004.

Sonnemann, Guido, Francesc Castells, Marta Schuhmacher, *Integrated Life-Cycle and Risk Assessment for Industrial Processes*, CRC Press/Lewis Publishers, Boca Raton, FL, 2003.

Townsend, Mardie, *Making Things Greener: Motivations and Influences in the Greening of Manufacturing,* Ashgate, Publishing, Aldershot, U.K., 1998.

Udo de Haes, Helias A., *Life-cycle Impact Assessment: Striving Towards Best Practice*, Society of Environment Toxicology and Chemistry, Pensacola, FL, 2002.

Van den Bergh, Jeroen C.J.M., and Marco A. Janssen, Eds., *Economics of Industrial Ecology: Materials, Structural Change, and Spatial Scales*, MIT Press, Cambridge, MA, 2004.

QUESTIONS AND PROBLEMS

1. Define industrial ecology.

2. Define an industrial ecosystem.

3. Name four major parts of an industrial ecosystem.

4. Give the name of the processes to which materials and components are subjected in industrial ecosystems.

5. Give a definition of wastes in terms of natural resources.

6. What is the general pathway of materials through industrial systems as they currently operate?

7. What is meant by "level of recycling" and how is it related to embedded utility?

8. Name the three kinds of analyses and three categories considered in a life-cycle assessment.

9. What are the three kinds of products, classified in part on their amenability to recycling, that are normally considered in life-cycle assessments?

10. Give three important useful characteristics of consumable products related to their potential environmental effects.

11. Name three key attributes of industrial ecosystems that largely determine the well being of the systems.

12. Given that an abundant source of energy can make almost anything possible in an industrial ecosystem, in what respects do vast reserves of coal, wind power, and solar energy fall short of being ideal energy sources?

13. Explain cogeneration of energy. What are its advantages?

14. Name three approaches to providing materials other than from virgin sources.

15. Consumable items and products cannot be recycled on a practical basis. Name three other categories of goods or products that can be recycled.

16. What is Kalundborg, Denmark, noted for?

17. Name several characteristics that facilitated development of the Kalundborg industrial ecosystem.

18. In biological ecosystems, a process called mineralization occurs as defined in this book. Name and describe a process analogous to mineralization that occurs in an industrial ecosystem.

19. How are the terms industrial metabolism, industrial ecosystem, and sustainable development related to industrial ecology?

20. How is industrial symbiosis related to industrial ecology?

21. Justify or refute the statement that in an operational industrial ecosystem only energy is consumed.

22. In what sense is the consumer sector the most difficult part of an industrial ecosystem?

23. In what sense might a "moon station" or a colony on Mars advance the practice of industrial ecology?

24. In what sense do modern solid state electronic devices illustrate both dematerialization and material substitution?

25. As applied to material resources, what is the distinction between dematerialization and material substitution? Use the automobile as an example.

26. How does "design for recycling" (DFR) relate to embedded utility?

27. Distinguish among consumable, durable (service), and recyclable products.

28. List some of the "environmentally friendly" criteria met by soap as a consumable commodity.

29. What are the enterprises that serve to underpin the Kalundborg industrial ecosystem? How might they compare to the basic enterprises of an industrial ecosystem consisting of rural counties in the state of Iowa?

30. Consider a university as an industrial ecosystem in which the ultimate "consumer" is society that utilizes and benefits from educated graduates. Describe ways in which the university fits the model of an industrial ecosystem and ways in which it does not. Is there any recycling? Can you suggest ways in which a university might become a more efficient ecosystem?

31. Suppose that it is proposed to construct a huge system to divert a significant amount of water from near the mouth of the Mississippi River and pump it with power provided by giant wind farms in Texas across the southern U.S. and into southern California and northern Mexico. Suggest how such a project might constitute an industrial ecosystem and what it would include. Suggest advantages and possible disadvantages.

32. The Mississippi River water that would be used in the project suggested in the preceding question contains algal (plant) nutrients in the form of phosphates, inorganic nitrogen, and potassium that cause excessive plant growth (eutrophication) in large regions of the Gulf of Mexico near the mouth of the river. The water also contains relatively high levels of oxygen-demanding organics, silt, and some industrial chemicals which, along with eutrophication, result in the formation of a "dead zone" at certain times of the year in the Gulf of Mexico. Suggest how ecological engineering could be applied to the proposed water project to mitigate these water pollution problems and deliver a clean water product to the end users.

33. Globalization of economies is a contentious issue. Suggest how globalization may relate to the practice of industrial ecology. Suggest ways in which globalization may help and may hurt the proper practice of industrial ecology.

18. ADEQUATE, SUSTAINABLE ENERGY: KEY TO SUSTAINABILITY

18.1. INTRODUCTION

With enough energy, all things are possible. Almost any sustainability or material resource problem can be solved if there is enough energy available. Water can be desalinated, wastes and low-grade ores can be processed to obtain scarce materials, and transportation needs can be met. Adequate, sustainable energy means that energy supplies are not only adequate for all needs, but also can be utilized sustainably. This means not only that supplies are sustainable, but energy sources must also be usable without causing major environmental harm.

Petroleum, the world's current leading energy source is neither adequate nor sustainable. Production is peaking, and supplies inevitably will become tighter. Adequate coal resources are available, but coal, as it is now used, is not sustainable because of greenhouse gas (carbon dioxide) emissions. Some enthusiasts advocate hydrogen as the fuel of the future, conveniently forgetting that elemental hydrogen is not an energy source, but is only a means of transporting, storing, and utilizing energy. The elemental hydrogen must be produced from fossil fuels or extracted from water using some other energy source.

There are energy alternatives to fossil fuels that can be developed, that are environmentally safe (or can be made so), and that, taken in total, can be adequate to supply energy needs. These include wind, solar, biomass, and nuclear energy sources. Some other miscellaneous sources, such as tidal energy, may contribute as well. Fossil fuels will continue to be used and may contribute sustainably for decades with sequestration of greenhouse gas carbon dioxide. And, of course, energy conservation and greatly enhanced efficiency of energy use will make substantial contributions. This chapter discusses the energy alternatives listed earlier with emphasis on energy sustainability.

18.2. NATURE OF ENERGY

Energy is the capacity to do work (basically, to move matter around) or **heat** in the form of the movement of atoms and molecules. **Kinetic energy** is contained in moving objects. One such is the energy contained in a rapidly spinning flywheel, a device of growing importance for energy storage. **Potential energy** is stored energy, such as in an elevated reservoir of water used as a means of storing hydroelectric energy for later use that can be run through a hydroelectric turbine to generate electricity as needed.

A very important form of potential energy is **chemical energy** stored in the bonds of molecules and released, usually as heat, when chemical reactions occur. For example, in the case of methane (CH_4), in natural gas, when the methane burns,

$$CH_4 + 2O_2 \rightarrow CO_2 + 2H_2O \tag{18.2.1}$$

the difference between the bond energies in the CO_2 and H_2O products and the CH_4 and O_2 reactants is released, primarily in the form of heat. If the heat is released by combustion of methane in a gas turbine, part of the heat energy can be converted to **mechanical energy** in the form of the rapidly spinning turbine and an electrical generator to which it is attached. The generator, in turn, converts the mechanical energy to **electrical energy**.

The standard unit of energy is the **joule**, abbreviated **J**. A total of 4.184 J of heat energy will raise the temperature of 1 g of liquid water by 1°C. This amount of heat is equal to 1 **cal** of energy (1 cal = 4.184 J), the unit of energy formerly used in scientific work. A joule is a small unit, and the kilojoule, kJ, equal to 1000 J is widely used in describing chemical processes. The calorie commonly used to express the energy value of food (and its potential to produce fat) is actually a kilocalorie, kcal, equal to 1000 cal.

Power refers to energy generated, transmitted, or used per unit time. The unit of power is the **watt** equal to an energy flux of 1 joule per second ($J \cdot sec^{-1}$). A compact fluorescent light bulb adequate to illuminate a desk area might have a rating of 21 W. A large power plant may put out electricity at a power level of 1000 **MW** (where 1 MW is equal to 1 million watts). Power on a national or global scale is often expressed in **gigawatts**, each one of which is equal to a billion watts or even **terawatts**, where a terawatt is equal to a trillion watts.

The science that deals with energy in its various forms and with work is thermodynamics. There are some important laws of thermodynamics. The first law of thermodynamics states that energy is neither created nor destroyed. This law is also known as the law of conservation of energy. The first law of thermodynamics must always be kept in mind in the practice of green technology, the best practice of which requires the most efficient use of energy. Thermodynamics enables calculation of the amount of usable energy. As described by the laws of thermodynamics, only a relatively small amount of the potential energy in fuel can be converted to mechanical energy or electrical energy with the remaining energy from the combustion of the fuel dissipated as heat. With the application of green technology, much of this heat is salvaged for applications such as district heating of homes.

18.3. SOURCES OF ENERGY USED IN THE ANTHROSPHERE

Before the 1800s, most of the energy used in the anthrosphere came from biomass produced during plant photosynthesis. Houses were heated with wood. Soil was cultivated and goods and people moved using the power of animals or of humans themselves, obtaining their energy from food biomass. Wind drove sailing ships and windmills, and falling water moved waterwheels. These sources were renewable and sustainable, with solar energy captured by photosynthesis to generate biomass, wind produced by differences in solar-heated masses of atmospheric air, and flowing water moved as part of the solar-driven hydrologic cycle.

Although coal from readily accessible deposits had long been used in small quantities for home heating, exploitation of this energy source grew rapidly after the invention of the steam engine around 1800. During the 1800s, coal became the predominant source of energy in the U.S., England, Europe, and other countries that had readily accessible coal resources, a major shift from renewable biomass, wind, and water to a depletable resource that had to be dug from the ground. By 1900, petroleum had become a significant source of energy and, by 1950, had surpassed coal as the leading energy supply in the U.S. Lagging behind petroleum, natural gas had become a significant energy supplier by 1950. By 1950, hydroelectric power was providing a large fraction of energy used in the anthrosphere and still is so. By around 1975, nuclear energy was supplying significant amounts of electricity and has maintained a share of several percent worldwide until the present. Although just a "blip" in the total energy picture, miscellaneous renewable sources including geothermal and, more recently, solar and wind energy, now make contributions to total energy supply. Biomass still contributes significantly to the total of the sources of energy used.

Figure 18.1 shows U.S. and world energy sources used annually as of the year 2000. The predominance of **fossil fuel** petroleum, natural gas, and coal are obvious. One of two major problems with fossil fuel is that sources are running out or will become impossibly expensive. Indications are that world petroleum production

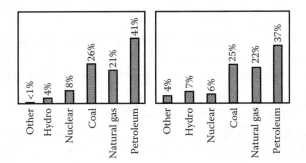

Figure 18.1.   U.S. (left) and world (right) sources of energy. Percentages of total rounded to the nearest 1%.

has peaked in the early 2000s. Resources of coal are considerably more abundant, but this gets into the second major problem with fossil fuels, their contribution to greenhouse gas — atmospheric carbon dioxide. As discussed in Chapter 9, Section 9.2, atmospheric carbon dioxide levels have grown by about 50% during the last 150 years. Now at about 380 ppm by volume in the atmosphere, these levels are increasing by about 1 ppm/year. Most reputable models for the effect this will have upon global temperatures project an increase of several °C within the next several decades as atmospheric CO_2 levels rise, a small increase but enough to have profound and perhaps even catastrophic effects on global climate.

The contributions of different fossil fuels to carbon dioxide emissions vary with the chemical nature of the fuel with greater contributions from those that have relatively less hydrogen. For example, the equation for the combustion of methane, CH_4,

$$CH_4 + 2O_2 \rightarrow CO_2 + 2H_2O + energy \qquad (18.3.1)$$

shows that two molecules of H_2O are produced for each molecule of CO_2. Because the conversion of chemically bound hydrogen to H_2O produces a large amount of heat, a relatively smaller amount of CO_2 is released per unit of heat generated. Petroleum hydrocarbons, such as those in gasoline or diesel fuel contain essentially only two atoms of H per atom of C. The combustion of such a molecule showing the conversion of one atom of C to CO_2 is represented as

$$CH_2 + {}^3/_2O_2 \rightarrow CO_2 + H_2O + energy \qquad (18.3.2)$$

demonstrating that only half as much hydrocarbon-bound H is burned per molecule of CO_2 produced, so significantly less energy is produced per C atom than in the combustion of natural gas. Coal is even worse. Coal is a black hydrocarbon with an approximate simple formula of $CH_{0.8}$, so the combustion of an atom of carbon in coal can be represented as the following:

$$CH_{0.8} + 1.2\ O_2 \rightarrow CO_2 + 0.4H_2O + energy \qquad (18.3.3)$$

Much less hydrocarbon-bound hydrogen is available to burn per atom of C in coal compared to petroleum or, especially, natural gas, so the amount of carbon dioxide emitted to the atmosphere per unit energy produced from coal is higher than with petroleum and much higher than with natural gas.

The problem that industrialized societies have become dependent on unsustainable fossil energy is clear, the solution less so. Alternatives must be developed and the transition to new sources will not be easy. The alternatives are discussed later in this chapter.

18.4. ENERGY DEVICES AND CONVERSIONS

Energy occurs in many forms and its utilization requires conversion to other forms. Many devices exist for the utilization of energy and its conversion to other forms. The most common of these are shown in Figure 18.2. The types of energy available, the forms in which they are utilized, and the processes by which they are converted to other forms have a number of implications for green technology and sustainability. For example, the wind turbine shown in Figure 18.2(1), once in place, continues to pump electricity into the power grid with virtually no harm to the environment (though some people regard them as unsightly whereas others think they are picturesque), whereas the steam power plant shown in Figure 18.2(2) requires mining depletable coal, combustion of the fossil fuel with its potential for air pollution, control of air pollutants, and means for cooling the steam exiting the turbine, with its potential for thermal pollution of waterways.

An important aspect of energy utilization is its conversion to usable forms. For example, gasoline burned in automobile engines comes originally from petroleum pumped from underground, the petroleum constituents are separated, molecules with properties suitable for engine fuel are produced chemically, the gasoline product is burned in an internal combustion engine converting chemical to mechanical energy, and the mechanical energy is transmitted to the wheels of the automobile in the form of kinetic energy that moves the automobile. Significantly less than half of the energy in the gasoline is actually converted to mechanical energy of the automobile's motion; the rest is dissipated as waste heat through the engine's cooling system.

Figure 18.3 illustrates major forms of energy and conversions between them. A significant point of this illustration is the very large ranges of energy conversion efficiencies from just a few percent or less to almost 100%. These differences suggest areas in which improvements may be sought. One of the most striking efficiencies is the less than 0.5% conversion of light energy to chemical energy by photosynthesis. Despite such a low conversion efficiency, photosynthesis has generated the fossil fuels from which industrialized societies now get their energy and provides a significant fraction of energy in areas where wood and agricultural wastes are used. Doubling photosynthesis efficiency with genetically engineered plants could be a major factor in making biomass a more desirable energy source. Replacement of woefully inefficient incandescent light bulbs with fluorescent bulbs that are 5 to 6 times more efficient in converting electrical energy to light can save large amounts of energy.

A particularly important energy conversion carried out in the anthrosphere is that of heat, such as from chemical combustion of fuel, to mechanical energy used to propel a vehicle or run an electrical generator. This occurs, for example, when gasoline in an engine burns, generating hot gases that move pistons in the engine connected to a crankshaft, which converts the up-and-down movement of the piston

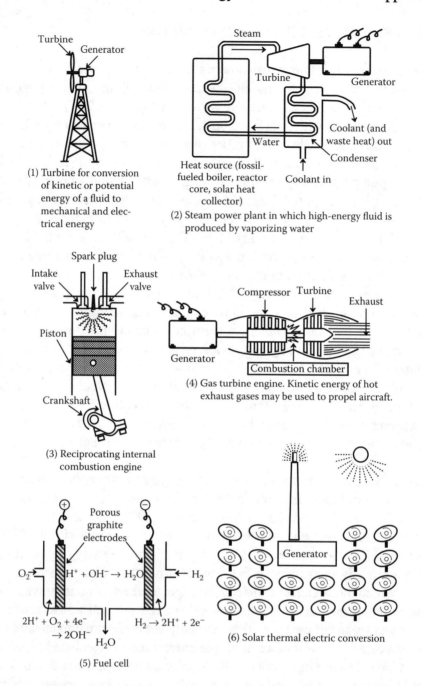

(1) Turbine for conversion of kinetic or potential energy of a fluid to mechanical and electrical energy

(2) Steam power plant in which high-energy fluid is produced by vaporizing water

(3) Reciprocating internal combustion engine

(4) Gas turbine engine. Kinetic energy of hot exhaust gases may be used to propel aircraft.

(5) Fuel cell

(6) Solar thermal electric conversion

$$2H^+ + O_2 + 4e^- \rightarrow 2OH^-$$

$$H_2 \rightarrow 2H^+ + 2e^-$$

$$H^+ + OH^- \rightarrow H_2O$$

Figure 18.2. Examples of many devices for the collection of energy and its conversion to other forms.

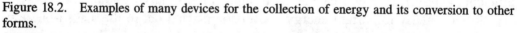

to rotary motion that drives a vehicle's wheels. It also occurs when hot steam generated at high pressure in a boiler flows through a turbine connected directly to an electrical generator. A device, such as a steam turbine, in which heat energy is converted to mechanical energy, is called a **heat engine**. Unfortunately, the laws of

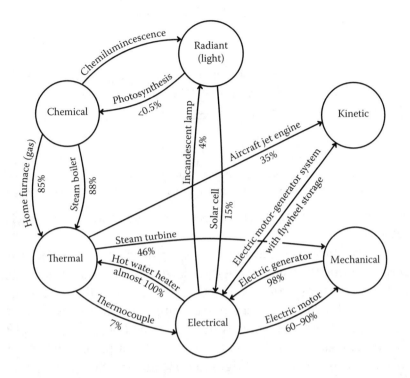

Figure 18.3. Major types of energy and conversions between them showing conversion efficiencies.

thermodynamics dictate that the conversion of heat to mechanical energy is always much less than 100% efficient. The efficiency of this conversion is given by the Carnot equation,

$$\text{Percent efficiency} = \frac{T_1 - T_2}{T_1} \times 100 \qquad (18.4.1)$$

in which T_1 is the inlet temperature (for example, of steam into a steam turbine) and T_2 is the outlet temperature, both expressed in Kelvin (°C + 273). Consider a steam turbine as shown in Figure 18.4. Substitution into the Carnot equation of 875 K for T_1 and 335 K for T_2 gives a maximum theoretical efficiency of 62%. However, it is not possible to introduce all the steam at the highest temperature and friction losses occur so that the energy conversion efficiency of most modern steam turbines is just below 50%. About 80% of the chemical energy released by combustion of fossil fuel in a boiler is actually transferred to water to produce steam so that the net efficiency for conversion of chemical energy in fossil fuels to mechanical energy to produce electricity is about 40%. The overall conversion of chemical energy to electricity is essentially the same because an electrical generator converts virtually all of the energy of a rotating turbine to electricity. Because nuclear reactor peak temperatures

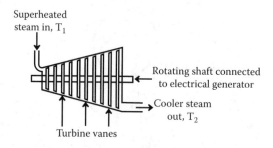

Figure 18.4.   In a steam turbine, superheated steam impinges on vanes attached to a shaft to produce mechanical energy. For generation of electricity, the shaft is coupled to an electrical generator.

are limited for safety reasons, their conversion of nuclear energy to electricity is only about 30%.

A particularly important machine for converting chemical to mechanical energy is the **internal combustion piston engine** shown in Figure 18.5. Most internal combustion engines operate on a cycle of four strokes. In the first of these, the piston moves downward drawing air or an air–fuel mixture into the cylinder. Next, with both valves closed, the air or air–fuel mixture is compressed as the piston moves upward. With the piston near the top of the cylinder (a point at which fuel may be injected if only air is compressed), ignition occurs, and the burning fuel creates a mass of highly pressurized combustion gas in the cylinder, which drives the piston down in the third stroke. The exhaust valve then opens and the exhaust gas is expelled during the exhaust stroke.

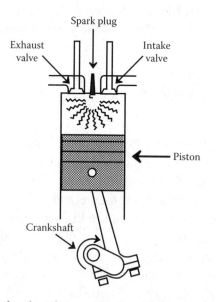

Figure 18.5.   An internal combustion piston engine in which a very rapidly burning mixture of air and fuel drives a piston downward during the power stroke and this motion is converted to rotary mechanical motion by the crankshaft.

The efficiency of the internal combustion engine increases with the peak temperature reached by the burning fuel, which increases with the degree of compression during the compression stroke (around 20:1 for a modern diesel engine). This temperature is highest for the diesel engine in which the compression is so high that fuel injected into the combustion chamber ignites without a spark plug ignition source. Whereas a standard gasoline engine is typically about 25% efficient in converting chemical energy in fuel to mechanical energy, a diesel engine is typically 37% efficient, with some reaching higher values.

Although highly superior from the standpoint of efficiency, diesel engines do have some disadvantages with respect to emissions. The first of these is that the combustion zone is not homogeneous because the fuel is injected into the highly compressed air at the top of the compression stroke resulting in incomplete combustion and production of carbon particles; improperly adjusted diesel engines are a major source of particle air pollution in urban areas. In addition, because of their very high combustion temperatures and high ignition pressures, diesel engines tend to produce elevated levels of air pollutant nitrogen oxides. Recent advances in diesel engine design, computerized control, and exhaust pollutant control devices have greatly reduced diesel engine emissions.

Fuel Cells

Fuel cells are devices that convert the energy released by electrochemical reactions directly to electricity without going through a combustion process and electricity generator. Fuel cells are the primary means for utilizing hydrogen fuel and are becoming more common as electrical generators. A fuel cell has an anode at which elemental hydrogen is oxidized, releasing electrons to an external circuit, and a cathode at which elemental oxygen is reduced by electrons introduced from the external circuit, as shown by the half-reactions in Figure 18.6. The H^+ ions generated at the anode migrate to the cathode through a solid membrane permeable to protons. The net reaction is

$$2H_2 + O_2 \rightarrow 2H_2O + \text{electrical energy} \tag{18.4.2}$$

and the only product of the fuel cell reactions is water.

Although elemental hydrogen is the ultimate fuel for fuel cells, it may be produced by the chemical breakdown of hydrogen-rich fuels, such as methane, methanol, or even gasoline, a process that also generates carbon dioxide. Tubular-style solid-oxide fuel cells, such as those manufactured by Siemens Westinghouse, operate at an elevated temperature of about 1000°C and produce an exhaust that is hot enough to drive a turbine or even to cogenerate steam. Such systems may be able to develop overall efficiencies of up to 80%.

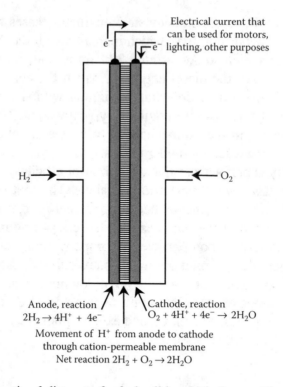

Figure 18.6.   Cross-sectional diagram of a fuel cell in which elemental hydrogen can be reacted with elemental oxygen to produce electricity directly with water as the only chemical product.

18.5. GREEN TECHNOLOGY AND ENERGY CONVERSION EFFICIENCY

One of the best ways to conserve fuel resources is through increasing the efficiency of energy conversion including that of chemical to mechanical energy with the intermediate step of production of heat energy. Many advances have been made in this area since the late 1800s. Part of the increase in conversion of fuel energy to electricity going from around 4% conversion in 1900 to more than 40% at present resulted from increasing the input temperature (T_1 in the Carnot equation) in the heat engines driving electrical generators. Energy use efficiency increased by more than fourfold when picturesque steam engines on railroads were replaced by diesel/electric locomotives during the 1940s and 1950s. Substitution of diesel engines for gasoline engines in trucks and farm and construction equipment have resulted in gains in energy efficiency.

Much of the increased efficiency in fuel utilization has come from improved materials that allow higher operating temperatures. In addition to high-temperature-tolerant metals in engines, a contribution has been made by lubricating oils that do not break down at high temperatures. Much of the progress has been achieved with better engineering, now greatly aided by computerized design, evaluation, and manufacturing of engines. Engineers of a century ago had never heard of green technology, and probably would not have cared had they known about it. But they did understand costs of fuel (which on the basis of constant value currency were often higher then than they are now) and they welcomed the greater efficiencies they achieved on the basis of costs.

A key aspect of the most efficient conversion of chemical to mechanical energy in engines is the precise control of operational aspects such as ignition timing, valve timing, and fuel injection. In modern engines, key operation parameters are controlled by a computer, leading to optimum efficiency in engine operation.

As an inevitable consequence of the thermodynamics described by the Carnot equation, engines that convert heat to mechanical energy cannot utilize much of the heat, which is carried away by an engine cooling system. Typically, a small portion of this heat is used in automotive heaters on cold days. On a broader scale, such as municipal electrical systems, this heat can be used for heating buildings. Such efficiencies are discussed under "Combined Power Cycles" in Section 18.18.

18.6. THE ENERGY PROBLEM

Since the first major "energy crisis," of 1973 to 1974, much has been said and written, many learned predictions have gone awry, and some concrete action has even taken place. Catastrophic economic disruption, people "freezing in the dark," and freeways given over to bicycles (perhaps a good idea) have not occurred. In the U.S., concern over energy supplies and measures taken to ensure alternate supplies reached a peak in the late 1970s. Significant programs on applied energy research were undertaken in the areas of renewable energy sources, efficiency, and fossil fuels. The financing of these efforts reached a peak around 1980, then dwindled significantly after that date. By the year 2000, a perceived abundance of fossil energy had resulted in a false sense of security regarding energy sources. However, since then, a true energy crisis has emerged and, as of 2006, several aspects of energy supply such as the following have become obvious:

- World economies have become too dependent upon petroleum.

- Discovery of new petroleum resources are not keeping pace with increases in consumption.

- Rapid development of new economies, especially in highly populated China and India, are adding tremendously to the demand for petroleum.

- The evidence is mounting that global warming caused by carbon dioxide emissions is in fact occurring, which would argue against exploitation of relatively abundant resources of coal and oil shale for energy.

The solutions to energy problems are strongly tied to environmental considerations. For example, a massive shift of the energy base to coal in nations that now rely largely on petroleum for energy would involve much more strip mining, potential production of acid mine water, use of scrubbers, and release of greenhouse gases (carbon dioxide from coal combustion and methane from coal mining). Similar examples could be cited for most other energy alternatives.

Dealing with the energy problem requires a heavy reliance on technology, especially green technology, which is discussed in numerous places in this book. Computerized control of transportation and manufacturing processes enables much more efficient utilization of energy. New and improved materials enable higher peak temperatures and therefore greater extraction of usable energy in thermal energy conversion processes. Innovative manufacturing processes have greatly lowered the costs of photovoltaic cells used to convert sunlight directly to energy. Technology will have a vital role to play in the shift away from CO_2-emitting fossil fuels that will be required to slow down the rate of global warming due to the greenhouse effect.

18.7. WORLD ENERGY RESOURCES

At present, most of the energy consumed by humans is produced from fossil fuels. Estimates of the amounts of fossil fuels available differ; those of the quantities of recoverable fossil fuels in the world before 1800 are given in Figure 18.7. By far,

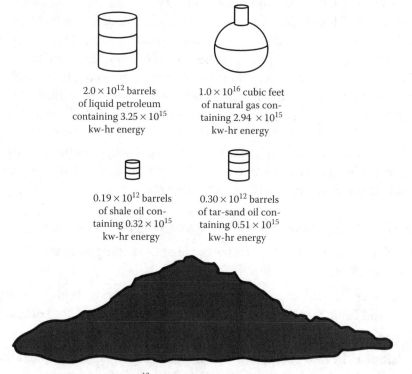

2.0 × 10¹² barrels of liquid petroleum containing 3.25 × 10¹⁵ kw-hr energy

1.0 × 10¹⁶ cubic feet of natural gas containing 2.94 × 10¹⁵ kw-hr energy

0.19 × 10¹² barrels of shale oil containing 0.32 × 10¹⁵ kw-hr energy

0.30 × 10¹² barrels of tar-sand oil containing 0.51 × 10¹⁵ kw-hr energy

7.6 × 10¹² metric tons of coal and lignite, containing 55.9 × 10¹⁵ kw-hr of energy

Figure 18.7.   Original amounts of the world's recoverable fossil fuels (quantities in thermal kilowatt hours of energy based upon data taken from M. K. Hubbert, "The Energy Resources of the Earth," in Energy and Power, W. H. Freeman and Co., San Francisco, 1971).

the greatest recoverable fossil fuel is in the form of coal and lignite. Furthermore, only a small percentage of this energy source has been utilized to date, whereas much of the recoverable petroleum and natural gas has already been consumed. Projected use of these latter resources indicates rapid depletion.

Although world coal resources are enormous and potentially can fill energy needs for a century or two, their utilization is limited by environmental disruption from mining and emissions of carbon dioxide and sulfur dioxide. These would become intolerable long before coal resources were exhausted. Assuming only uranium-235 as a fission fuel source, total recoverable reserves of nuclear fuel are roughly about the same as fossil fuel reserves. These are several orders of magnitude higher if the use of breeder reactors that convert unfissionable uranium-238 (which composes around 99% of natural uranium) to fissionable plutonium is assumed. Extraction of only 2% of the deuterium present in the Earth's oceans would yield about a billion times as much energy by controlled nuclear fusion as was originally present in fossil fuels! This prospect is tempered by the lack of success in developing a controlled nuclear fusion reactor. Geothermal power, currently utilized in northern California, Italy, and New Zealand, has the potential for providing a significantly greater percentage of energy worldwide. The same limited potential is characteristic of several renewable energy resources, including hydroelectric energy, tidal energy, and especially wind power. All of these will continue to contribute significant amounts of energy. Renewable, nonpolluting solar energy comes close to being an ideal energy source and it almost certainly has a bright future.

18.8. ENERGY CONSERVATION AND RENEWABLE ENERGY SOURCES

Any consideration of energy needs and production must take energy conservation into consideration. This does not have to mean cold classrooms with thermostats set at 60°F in mid-winter, nor swelteringly hot homes with no air-conditioning, nor total reliance on the bicycle for transportation, although these and even more severe conditions are routine in many countries. The fact remains that the U.S. and several other industrialized nations have wasted energy at a deplorable rate. For example, U.S. energy consumption is higher per capita than that of some other countries that have equal, or significantly better, living standards. Obviously, a great deal of potential exists for energy conservation that will ease the energy problem.

Efficient use of energy can in fact correlate positively with higher economic standards. Figure 18.8 shows a plot of the ratio of energy use per unit of gross domestic product in developed industrialized nations and illustrates a steady and favorable decrease of energy required relative to economic output. Whereas in 2000, 1.7 barrels of oil equivalents were required per $1000 gross domestic product in developed nations, the corresponding figure for developing nations, which tend to lack advanced means of using energy efficiently, was 5.2 barrels, or 3 times as much. These figures indicate the substantial potential for decreased energy consumption

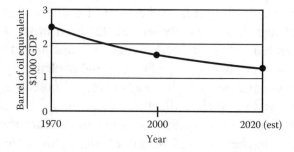

Figure 18.8.   Plot of barrel of oil equivalent required per $1000 gross domestic product (GDP) as a function of year in industrialized nations.

by energy conscious development of the less industrially advanced nations as well as the additional conservation that can be achieved if citizens of industrialized nations can be persuaded to forego wasteful energy practices, such as excessively large and inefficient vehicles and overly large dwellings.

Transportation is the economic sector with the greatest potential for increased efficiencies. The private auto and airplane are only about one-third as efficient as buses or trains for transportation. Transportation of freight by truck requires about 3800 Btu/t-mi, compared to only 670 Btu/t-mi for a train. Truck transport is terribly inefficient compared to rail transport (as well as dangerous, labor-intensive, and environmentally disruptive). Major shifts in current modes of transportation in the U.S. will not come without anguish, but energy conservation dictates that they be made.

Figure 18.9 shows the trend in U.S. automobile fuel economy during recent decades. The gains through about 1990 were very impressive, then dropped off as less fuel efficient vehicles became more popular. If the same trends from this period would have been maintained, the U.S. automobile fleet would by now average close to 40 mi/gal (MPG). Such a figure is readily achievable without seriously compromising safety or comfort and, as is obvious from the figure, with much lower emissions from pollutants compared to 1970.

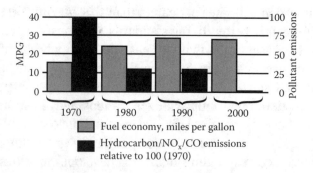

Figure 18.9.   U.S. auto fleet fuel economy and emissions over 3 decades. Fuel economy has improved markedly while emissions have been greatly reduced.

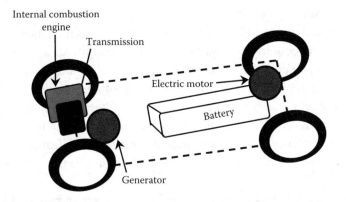

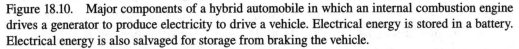

Figure 18.10.   Major components of a hybrid automobile in which an internal combustion engine drives a generator to produce electricity to drive a vehicle. Electrical energy is stored in a battery. Electrical energy is also salvaged for storage from braking the vehicle.

Household and commercial uses of energy are relatively efficient. Here again, appreciable savings can be made. The all-electric home requires much more energy (considering the percentage wasted in generating electricity) than a home heated with fossil fuels. The sprawling ranch-house style home uses much more energy per person than does an apartment unit, row house, or even a home of comparable floor area built in a compact format (more like a square box). Improved insulation, sealing around the windows, and other measures can conserve a great deal of energy. Electric generating plants centrally located in cities can provide waste heat for commercial and residential heating and cooling and, with proper pollution control, can use municipal refuse for part of their fuel, thus reducing quantities of solid wastes requiring disposal.

One of the greatest contributions to energy conservation and energy use efficiency in very recent years has been the **hybrid vehicle** that uses an internal combustion engine to produce electricity that is stored for propulsion of the vehicle in a nickel-metal-hydride battery (Figure 18.10). Although not greatly more efficient for prolonged driving at highway speeds, these vehicles have achieved improvements up to 50% in stop-and-go driving in traffic. For routine operation, the internal combustion engine supplies all the power needed plus additional power, if required, to run the generator to recharge a battery (larger than the battery in a conventional automobile but significantly smaller than the battery in an all-electric vehicle). When a surge of power is required, electricity from the storage battery drives the electric motor to produce the additional power. The braking system also generates electricity that is stored in the battery. When the vehicle is stopped, the internal combustion engine does not run, which also saves fuel.

Although gasoline engines are now employed in hybrid vehicles, the ultimate in fuel economy could be achieved with an inherently more efficient diesel engine as the internal combustion engine component. By allowing the diesel engine to run at a generally steady rate, the output of exhaust pollutants, which are produced at higher levels by diesel engines as the engine speed is changed, could be greatly reduced.

Furthermore, diesel engines idle with remarkably little fuel consumption, so that the diesel engine would not need to be turned off when the vehicle is stopped, thus staying hot and further reducing emissions when it is brought up to speed.

As scientists and engineers undertake the crucial task of developing alternative energy sources to replace dwindling petroleum and natural gas supplies, energy conservation must receive proper emphasis. In fact, zero energy-use growth, at least on a per capita basis, is a worthwhile and achievable goal. Such a policy would go a long way toward solving many environmental problems. With ingenuity, planning, and proper management, it could be achieved while increasing the standard of living and quality of life.

Closely related to energy conservation is the concept of **renewable energy** from sources that do not run out. Essentially all of these depend upon energy from the sun, including direct solar energy, wind driven by the solar heating of air masses, falling water from the solar-powered hydrologic cycle, and biomass formed from photosynthesis. For most of its lifetime on Earth, humankind has depended entirely upon renewable sources of energy, and essentially all countries are again emphasizing these sources.

Enlightened sustainable energy policies are being implemented in some developing countries. China implemented a renewable energy law in 2006. This policy encourages renewable energy alternatives including wind power, biomass energy, and biomethane generation. Long known for its utilization of wastes (including even use of human wastes as fertilizer for growing vegetables), China has constructed many waste-to-methane generators in rural areas with 17 million families served by such facilities by 2006. Experimental biopower projects burning crop biomass by-products have been undertaken. As of 2006, China had 80 million square meters of solar collectors to heat water, equivalent to the energy from 10 million tons of coal per year. The total renewable energy capacity of China in 2006 was 7% of China's energy use, equivalent to 160 million tons of coal per year.

18.9. PETROLEUM AND NATURAL GAS

Liquid **petroleum** occurs in rock formations ranging in porosity from 10 to 30%. Up to half of the pore space is occupied by water. The oil in these formations must flow over long distances to an approximately 15-cm diameter well from which it is pumped. The rate of flow depends on the permeability of the rock formation, the viscosity of the oil, the driving pressure behind the oil, and other factors. Because of limitations in these factors, **primary recovery** of oil yields an average of about 30% of the oil in the formation, although it is sometimes as little as 15%. More oil can be obtained using **secondary recovery** techniques, which involve forcing water under pressure into the oil-bearing formation to drive the oil out. Primary and secondary recovery together typically extract somewhat less than 50% of the oil from a formation. Finally, **tertiary recovery** can be used to extract even more oil, normally through the injection of pressurized carbon dioxide, which forms a mobile solution with the oil and allows it to flow more easily to the well. Other chemicals, such as detergents, may be used to aid in tertiary recovery. Currently, about 300 billion barrels of U.S. oil are not available through primary recovery alone. A recovery efficiency of 60% through secondary or tertiary techniques could double the amount of available petroleum. Much of this would come from fields which have already been abandoned or essentially exhausted using primary recovery techniques.

Shale oil is a possible substitute for liquid petroleum. Shale oil is a pyrolysis product of oil shale, a rock containing organic carbon in a complex structure of biological origin from eons past called *kerogen*. Oil shale is believed to contain approximately 1.8 trillion barrels of shale oil that could be recovered from deposits in Colorado, Wyoming, and Utah. In the Colorado Piceance Creek basin alone, more than 100 billion barrels of oil could be recovered from prime shale deposits. However, the environmental implications of recovering shale oil by heating oil shale including production of vast amounts of carbon dioxide and water-soluble salt residues of the pyrolysis of oil shale make it unlikely that this resource will ever be developed on a large scale.

Natural gas, consisting almost entirely of methane, is a very attractive fuel that produces few pollutants and less carbon dioxide per unit energy than any other fossil fuel. In addition to its use as a fuel, natural gas can be converted to many other hydrocarbon materials. It can be used as a raw material for the Fischer-Tropsch synthesis of gasoline. As of 2004, increased demand for natural gas had led to tight supplies in the U.S. and the situation has become worse since then. Production of natural gas from coal seams in Wyoming has required pumping saline, alkaline water from the seams, which has caused water pollution problems. New unconventional sources of natural gas, such as may exist in geopressurized zones, and importation of liquified methane held at low temperatures could provide abundant energy reserves for the U.S., though at substantially increased prices.

18.10. COAL

From Civil War times until World War II, **coal** was the dominant energy source behind industrial expansion in most nations. However, after World War II, the greater convenience of lower-cost petroleum resulted in a decrease in the use of coal for energy in the U.S. and in a number of other countries. Annual coal production in the U.S. fell by about one-third, reaching a low of approximately 400 million tons in 1958, but, since then, coal production for electricity generation has reached about 1 billion tons per year in the U.S. About one-third of the world's energy and around 50% of electrical energy is provided by coal.

The general term coal describes a large range of solid fossil fuels derived from partial degradation of plants. Coal is differentiated largely by **coal rank** based upon percentage of fixed carbon, percentage of volatile matter, and heating value. The approximate average empirical formula of coal is $CH_{0.8}$, coal typically contains from 1 to several percent sulfur, nitrogen, and oxygen. Of these elements, sulfur bound to the organic coal molecule and mixed with coal as mineral pyrite (FeS_2) presents major environmental problems because of production of air pollutant sulfur dioxide during combustion. Much of the FeS_2 can be removed physically from coal prior to combustion and sulfur dioxide can be removed from stack gas by various scrubbing processes.

Coal Conversion

As shown in Figure 18.11, coal can be converted to gaseous, liquid, or low-sulfur, low-ash solid fuels such as coal char (coke) or solvent-refined coal (SRC). Coal conversion is an old idea; a house belonging to William Murdock at Redruth, Cornwall, England, was illuminated with coal gas in 1792. The first municipal coal-gas system was employed to light Pall Mall in London in 1807. The coal-gas industry began in the U.S. in 1816. The early coal-gas plants used coal pyrolysis (heating in the absence of air) to produce a hydrocarbon-rich product particularly useful for illumination. Later in the 1800s, the water-gas process was developed, in which steam was impinged upon hot coal coke to produce a mixture consisting primarily of H_2 and CO. It was necessary to add volatile hydrocarbons to this "carbureted" water-gas to bring its illuminating power up to that of gas prepared by coal pyrolysis. The U.S. had 11,000 coal gasifiers operating in the 1920s. At the peak of its use in

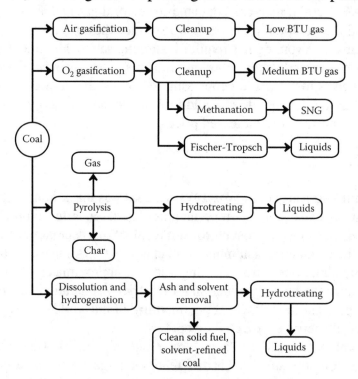

Figure 18.11.   Routes to coal conversion. BTU refers to British thermal units, a measure of the heat energy that can be obtained from a fuel. Methanation means synthesis of CH_4 gas. Hydrogenation and hydrotreating refer to reaction with elemental H_2 gas.

1947, the water-gas method accounted for 57% of U.S.-manufactured gas. The gas was made in low-pressure, low-capacity gasifiers that by today's standards would be inefficient and environmentally unacceptable (many sites of these old plants have been designated as hazardous waste sites because of residues of coal tar and other wastes). This was definitely not a green technology because of the high toxicity of carbon monoxide in the gas product, and many people were killed by release of CO in their houses. During World War II, Germany developed a major synthetic petroleum industry based on coal, which reached a peak capacity of 100,000 barrels per day in 1944. A synthetic petroleum plant operating in Sasol, South Africa, reached a capacity of several tens of thousands of tons of coal per day in the 1970s.

A number of environmental implications are involved in the widespread use of coal conversion. These include strip mining, water consumption in arid regions, lower overall energy conversion compared to direct coal combustion, and increased output of atmospheric carbon dioxide. These plus economic factors have prevented coal conversion from being practiced on a very large scale. However, coal conversion does enable relatively facile carbon sequestration (see following section), which could enable much more sustainable coal utilization.

18.11. CARBON SEQUESTRATION FOR FOSSIL FUEL UTILIZATION

Carbon sequestration, which prevents carbon dioxide generated by fossil fuels from entering the atmosphere, holds the promise of enabling utilization of fossil fuels without contributing to greenhouse warming. Basically, the various schemes that have been proposed entail capturing carbon dioxide from a product or waste stream and sequestering it in a place where it cannot enter the atmosphere. Several approaches have been suggested or tried for capturing carbon dioxide, and there are several possibilities for sequestration.

The easiest sources of carbon dioxide that can be captured are those from industrial processes that generate the gas in high concentrations. An example of such a process is the fermentation of carbohydrates to produce ethanol for fuel or other uses. This source provides much of the carbon dioxide that is used commercially at present. The largest source of carbon dioxide now discharged to the atmosphere is generated in power plants fueled with fossil fuels. These sources present a substantial challenge for carbon dioxide removal because they are so dilute. A power plant fueled with carbon-rich coal produces an exhaust stream that is 13 to 15% carbon dioxide, whereas one burning hydrogen-rich methane produces only 3 to 5% carbon dioxide. A third possibility is to capture carbon dioxide released from the gasification of fossil fuels, particularly coal (see Section 18.10). Normally, gasification is performed using oxygen as an oxidant, and the initial product consists of carbon dioxide and combustible gases H_2 and CO. Carbon monoxide in the synthesis gas product can be reacted with steam,

$$CO + H_2O \rightarrow H_2 + CO_2 \qquad (18.11.1)$$

to produce elemental hydrogen, a nonpolluting fuel for fuel cells, and carbon dioxide.

There are several possible sinks in which carbon dioxide can be sequestered. The largest of these, a natural sink for the gas, is the ocean. Earth's oceans have an almost inexhaustible capacity for carbon dioxide. However, lowering the average pH of the oceans by as little as 0.1 pH unit from acidic carbon dioxide could have a serious adverse effect upon ocean life and productivity. Deep saline formations also have a very high capacity for carbon dioxide sequestration. Depleted oil and gas reservoirs and unmineable coal seams have much lower, but still significant, carbon dioxide capacities.

Geological carbon dioxide sequestration can be accomplished by injecting the gas into porous sedimentary formations at depths exceeding approximately 1000 m. Experience in the petroleum industry with underground disposal of carbon dioxide and injection of the gas into oil-bearing formations for petroleum recovery have provided the technology required for geological carbon dioxide sequestration. The carbon dioxide injected into sedimentary formations rises and is confined by poorly permeable cap-rock. Breaches in cap rock, such as those from abandoned oil wells can result in carbon dioxide release. Eventually, the carbon dioxide dissolves in the generally saline pore waters in the sedimentary formation into which it is injected. Chemical reactions in the water and with the surrounding geological strata can result in long-term stability of the carbon dioxide.

The first commercial application of carbon dioxide sequestration has operated since 1996 in the North Sea, about 240 km from the Norwegian coast, in a region known as the Sleipner oil and gas field. The natural gas product from this field is about 9% carbon dioxide, a value that must be reduced to 2.5% for commercial distribution of the gas. Whereas all other gas-producing operations simply discharge the carbon dioxide removed to the atmosphere, at Sleipner it is pumped under pressure into a 200-m thick layer of sandstone, the Utsira formation that is about 1000 m below the seabed.

An interesting possibility for carbon sequestration is the sequestration of biomass produced by the removal of carbon dioxide from the atmosphere by photosynthesis such that it does not burn or decay to produce carbon dioxide. Burning biomass for fuel is a CO_2-neutral technology in that all of the carbon dioxide released was removed from the atmosphere by photosynthesis. Burial of biomass in a secure location would result in net removal of carbon dioxide from the atmosphere. This is, of course, what happened during times that coal and similar fossil fuels were formed from plant matter.

18.12. NUCLEAR ENERGY

The awesome power of the atom nucleus revealed at the end of World War II held out enormous promise for the production of abundant, cheap energy. This promise has never really come to full fruition, although nuclear energy currently provides a significant percentage of electric energy in many countries, and it may be the only source of electrical power that can meet world demand without unacceptable environmental degradation, particularly through the generation of greenhouse gases.

Nuclear Fission

Nuclear fission for power production is carried out in nuclear power reactors in which the fission (splitting) of uranium-235 or plutonium nuclei occurs. Each such event generates two radioactive fission product atoms of roughly half the mass of the nucleus fissioned, an average of 2.5 neutrons, plus an enormous amount of energy compared to normal chemical reactions. The neutrons, initially released as fast-moving, highly energetic particles, are slowed to thermal energies in a moderator medium. For a reactor operating at a steady state, exactly one of the neutron products from each fission is used to induce another fission reaction in a chain reaction (Figure 18.12).

The energy from these nuclear reactions is used to heat water in the reactor core and produce steam to drive a steam turbine, as shown in Figure 18.13. As noted in Section 18.4, temperature limitations make nuclear power less efficient in converting heat to mechanical energy and, therefore, to electricity, than fossil energy conversion processes.

A limitation of fission reactors is the fact that only 0.71% of natural uranium is fissionable uranium-235. This situation could be improved by the development of **breeder reactors**, which convert uranium-238 (natural abundance 99.28%) to fissionable plutonium-239.

A major consideration in the widespread use of nuclear fission power is the production of large quantities of highly radioactive waste products. These remain lethal for thousands of years. They must either be stored in a safe place or disposed of permanently in a safe manner. At the present time, spent fuel elements are being stored under water at the reactor sites. Under current regulations in most countries, the wastes from this fuel will eventually have to be buried. An alternative

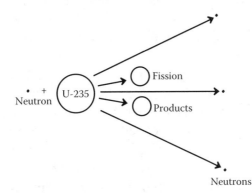

Figure 18.12.   Fission of a uranium-235 nucleus.

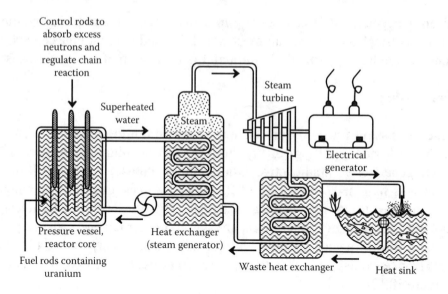

Figure 18.13.   A typical nuclear fission power plant.

favored by many investigators is to process the material in the spent fuel elements to remove radioactive products from uranium fuel, isolate the relatively short-lived fission products that decay spontaneously within several hundred years, and bombard the longer-lived nuclear wastes with neutrons in nuclear reactors. The absorption of neutrons by the nuclei of the nuclear waste elements causes **transmutation** in which the elements are converted to other elements or fission products with shorter half lives resulting in relatively rapid production of stable isotopes. Radioactive waste elements for which transmutation is feasible include plutonium, americium, neptunium, curium, technetium-99, and iodine-129. Plutonium, americium, neptunium, and curium are heavy actinide elements that are fissionable and add fuel value in a nuclear reactor.

Another problem to be faced with nuclear fission reactors is their eventual decommissioning. There are three possible solutions. One is dismantling soon after shutdown, in which the fuel elements are removed, various components are flushed with cleaning fluids, and the reactor is cut up by remote control and buried. "Safe storage" involves letting the reactor stand 30 to 100 years to allow for radioactive decay, followed by dismantling. The third alternative is entombment, encasing the reactor in a concrete structure.

The course of nuclear power development was altered drastically by two accidents. The first of these occurred on March 28, 1979, with a partial loss of coolant water from the Metropolitan Edison Company's nuclear reactor located on Three Mile Island in the Susquehanna River, 28 miles outside of Harrisburg, PA. The result was a loss of control, overheating, and partial disintegration of the reactor core. Some radioactive xenon and krypton gases were released and some radioactive water was dumped into the Susquehanna River. Eventually the reactor building was sealed. A much worse accident occurred at Chernobyl in the Soviet Union in

April of 1986 when a reactor blew up spreading radioactive debris over a wide area and killing a number of people (officially 31, but certainly many more). Thousands of people were evacuated and the entire reactor structure had to be entombed in concrete and steel plate. Food was seriously contaminated as far away as northern Scandinavia.

As of 2006, 28 years had passed since a new nuclear electric power plant had been ordered in the U.S., in large part because of the projected high costs of new nuclear plants. Although this tends to indicate hard times for the nuclear industry, pronouncements of its demise may be premature. Properly designed nuclear fission reactors can generate large quantities of electricity reliably and safely and have done so for decades in U.S. naval submarines and carrier ships. The single most important factor that may lead to renaissance of nuclear energy is the threat to the atmosphere from greenhouse gases produced in large quantities by fossil fuels. It can be argued that nuclear energy is the only proven alternative that can provide the amounts of energy required within acceptable limits of cost, reliability, and environmental effects.

New designs for nuclear power plants can enable power reactors that are much safer and environmentally acceptable than those built with older technologies. The proposed new designs incorporate built-in passive safety features that work automatically in the event of problems that could lead to incidents such as TMI or Chernobyl with older reactors. These devices — which depend upon passive phenomena such as gravity feeding of coolant, evaporation of water, or convection flow of fluids — give the reactor the desirable characteristics of **passive stability**. They have also enabled significant simplification of hardware, with only about half as many pumps, pipes, and heat exchangers as are contained in older power reactors.

Nuclear Fusion

The fusion of a deuterium nucleus and a tritium nucleus releases a lot of energy as shown below, where Mev stands for million electron volts, a unit of energy:

$$_1^2\text{H} + {}_1^3\text{H} \rightarrow {}_2^4\text{He} + {}_0^1\text{n} + 17.6 \text{ Mev (energy released per fusion)} \qquad (18.12.1)$$

This reaction is responsible for the enormous explosive power of the "hydrogen bomb." So far it has eluded efforts at containment for a practical continuous source of energy. And because physicists have been trying to make it work on a practical basis for the last approximately 50 years, it will probably never be done. (Within about 15 years after the discovery of the phenomenon of nuclear fission, it was being used in a power reactor to power a nuclear submarine.) However, the tantalizing possibility of using the essentially limitless supply of deuterium, an isotope of hydrogen, from Earth's oceans for nuclear fusion still give some investigators hope of a practical nuclear fusion reactor.

Nuclear fusion was the subject of one of the greatest scientific embarrassments of modern times when investigators at the University of Utah in 1989 announced that they had accomplished so-called cold fusion of deuterium during the electrolysis of deuterium oxide (heavy water). This resulted in an astonishing flurry of activity as scientists throughout the world sought to repeat the results, whereas others ridiculed the idea. Unfortunately, for the attainment of a cheap and abundant source of energy, the skeptics were right, and the whole story of cold fusion stands as a lesson in the (temporary) triumph of wishful technological thinking over scientific good sense.

18.13. GEOTHERMAL ENERGY

Underground heat in the form of steam, hot water, or hot rock used to produce steam has been used as an energy resource for about a century and can be regarded as largely renewable. This energy was first harnessed for the generation of electricity at Larderello, Italy, in 1904, and has since been developed in Japan, Russia, New Zealand, the Philippines, and at the Geysers in northern California.

Underground dry steam is relatively rare, but is the most desirable from the standpoint of power generation. More commonly, energy reaches the surface as superheated water and steam. In some cases, the water is so pure that it can be used for irrigation and livestock; in other cases, it is loaded with corrosive, scale-forming salts. Utilization of the heat from contaminated geothermal water generally requires that the water be reinjected into the hot formation after heat removal to prevent contamination of surface water.

The utilization of hot rocks for energy requires fracturing of the hot formation, followed by injection of water and withdrawal of steam. This technology is still in the experimental state, but promises approximately ten times as much energy production as steam and hot-water sources.

Land subsidence and seismic effects, such as the mini-earthquakes that occur when water is pumped under extreme pressure into hot rock formations that fracture as a consequence, are environmental factors that may hinder the development of geothermal power. However, this energy source holds considerable promise, and its development continues.

18.14. THE SUN: AN IDEAL, RENEWABLE ENERGY SOURCE

Solar power is an ideal source of energy that is unlimited in supply, widely available, and inexpensive. It does not add to the Earth's total heat burden or produce chemical air and water pollutants. On a global basis, utilization of only a small fraction of solar energy reaching the Earth could provide for all energy needs. In the U.S., for example, with conversion efficiencies ranging from 10 to 30%, it would only require collectors ranging in area from one tenth down to one thirtieth that of the state of Arizona to satisfy present U.S. energy needs. (This is still an enormous amount of land, and there are economic and environmental problems related to the use of even a fraction of this amount of land for solar energy collection. Certainly, many residents of Arizona would not be pleased at having so much of the state covered by solar collectors, and some environmental groups would protest the resultant shading of rattlesnake habitat.)

Solar power cells (photovoltaic cells) for the direct conversion of sunlight to electricity have been developed and are widely used for energy in space vehicles. With present technology, however, they remain too expensive in most places for large-scale generation of electricity, although the economic gap is narrowing. Some schemes for the utilization of solar power depend upon the collection of thermal energy followed by conversion to electrical energy. The simplest such approach involves focusing sunlight on a steam-generating boiler (see Illustration 6 in Figure 18.2). Parabolic reflectors can be used to focus sunlight on pipes containing heat-transporting fluids. Selective coatings on these pipes can be used so that most of the incident energy is absorbed.

The direct conversion of energy in sunlight to electricity is accomplished by special solar voltaic cells. A common type of photovoltaic cell depends on the special electronic properties of silicon atoms containing low levels of other elements. In a typical photovoltaic cell, the cell consists of two layers of silicon, a donor layer that is doped with about 1 ppm of arsenic atoms and an acceptor layer doped with about 1 ppm of boron. Recall from Chapter 3, Section 3.3, that Lewis symbols use dots to represent the outermost valence electrons of atoms, those that can be lost, gained, or shared in chemical bonds. Examination of the Lewis symbols of silicon, arsenic, and boron,

$$\cdot \text{Si} \colon \qquad \cdot \text{As} \colon \qquad \text{B} \colon$$

shows that substitution of an arsenic atom with its five valence electrons for a silicon atom with its four valence electrons in the donor layer gives a site with an excess of one electron, whereas substitution of a boron atom with only three electrons for a silicon atom in the acceptor layer gives a site "hole" that is deficient in one electron. The surface of a donor layer in contact with an acceptor layer contains electrons that are attracted to the acceptor layer. When light shines on this area, the energy of the photons of light can push these electrons back onto the donor layer, from which they can go through an external circuit back to the acceptor layer as shown in Figure 18.14. This flow of electrons constitutes an electrical current that can be used for energy.

Solar voltaic cells based on crystalline silicon have operated with a 15% efficiency for experimental cells and 11 to 12% for commercial units, at a cost of around 20 cents per kilowatt-hour (kWh), several times the cost of conventionally generated electricity. Part of the high cost results from the fact that the silicon used in the cells must be cut as small wafers from silicon crystals for mounting on the cell surfaces. Significant advances in costs and technology are being made with thin-film photovoltaics, which use an amorphous silicon alloy. A newer approach to the design and

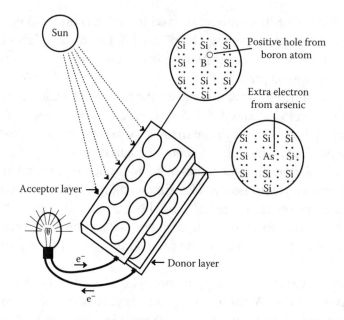

Figure 18.14.   The operation of a photovoltaic cell.

construction of amorphous silicon film photovoltaic devices uses three layers of amorphous silicon to absorb, successively, short wavelength ("blue"), intermediate wavelength ("green"), and long wavelength ("red") light as shown in Figure 18.15. Thin-film solar panels constructed with this approach have achieved solar-to-electricity energy conversion efficiencies just over 10%, lower than those using crystalline silicon, but higher than other amorphous film devices. The low cost and relatively high conversion efficiencies of these solar panels should enable production of electricity at only about twice the cost of conventional electrical power, which would be competitive in some situations.

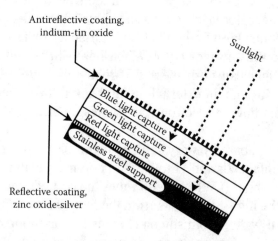

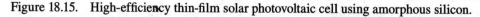

Figure 18.15.   High-efficiency thin-film solar photovoltaic cell using amorphous silicon.

A major disadvantage of solar energy is its intermittent nature. However, flexibility inherent in an electric power grid would enable it to accept up to 15% of its total power input from solar energy units without special provision for energy storage. Existing hydroelectric facilities may be used for pumped-water energy storage in conjunction with solar electricity generation. Heat or cold can be stored in water, in a latent form in water (ice) or eutectic salts, or in beds of rock. Enormous amounts of heat can be stored in water as a supercritical fluid contained at high temperatures and very high pressures deep underground. Mechanical energy can be stored with compressed air or flywheels. Utilization of solar energy to produce elemental hydrogen as a means to store, transfer, and utilize energy as discussed in Section 18.15 below will probably come into widespread use.

No really insurmountable barriers exist to block the development of solar energy, such as might be the case with fusion power. In fact, the installation of solar space and water heaters became widespread in the late 1970s, and research on solar energy was well supported in the U.S. until after 1980, when it became fashionable to believe that free-market forces had solved the "energy crisis." With the installation of more heating devices and the probable development of some cheap, direct solar electrical generating capacity, it is likely that during the coming century solar energy will be providing an appreciable percentage of energy needs in areas receiving abundant sunlight.

18.15. HYDROGEN AS A MEANS TO STORE AND UTILIZE ENERGY

Hydrogen gas, H_2, is an ideal chemical fuel in some respects that may serve as a storage medium for solar energy. Solar-generated electricity can be used to electrolyze water:

$$2H_2O + \text{electrical energy} \rightarrow 2H_2\ (g) + O_2\ (g) \qquad (18.15.1)$$

The hydrogen fuel product, and even oxygen, can be piped some distance and the hydrogen burned without pollution, or it may be used in a fuel cell (Figure 18.6). This may, in fact, make possible a "hydrogen economy." Disadvantages of using hydrogen as a fuel include its low heating value per unit volume and the wide range of explosive mixtures it forms with air. Although not yet economical, photochemical processes can be used to split water to H_2 and O_2 that can be used to power fuel cells.

Fuel-cell-powered vehicles are now practical in some applications. One of the greatest barriers to their widespread adoption has been their inability to carry sufficient hydrogen for an acceptable range. Several solutions to this problem are now being investigated. One is the potentially problematic use of very cold liquid hydrogen as a fuel source. Another is the use of very-high-pressure containers composed of multilayer cylinders wrapped with carbon composite and filled with hydrogen at pressures up to 10,000 psi (about 670 × atmospheric pressure!) reputed to contain sufficient hydrogen to propel an automobile 300 mi. Other systems use catalysts to break down liquid fuels, such as methanol or gasoline, to generate hydrogen for fuel cells.

Although there has been much enthusiasm in some quarters for hydrogen fuel and predictions of a new "hydrogen economy," some of the more extravagant arguments for hydrogen fuel may be too optimistic. The most important point is that, unlike fossil fuels, hydrogen is not a primary source of energy and has to be made by processes such as the electrolysis of water (Equation 18.15.1) that use other sources of energy. Most of the 6 million tons of elemental hydrogen used in the U.S. each year is made from steam reforming of methane from natural gas:

$$CH_4 + H_2O \rightarrow 3H_2 + CO \qquad (18.15.2)$$

The carbon monoxide product can be reacted with steam,

$$CO + H_2O \rightarrow H_2 + CO_2 \qquad (18.15.3)$$

to produce additional H_2 and the CO_2 can be sequestered as discussed in Section 18.11.

In principle, the process described above and the utilization of elemental hydrogen in fuel cells can provide a transport fuel that is pollution-free. However, methane gas is easier to store than elemental hydrogen and the modern internal combustion engine with associated emissions control equipment is virtually pollution-free. So the intermediate production of elemental hydrogen is unlikely to be the greenest approach. Production of elemental hydrogen by electrolysis of water using electricity from renewable sources, such as photovoltaics and wind power, is essentially nonpolluting, but it requires balancing the relatively inefficient electrolysis process against the value of the electricity that it consumes.

18.16. ENERGY FROM MOVING AIR AND MOVING WATER

The Surprising Success of Wind Power

Wind power using huge turbines mounted on high towers and coupled to electrical generators is emerging at a somewhat surprising rate as a source of renewable energy. Although used for centuries with windmills that drove grain grinding and water pumping operations, and, during the early 1900s, for small-scale electricity generation, modern large-scale wind powered electrical generators emerged during the 1900s as economical means of generating electrical power. Wind power is completely renewable and nonpolluting. It is an indirect means of utilizing solar energy, because winds are caused by the movement of air masses heated by the sun.

Wind power has become a major factor in the energy supply of European Union countries, which have more wind power capacity than all the rest of the world combined. By the end of 2005, EU countries had an installed capacity of 40,504 MW of

Figure 18.16.   Wind-powered electrical generators mounted on towers are becoming increasingly common sights in the world in areas where consistent wind makes this nonpolluting source of renewable energy practical.

wind generators, which had increased by 18% during the year. The leading countries are Germany (18,428 MW), Spain (10,027 MW) and Denmark (3122 MW), the country with the largest percentage of its energy produced by wind of all countries in the world. By the year 2006, China had 2000 MW of wind energy capability and was planning for much more.

The impetus for increasing wind power capacities in Europe were heightened in 2005/2006 because uncertainties regarding Russia's reliability as a supplier of natural gas. As of 2006, of the 6% of the energy in the European Union's 15 countries that comes from renewable sources other than hydroelectric, half came from wind power. The EU has set a goal of 12% of energy from renewables by 2010 and the European Parliament has mandated a benchmark of 20% of energy from renewable sources by 2010.

Wind energy is likewise gaining popularity in the U.S. The year 2005 set a record for U.S. wind energy installations with an additional 2500 MW of capacity installed in 22 states at a cost of around $3 billion. It was anticipated that an additional 3000 MW of capacity would be installed in 2006. As of the beginning of 2006, the U.S. had just over 9000 MW of wind power capacity installed with plants in 30 states. California leads the U.S. in wind power capacity at 2150 MW in 2006, but is likely to be overtaken by Texas at 1995 MW. The next three states are Iowa (836 MW), Minnesota (744 MW), and Oklahoma (475 MW). Wind energy in the U.S. is now sufficient to power 2.3 million average households eliminating as much as 15 million tons of carbon dioxide emissions per year.

Wind power is especially attractive for some agricultural regions. One reason that this is so is that electricity generated by wind energy can be used to electrolyze water to produce elemental H_2 and O_2 (Reaction 18.15.1), an application not handicapped by wind's intermittent nature. The H_2, currently produced from natural gas (Reaction 18.15.2), is the most expensive component in the synthesis of ammonia (NH_3) required by all agricultural crops as fertilizer and its production from water by inexpensive wind energy should keep the price of ammonia fertilizer at reasonable levels. Furthermore, both elemental H_2 and O_2 can be used to convert crop by-product biomass to hydrocarbon fuels as discussed in Section 18.17.

Northern regions, including parts of Alaska, Canada, the Scandinavian countries, and Russia often have consistently strong wind conditions conducive to the generation of wind power. Isolation from other sources of energy makes wind power attractive for many of these regions. Severe climate conditions in these regions pose special challenges for wind generators. One problem can be the buildup of rime consisting of ice condensed directly on structures from supercooled fog in air. (In warmer regions, the remains of insects zapped by the rotating turbine blades have built up to the point of reducing the aerodynamic efficiency of the blades.)

Energy from Moving Water

Water flowing in contact with a device called a waterwheel is one of the oldest sources of power other than humans or animals. Grain mills driven by waterpower existed in ancient Greece and Rome, and large waterwheels developing up to 50 hp were constructed in the Middle Ages. In colonial North America, waterwheels drove grist mills and sawmills and were further applied to leather, textile, and machine shop operations. Because of problems with low water flow in the summer and ice formation in the winter, these operations were rapidly displaced when steam engines became available in the early 1800s.

With the development of electric power in the late 1800s, water power underwent a spectacular renaissance to drive electrical generators. The first practical hydroelectric plant went into operation on the Fox River near Appleton, Wisconsin, in 1882. Hydroelectric power grew rapidly as an energy source from that time and, by 1980, accounted for about 25% of world electrical production and 5% of total world energy use. The potential to construct hydroelectric plants is favored by mountainous terrain with large river valleys and is distributed relatively evenly around the world. China has about 1/10 of the world's potential for hydroelectric power. About 99% of Norway's electric power is hydroelectric accounting for about 50% of that country's energy use.

The largest hydroelectric project is the Three Gorges installation on the huge Yangtze River in China. Located at the end of a number of steep canyons, the dam spans 2.3 km across the river valley and reaches a height of 185 meters. When filled, the reservoir formed by the dam will extend for 630 km with an average width of 1.3 km. When the dam is finished in 2009, there will be 26 generating units each capable of generating 700 MW, a total capacity of 18.2 GW. After 2009, 6 more units are to be constructed in a subterranean power house bringing the total generating capacity to 22.4 GW.

The sustainability and environmental acceptability of hydroelectric power present a mixed picture. In the modern age, construction of water impoundments tends to displace significant numbers of people, more than 1 million for the Chinese Three Gorges project. Altering the flow of rivers can change their aquatic ecology. Esthetics can be harmed by filling scenic river valleys with impounded water. In some cases in the U.S., dams are being dismantled to restore river valleys to their former state.

However, hydroelectric power prevents release of greenhouse gases from equivalent fossil energy powered plants. Reservoirs can provide recreational facilities and serve as sources of fish.

18.17. BIOMASS ENERGY

Fossil fuels originally came from photosynthetic processes. Photosynthesis does hold some promise of producing combustible chemicals to be used for energy production and could certainly produce all needed organic raw materials to substitute for petroleum in the current petrochemicals industry. It suffers from the disadvantage of being a very inefficient means of solar energy collection (a collection efficiency of only a fraction of a percent by photosynthesis is typical of most common plants). However, the overall energy conversion efficiency of several plants, such as sugarcane, is around 0.6%. Furthermore, some plants, such as *Euphorbia lathyrus* (gopher plant), a small bush growing wild in California, produce hydrocarbon emulsions directly. The fruit of the Philippine plant, *Pittosporum resiniferum*, can be burned for illumination because of its high content of hydrocarbon terpenes primarily a-pinene and myrcene. Conversion of agricultural plant residues to energy could be employed to provide much of the energy required for agricultural production. Indeed, until about 80 years ago, virtually all of the energy required in agriculture — hay and oats for horses, home-grown food for laborers, and wood for home heating — originated from plant materials produced on the land. (An interesting exercise is to calculate the number of horses required to provide the energy currently used for transportation in the Los Angeles basin. It can be shown that such a large number of horses would fill the entire basin with manure at a rate of several feet per day.)

Annual world production of biomass is estimated at 146 billion metric tons, mostly from uncontrolled plant growth. Many farm crops and trees can produce around 2 metric tons per acre per year of dry biomass, and some algae and grasses can produce significantly more. The heating value of this biomass is 5000 to 8000 Btu/lb, about half of typical values for coal. However, biomass contains virtually no ash or sulfur, both problems with coal. Another sustainability advantage of biomass is that all of the carbon in it is taken from carbon dioxide in the atmosphere so that biomass combustion does not add any net quantities of carbon dioxide to the atmosphere. Indeed, use of biomass to produce hydrogen-rich methane or elemental hydrogen along with sequestration of by-product carbon dioxide would result in an overall loss of carbon dioxide from the atmosphere.

As it has been throughout history, biomass is significant as heating fuel, and, in some parts of the world, is the fuel most widely used for cooking. Scavenging wood for cooking fuel has been a major contributor to deforestation in some areas. About 15% of Finland's energy needs are provided by wood and wood products (including black liquor by-product from pulp and paper manufacture), about ⅓ of which is from solid wood. Despite the charm of a wood fire and the sometimes pleasant odor of wood smoke, air pollution from wood-burning stoves and furnaces is a significant problem in some areas. Currently, wood provides about 8% of world energy needs. This percentage could increase through the development of energy plantations consisting of trees grown solely for their energy content.

Seed oils show promise as fuels, particularly for use in diesel engines. The most common plants producing seed oils are sunflowers and peanuts. More exotic species include the buffalo gourd, cucurbits, and Chinese tallow tree. Vegetable oils from soybeans and other biological sources are used to make biodiesel fuel as discussed below.

Biomass could be used to replace much of the 100 million metric tons of petroleum and natural gas currently consumed in the manufacture of primary chemicals in the world each year. Among the sources of biomass that could be used for chemical production are grains and sugar crops (for ethanol manufacture), oilseeds, animal by-products, manure, and sewage (the last two for methane generation). The biggest potential source of chemicals is the lignocellulose making up the bulk of most plant material. For example, both phenol and benzene might be produced directly from lignin. Brazil has had a program for the production of chemicals from fermentation-produced ethanol.

Ethanol Fuel

A major option for converting photosynthetically produced biochemical energy to forms suitable for internal combustion engines is the production of ethanol, C_2H_6O, by fermentation of sugars from biomass. Suitably designed internal combustion engines can burn pure ethanol or a mixture of 85% ethanol and 15% gasoline called E85. More commonly, ethanol is blended in proportions of around 10% with gasoline to give **gasohol**, a fuel that can be used in existing internal combustion engines with little or no adjustment.

Gasohol boosts octane rating and reduces emissions of carbon monoxide. From a resource viewpoint, because of its photosynthetic origin, alcohol may be considered a renewable resource rather than a depletable fossil fuel. Ethanol is most commonly produced biochemically by fermentation of carbohydrates. Brazil, a country that produces copious amounts of fermentable sugar from sugarcane, has been a leader in the manufacture of ethanol for fuel uses, with about 16 billion liters produced in 2006. All motor fuels in Brazil contain at least 24% ethanol and some fuel is essentially pure ethanol. Significant amounts of gasoline in the U.S. are supplemented with ethanol, more as an octane-ratings booster than as a fuel supplement.

Although most of the ethanol that has been produced for fuel has been made from the fermentation of grain or sugar, there is legitimate concern that, considering the energy that goes into producing grain ethanol, there is no net energy gain. A potentially much more abundant and cheaper source of ethanol consists of biomass generated as a by-product of crop production, including straw from wheat or rice production or cornstalks from growing corn. In the past, much rice straw from commercial production in the U.S. was simply burned to save the cost of cultivating it back into the soil. Straw cannot be fermented directly, but must be broken down to hexose and pentose sugars for fermentation. This has traditionally been done with acid treatment, which is expensive, although technologies exist for recycling acid. It is now generally agreed that production of ethanol from plant biomass by-products will require enzymatic hydrolysis with cellulase enzyme to produce the required sugars. The Canadian Iogen Corporation has a means for obtaining fermentable sugars from wheat straw and other plant materials and has attempted to develop a cost-effective commercial process.

Biodiesel Fuel

Biodiesel fuel is a growing source of renewable liquid hydrocarbon fuels. Rudolf Diesel developed the high-compression, compression-ignited diesel engine in the late 1800s and first operated the engine in Augsburg, Germany in 1893. He demonstrated his invention at the World Fair in Paris in 1900, receiving the "Grand Prix" (highest prize) for his invention. Interestingly, the fuel used in this and other demonstrations of Diesel's engine was peanut oil, and vegetable oils were the main source of fuel for diesel engines during the first two decades of their use. Vegetable oils were eventually replaced by petroleum-based hydrocarbons that did not solidify in cold weather. More recently, diesel fuels from vegetable oil sources have been developed that are derivatives of the fatty acids in the oils.

Vegetable oils are fatty acid esters of glycerol, a 3-carbon alcohol with 3-OH groups attached. To produce biodiesel fuel, the glycerol esters are hydrolyzed by strong base (NaOH) in the presence of methanol alcohol ($HOCH_3$), and the fatty acids are converted to their methyl esters, the molecules composing biodiesel fuel:

$$(18.17.1)$$

In this reaction, R stands for a long hydrocarbon chain in one of a number of fatty acids including stearic acid, linoleic acid, oleic acid, lauric acid, and behenic acid. For example, in stearic acid, R is a straight chain with 17 carbon atoms, $C_{17}H_{35}$.

Major oils that are used for biodiesel fuel production are rapeseed, sunflower, soybean, palm, coconut, and jatropha. Rapeseed, long grown for animal feed, is the largest source of oil for biodiesel manufacture and is widely produced in Europe, whereas soybean oil predominates in the U.S. Both of these oils offer the advantage of providing a protein-rich animal feed after the oil has been squeezed from the seed. In terms of sustainability, both coconut oil (from coconut trees) and jatropha (from *Jatropha curcus*, planted for hedges) are especially attractive because they are from perennial plants that thrive in the tropics.

There have been some problems introducing biodiesel fuel to the market. Under a law passed by the Minnesota legislature, that state mandated that all diesel fuel sold there should contain at least 2% biodiesel. Implemented in late 2005, the regulation was quickly suspended because of problems attributed to bad batches of biodiesel fuel leading to clogged fuel filters and shutting down whole fleets of trucks.

An interesting possibility for biodiesel fuel production is algae that have an oil content exceeding 50%. Such algae can grow profusely in ponds fed with nutrient-rich effluent from wastewater treatment plants. Oil-producing algae have also been grown in the carbon-dioxide-rich atmosphere provided by powerplant stack gases. By growing algae in treated wastewater within a carbon-dioxide-rich atmosphere from power plants, nutrients can be removed from wastewater, thus reducing eutrophication in receiving waters, and some of the carbon dioxide can be removed from stack gas emissions.

The Unrealized Potential of Lignocellulose Fuels

Both grain- and sugar-based ethanols, as well as vegetable oils used to make biodiesel fuel are not the best candidates for biomaterial fuels, because they use only relatively small fractions of the plants consisting of the parts that have the most value for food, animal feed, and raw materials. A much more abundant source of fuel consists of the **lignocellulose** parts of plants, a material with an approximate empirical formula of CH_2O that composes the structural members of plants including stalks, straw, corncobs, and leaves.

Lignocellulose is produced naturally as an essential part of all plant growth. One major source of it is the biomass by-product from crop growth — corn stover (stalks, leaves, cobs), wheat straw, and rice straw. The other source is from trees and other plants that, with the exception of nuts and fruit from trees, do not produce products useful for food or other applications.

Large amounts of crop by-product biomass are generated annually. Assuming conservatively that the amount of this material available in the U.S. is equal to the mass of corn grain produced, about 230 million tons of crop by-product biomass could be made available for fuel each year; the actual figure might be much higher.

The amount of biomass that could be generated from dedicated trees and grass is very high, an estimated 2,240 million tons per year in the U.S. alone. A major advantage of this source is that it comes from perennial plants that can be grown on erodable land, much of which has been taken from agricultural production as the result of government programs. One of the plants that is remarkably productive of biomass consists of hybrid poplars, from the genus *Populis* that includes cottonwoods and aspens. These trees may grow more than 2 m/year and will establish new growth from the stumps of harvested trees.

The grass most commonly considered for its biomass productivity is switchgrass. Native to North America, switchgrass is disease- and pest-resistant and requires little fertilizer. It tolerates both drought and flooding very well. Upland varieties of switchgrass grow up to 2 m tall in one growing season on well-drained soils. Lowland varieties can reach heights of 4 m and grow best on heavy soils in bottomlands. Improved varieties of switchgrass have been developed for animal forage and yield around 8 tons per acre of biomass each year. Average biomass yields of forests are only about half as high.

Another high-yielding grass native to swampy regions, such as the Florida everglades, is saw grass (*Cladium jamaicense*), which gets its name from the sawlike serrations on its leaves. It is well adapted to cultivation in wet areas where other crops cannot be grown and has the additional advantage of providing good wildlife cover.

Biomass can be used as a fuel with a heating value on a dry mass basis about half that of coal. It is extremely low in sulfur and ash and its mineral ash components do not contain toxic elements, such as the arsenic that occurs in some coals. The most direct way to use biomass fuel is direct combustion to produce heat. Biomass can be converted to other high-value fuels including hydrocarbons by gasification. One approach to biomass gasification begins with combustion of part of the biomass (represented by the formula $\{CH_2O\}$) with pure molecular oxygen oxidant (to avoid diluting the product gas with N_2 from air),

$$\{CH_2O\} + O_2 \rightarrow CO_2 + H_2O + heat \qquad (18.17.2)$$

yielding heat required for the rest of the gasification process. Under the oxygen-deficient conditions through which gasification is carried out, part of the biomass is partially oxidized to combustible carbon monoxide, CO:

$$\{CH_2O\} + \tfrac{1}{2}O_2 \rightarrow CO + H_2O + heat \qquad (18.17.3)$$

Part of the biomass is pyrolyzed by the heat produced by Reaction 18.17.2,

$$\{CH_2O\} + heat \rightarrow C + H_2O \qquad (18.17.4)$$

yielding hot carbon. The hot carbon reacts with steam,

$$C + H_2O + heat \rightarrow CO + H_2 \qquad (18.17.5)$$

yielding a **synthesis gas** mixture of CO and H_2. Biomass may also react as it is heated,

$$\{CH_2O\} + heat \rightarrow CO + H_2 \qquad (18.17.6)$$

yielding synthesis gas. The carbon monoxide in synthesis gas can be subjected to the water-gas shift reaction,

$$CO + H_2O \rightarrow CO_2 + H_2 \qquad (18.17.7)$$

giving elemental H_2 as the only gaseous product. The CO_2 generated in Reaction 18.17.2 and Reaction 18.17.7 can be separated and sequestered, such as by pumping into underground petroleum formations to enable recovery of petroleum, thereby preventing the release of this greenhouse-warming gas to the atmosphere.

Elemental hydrogen can be used as an end product of biomass gasification directly as a fuel in gas turbines and other heat engines or to generate electricity in fuel cells. As noted earlier in this chapter, elemental hydrogen can be used to synthesize ammonia (NH_3), an important industrial chemical and fertilizer. A mixture of CO and H_2 can be reacted over a catalyst in a methanation reaction

$$CO + 3H_2 \rightarrow CH_4 + H_2O \qquad (18.17.8)$$

to produce methane, which, made by this method, is called **synthetic natural gas** (SNG). With different proportions of hydrogen and CO reactants and a different catalyst, a mixture of CO and H_2 can react according to the Fischer–Tropsch reaction to yield a variety of hydrocarbons including gasoline, jet fuel, and diesel fuel as shown by the reaction below for the synthesis of octane, one of the hydrocarbons in gasoline:

$$8CO + 17H_2 \rightarrow C_8H_{18} + 8H_2O \qquad (18.17.9)$$

A similar reaction can also be used to make methanol, CH_3OH, which can be used as a fuel, gasoline additive, and material to produce H_2 for fuel cells.

An interesting possibility for increasing the amount of hydrocarbon fuel that can be obtained from biomass is to use H_2 and O_2 produced by the electrolysis of water with electricity generated by wind power,

$$2H_2O + electrical\ energy \rightarrow 2H_2\ (g) + O_2\ (g) \qquad (18.14.1)$$

to react with biomass for gasification. The pure elemental oxygen can be used to produce energy from biomass as shown by Reaction 18.17.2 above without diluting the gas product with elemental nitrogen, N_2, as would be the case with air oxidant. The elemental hydrogen generated by electrolysis of water can be reacted directly with biomass to produce hydrocarbons, such as by the following reaction:

$$\{CH_2O\} + 2H_2 \rightarrow CH_4 + H_2O \qquad (18.17.10)$$

Such a process provides a means for using the energy originally generated by wind power to make a high-energy hydrocarbon fuel that can be used for transportation or other purposes.

At several barrels of oil per ton of biomass, the use of biomass to produce synthetic fuels and substitute petrochemicals looks like an interesting possibility worthy of further development. This is especially so, considering that the biomass raw material may be produced by fast-growing grasses and trees that would have little application for agricultural uses grown on land that is not very suitable for sustained production of annual food and feed crops.

Biogas

A significant source of clean-burning methane can be obtained from the anoxic (oxygen-free) bacterial fermentation of biomass of a variety of kinds. Representing biomass as $\{CH_2O\}$ the biochemical reaction is the following:

$$2\{CH_2O\} \rightarrow CH_4 + CO_2 \qquad (18.17.11)$$

This reaction has long been used in the anoxic digesters of sewage treatment plants to reduce the amount of degradable organic matter in excess sewage sludge. A well-balanced plant makes enough methane to provide for all its energy needs. Large livestock feeding operations may have digesters to produce methane from livestock manure and other biological wastes associated with the feeding operation. Another source of methane generated by anoxic fermentation is obtained by burying collector pipes in mounds of municipal solid wastes and collecting the offgas.

18.18. COMBINED POWER CYCLES

Combined power cycles, as illustrated in Figure 18.17, enable much more efficient utilization of combustible fuels by first using the heat of combustion in a turbine coupled to an electrical generator, raising steam in a boiler with the hot exhaust gas from this turbine, using the steam to power a second turbine linked to a generator, and finally using the steam and hot water from the steam turbine for applications, such as processing in the chemical industry, heating commercial buildings, or heating homes. The water condensed from the steam used for heating is pure and is recycled to the boiler, thus minimizing the amount of makeup boiler feed water, which requires expensive treatment to make it suitable for use in boilers. The use of steam leaving a steam turbine for heating, a concept known as **district heating**, is commonly practiced in Europe (and many university campuses in the U.S.) and can save large amounts of fuel otherwise required for heating. Such a system as the one described is in keeping with the best practice of industrial ecology and should be employed whenever it is practical to do so.

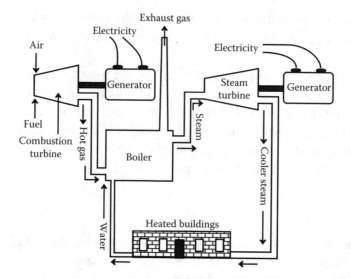

Figure 18.17.   A combined power cycle in which combustible gas or oil is first used to fire a gas turbine connected to an electrical generator. The hot gases from this turbine are fed to a boiler to raise steam, which drives a steam turbine, also connected to a generator. The still hot exhaust steam from the steam turbine is used for process heat or conveyed to commercial or residential buildings for heating. The water condensed from the steam in this final application is returned to the power plant to generate more steam, thus conserving water and avoiding the necessity to treat more water to the high purity standards required by the boiler.

SUPPLEMENTARY REFERENCES

Amos, Salvador, *Energy: A Historical Perspective and 21st Century Forecast*, The American Association of Petroleum Geologists, Tulsa, OK, 2005.

Archer, Mary D., and Robert Hill, *Clean Electricity from Photovoltaics*, Imperial College Press, London, 2001.

Brenes, Michael D., Ed., *Biomass and Bioenergy: New Research*, Nova Science Publishers, New York, 2006.

Cothran, Helen, *Energy Alternatives: Opposing Viewpoints*, Greenhaven Press, San Diego, CA, 2002.

Dickson, Mary and Mario Fanelli, Eds., *Geothermal Energy: Utilization and Technology*, Earthscan, Sterling, VA, 2005.

Fay, James A. and Dan S. Golomb, *Energy and The Environment*, Oxford University Press, New York, 2002.

Gipe, Paul, *Wind Power: Renewable Energy for Home, Farm, and Business*, Chelsea Green Pub. Co., White River Junction, VT, 2003.

Goetzberger, A. and V. U. Hoffmann, *Photovoltaic Solar Energy Generation*, Springer-Verlag, Berlin, 2005.

Gonzalo, Roberto and Karl J. Habermann, *Energy Efficient Architecture: Basics for Planning and Construction*, Birkhauser-Publishers for Architecture, Basel, Switzerland, 2006.

Hau, Erich, *Windturbines: Fundamentals, Technologies, Application, and Economics*, 2nd ed., Springer-Verlag, Berlin, 2006.

Harvey, L.D. Danny, Ed., *A Handbook on Low-Energy Buildings and District-Energy System: Fundamentals, Techniques and Examples*, Earthscan, Sterling, VA, 2006.

Hoffmann, Peter, *Tomorrow's Energy: Hydrogen, Fuel Cells, and the Prospects for a Cleaner Planet*, MIT Press, Cambridge, MA, 2001.

Jamasb, Tooraj, William J. Nuttall, and Michael G. Pollitt, Eds., *Future Electricity Technologies and Systems*, Cambridge University Press, Cambridge, U.K., 2006.

Kanninen, Barbara, Ed., *Atomic Energy*, Green Haven Press, San Diego, CA, 2006.

Krauter, Stefan C.W., *Solar Electric Power Generation Photovoltaic Energy Systems*, Springer-Verlag, New York, 2006.

Larminie, James and Andrew Dicks, *Fuel Cell Systems Explained*, 2nd ed., John Wiley & Sons, New York, 2003.

Manwell, J.F., J.G. McGowan, and A.L. Rogers, *Wind Energy Explained: Theory, Design and Application,* John Wiley & Sons, New York, 2002.

Mori, Y.H. and K. Ohnishi, Eds., *Energy and Environment Technological Challenges for the Future*, Springer-Verlag, New York, 2001.

Nutall, W.J., *Nuclear Renaissance: Technologies and Policies for the Future of Nuclear Power*, Bristol, Philadelphia, 2005.

Organisation for Economic Co-operation and Development, *Nuclear Energy Today*, OECD Publications, Paris, 2003.

Patel, Mukund R., *Wind and Solar Power Systems: Design, Analysis, and Operation*, 2nd ed., Taylor & Francis, London, 2006.

Romm, Joseph J., *The Hype About Hydrogen: Fact and Fiction in the Race to Save the Climate*, Island Press, Washington, D.C., 2004.

Silveira, Semida, Ed., *Bioenergy, Realizing the Potential*, Elsevier, Amsterdam, 2005.

Smith, Trevor, *Renewable Energy Resources*, Weigl Publishers, Mankato, MN, 2003.

Snedden, Robert, *Energy Alternatives*, Heinemann Library, Chicago, IL, 2002.

QUESTIONS AND PROBLEMS

1. The tail of a firefly glows, although it is not hot. Explain the kind of energy transformation that is most likely involved in the firefly's producing light.

2. What is the standard unit of energy? What unit did it replace? What is the relationship between these two units?

3. Which law states that energy is neither created nor destroyed.

4. What is the special significance of 1340 W?

5. What is the reaction in nature by which solar energy is converted to chemical energy?

6. In what respects is wind both one of the oldest, as well as one of the newest, sources of energy?

7. What are two major problems with reliance upon coal and petroleum for energy?

8. Why does natural gas contribute much less to greenhouse warming than does coal?

9. How might coal be utilized for energy without producing greenhouse gas carbon?

10. What is a large limiting factor in growing biomass for fuel, and in what respect does this limit hold hope for the eventual use of biomass fuel?

11. What relationship describes the limit to which heat energy can be converted to mechanical energy?

12. Why does a diesel-powered vehicle have significantly better fuel economy than a gasoline-powered vehicle of similar size?

13. Why is a nuclear power plant less efficient in converting heat energy to electricity than is a fossil-fueled power plant?

14. Instead of having a spark plug that ignites the fuel, a diesel engine has a glow plug that only operates for ignition during engine startup. Explain how ignition occurs during normal operation of the engine.

15. Cite two examples of significantly increased efficiency of energy utilization that are very useful in conserving energy.

16. Describe a combined power cycle. How may it be tied with district heating?

17. What are three reactions used in biomass gasification?

18. What is a major proposed use of liquid methanol as a fuel for the future?

19. Describe a direct and an indirect way to produce electricity from solar energy.

20. What is the distinction between donor and acceptor layers in photovoltaic cells?

21. Using Internet resources for information list some possible means for storing energy generated from solar radiation.

22. What are the advantages of *Pittsosporum reiniferum* and *Euphorbia lathyrus* for the production of biomass energy?

23. Corn produces biomass in large quantities during its growing season. What are two potential sources of biomass fuel from corn, one that depends upon the corn grain and the other that does not?

24. Does biomass contribute to greenhouse gas carbon dioxide? Explain.

25. What fermentation process is used to generate a fuel from wastes, such as animal wastes?

26. What are two potential pollution problems that accompany the use of geothermal energy to generate electricity?

27. What basic phenomenon is responsible for nuclear energy? What keeps the process going?

28. What is the biggest problem with nuclear energy? Why is it not such a bad idea to store spent nuclear fuel at a reactor site for a number of years before moving it?

29. What is meant by passive stability in nuclear reactor design?

30. What is the status of thermonuclear fusion for power production?

31. Arrange the following energy conversion processes in order from the least to the most efficient: (a) electric hot water heater, (b) photosynthesis, (c) solar cell, (d) electric generator, and (e) aircraft jet engine.

32. Considering the Carnot equation and common means for energy conversion, what might be the role of improved materials (metal alloys, ceramics) in increasing energy conversion efficiency?

33. As it is now used, what is the principle or basis for the production of energy from uranium by nuclear fission? Is this process actually used for energy production? What are some of its environmental disadvantages? What is one major advantage?

34. What would be at least two highly desirable features of nuclear fusion power if it could ever be achieved in a controllable fashion on a large scale?

35. Justify describing the sun as "an ideal energy source." What are two big disadvantages of solar energy?

36. What are some of the greater implications of the use of biomass for energy? How might such widespread use affect greenhouse warming? How might it affect agricultural production of food?

INDEX

C

I

RELATED TITLES

Environmental Chemistry, Eighth Edition
Stanley E Manahan
ISBN: 1566706335

*Introduction to Environmental Toxicology: Impacts of Chemicals
upon Ecological Systems, Third Edition*
Wayne G Landis and Ming-Ho Yu
ISBN: 1566706602

*Environmental Toxicology: Biological and Health Effects
of Pollutants, Second Edition*
Ming-Ho Yu
ISBN: 156670670X

Principles of Ecotoxicology, Third Edition
C. H. Walker
ISBN: 084933635X

Fundamentals of Ecotoxicology, Second Edition
Michael C Newman and Michael A Unger
ISBN: 1566705983

Air Quality, Fourth Edition
Thad Godish
ISBN: 156670586X

Biological and Bioenvironmental Heat and Mass Transfer
Ashim K Datta
ISBN: 0824707753